Evolution: an introduction

D0147815

Evolution:
an introduction

SECOND EDITION

Stephen C. Stearns
Department of Ecology and Evolutionary Biology
Yale University

Rolf F. Hoekstra
Laboratory of Genetics, Department of Plant Sciences
Wageningen University

UNIVERSITY PRESS

OXFORD

UNIVERSITY PRESS

Great Clarendon Street, Oxford OX2 6DP

Oxford University Press is a department of the University of Oxford.
It furthers the University's objective of excellence in research, scholarship,
and education by publishing worldwide in

Oxford New York

Auckland Cape Town Dar es Salaam Hong Kong Karachi
Kuala Lumpur Madrid Melbourne Mexico City Nairobi
New Delhi Shanghai Taipei Toronto

With offices in

Argentina Austria Brazil Chile Czech Republic France Greece
Guatemala Hungary Italy Japan Poland Portugal Singapore
South Korea Switzerland Thailand Turkey Ukraine Vietnam

Oxford is a registered trade mark of Oxford University Press
in the UK and in certain other countries

Published in the United States
by Oxford University Press Inc., New York

© Stephen C. Stearns & Rolf F. Hoekstra, 2005

The moral rights of the authors have been asserted
Database right Oxford University Press (maker)

First edition published 2000
Second edition published 2005

All rights reserved. No part of this publication may be reproduced,
stored in a retrieval system, or transmitted, in any form or by any means,
without the prior permission in writing of Oxford University Press,
or as expressly permitted by law, or under terms agreed with the appropriate
reprographics rights organization. Enquiries concerning reproduction
outside the scope of the above should be sent to the Rights Department,
Oxford University Press, at the address above

You must not circulate this book in any other binding or cover
and you must impose the same condition on any acquirer

British Library Cataloguing in Publication Data

Data available

Library of Congress Cataloging in Publication Data

Data available

Typeset by Newgen Imaging Systems (P) Ltd., Chennai, India
Printed in Great Britain
on acid-free paper by
Ashford Colour Press, Gosport, Hampshire

ISBN 0–19–925563–6   978–0–19–925563–4

3 5 7 9 10 8 6 4 2

PREFACE

To the students

All living things have evolved. Evolutionary biology studies the history of that process and the mechanisms that cause it to happen. Those mechanisms include inheritance, development, and natural selection. Organisms inherit their traits from their ancestors, they express their traits as they develop into functioning adults, and natural selection causes some of their traits to change over generations. Genetics has shed much light on inheritance, developmental biology has revealed many of the mechanisms by which organisms are built, and evolutionary biology integrates the interactions of genetics and development in populations of organisms changing over time. Evolutionary biology is not a specialism that focuses on one part of biology, as do biochemistry, development, genetics, and cell biology. It is the integrative explanation of how all those processes and structures came to be. Thus, biology only makes sense in the light of evolution.

This book aims to introduce what is essential and exciting in evolutionary biology to students taking a first course in the subject. We have not tried to cover everything, for several recent, and much larger books do that well. By featuring concepts and examples prominently throughout the text, we have tried to illustrate how evolution explains the biological world.

In writing this new edition, our aim has been to clarify and communicate what we know about evolution. That aim, however, is a double-edged sword, for in communicating what we know in compact form, it is easy to give two misleading impressions. The first is that there is not much left to be discovered. The second is that there is not much point in questioning what is presented. Neither is the case. There is a great deal still to discover, and it is only healthy to question what we know about evolutionary biology. We have not written with the attitude that we are describing immutable truths; we have simply tried to present what is most reliable about what has been discovered to date. There is a world of difference between the two.

There is no controversy about the fact of evolution; all scientists accept that it occurred. But there are still disagreements about the interpretation of the details. To appreciate this, we recommend that you read papers in the recent original literature to get different views on issues that remain clearly open. Some controversial topics and papers are listed on this book's web site. Others can be discovered in almost any issue of the leading research journals, such as *Science* and *Nature*. For example, during the writing of this preface, it was announced that a fossil had been discovered that documents the existence of a previously unknown, dwarf species of *Homo* on the island of Flores in Indonesia

(Brown et al. 2004). Because the fossil is only 18 000 years old, that species must have co-existed with our own. It illustrates an evolutionary process that also made dwarves out of other large mammals that colonized small islands, such as the extinct dwarf elephants that used to live on islands in the Mediterranean. Nothing about the discovery of this fossil changes our views of the evolutionary process, but it does add an important detail to our view of human history. This one example illustrates the dynamic nature of our knowledge of evolution: the central structure is solid, but details are being changed continuously on many fronts.

To the faculty

We have tried to present the material in a manner understandable by students taking a first course in evolution. To meet this challenge, which faces all textbook writers, we have restructured the book, revised the text, added new chapters, and introduced new pedagogy.

In place of marginal summaries, found in the first edition, we now use subheadings to communicate the main messages as sentences. The margins are used to emphasize key concepts in evolution and to provide definitions of new terminology at the point it is first introduced. All terms printed in green are also defined in the Glossary, which has been expanded.

The division of the book into five parts is new to the second edition; it is intended to help readers see the big picture. We reorganized the introductory matter, moving a shortened Prologue into a new first chapter that introduces evolution and discusses and counters Creationist criticisms of evolution. We wrote new chapters on *The importance of development in evolution* (Chapter 6), *The fossil record and life's history* (Chapter 17), *Coevolution* (Chapter 18), and *Human evolution and evolutionary medicine* (Chapter 19). In so doing we moved some material from other chapters in the first edition into the new chapters.

The chapters on *Speciation* (Chapter 12), *Phylogeny and systematics* (Chapter 13), and *Comparative methods: trees, maps, and traits* (Chapter 14) were completely reorganized and rewritten.

Readers would benefit from a prior or parallel course in genetics. For those without that background, we provide a summary of the essentials in a *Genetic appendix*.

Mathematics has been restricted to little more than algebra; calculus is not required. Because the consequences of variation are a major theme, some understanding of probability and statistics is needed, but we do not assume much.

Structure of the book

In organizing the book, we have chosen to move from microevolution and macroevolution to the integration of the two.

Part I Microevolutionary concepts

Microevolution deals with short-term evolutionary dynamics occurring within populations and species—the nuts and bolts of evolutionary change—such as natural selection acting on whole organisms, the genetic response to selection, genetic drift, the role of development in evolution, and the expression of genetic variation.

Part II Design by selection for reproductive success

In producing adaptations, natural selection designs organisms for reproductive success. The issues are many. Here we have included those with the broadest implications: the evolution of sex, the role of genetic conflict, the evolution of life histories, the evolution of sex allocation, and sexual selection.

Part III Principles of macroevolution

Macroevolution deals with evolution above the species level, with deep time, broad relationships, and big patterns. Speciation links micro- and macroevolution. After discussing speciation, we examine how systematics helps us to infer the relationships among species. We then use that information to analyze historical biogeography and the macroevolution of traits.

Part IV The history of life

The history of life is described from three perspectives: as changes in the nature of the evolving units (key events), as the interaction of life with geology (history of life and the planet), and as the fossil record of evolutionary change producing a history of increasing diversity and complexity of organisms.

Part V Integrating micro- and macroevolution

We use two foci to illustrate the integration of micro- with macroevolution: coevolution and evolutionary medicine.

Conclusion and prospect

We conclude with reflections on the reality of evolution and the reliability of claims about evolution, the scope of evolutionary explanation and the controversies resulting from current attempts to expand that scope, major puzzles and unsolved problems, and the limits to evolutionary prediction.

We hope that readers who finish this book:

- will understand, and be able to explain, why biologists view evolution as a fact for which there is a good explanation;
- will be able to see how both micro- and macroevolution have contributed to the biology in which they are interested;
- will be comfortable navigating the original literature; and
- will see unsolved problems in need of solutions that they want to discover.

Companion web site

Companion web sites provide students and lecturers with ready-to-use teaching and learning resources. They are free-of-charge, designed to complement the textbook, and offer additional materials which are suited to electronic delivery.

All these resources can be downloaded and are fully customizable, allowing them to be incorporated into your institution's existing virtual learning environment.

You will find this material at: http://www.oup.com/uk/booksites/biosciences/

Lecturer resources

• Illustrations from the book

Student resources

• Active learning exercises
• Evolutionary timeline
• Weblinks

ACKNOWLEDGEMENTS

As with the first edition, the second edition has benefited greatly from comments from many reviewers. We thank them profusely for the time they took from their busy schedules to help us make this a better book. In particular we thank Brett Edkins, Kent Garber, Harry Flaster, Vivian Irish, Talia Lerner, Gunter Wagner, and Alex White.

Dr Ross Coleman, School of Biological Sciences, Plymouth University

Dr Patrick Doncaster, School of Biological Sciences, University of Southampton

Dr Dmitry Filatov, School of Biological Sciences, University of Birmingham

Dr Chris Gliddon, School of Biological Sciences, University of Wales, Bangor

Dr John Grahame, School of Biological Sciences, University of Leeds

Dr David Harper, School of Life Sciences, University of Sussex

Dr Rus Hoelzel, School of Biological & Biomedical Sciences, University of Durham

Professor Allen Moore, Faculty of Life Sciences, University of Manchester

Dr Richard Preziosi, Faculty of Life Sciences, University of Manchester

Dr Liz Somerville, School of Life Sciences, University of Sussex

Dr Steve Waite, Division of Pharmacy & Biomolecular Sciences, Brighton University

Dr Charles Wellman, Department of Animal & Plant Sciences, University of Sheffield

Dr Mark Willcox, School of Biological & Earth Sciences, Liverpool John Moores University

Dr Mark Jervis, Cardiff School of Biosciences, Cardiff University

Dr Durr Aanen, Department of Population Ecology, University of Copenhagen

Professor Kuke Bijlsma, Evolutionary Genetics, University of Groningen

Dr Gerdien de Jong, Evolutionary Population Biology, Utrecht University

Professor Jacob Höglund, Evolutionary Biology Centre, Uppsala University

Professor Kerstin Johannesson, Department of Marine Ecology, University of Goteborg

Professor Soren Nylin, Department of Zoology, Stockholm University

Professor Dave Parker, Department of Genetics and Ecology, Aarhus University

Professor Outi Savolainen, Department of Biology, University of Oulu

Professor David Archibald, Department of Biology, San Diego State University

Dr Christopher Beck, Department of Biology, Emory University

Dr David King, School of Medicine, Department of Anatomy and College of Science, Southern Illinois University

Professor John Olsen, Department of Biology, Rhodes College

Professor Nathan Rank, Department of Biology, Sonoma State University

We also thank our editor at Oxford University Press, Ruth Hughes, for her attention to detail, insistence on clarity, and exemplary engagement with the process of revision.

Steve Stearns
Rolf Hoekstra
March 2005

BRIEF CONTENTS

CONTENTS

CHAPTER 1

Introduction

What evolution is about

At its best science efficiently converts the mysterious into the obvious. It does so by objectively criticizing ideas and data to achieve reliable understanding of the evidence available. The study of evolution has converted much that was mysterious into things every evolutionary biologist now finds obvious. Doing so has taken two centuries of intense work—the gathering of an enormous amount of data, the comparison of many possible interpretations and the rejection of quite a few of them, and the resolution of numerous controversies. Our understanding of evolution has not been bought cheaply. The ideas and observations that we now consider reliable have survived intense scrutiny and explicit consideration of alternatives. We sketch the history of that scrutiny later in this chapter, then conclude by considering and rejecting the Creationist alternative.

> ● **KEY CONCEPT**
> Evolutionary biology explains the history, diversity, adaptation, and complexity of life.

So what is evolutionary biology?

Evolutionary biology aims to understand the diversity of life and the processes responsible for shaping it. It does so at two levels and on two timescales. At the level of populations and on a relatively short timescale, it investigates how traits become well adapted to the job of survival and reproduction. On a longer timescale, it investigates the processes that produced the diversity of life now and throughout the fossil record. The evolutionary processes that occur rapidly within populations are called microevolution, evolution in the here and now, where change is readily observed. The evolutionary processes that occur slowly, where the history of life is written both in the fossil record and in the relationships of living species, are called macroevolution.

> **Microevolution:** The process of evolution within populations, including adaptive and neutral evolution.

> **Macroevolution:** The pattern of evolution at and above the species level, including most of fossil history and much of systematics.

Microevolution studies rapid change within populations

Adaptive evolution: *The process of change in a population driven by variation in reproductive success that is correlated with heritable variation in a trait.*

Gene: *A gene is a unit of heredity, a segment of DNA transcribed to produce a messenger RNA that is translated to produce a protein. See the Genetic appendix for details.*

Adaptation: *A state that evolved because it improved relative reproductive performance; also the process that produces that state.*

Natural selection: *The differential survival and reproduction of individuals that differ in a heritable trait. In responding to selection, traits that improve reproductive success increase in frequency over time.*

Heritable variation: *Differences in traits between individuals that are determined by genes and can therefore be passed from individuals to their offspring.*

Neutral evolution: *Changes in the genetic composition of populations that occur because genes that are not correlated with reproductive success are still influenced by random processes.*

Genetic drift: *Random change in allele frequencies due to chance factors.*

Some microevolutionary change is adaptive, and some is neutral. **Adaptive evolution** occurs in traits that fulfill three conditions:

(1) they vary among individuals;

(2) their variation affects how well those individuals survive and reproduce;

(3) some of their variation is determined by **genes** that also vary among individuals.

The reproduction of the successful individuals then causes the frequency of the genes and the traits that they determine to increase in the next generation. As this process continues over generations, the population becomes better **adapted**, more capable of successfully surviving and reproducing, and the inheritance of the adaptive change is reflected in changes in the genetic composition of the population. **Natural selection** is the correlation of traits with reproductive success; the *response to selection* is the change in the genetic composition of the population caused by variation in reproductive success. Natural selection thus works on **heritable variation** to produce adaptive change. It is described in Chapters 2 and 4 (pp. 27–53, 70–98).

Consider, for example, a flock of seed-eating birds that are the first immigrants of their kind to an isolated island, such as one of the Hawaiian islands. The plants on that island produce seeds that are larger and harder than the seeds on the mainland, and the birds vary in their ability to crack the larger seeds. Those with deep, strong bills do the best. The birds that crack seeds more efficiently survive better and have more surviving offspring than the less-efficient seed-crackers, and some of the variation in bill size and shape is inherited. As the more efficient seed-crackers continue to produce more surviving offspring, their bill characteristics are inherited and spread through the population, driven by the increment in reproductive success that is associated with bill shape. On another island, flowers and nectar are abundant, and the first immigrants switch from eating seeds to eating nectar. After some time, the bill shape of the birds on the second island becomes well adapted to eating nectar. The process continues, eventually producing a large variety of bill shapes, each adapted to a particular diet (Figure 1.1).

At no point in this process was the 'best' version of a bill ever evaluated; all along the way better versions were substituted for less efficient ones from the variants then available. And there is no guarantee that at the end of the process the version that prevailed is the best one conceivable; it is simply the best of those that were tested in terms of their contribution to lifetime reproductive success.

Changes in the genetic composition of populations also occur through **neutral evolution**, the aimless fluctuation of genes that vary but have no correlation with reproductive success. This kind of evolution is called neutral because the variation is neutral with respect to selection; no variant has any systematic advantage over any other. It is also called **genetic drift** to communicate the lack of direction of neutral genes drifting through the population over many generations. Drift

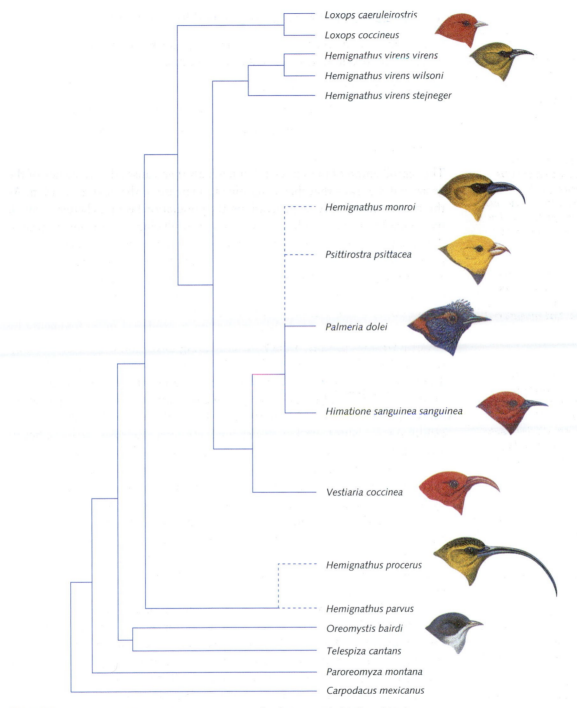

Figure 1.1 The ancestor of the Hawaiian Honeycreepers (family Drepaniidae) is thought to have been a seed-eating finch from Central America. Selection for seed-eating produced the large, thick beaks of the *Psittirostra* group; selection for nectar-feeding produced the long, down-curved beaks of *Vestiaria* and *Hemignathus*. These relationships are based on molecular data (solid line) and morphological similarity (dotted line). (Adapted from Tarr and Fleischer 1995.)

produces random change in both large and small populations, but it works more rapidly and over a broader range of conditions in small populations.

To see one of its important effects, consider an example of drift simulated on a computer (see Hartl 1994). We start 12 populations, each with eight individuals. Each individual has two versions of the same gene, one from its father and one from its mother, and at the start those versions—A_1 and A_2—are different. Thus the process starts with equal numbers of A_1 and A_2; every individual has one of each. Then mating occurs at random, which means that the genes have probabilities of occurring in offspring that are directly proportional to their frequencies in the parents. The number of offspring is always the same as the number of parents; each population always consists of eight adults in every generation. At the beginning the frequencies of A_1 and A_2 are 0.5 and 0.5, but any one mating in the first generation could produce A_1A_1, A_1A_2, or A_2A_2 offspring; the probabilities of each would be 0.25, 0.50, and 0.25. The process is exactly like flipping a coin twice; in 25% of the cases you will get two heads, in 50% of the cases you will get one head and one tail, and in 25% of the cases you will get two tails. So is it also with drifting genes. In some of the populations, A_1 will increase because the A_1A_1 combination is produced a bit more frequently by chance; in others it will decrease because the A_2A_2 combination is produced more frequently by chance. One run of a computer program yielded the results in Table 1.1.

Note the random variation among populations in the outcome in any generation. For example, after 20 generations five of the populations consisted only of A_1, three consisted only of A_2, and four still contained both. Thus random changes had, just by chance, caused A_1 to increase in five of the populations until it was the only gene left, and random changes had, just by chance, caused A_2 to increase in three of the populations until it was the only gene left. The populations had diverged from each other. They had changed not because genes were selected in different directions in different populations, but because they experienced random changes that accumulated over generations.

Table 1.1 A computer simulation of genetic drift. The 12 populations are labeled a–l. Each population of eight individuals starts with eight copies of A_1 and eight copies of A_2. Random drift causes the numbers of copies to change aimlessly. As soon as all the copies are A_1 (16) or all are A_2 (0), the process stops because one of the copies has taken over the population, which thereafter remains in the same state. (After Hartl 1994.)

Generation	Population...	a	b	c	d	e	f	g	h	i	j	k	l
	Number of Copies of A_1												
0		8	8	8	8	8	8	8	8	8	8	8	8
5		8	8	5	3	13	2	8	12	9	5	15	6
10		16	6	5	9	8	0	9	13	3	10	16	0
15		16	0	3	14	7	0	8	3	12	13	16	0
20		16	0	1	16	2	0	15	16	9	16	16	0

In the short term, neutral evolution is disorderly, but in the long term it produces an important regularity: particular versions of neutral genes accumulate in a lineage at a nearly constant rate, generating measurable differences among branches on the tree of life. Neutral evolution is described further in Chapter 3 (pp. 54–69) and its role in systematics is discussed in Chapters 13 and 14 (pp. 303, 331).

Macroevolution studies long-term changes among species

Macroevolution deals with broad patterns and deep time, time measured in millions and billions of years and space at scales as large as the planet. One key process in macroevolution is speciation, the splitting of one species into two; it is described in Chapter 12 (p. 277). And one key method in macroevolution is phylogenetics, the part of evolutionary biology that infers relationships among species (phylogenies); it is described in Chapter 13 (p. 303). The history of life and the planet are approached from three different angles in Chapters 15 (p. 355), on key events, 16 (p. 375), on the geological theater, and 17 (p. 403), on the fossil record.

Macroevolution resembles microevolution in several ways. Speciation is like reproduction; extinction is like death; and species can disappear in mass extinctions for reasons that have nothing to do with their performance under normal conditions, just as genes can disappear randomly when their effects are not correlated with reproductive success, this being one way that chance is introduced into history.

The biogeography of ratite birds recalls the breakup of Gondwana more than 80 million years ago

A good example of macroevolution is the distribution of ratites, a group of related birds consisting of ostriches, their living relatives the rheas, emus, cassowaries, kiwis, and tinamous, and the extinct elephant birds of Madagascar and moas of New Zealand. The ratites are large, incapable of flight, and they do not swim far. Before continental drift was firmly established with geological evidence, the distribution of the ratites was a puzzle. It was sometimes suggested simply on the basis of their geographic distribution that the southern continents must previously have been in contact. We now know from geology that South America, Africa, Madagascar, Australia, and New Zealand were indeed part of the same landmass, Gondwana, for several hundred million years, and that Gondwana began to break up about 200 million years ago. There followed a period when Africa was isolated but South America and Australia were still in contact via Antarctica.

Recently Cooper et al. (2001) sequenced fossil DNA from two species of extinct New Zealand moas and combined that with similar sequence information from other ratites, including a short sequence from the extinct elephant bird of Madagascar. With that information they could establish the relationships of the different ratite species, both living and extinct. When we plot their

Speciation: *The process by which new species originate and thereafter remain separate. Speciation is ultimately responsible for the diversity of much of life.*

Phylogenetics: *The branch of biology that reconstructs the evolutionary history of species.*

Phylogeny: *The history of a group of taxa described as an evolutionary tree with a common ancestor as the base and descendent taxa as branch tips.*

DNA: *Deoxyribonucleic acid or DNA is the hereditary material, the substance of genes, in all organisms except the RNA viruses. See the Genetic appendix for details.*

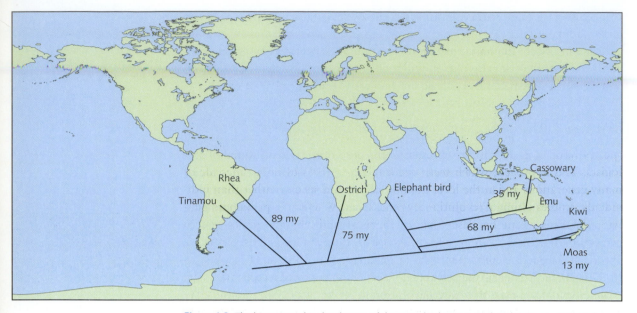

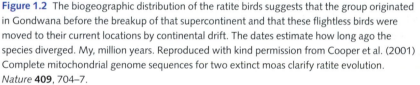

Figure 1.2 The biogeographic distribution of the ratite birds suggests that the group originated in Gondwana before the breakup of that supercontinent and that these flightless birds were moved to their current locations by continental drift. The dates estimate how long ago the species diverged. My, million years. Reproduced with kind permission from Cooper et al. (2001) Complete mitochondrial genome sequences for two extinct moas clarify ratite evolution. *Nature* **409**, 704–7.

relationships on to the map of the world (Figure 1.2), we see clear traces of Gondwana, and the estimated dates of divergence correspond reasonably well with the geological sequence of events.

Evolution combines with physics and chemistry to explain biology

Evolution contains the general principles that we need to understand biology that are not already established by physics and chemistry. Physics and chemistry do a fine job of explaining particular mechanisms in molecules and cells, but they cannot explain how those molecules and cells came to be, how life diversified, or how it became more complicated and better adapted. Dobzhansky put it well: 'Nothing in biology makes sense except in the light of evolution.' Thus all biologists are in some sense evolutionary biologists, for evolutionary biology is not a specialty, like genetics or development—it is an explanation of what is investigated by all biological specialties.

Some biologists emphasize explanations couched primarily in chemical and physical terms; they concentrate on immediate, mechanical causes. How does photosynthesis work?, what determines the sex of an organism?, and what causes disease? are questions about immediate causes whose answers can be sought in fields like physiology, genetics, cell biology, biochemistry, and development. This approach, which seeks causes acting within a period less than the lifetime of a single organism, studies **proximate** (mechanical) **causation**. It constitutes much of biology.

> **Proximate causation:** *The mechanical causes of a biological effect, couched in terms of chemistry and physics.*

Other biologists emphasize explanations couched in evolutionary terms; they study the impact of selection, drift, and history and ask different questions and investigate different kinds of causes in search of answers. Why does photosynthesis occur in the chloroplasts and not in the cell nucleus?, why do many species have approximately equal numbers of males and females, and why are there dramatic exceptions?, why do organisms senesce?, and why do some small species have large relatives? represent questions asked by evolutionary biologists. This approach, which seeks the causes of processes on a timescale of many generations and at the level of populations and species rather than individuals, studies **ultimate** (evolutionary) **causation**. Whereas in proximate analysis the causes can be described as chemical and physical processes, in evolutionary analysis one describes the causes as how natural selection, chance events, and descent from shared ancestors shaped the outcome.

> **Ultimate causation:** *The evolutionary causes of a biological effect, couched in terms of selection and history.*

All traits have both proximate and ultimate causes; a balanced explanation requires understanding of both types. Isolating the two kinds of analysis from each other is a strategic error because it reduces the number of interesting questions that can be asked. A biologist should be able to see the world both ways—from the bottom up, from molecules to species, and from the top down, from selection and history to molecules.

Evolution yields striking insights

The following four stories each reveal an important point about evolution.

> ● **KEY CONCEPT**
> Selection and history combine to explain evolution.

The focus of selection is not on survival but on reproductive success

The Australian redback spider belongs to a genus (*Latrodectus*) whose bite is so poisonous that it can kill a human. After a long and dangerous search for a mate, a male redback spider finally encounters a female. She is suspended upside down in the middle of her web, waiting for prey, and is much larger than he is. After signaling that he is a male of her species and not prey, he approaches the female and copulates with her by inserting one of his two sperm-laden palps into one of her two genital openings. As soon as sperm is being transferred, and with his copulatory organ still inserted, he somersaults, positioning his abdomen directly under the female's mouth, from which digestive fluids almost immediately appear. While insemination proceeds, the female chews the male's abdomen and may stab it with her poison-laden fangs. When copulation with the first palp is over, the male moves a few centimeters from the female, re-enacts some courtship maneuvers, and re-engages, inserting his second palp despite his mutilated abdomen. Again he does his somersault, and again the

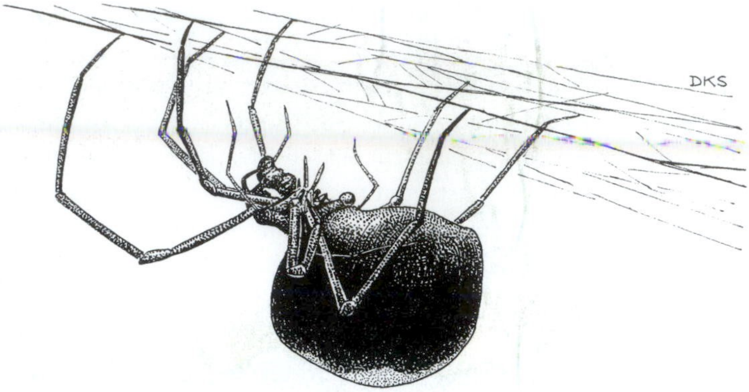

Figure 1.3 A female Australian red-backed spider, *Latrodectus hasselti*, eating a male after copulation. The male is much smaller than the female. (By Dafila K. Scott.)

female chews on his abdomen. After he withdraws the second time, the female wraps her silk around him, storing him for later consumption (Figure 1.3).

If the female is not hungry, she might not eat him, but even males that escape die of their injuries within 2 days. The copulatory behavior of the male Australian redback spider is stereotypic, which means that its sequence of elements unfold in rigid order. The somersault that brings his abdomen into contact with the female's mouth is unique within the genus *Latrodectus*. If a male redback spider is mated to a female of another species in the same genus, he makes the somersault, but she does not wound him. The pattern of the male's movements not only increases the probability that the female will eat him; it also appears to be genetically programmed (Forster 1992).

How could evolution have produced behavior that reduces the chance of survival? By being eaten during copulation males contribute to the nourishment of their offspring and thereby increase their lifetime reproductive success more than they would, on average, if they survived the first mating attempt, wandered off in search of another mate, and probably died before finding that mate. Sexual cannibalism is found in snails, crustacea, insects, and arachnids and occurs most frequently in scorpions and spiders (Elgar 1992). It is not expected to evolve when males have a good chance of finding another mate.

Strong selection produces rapid evolution in large populations

Guppies (*Poecilia reticulata*) are small freshwater fish popular in the aquarium trade. In nature, they often encounter two species of predatory fish. One is a large cichlid, *Crenicichla alta*, that eats guppies of all sizes, juvenile or adult. The other is a small killifish, *Rivulus hartii*, that preys primarily on small, juvenile guppies. In northern Trinidad many streams flow down a mountain range into the Aripo River. The large cichlid occurs primarily in the river itself and in the lower reaches of the streams that feed it; the small killifish occurs in many of the smaller streams. By raising guppies from populations that occur only with

Figure 1.4 A male and female guppy, *Poecelia reticulata* (bottom), with the killifish *Rivulus* (middle) and the cichlid *Crenicichla* (top). (Photo courtesy of David Reznick.)

10 mm

the large cichlid or only with the small killifish, it was discovered that where guppies had coexisted with the large cichlid, they matured earlier, at a smaller size, and gave birth to more, smaller offspring. They also showed greater innate tendencies to avoid predators, and the males were less colorful and had less elaborate courtship behavior than did males from populations that had coexisted with the less dangerous killifish (Figure 1.4).

Thus the populations differed genetically in important and interesting traits—they had evolved adaptations to local conditions. How rapidly might such differences evolve? To answer that question, guppies were taken from the Aripo River, where they had long coexisted with the large, dangerous predator, and introduced to a stream that contained the less dangerous predator but no other guppies (it was above a waterfall). To follow genetically based changes guppies were regularly taken to the laboratory and raised under the same controlled conditions. After just 11 years, or about 20 guppy generations, the introduced guppies had evolved traits like those in guppies that had coexisted with the killifish for a long time (Reznick et al. 1990). Selection was strong on the traits directly involved with reproduction: age and size at maturity and size and number of offspring.

Old mysteries can be resolved with molecular systematics

It is hard to determine the relationships of highly specialized parasites, for in adapting to their hosts they often lose many of the features that they had shared with their relatives. For example, the tongue worms, or pentastomids, are long, glassy, worm-like animals that occur mostly in the tropics or subtropics, where they live in the noses and lungs of their primary hosts, dogs and crocodiles.

They usually have three larval stages. The first stage develops within the unhatched eggs in the stomach of a herbivore like a rabbit or a fish. The second stage hatches into the stomach of that intermediate host and looks like a tardigrade, a group of minute arthropods called water bears. From the stomach, the tiny larvae swim to and bore into the lungs or liver of their intermediate host and enter a third stage, encapsulating to form cysts. There they wait for their host to be eaten by their definitive host, a carnivore. They leave their cyst in the mouth or stomach of the carnivore, embed themselves again in nasal sinuses or lung tissues, grow and mature, and the cycle begins again (Margulis and Schwartz 1982; Figure 1.5).

In the simplified morphology of pentastomids there are not many clues to their phylogenetic relations. Like arthropods, they have a segmented, chitinous exoskeleton, but there is no internal segmentation. Otherwise, they look like a transparent worm with some hooks for holding on to the host located near the mouth, a simple straight intestine, and a body filled with reproductive organs. Sexes are separate. Except for the hooks and mouth parts, there are few useful morphological features in the adults. Various authors have treated them as relatives of tardigrades, mites, onychophorans, annelids, and myriapods. In most zoology texts they are placed in their own phylum related to the arthropods,

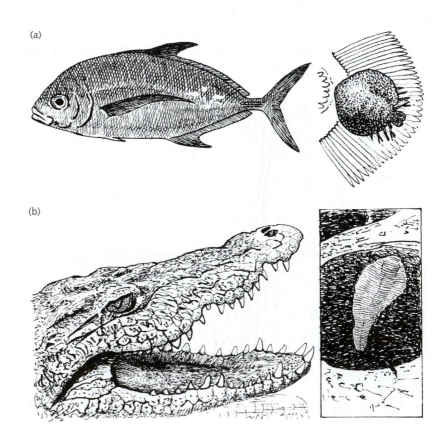

(a)

(b)

Figure 1.5 (a) A branchiuran fish louse attached to the side of a triggerfish. (b) Pentastomid parasites crawling on a crocodile's palate. (By Dafila K. Scott.)

and parasitology texts describe them variously as an independent phylum, a class of the Mandibulata, or an order of the Arachnida. Only the morphology of their sperm suggests a relationship to branchiuran crustaceans.

The mystery of pentastomid relationships was cleared up by comparisons of DNA sequences from pentastomids, branchiuran fish lice, other crustaceans, and representative annelids, chelicerates, myriapods, and insects (Abele et al. 1989). They are relatives of branchiuran fish lice and belong well within the crustacea; they are definitely not in a phylum on their own.

Some evolutionary change is irreversible

The plethodontids are a family of salamanders found in Central and North America. Their ancestor had aquatic larvae, metamorphosis, and semi-terrestrial adults, a life cycle still found in some plethodontids and in many related salamanders. Some plethodontids now mature as larvae; others have embryos that develop directly into adult forms. All have lost their lungs and breathe through their skin. The bones and muscles previously associated with lung breathing have moved forward in the thorax where they help to construct a protrusible tongue used in capturing food. In species that switched from larval to direct development, the structures previously used in larval gills have also found new application in adult structures. It would be hard to select for reversal to lung breathing and larval development because the structures that ancestral salamanders used for those two processes have been employed elsewhere and could not be recovered without killing the developing animal (Wake and Larson 1987; Figure 1.6).

Figure 1.6 The plethodontid salamander capturing a spider with a tongue-flip. (Photo courtesy of Steve Deban.)

Similarly, the bones of the mammalian inner ear were originally breathing aids as gill arches in fish, evolved into feeding aids in the jaws of amphibians and reptiles, then became hearing aids in mammals (Romer 1962). To select mammals to relocate the incus, malleus, and stapes from the inner ear into the jaw again would cause drastic hearing loss and probably death as embryos.

Thus key innovations, such as lunglessness and direct development, imply irreversibility because they allow morphological elements to be used in other structures with different functions, functions so important that a return to the original state would involve costs too high to pay.

Organisms are mosaics of parts with different ages

The first two examples—sexual cannibalism in spiders and rapid evolution in guppies—illustrate the power of natural selection. The second two examples—the puzzle of pentastomid relationships and irreversibility in the evolution of salamanders—illustrate the importance of history. Natural selection and history are the two great themes of evolution that combine to explain the evolution of biological patterns. Every organism is a mosaic, some of whose parts reflect the role of recent selection, others of which recall its phylogenetic history. One of our goals is to understand the roles of selection and history in determining any biological process that interests us.

These two great themes of evolution were established over the last two centuries and were made possible by the work of natural historians that started more than 2000 years ago.

Ideas about evolution have a history

● **KEY CONCEPT**

Every major concept in evolutionary biology has survived challenges and controversies.

While Charles Darwin is the person most strongly identified with the discovery of evolution, he could not have made his creative synthesis without the observations and ideas of many predecessors. In particular, evolutionary biology is based on and stimulated by the accurate observation of nature. Aristotle wrote the first comprehensive natural history in the 4th century BC, and Pliny the Elder, whose curiosity about volcanoes lead to his death in the eruption of Vesuvius in 79 AD, wrote another. Their works were the standard references until the Renaissance in the 15th and 16th centuries. The early naturalists established that organisms can be sorted into groups, which we now call species. They realized that the differences among individuals *within* groups are smaller than are the differences among individuals *between* groups. They also accepted the idea that species have some kind of relationship to one another, birds being more closely related to other birds, for example, than they are to fish. The great figure in developing a classification of living things that reflected their relatedness was the Swede Carolus Linnaeus (1707–78). But even Linnaeus, who attempted the first classification of all living things, did not yet understand why organisms were related, or complicated, or adapted. Answers to those questions had to await the discovery of evolution.

Others prepared the way for Darwin

The transformation of scientific thought about life and the planet that has occurred since the 18th century has been revolutionary. At the end of that century many educated westerners thought that the Earth was thousands, not billions, of years old; that species could not change; that all species that had ever existed were currently living somewhere on the planet, waiting to be found by explorers; and that adaptations were produced by divine intervention. To get to where we are today required the discovery of deep time, the recognition that fossils represent extinct species with living descendants that differ from their ancestors, that species are thus mutable, that most species are extinct, and that adaptations are produced by natural selection. Each step was a major event in the assembly of evolutionary thought.

Geologists discovered deep time early in the 19th century. A key figure was Charles Lyell, who argued that we can understand major geological features as the product of processes observable today. Using observations of the rate at which erosion occurs today, Lyell determined that the Earth must be hundreds of millions of years old to account for the amount of erosion that has occurred in the past. The accurate dating of the Earth had to wait for the discovery of radioactive elements in the early 20th century and the development of atomic clocks in the middle of the 20th century. The nuclear decay of uranium into lead occurs slowly enough to allow estimates of the age of the oldest rocks on Earth. Currently the best estimate of the age of the planet is 4.6 billion years, and the oldest surviving rocks are thought, based on the geochemistry of isotope ratios, to be 3.8 billion years old. The shift from a few thousand years to several billion years in the estimated age of the Earth had enormous impact on ideas about the history of life.

The notion that species are related through descent from common ancestors was discussed at the end of the 18th and the beginning of the 19th century. Erasmus Darwin (Charles Darwin's grandfather) mentioned it in *Zoonomia* (1796), as did Jean-Baptiste Lamarck in *Zoological Philosophy* (1809). The idea stimulated a controversy that culminated in a famous 1830 debate in Paris between Baron Charles Cuvier and Lamarck's younger colleague and defender, Etienne Geoffrey St. Hilaire. At issue was the question of whether the body parts of one species could be understood as having a relationship to those of another species because the species themselves were related through a common ancestor that had had the feature in question. For example, are the forelimbs of horses, bats, and whales similar because all mammals are descended from a common ancestor who had a forelimb with the same basic layout? If they are,

Homology: *A hypothesis that similarity of a trait in two or more species indicates descent from a common ancestor.*

then we can call them homologous (see Chapter 13, p. 315). That was the question at stake, and there was a lot riding on the outcome, for if such relationships could be drawn, then the history of life could be reconstructed as a tree of relationships based on the accurate identification of homologies. If they could not, then species would stand independent of each other, unrelated. The debate was followed in the entire scientific community with great interest. It was resolved with the elaboration of evolutionary biology started by Darwin and continuing up to the present. How to recognize homologies remains an important issue on which the most sophisticated methods of molecular and developmental biology are now brought to bear.

Whereas Cuvier opposed the idea that groups of animals shared common ancestors in the 1830 debates, it was he, with his investigations of fossil mammals from the Paris Basin, who did more than anyone else to show that fossils were the remains of extinct species. By the 1820s that idea had become well established.

Darwin saw that species change and that time is deep

Thus by the time Charles Darwin went on his voyage of exploration in the *Beagle* (1831–6), taking Lyell's recently published *Principles of Geology* with him, he knew that the Earth was probably very old, that fossils represented extinct species, and that living organisms were related in groups, probably because they had shared common ancestors. He was an ambitious young naturalist, and it was clear that whoever solved the problem of how new species originated would become famous. On the voyage he saw fossil giant armadillos in Argentina buried a few feet below living modern armadillos; he saw vast geological sections in the Andes that argued for a very dynamic history of the Earth, as did the uplift he viewed in the harbor at Valparaiso in Chile shortly after a large earthquake. In the Galápagos Islands he saw different species of closely related birds and lizards on the different islands (notably, he did not notice that the finches differed until a specialist pointed that out to him after his return to London). He experienced evidence that strongly suggested that species can change and that time is deep.

After his return to London he read Malthus, who argued that population growth would rapidly bring individuals into competition for scarce resources, and he began a journal to record his thoughts as he assembled his materials for publication. In that journal, in 1838, he first described natural selection. He soon saw that it could explain the origin of adaptations, and when he combined it with geographic separation he could explain the origin of species. But he knew the idea would be controversial, for the consequences were enormous, and he refrained from publishing for 20 years while continuing to gather evidence. Then, in 1858, a bombshell arrived in London, a letter from Alfred Russel Wallace, a naturalist then working in Indonesia. In a malaria-induced period of bed rest, Wallace had also had the idea of natural selection.

Darwin's initial reaction was despair, for he thought that the honorable course was to let Wallace publish alone, but prominent friends who knew what he had in his journals encouraged him to arrange a joint publication. Thus in 1858 Darwin and Wallace published back-to-back papers on natural selection in the *Journal of the Linnean Society of London*. Working rapidly, Darwin then rushed into press with a book summarizing part of the evidence he had collected and some of the interpretations he had worked out: *The Origin of Species by Means of Natural Selection* (1859). The first print run sold out on the first day.

Darwin's book was a sensation that set off a controversy. For some, its most controversial implication was that humans are descended from apes, but the book implied something far more important and general than that. Darwin had worked out most of the details of a process by which the astonishing diversity and precise adaptations of all living things could be explained with material processes operating in the here and now. As he put it in his famous closing paragraph,

It is interesting to contemplate a tangled bank, clothed with many plants of many kinds, with birds singing on the bushes, with various insects flitting about, and with worms crawling through the damp earth, and to reflect that these elaborately constructed forms, so different from each other, and dependent upon each other in so complex a manner, have all been produced by laws acting around us. . . . Thus, from the war of nature, from famine and death, the most exalted object which we are capable of conceiving, namely, the production of the higher animals, directly follows. There is grandeur in this view of life, . . . that, whilst this planet has gone cycling on according to the fixed law of gravity, from so simple a beginning endless forms most beautiful and most wonderful have been, and are being, evolved.

Darwin was criticized for his faulty genetics and his assumption that the Earth is ancient

The Origin of Species was a triumph, but it was not without problems. Two became famous. The first concerned inheritance. Darwin never became aware of Mendel's demonstration, published in 1865, that genes are material particles that exist in most plants and animals in two copies, one from each parent (see the Genetic appendix). Instead Darwin suggested a model of blending inheritance that assumed that genes behaved like a fluid. His model simply did not work, for it led to the disappearance of the very differences among individuals necessary for a response to natural selection, differences that were observed to be maintained. Thus Darwin's model of inheritance was logically flawed and did not fit the facts.

To discover how inheritance works and whether it is consistent with natural selection were not simple problems to solve. They required contributions from a series of scientists whose discoveries earned each of them a place in history. The first mistake they corrected was Lamarck's idea of the inheritance of

Acquired characteristics: *The states of traits acquired by individuals as they develop through their interaction with the environment; they are not inherited and do not respond to natural selection.*

Germ line: *The cell lineage of gametes, zygotes, stem cells, and gonadal cells that stretches continuously back from each organism to the origin of life.*

Soma: *The cells of the body that carry out all the functions of life except gamete production.*

Central dogma: *A key concept in molecular biology that states that DNA makes RNA, and RNA makes proteins. Thus information flows from DNA out to the cell, not from the cell back into the DNA.*

acquired characteristics. Darwin resorted to this idea in later editions of *The Origin of Species* as he reacted to criticism of his model of blending inheritance. Lamarck had claimed that offspring could inherit the states that their parents acquired by interacting with the environment. For example, if a giraffe stretched higher and higher to reach leaves at the top of trees, it would become slightly taller. Lamarck thought that its offspring could inherit that condition, and that this would explain how giraffes had evolved their great height. In later editions of *The Origin of Species* Darwin adopted the idea of Lamarckian inheritance, for he thought that the use and disuse of parts could produce heritable modification, but he was mistaken. For example, Darwin thought ostriches lack wings in part because their ancestors did not use wings and in part as a response to natural selection.

Weismann destroyed Lamarck's logic by distinguishing **germ line** from **soma** in 1892 and supporting his distinction with convincing experimental evidence. The germ line consists of the gonads and their egg and sperm cells. These are normally sequestered early in animal development; cells from the rest of the body, the soma, do not later move into the gonads. Now we also know, as the **central dogma** of molecular biology puts it, that DNA makes RNA makes proteins. The central dogma establishes at the molecular level for all organisms the idea that Weismann established at the level of cells and organs for some organisms. Information does not flow from the organism back into the genes, where it could be transmitted to offspring; it flows from the genes out to the organism. Communication is one-way, from genes to organisms. (There are a few exceptions, but they do not affect the general conclusion.) Thus there is no mechanism by which developmental changes in neck length in giraffes could be communicated to the giraffe genes and passed on to their offspring.

Weismann's insight was brilliant, but he did not know what genes are made of. Part of that answer came when Correns, Tschermak, and de Vries rediscovered Mendel's laws of inheritance in 1900. They established that genes are material particles, not fluids, present in most organisms in two copies that segregate when gametes are formed. Gametes thus contain just one copy of each gene, and gamete fusion re-establishes the two-copy condition. By 1915 Morgan and his group had shown that genes are located on chromosomes and that the behavior of chromosomes in meiosis explains Mendel's laws. Between 1918 and 1932 Fisher, Haldane, and Wright showed that Mendelian genetics is consistent with natural selection. Only then, more than 60 years after the publication of *The Origin of Species*, was the genetic objection to natural selection finally removed. Modern molecular and developmental genetics have confirmed in exquisite chemical detail the key aspects of genetics necessary for Darwin's ideas to work: that the genetic material is DNA, that DNA has a sequence that can be replicated precisely, that DNA mutates, and that DNA contains information in its genetic code that determines the precise structure of the proteins whose structures and activities form the materials and control the processes out of which organisms are built.

The second big problem with *The Origin of Species* was the age of the Earth. In the 1860s it was thought that the Earth had formed from a ball of liquid lava that had cooled down continuously since it was formed. In 1862 William Thomson calculated from the current temperature of the Earth that the planet must be much younger—somewhere between 20 and 400 million years—than the span of time that Darwin thought was needed for evolution to have produced the diversity of life. Later Helmholtz and Newcomb estimated the age of the Sun at 100 million years, assuming a constant rate of condensation from a ball of gas, and Joly estimated the age of the oceans at 90 million years, assuming a constant accumulation of salt flowing in from rivers. Respected physicists thus thought that there was not enough time for Darwin's process to have accomplished what he claimed. They were wrong.

Darwin died, in 1882, before solutions were found. Becquerel's (1896) and Curie's (1897) discovery of radioactivity greatly extended the age of the Earth, for naturally radioactive minerals deliver heat continuously to the Earth's interior and can account for the current temperature of the Earth even if the planet is several billion years old. With the later recognition that the Sun runs on thermonuclear reactions came the realization that the constant-condensation model was unrealistic and that the Sun is much older than Helmholtz and Newcomb had thought. And when it was discovered that much of the salt that entered the ocean precipitated and was stored in rocks, some of which are carried down into the mantle at the tectonic boundaries where plates are subducted (pushed down below the continental margins), it became clear that the accumulation of salt from rivers seriously underestimates the age of the oceans. There was plenty of time for evolution to have taken place after all.

By the middle of the 20th century the major elements of evolutionary thought had been tested rigorously against alternatives and remained standing as the best available explanations of biological diversity and adaptation. Subsequent work has confirmed evolution in great detail. In the early 21st century no scientist doubts the reality of biological evolution. While it is no longer controversial within science, it remains controversial within society as a whole.

Creationists object to evolution for several reasons

Religions contain a wealth of wisdom about the human condition; they are the source of much of our ethics and morality. There is much in religion to respect, and certainly all educated persons, whether they believe in a religion or not, should be acquainted with the doctrines and texts of the major religions, for they have been a key element of the human experience and a major force in shaping our cultures and histories.

Some religious groups object to evolution because they think that its science undermines their religion. They include fundamentalist Protestants, some of

● **KEY CONCEPT**

The nature of reality is not subject to the decrees of human institutions.

whom believe that the world is 6400 years old and that each species was created separately, and fundamentalist Muslims. Because some of you may come from such a background, and all of you are likely to encounter such people, we discuss some of the principal Creationist objections to evolution below.

Before dealing with those objections, we note two things. First, there is a major difference between religion and science. Religion is based on belief; science is based on observation and experiment. Religions are not rejected when they are tested and found wanting; no believing Christian has abandoned Christianity when prayers are not answered. But there is no idea in science so important that it would not be thrown out if it were inconsistent with the facts and could be replaced by a better alternative. We can state the conditions under which scientists would be willing to accept an alternative to evolution as an explanation of biological diversity and adaptation. The alternative would have to explain everything that evolution explains just as well as evolution explains it, would have to be consistent with all the many parts of science with which evolution is consistent, and would in addition have to be able to explain some important empirical observation that evolution cannot explain. At the moment, no alternative—including Creationism—comes close to doing that. Evolution is a scientific fact that has been demonstrated so many times that it would be perverse to withhold provisional assent (Gould 1999).

Second, we note that evolution now poses no problem for the Catholic Church. In 1996 the Pope officially stated that evolution has occurred and is compatible with Catholicism. In so doing he reacted to the Church's experience following its confrontation with Galileo in the early 16th century. Then the controversy was over the structure of the universe—whether the Earth was the center of the universe, or whether, as Copernicus had argued, the Earth revolved around the Sun and other non-solar bodies also moved. Through his new telescope Galileo observed four moons moving around Jupiter. He realized that he was looking at a miniature model of the solar system, and that Copernicus was right. He then proclaimed his discovery in public lectures and in a provocative book, *Dialogue Concerning the Two Chief World Systems*.

The Church forced Galileo to recant, but the victory was short-lived, for the nature of reality is not subject to the decrees of human institutions. A few decades later Newton discovered gravitation and the laws of motion, which predicted the orbits of the planets with astonishing accuracy. Continuing advances in physics, astronomy, and cosmology made the position of the Church increasingly embarrassing. In the late 20th century it changed its position on Galileo retroactively and, having learned the difficulties of dealing with a position that contradicts well-established science, it accepted evolution. A similar scenario was acted out in the Soviet Union when Stalin accepted the arguments of Lysenko for the inheritance of acquired characteristics, denied the established facts of genetics, destroyed genetic research and the agriculture on which it was based, and caused famines that killed millions. And when the president of South Africa recently denied that AIDS is caused by the HIV virus, thousands were denied appropriate treatment. People can pay with their lives when authorities ignore or deny what science has learned about nature. And authorities who deny science risk embarrassment or worse.

With that as background, we now consider some of the more frequently encountered Creationist objections to evolution.

Creationist objections misconstrue science

Antolin and Herbers (2001) include the following among the chief Creationist objections to evolution.

Evolution is just a theory

To say that evolution is 'just' a theory is to miss a key point about science. Evolution is a theory in the same sense that Newton's theory of motion, Einstein's theory of relativity, or the atomic theory of matter are theories. The major ideas in science start out as working hypotheses, and they have to survive test and controversy before they are accepted by the scientific community. During that period of test and controversy they are referred to as theories, but as a result of confirmation and long use as reliable tools they acquire the status of well-supported ideas that all scientists accept. That happened to Newton's ideas, it happened to Einstein's, and it happened to Darwin's. To call evolution a theory does not imply to a scientist that it is unreliable. Evolution is a major scientific theory, as well confirmed and as reliable as any.

The scientific explanation of the origin of life remains weak

That this objection is true does not detract from the status of evolution as a major scientific insight. We do not understand the origin of life nearly as well as we understand speciation, or selection, or many other aspects of evolution. But our failure is not one of principle or logic; it is a failure of imagination and technique. We have not yet been able to imagine and test all of the intermediate states between inanimate matter and organisms capable of metabolism and reproduction. We have a good account of many of those stages, and those that are not understood are interesting scientific challenges. A good description of what we do—and do not—understand about the origin of life can be found in Maynard Smith and Szathmary (1995). It would be very risky to claim that these problems will not eventually be solved, for experience suggests that they will.

Evolved structures are too complex to have evolved from random mutations

This claim is often encountered not only among Creationists but also among people, some of them scientists, who have not understood the extraordinary power of natural selection to rapidly create complex order out of chaotic variation. Natural selection is in fact the only natural process known that can create and maintain order against the tendency of all physico-chemical systems to become disordered. It does so by creating and then improving closed systems that import energy to maintain and reproduce themselves. The power of natural selection efficiently to create structures that at first glance seem highly improbable is discussed in more detail in Chapter 2 (p. 32).

The evolutionary view of nature is unacceptably grim

Early interpretations of evolution did conclude that evolved behavior should be unremittingly selfish so that individuals could maximize the number of off-spring they produced. That view resembled Thomas Hobbes' famous description of life in *Leviathan*: 'nasty, brutish, and short.' Our first response is that even if evolution's view of nature were grim, it could still be true. However, there is more to it than that. In the 1960s, evolutionary theories predicted altruistic (self-sacrificial) behavior for relatives, such as worker bees foregoing reproduction to raise sisters; in the 1980s they predicted cooperative behavior among non-relatives, such as vampire bats feeding unrelated vampire bats who had not found food that night. Since then we have seen a steady development of ideas and evidence that has broadened the conditions under which we would expect to see cooperation evolve. Thus if religion and evolution differ on altruism and cooperation, then the difference is not about whether they are expected to exist, but what produced them and when we should expect them to be elicited and used. Evolved behavior may have been shaped by reproductive success, but it sometimes has consequences that we would describe as admirable and benign.

Evolution means survival of the fittest, a doctrine that has unacceptably justified Social Darwinism, eugenics, and genocide

This accusation is a striking instance of the Naturalistic Fallacy—the idea that if it happens in nature, it is justified in human relations—that what is, is good. The Naturalistic Fallacy is just that—a fallacy. There are no values in nature. Evolved behavior is neither good nor evil. It just is. Humans invent values and use them in their interactions with each other. Sometimes they are misused. There is a long history of one group or another using some analogy with nature to justify some human policy or action. Such interpretations are propaganda used to advance particular agendas and should be viewed with suspicion. We need not accept such analogies, for they are not based on logic. We are free to consider alternatives, one of which is that we should not always behave in ways that evolution has shaped.

Evolution is atheism

Some evolutionary biologists are atheists, some are religious, and some atheists do not base their stance on evolution. Evolution is compatible with many religions, including some versions of Christianity. It is not compatible with a literal interpretation of parts of the Christian Bible, particularly those concerning the age of the Earth and the creation of life. Anyone who accepts those passages has difficulties not just with evolution but with physics, cosmology, and geology as well.

Scientific Creationism is a valid alternative hypothesis, the evidence for evolution is weak, and therefore Creationism must be right

Creationism is not scientific because it is not subject to change when tested and found wanting. It is not a valid alternative hypothesis because it cannot explain what evolution explains and because it is not consistent with the other aspects of science—physics, chemistry, geology—with which evolution is consistent. The evidence for evolution is not weak, it is strong; you can see that both later in this book and in the vast literature on evolutionary biology. And it is illogical to claim that there are only two alternative explanations for anything—that is a failure of imagination. Experience shows that new alternatives are always being thought up. The issue in science is not to think of an alternative, which is relatively easy, but to think of one that works better than the existing explanation, which can be very hard when that explanation has survived many tests. Such is the case with evolution.

Teaching Creationism is only fair

Fair to whom? We do not teach that the Earth is flat, or that the Sun rotates around the Earth, or that heat consists of a substance called phlogiston, or that infectious disease is caused by bad air. We have learned that those ideas are incorrect. It would be a disservice to students to take up time better used discussing what we do know to be correct by discussing incorrect ideas that the scientific community has long since discarded. Worse than that, it would be dishonest.

Scientific disagreement is a sign of weakness

This criticism gets the issue exactly backwards. In science disagreement and debate are a sign of strength and part of the mechanism by which science makes progress. They are the means by which every important scientific issue is examined from all angles until we are sure it is reliable. Anyone making this criticism has not understood how science works.

Evolution has never been tested

Experimental microevolution is a flourishing field in which many papers are published each year; it has frequently observed and confirmed evolution in laboratory experiments. Scientific inference is also based on observations and comparisons without experimentation. Astronomy and geology are two fields whose major conclusions are not based on experiments, but they have made progress and yielded a great deal of reliable knowledge. The same is true of macroevolution. The analysis of fossils, of the relationships of living organisms, and of their geographic distribution has allowed us reliably to reconstruct the major features of the history of life. Evolution has been tested many times in great detail, both experimentally and with the comparative method. It has not been found wanting.

Macroevolution has never been experimentally confirmed

The key process in macroevolution is speciation. One project within experimental evolution is to produce speciation in the laboratory. Some important elements of speciation—the evolution of mate choice leading to reproductive isolation coupled with subsequent divergence of the two populations—have already been achieved (see Chapter 12, pp. 299 f.). Moreover, for more than 50 years biologists have recreated in the laboratory the speciation of many plants, among them those in the cabbage family, through hybridization. It would be unwise to claim that speciation will not someday be based on a substantial and convincing program of laboratory experiments. It already is in part.

Ignoring the reality of evolution would be dangerous

The evidence and logic summarized and cited in this book confirm the reality of evolution. If we denied evolution and ignored its consequences, we would place ourselves at great risk. We would not then know how to deal with the evolution of antibiotic resistance; millions would die of infections. We would not know how to deal with the evolution of herbicide and pesticide resistance; crops would fail and millions would starve. We would not understand how virulence evolves, and why some diseases become more, others less deadly. We would not know how to understand the history and distribution of human genetic diversity and thus would not realize that the claims of racists are without scientific foundation. We would not understand the history of life, marked by mass extinctions, and would not know how to think about the consequences of the mass extinction occurring right now.

The reality of evolution places these and many other issues of vital significance in an importantly different light. If we ignored evolution, we would miss as much wonder and awe at the nature of life and the human condition as were ever produced by a religion. The history and diversity of life, the power of natural selection to shape precise adaptations, and the implications of our evolved condition are absolutely astonishing.

● SUMMARY

- Microevolution occurs within populations in the short term; it operates both through evolution driven by natural selection, which produces adaptations, and through neutral evolution driven by genetic drift, which produces random divergence among populations.

- Natural selection results from the correlation between traits and reproductive success; a response to selection follows if the variation in the traits is heritable.

- Genetic drift results when genes that do not affect fitness occur in families of different sizes and when such genes are placed at random into gametes.

- Macroevolution occurs among species in the long term; the patterns it produces are driven by speciation and extinction.

- Evolution adds natural selection, drift, and history to the laws of physics and chemistry, thereby creating an explanatory framework that is sufficient to understand the complexity, diversity, and adaptation of living things.

- Proximate causation results from chemical and physical mechanisms operating within the lifetime of a single organism.

- Ultimate causation results from micro- and macroevolution operating over many generations.

- Adaptation is not primarily about survival; it is essentially about reproductive success. If death will increase reproductive success, death will evolve.

- Microevolution is fast when selection is strong in large populations that contain much genetic variation.

- Macroevolution can be irreversible when organs and structures lose old functions and acquire new ones.

- Our view of the relationships among species has often changed when new evidence has appeared, and it will continue to do so for some time to come.

- The ideas of evolution have survived many controversies and tests and are now considered as reliable as any ideas in science. The facts of evolution are well established, by both experiments and historical inference.

- Creationist objections to evolution are invalid, but evolution is not in conflict with many interpretations of religion.

● RECOMMENDED READING

Darwin, C. (1859) *On the origin of species by means of natural selection or the preservation of favoured races in the struggle for life*. John Murray, London.

Eiseley, L. (1961) *Darwin's century*. Anchor Books, Garden City, NY.

Futuyma, D.J. (1997) *Science on trial*. Sinauer Associates, Sunderland, MA.

The Pope's message on evolution and four commentaries. (1997) *Quarterly Review of Biology* **72**, 375–406.

PART 1

Microevolutionary concepts

Microevolution operates within populations. It includes adaptive evolution, which consists of natural selection and the genetic response to it, and neutral evolution, which consists of the random drift of traits and genes that have no effect on reproductive success. It also includes the short-term effects within populations of ancient developmental programs. Each of these is explored in more detail in the following chapters, and each describes a key mechanism of evolution. These are the nuts and bolts of the process.

Chapter 2 describes adaptive evolution: how natural selection works, what forms it takes, and how strong it is in nature. Chapter 3 discusses neutral evolution: how genetic drift comes about, why it is significant for molecular evolution, and how analogous processes generate patterns in biodiversity and history. Chapters 4 and 5 discuss how genetic variation responds to selection and how it originates and is maintained in populations.

Developmental biology studies how information stored in genes is used to build organisms. Development is central in evolution because all changes in organisms require changes in development. Chapter 6 focuses on ancient developmental programs that control the construction of organisms. Chapter 7 discusses how variation is expressed in the phenotype. One key idea is the reaction norm, which describes how one genotype can interact with the environment to produce different phenotypes.

Thus the central ideas of microevolution are selection, drift, genetics, and development. Each of those big ideas has many parts; much of the action in microevolution takes place at their intersection.

CHAPTER 2
Adaptive evolution

This chapter discusses adaptive evolution, introduced in Chapter 1, in greater detail. Adaptive evolution is driven by natural selection, and natural selection is one consequence of variation in lifetime reproductive success. (Variation in reproductive success is also one cause of neutral evolution, taken up in Chapter 3, p. 54.) Variation in lifetime reproductive success arises among *material* organisms. The response to selection, however, is recorded not in the organisms, all of which eventually die, but in the genes transmitted to the next generation—the genes that were in the organisms that had reproductive success. And the essential feature of genes is that they store *information* on how to build organisms. That is why we distinguish clearly between organisms in the definition of natural selection and genes in the definition of the response to selection. Natural selection occurs in *material* organisms; the response to selection is recorded in *information* stored in the genes. Both are necessary for adaptive evolution to occur.

● **KEY CONCEPT**

Selection occurs in material organisms; the response to selection occurs in genetic information.

When is evolution adaptive, and when is it neutral?

Either adaptive or neutral evolution may occur when two conditions are met

Natural selection causes adaptive evolution in traits and genes; neutral evolution causes non-adaptive change in traits and genes; and two conditions are necessary for both adaptive and neutral evolution to occur. These two conditions are:

(1) *variation among individuals in lifetime reproductive success*, in the number of their offspring that survive to reproduce;

(2) *heritable variation in the trait*; that is, whether the state of the trait in the parent—at least whether it is above or below average—is inherited by the offspring. Only if at least some of the variation in the trait is heritable will the genes that are responsible for the trait change in frequency and record the action of selection or drift.

● **KEY CONCEPT**

For a trait to undergo evolutionary change, it must vary heritably among the organisms in the population, and those organisms must also vary in their reproductive success.

● **KEY CONCEPT**

The correlation between traits and reproductive success determines whether evolution will be adaptive or neutral.

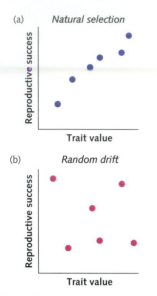

(a) *Natural selection*

Reproductive success

Trait value

(b) *Random drift*

Reproductive success

Trait value

Figure 2.1 The distinction between selection and drift arises in the correlation between lifetime reproductive success and trait value. If the correlation is positive (as it is in panel a), selection acts. If it is zero (as it is in panel b), drift acts.

It is not these two conditions but the link between them that determines whether evolution is adaptive or neutral. The link is *the correlation of the trait or gene with lifetime reproductive success*; that is, a consistent relationship between a state of the trait or gene and whether or not the individual carrying it has more or fewer than the average number of offspring that survive to reproduce.

Thus two conditions and the link between them are the key elements of microevolution. The conditions are heritable variation in traits and variation in lifetime reproductive success. The link is the correlation between the two. The difference between adaptive and neutral evolution lies in the correlation (Figure 2.1). When that correlation is positive or negative, natural selection is operating on the trait, and evolutionary change will move the trait from genera-tion to generation in the direction of increasing adaptation. When the correla-tion is zero, natural selection disappears, and then the things that are inherited and that vary, whether genes or traits, fluctuate randomly in the population.

Consider two traits in the same population. Variation in the first trait is cor-related with reproductive success, and some of that variation is genetic (Figure 2.1a). Variation in the second trait is not correlated with reproductive success, and some of that variation is also genetic (Figure 2.1b). The same vari-ation in reproductive success that causes adaptive evolutionary change in the first trait will cause neutral drift in the second. To see this, consider a neutral gene or trait in a population in which there is much variation in reproductive success, where some females have many surviving offspring whereas others have few or none. The neutral gene or trait increases or decreases erratically depending on whether it occurs in a large or a small family. Because it is not cor-related with family size, occurring in many different family sizes at random over the generations, it does not change steadily in any particular direction.

This way of distinguishing between adaptive and neutral evolution unifies the two basic microevolutionary processes. It is the strength of the correlation between trait variation and reproductive success that determines whether a trait or gene will undergo adaptive change or perform a random walk through time. In that correlation, in the reasons for it being positive, negative, or zero, lies much of the causal explanation of microevolution.

Ecological interactions guarantee variation in lifetime reproductive success

Organisms in natural populations always vary in lifetime reproductive success, for such variation is produced by the ecological interactions that always occur among individuals. One important ecological interaction occurs as population density increases and the amount of food per individual decreases. Individuals then grow more slowly to become smaller adults, have fewer offspring, and suffer higher mortality rates. However, because of genetic variation between individuals, some are better able to cope with living at high density than others. These individuals will grow faster, become larger, survive better, and have more offspring than others

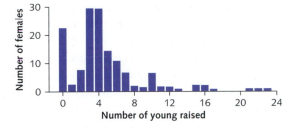

Figure 2.2 Lifetime reproductive success measured as offspring that fledged for 142 female sparrowhawks from 1971 to 1984. Twenty-three females produced no fledglings, one produced 21, one produced 22, and one produced 23. The potential for natural selection was great. (From Newton 1988.)

because they competed successfully for the limited resources. In this manner responses to increasing density generally produce a great deal of individual variation in the number of offspring produced per lifetime. Similarly, individuals react differently to other environmental challenges, including temperature, food supply, diseases, and predators. Such ecological interactions ensure that all natural populations contain individuals that vary in their reproductive success.

When many individuals in a population can be followed from birth to death, it is possible to quantify how much variation in reproductive success exists in nature (Clutton-Brock 1988). For example, in Newton's (1988) study of sparrowhawks in southern Scotland, 72% of the females that fledged died before they could breed, 4.5% tried to breed but produced no young, and the remaining 23.5% produced between 1 and 23 young apiece (Figure 2.2).

Large cactus finches on Isla Genovesa in the Galápagos Islands (Figure 2.3) also vary greatly in reproductive success. From 1978 to 1988 Grant and Grant

Figure 2.3 A cactus finch foraging on *Opuntia* cactus in the Galápagos Islands. (Photo courtesy of Dr. Robert H Rothman, Department of Biological Sciences, Rochester Institute of Technology.)

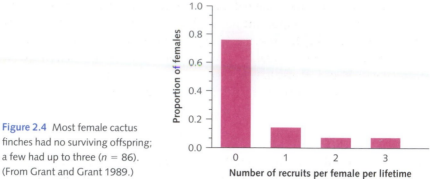

Figure 2.4 Most female cactus finches had no surviving offspring; a few had up to three (*n* = 86). (From Grant and Grant 1989.)

(1989) caught and individually banded most of the birds on the island, constructed detailed genealogies, and followed the fates of offspring.

During this period conditions were difficult and the population declined markedly: 79% of breeding males and 78% of breeding females produced no offspring that survived to breed. Most successful adults contributed only one offspring to the next generation, and a few contributed two or three (Figure 2.4).

The most successful parent was a male who bred from 1978 through 1982 and produced five surviving offspring that bred and in turn produced surviving offspring before the study ended. Females varied from 0 to 3 recruits (the offspring that survive to maturity) per lifetime. The most successful female also lived the longest. Between 1978 and 1987 she laid 31 clutches in 8 breeding years. Her 110 eggs produced 58 fledglings, only three of which survived to breed themselves. Such large individual variation in reproductive success translates into strong selection on any traits correlated with it.

Heritable variation of traits in natural populations is common

Even with great variation in reproductive success, there will be no evolutionary change in a trait that does not vary heritably, for the differences in performance exhibited by the parents will not be inherited by and expressed in the offspring. Many traits in natural populations display heritable variation. To determine whether variation is heritable we can ask whether offspring resemble their parents, as O'Neil (1997) did for a plant, the purple loosestrife (*Lythrum salicaria*). To see whether the variation in number of seeds per capsule was heritable, she plotted the values for the offspring against the average for the two parents (the mid-parent value; Figure 2.5).

Parents that produced above-average numbers of seeds per capsule had offspring that also produced above-average numbers of seeds per capsule, and parents with below-average numbers of seeds per capsule had offspring that also produced below-average numbers of seeds per capsule. Given such heritable variation selection could elicit a rapid response.

Purple loosestrife is but one example among thousands. Many traits in natural populations vary heritably, and many genes in natural populations are present

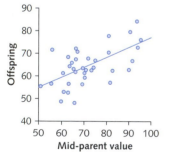

Figure 2.5 The resemblance of offspring to parents for number of seeds per capsule in purple loosestrife in a population growing along a river in Maine. The abundant heritable variation would enable a strong response to selection. Mid-parent means the average of the two parents. (Data courtesy of Pamela O'Neil.)

in several versions. That the potential for a response to selection is thus the rule, not the exception, is an important point based on decades of work by many biologists. Dobzhansky, for insects, Mayr, for birds, Stebbins, for flowering plants, and Lewontin and Ayala, for gene products detectable by electrophoresis, along with many others, have shown that a great deal of genetic variation is present in natural populations for many types of trait and organism.

Selection is not just the survival of the fittest

It is a misleading half-truth to claim that evolution is concerned only with the survival of the fittest. Survival is important, but only insofar as it contributes to reproductive success, to the number of offspring produced per lifetime that survive to reproduce. The action of selection is located in reproductive success and in the many ways that variation in genes and traits connects to reproductive success, creating the key correlation at the heart of selection.

Nothing consciously chooses what is selected

Nature is not a conscious agent who chooses what will be selected. Genes and traits increase or decrease in frequency because of their correlation with the reproductive success of the organisms that carry them. There is no long-term goal, for nothing is involved that could conceive of a goal. There is only short-term relative reproductive success producing change from one generation to the next.

The ordinary causes of selection have extraordinary adaptive consequences

The subtlety and power of natural selection are hidden in its ordinary causes and revealed in its extraordinary consequences. Its causes are the reasons for individual variation in survival rates, in mating success, in offspring number, in offspring survival, in the survival and reproduction of relatives, and all the many reasons why variation in traits is correlated with variation in reproductive success; for example, why being able to run faster leads to improved survival in zebras, and why producing colorful, long-lived flowers leads to improved pollination in orchids. Its consequences include the mechanisms of development and physiology, the structure and performance of flowers, brains, eyes, hearts, and other organs, much of behavior, much of population dynamics, and much of the ecological interactions of species: in short, much of biology. Do not let the apparent simplicity of its ordinary causes lead you to underestimate natural selection's scope and power.

Because natural selection is the correlation of a trait with reproductive success, survival is only important if it contributes to reproductive success. That is why male red-backed spiders commit suicide during copulation, that is why senescence evolves, that is why selection for mating success (sexual selection) can reduce survival, and that is why worker bees will commit suicide to defend their nest. Survival is only a means to achieve reproductive success. If death will bring the individual greater reproductive success than survival, death will evolve.

Natural selection can rapidly produce improbable states

● **KEY CONCEPT**

Selection rapidly finds improvements; inheritance accumulates them.

Selection creates order out of disorder. Under strong selection populations rapidly produce combinations of genes that are at first glance extremely unlikely. We demonstrate this with two examples. The first, an analogy using the letters of the alphabet, exaggerates the power of selection because it describes a process that aims to produce a very specific, concrete goal. This is not how natural selection works, but the example makes an important principle clear. We use the first example to prepare the more realistic, and more complicated, second example.

Inheritance preserves the improvements sorted by selection

The 31 letters in THEREISGRANDEURINTHISVIEWOFLIFE are analogous to a sequence of 31 genes, each with 26 different possible versions. If evolution assembled such sequences completely at random, it would have to sort through 26^{31} different possible combinations of letters to hit on this one. If it sorted through a hundred million sequences per second, it would take more than a hundred million years to do so. However, natural selection causes successful gene combinations to increase in frequency, and accurate replication preserves those increases—it remembers what worked before. In this artificial example let us suppose that strong selection retains the correct letter whenever it occurs. If we start with any random sequence of 31 letters and retain all the letters that happen to be correct, then repeat the process by generating new letters at random for the ones that are not yet correct, we get to the right sequence in about 100 trials, about 30 orders of magnitude faster than a random search (Dawkins 1986, Ewens 1993)—in other words, incredibly rapidly. This example is somewhat misleading because natural selection does not aim at any particular final state. It doesn't aim at anything. It just produces something that works better from among the variants currently available. And in this example the mutation rates change from very high to very low as soon as a correct letter is found, which is not biologically realistic. The next example moves from letter-play to meaningful experimentation.

Point mutation: *A change in a single DNA nucleotide, e.g. adenine mutates to thymine, or an insertion or deletion of a single nucleotide.*

A random process would be very unlikely to produce an adaptation

This example requires that you know that the **point mutation** rate is the rate at which mistakes are made when sequences are copied—for example when an AAT sequence is miscopied to produce ATT. It is the high point mutation rate of ribonucleic acid (RNA), a macromolecule that stores genetic information, which allows RNA to evolve rapidly in test tubes. RNA is the genetic molecule of many viruses, and its high mutation rate has important consequences. For example, HIV is an RNA virus, and its high mutation rate makes it hard to find a vaccine against it. The high mutation rate of RNA is what allowed the following experiment to work quickly.

From a virus that infects bacteria, one can prepare an enzyme that copies RNA. Given an RNA molecule as substrate, an energy source, and a supply of the four necessary building blocks—the nucleotides—from which RNA is made, this enzyme rapidly produces a large population of RNA copies in solution in a test tube. By transferring a drop of the solution into a new test tube every 30 minutes, one selects the copies present at highest frequency and most likely to be present in the drop. The molecules at highest frequency are those that are copied most rapidly. Replication is good but not exact: in about 1 in 10 000 cases one nucleotide is substituted for another—a mutation occurs. The mutations that are copied faster have greater reproductive success. Two types of molecule have an advantage: small ones, rapidly copied, and larger ones that fit well to the replicating enzyme.

After more than 100 transfers, a large, complex molecule dominates the population. Which one depends on details; one that occurs frequently is 218 nucleotides long. Hitting on such a molecule at random has a probability of 1 in 4^{218} or 1 in 10^{131}. Since there are about 10^{16} molecules in the test tube just before transfer, the procedure screens about 10^{16} molecules every half hour. If the process were random, it would take about 10^{115} years to find the one that is best at being rapidly copied. Instead, the procedure produces something close to the best one in about 2 days. The response to selection is efficient because each step leads to a molecule that is better than the previous one. Because the improvements are inherited, they accumulate (Maynard Smith 1998).

Remember this example if you encounter the false argument that natural selection cannot make precise, complicated organs like the vertebrate eye because it starts with random variation. Natural selection can rapidly convert initially random variation into highly adapted states. They only *appear* to be improbable because we have not watched the process at work. The efficiency of natural selection makes even extremely precise and complicated structures probable given enough time.

> **RNA:** *Ribonucleic acid or RNA is a macromolecule that, like DNA, has a sugar-phosphate backbone to which are attached nucleotides. It is the genetic molecule of RNA viruses. In other organisms it occurs as messenger RNA (mRNA), transfer RNA (tRNA), and ribosomal RNA (rRNA).*

Adaptations increase reproductive success

Whenever heritable variation in a trait is correlated with reproductive success, the trait changes from one generation to the next, and the result is improved reproductive performance. As this improvement continues, a *process* called adaptation, it results in a *condition* in the trait that we also call an adaptation.

● **KEY CONCEPT**

The remarkable precision of some adaptations demonstrates the power of selection.

An adaptation is the state of a trait, or a change in that state, that increases the reproductive success of the organisms that carry it.

The striking precision of adaptations is demonstrated by the accurate, coordinated timing of reproduction in marine organisms that rely on the predictability of the moon and the tides. For example, palolo worms (polychaete worms of several species in two families) live most of the year in the sea floor in shallow water in the western Pacific Ocean. As the reproductive season approaches, they grow a special reproductive organ on the back end of their body that looks like an individual worm. Reproduction is triggered by a specific phase of the moon, detected by millions of individuals scattered across a large area. On just a few nights of the year, the reproductive organ splits off the worm and swims to the surface, where it encounters millions of others, forming a massive swarm that spawns and dies. If the timing were not precise, reproduction would often fail, for lack of synchrony would result in lack of partners. Meanwhile, the adults survive and grow another reproductive organ for the next year.

Similarly, the grunion (*Leuresthes tenuis*), a member of the silversides family, is a fish found in coastal Californian waters that spawns with the highest tide of the month. It rushes in with the waves just as the tide is turning and throws itself out of the water, depositing eggs and sperm in pockets in the wet sand where the eggs will not be disturbed again by waves until a month later, at the time of the next spring tide. When that tide arrives and the waves disturb the eggs, the young hatch explosively and swim out with the receding water. The timing of reproduction, development, and hatching are coordinated precisely with the rhythm of the tides.

The precision of adaptation is also illustrated by the ability of bats to find prey in the dark using echolocation. Bats produce high-frequency cries whose energy is reflected by objects in their environment. There are physical constraints on this method of 'seeing.' Sound waves rapidly lose energy with distance. Lower-frequency waves penetrate further but can only detect large objects, and many bats use echolocation to detect small, rapidly moving, flying insects (Figure 2.6).

Griffin (1958) discovered that when a bat approaches to within 1 or 2 m of an obstacle, it increases the number and raises the frequency of the ultrasonic pulses that it emits. Griffin also tested the ability of bats to fly through grids of fine wires. They had no problem with wires that were 0.4 mm in diameter and had some success with wires down to 0.2 mm in diameter, which reflect very little sound energy. Later Simmons (1973) showed that a bat could discriminate target distances as small as 1 cm by detecting differences of as little as 60 μs in the pulse-to-echo interval. It could also discriminate a stationary from a vibrating target where the vibration was as small as 0.2 μs, implying a difference in pulse-to-echo interval of just 1 μs. For comparison, the duration of a single action potential in the bat's auditory nerve is about 1 ms, a thousand times longer. These astounding abilities to discriminate the distance and nature of an

Figure 2.6 A bat catching a moth in complete darkness. Bats have exquisite adaptations for catching prey, and moths have similarly exquisite adaptations for avoiding bats. (Photo copyright Stephen Dalton/NHPA.)

object in complete darkness result from selection for the ability to locate flying-insect prey whose wing beats convey information on distance and direction of movement.

Similarly impressive abilities to detect faint signals that carry critical information are found in the noses of migrating salmon, which can detect as little as a single molecule characteristic of their native stream, and in the dark-adapted eyes of nocturnal mammals, which can detect as little as a single photon of light. Natural selection has great power to shape precise adaptations.

Selection has been demonstrated in natural populations

Birds lay the number of eggs that gives them greatest reproductive success

When a trait has evolved to a state of adaptation, the version found in nature has greater reproductive success than the alternatives against which selection has tested it. This observation suggests a way of testing for adaptation. To see whether a trait is adapted, we can manipulate it and note the consequences. If it is adapted, the manipulated forms should have lower reproductive success than the natural state. Such a test would be most risky for the selection hypothesis if the manipulated trait were directly connected to reproductive success, for then selection should certainly have brought it into an adapted state. That is the case for clutch size in birds, a classic life-history trait discussed in greater detail in Chapter 10 (p. 214).

● **KEY CONCEPT**

Experimental manipulations reveal the impact of selection.

Figure 2.7 A kestrel feeding a lizard to its nestlings. Kestrels work hard to raise their chicks; increased reproduction decreases adult survival. (Photo courtesy of Gary Weber, University of Newcastle, Australia.)

Reproductive value: *The expected contribution of organisms in that stage of life to lifetime reproductive success.*

Residual reproductive value: *The remaining contribution to lifetime reproductive success after the current activity has made its contribution.*

For birds that naturally nest in tree holes and readily build nests in artificial boxes, a clutch can be increased or decreased by adding or removing an egg or a nestling. Daan et al. (1990) did such clutch-size manipulations on a falcon, the kestrel (Figure 2.7), at a site in the Netherlands where survival was good and clutches were large. After their third year these kestrels had survival rates of about 70% per year, and after their first year their clutch sizes were about five eggs per clutch. However, survival to the third year, the size of the first clutch produced by the offspring, and the probability that a clutch succeeded varied with laying date and manipulation.

The experimentally enlarged clutches fledged more offspring than the control or the reduced clutches, and the **reproductive value** of those clutches—the number that survived to reproduce, multiplied by the number of young they had—was also greater than the reproductive value of the control clutches. If that were the whole story, it would be better for kestrels to lay more eggs than they chose to lay. However, the parents of enlarged clutches had poorer survival than did the parents of control or reduced clutches. This effect on survival substantially reduced the number of grand-offspring that the parents of enlarged clutches could expect to have during the rest of their life—their **residual reproductive value**. When both the reproductive value of the manipulated clutch and the subsequent reproductive performance of the parents were combined into a single measure, total reproductive value, it became clear that natural clutch sizes are adaptations. The controls, kestrels with clutches where young had been removed and then returned to the same nest without changing their number, produced the most surviving offspring per lifetime: more than half a chick more than the reduced clutches, and a full chick more than the enlarged clutches. Clutch size had evolved to maximize lifetime reproductive performance, taking into account the dependence of adult mortality on the level of parental care.

Manipulations of corolla tubes in orchids reveal adaptations to pollinators

Darwin was fascinated by the natural history of pollination. His study of the morphology of red clover and the bees that pollinate it convinced him that the length of the flower's corolla and the length of the bee's tongue had evolved so that the bee had to extend its tongue as far as possible while bending forward into the flower, bringing its head into proper position to transfer pollen. Thus greater corolla depth selects for longer bee tongues, and longer bee tongues select for greater corolla depth. If the length of the pollinator's tongue and the depth of the corolla were not limited by other factors, the process could result in very deep flowers and very long tongues. Later Darwin had the opportunity to examine the Madagascar star orchid, which has an extremely long floral tube of 28–32 cm (Figure 2.8).

Figure 2.8 The Madagascar star orchid (*Angraecum sesquepedale*) is pollinated by a hawkmoth (*Xanthopan morgani praedicta*) whose existence and long proboscis were predicted by Darwin from the morphology of the orchid. The length of the orchid's nectary has evolved to fit the length of the hawk's proboscis. (Photo courtesy of L. T. Wasserthal.)

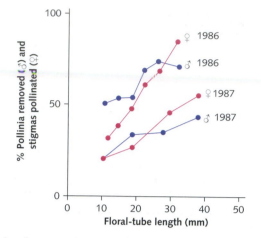
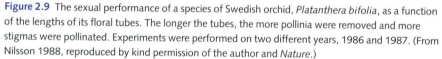

Figure 2.9 The sexual performance of a species of Swedish orchid, *Platanthera bifolia*, as a function of the lengths of its floral tubes. The longer the tubes, the more pollinia were removed and more stigmas were pollinated. Experiments were performed on two different years, 1986 and 1987. (From Nilsson 1988, reproduced by kind permission of the author and *Nature*.)

Reasoning that the same selective forces would drive the coevolution of the orchid and its pollinator, Darwin predicted that it would be pollinated by a giant hawkmoth with a tongue about 30 cm long (Darwin 1859, p. 202). Forty years later, the hawkmoth (*Xanthopan morgani praedicta*) was discovered. It had a tongue of 30 cm.

Darwin's elegant explanation of floral-tube depth remained untested until Nilsson (1988) manipulated the depth of the floral tubes of orchids on Baltic islands. He experimentally shortened the floral tubes by taping or tying them shut, squeezing the nectar above the constriction so that the pollinator's reward remained in place, then measuring the number of pollinia removed (male function) and the number of stigmas pollinated (female function) for each of the experimental treatments. Orchids with artificially shortened corollas had reduced fitness through both male and female function (Figure 2.9).

It now appears likely that the Madagascar orchids evolved their long floral tubes in response to pollinating moths that were evolving long tongues for another purpose: to keep their bodies well away from flowers on which spiders perch to attack them (Wasserthal 1997).

Bacteria rapidly evolve resistance to antibiotics

Staphylococcus aureus is a bacterium that causes dangerous infections in humans, including burn victims, patients suffering from wounds, and people recovering from operations in hospitals. In 1941, *S. aureus* could be treated with penicillin G, which inhibits the synthesis of the bacterial cell wall, but by 1944 some strains had already acquired resistance to penicillin, and by 1992 more than 95% of all

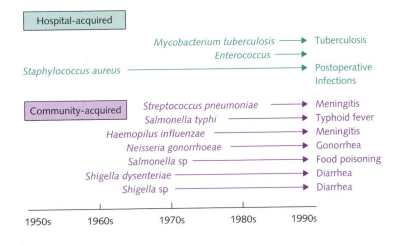

 of the figure area shows:

Hospital-acquired

Mycobacterium tuberculosis ⟶ Tuberculosis
Enterococcus ⟶
Staphylococcus aureus ⟶ Postoperative Infections

Community-acquired

Streptococcus pneumoniae ⟶ Meningitis
Salmonella typhi ⟶ Typhoid fever
Haemopilus influenzae ⟶ Meningitis
Neisseria gonorrhoeae ⟶ Gonorrhea
Salmonella sp ⟶ Food poisoning
Shigella dysenteriae ⟶ Diarrhea
Shigella sp ⟶ Diarrhea

1950s   1960s   1970s   1980s   1990s

Figure 2.10 Bacteria that cause many dangerous diseases and infections have rapidly evolved resistance to antibiotics. (From Cohen 1992.)

S. aureus strains worldwide were resistant to penicillin and related drugs (Neu 1992). The pharmaceutical industry responded by synthesizing methicillin, but by the late 1980s resistance to methicillin was common, and strains of *S. aureus* resistant to methicillin were also resistant to all drugs structurally similar to penicillin (Figure 2.10). Resistance was caused by a mutation in a gene that controls the structure of one element of the bacterial cell wall: the mutated gene alters the structure of the cell wall so that the drugs no longer bind to it.

Bacteria acquire resistance to antibiotics through several genetic mechanisms. The simplest, a new chromosomal mutation, was responsible for the original resistance of *S. aureus* to penicillin. Some bacteria carry previously evolved resistance genes that are not normally expressed; they are only induced when the antibiotic is encountered. Resistance genes can be transferred to other bacteria through transformation, a process in which bacteria acquire genes by taking up free DNA molecules released by dying bacteria into the surrounding medium. The viruses that parasitize bacteria can also pick up a bacterial resistance gene and carry it with them into their next host, where it may be inserted into the bacterial chromosome. If the bacterium survives the viral infection, it will have acquired antibiotic resistance. This mechanism is called transduction. Like transformation, it moves chromosomal genes between bacteria. Bacteria also contain small, circular pieces of extra-chromosomal DNA called plasmids, some of which contain genes for antibiotic resistance, and they occasionally exchange plasmids and chromosomal genes in a third process, conjugation, in which they fuse with each other. Furthermore, within the bacterial genome there are mobile genetic elements, transposons or so-called jumping genes, that can pick up a resistance gene and move it along the main chromosome or into a plasmid. Transposons and plasmids may induce bacteria to conjugate, creating opportunities for horizontal transmission of themselves and the resistance gene.

Because of strong selection, transposition, and integration of resistance genes into the main chromosome of *S. aureus*, some strains have become resistant to many antibiotics, including erythromycin, fusidic acid, tetracycline, minocycline,

Transformation: *Bacteria take up DNA from the medium and incorporate it into their circular chromosome.*

Transduction: *A virus that infects bacteria picks up some bacterial DNA from one host and transfers it to the next host, which may incorporate the DNA if it survives the infection.*

Plasmid: *A small piece of circular DNA that may exist in multiple copies in bacterial cells. Plasmids are involved in inducing conjugation, during which they carry genetic information from one bacterial cell to another.*

Conjugation: *A form of mating in which two bacteria build a tube between them through which both plasmids and chromosomal DNA can be exchanged. Bacteria can be induced to conjugate by plasmids, which thereby achieve horizontal transmission.*

Transposon: *A segment of DNA that can move from one site on a chromosome to another. If it leaves a copy behind, it can multiply, increasing in copy number. Transposons are also called jumping genes. They do not code for proteins that are expressed in the cells of their hosts, behaving instead as genetic parasites. Their movements are responsible for some mutations, for when they insert into the DNA at a new site they can disturb local gene expression.*

streptomycin, spectinomycin, and sulfonamides, and even to disinfectants and heavy metals, such as cadmium and mercury. In the mid-1980s new drugs were discovered, the fluoroquinolones, that killed *S. aureus* at low concentrations and cured serious infections. Some thought the problem of antibiotic resistance had been solved. However, around 1990 strains of *S. aureus* that resisted the new drugs increased from less than 5% to more than 80% within 1 year in a hospital in New York, and resistance to fluoroquinolones spread around the world. Recently a drug, mupirocin, was found that killed strains of *S. aureus* resistant to fluoroquinolones. It was first used in the UK, where strains have already evolved resistance to mupirocin. The genes for resistance to mupirocin are carried on plasmids and therefore can spread horizontally and rapidly.

S. aureus illustrates some properties of the evolution of antibiotic resistance that are true for many other dangerous bacteria (see Figure 2.10). Bacteria rapidly evolve resistance to virtually all antibiotics (Neu 1992). Resistance that evolves in one bacterium spreads horizontally to other bacteria through transduction, transformation, and conjugation. Hospitals are especially good breeding grounds for new forms of resistance because that is where antibiotics are used most intensively. Moreover, it costs much more to treat resistant forms. In the United States in 1992, it cost about $12 000 to treat one case of nonresistant tuberculosis, including drugs, procedures, and hospitalization; treatment for one case of multidrug-resistant tuberculosis cost about $180 000 (Cohen 1992).

At human body temperatures, bacteria have generation times of about an hour, and they can rapidly multiply to form populations of billions of cells in which many mutations occur every generation. Strong selection applied by antibiotics to large populations with short generation times produces rapid evolution. Since the invention of antibiotics, about 5 million tons of these drugs have been used in humans and domestic animals, exerting massive selection on the world's bacteria. This cure contained the seeds of its own evolutionary destruction. The more people whom doctors cure with antibiotics, the more rapidly new forms of resistance evolve. This open-ended coevolutionary arms race with bacteria is profitable for the pharmaceutical industry but risky for mankind. An expert said recently, '. . . the post-anti-microbial era may be rapidly approaching in which infectious disease wards housing untreatable conditions will again be seen' (Cohen 1992).

Many studies have demonstrated natural selection

Endler (1986) discusses methods of detecting natural selection. In at least 99 species of animals and at least 42 species of plants, adaptive evolution has been demonstrated in morphological, physiological, or biochemical traits. However, few of those studies measured lifetime reproductive success, few dealt with more than one or two traits, and in most one could not pinpoint the mechanism of selection. The most convincing method remains the following of large numbers of individuals through several generations to document reproductive success and its correlates.

When selection is strong evolution can be fast

Selection in natural populations can be strong

Darwin and the early population geneticists thought that selection was weak and that evolution was slow and gradual. This view lasted into the 1970s, when studies began to accumulate demonstrating strong selection and rapid evolution. Endler's (1986) summary of studies of selection in natural populations made clear that selection was quite strong on some species, traits, and genes and weak or absent on others. From studies on 57 species, Endler calculated 592 estimates of the strength of selection: 180 of them were significantly greater than zero, and many were impressively large. Because the strength of selection varies from strong to zero, we cannot assume it *a priori*: if we need to know it, we have to measure it. Endler's survey may have been biased towards strong selection, as he acknowledged, for such results are more likely to be published, but that criticism does not affect the main conclusions: selection can be strong, and microevolution can be fast.

Large populations, abundant genetic variation, and strong selection promote rapid evolution

The rate of evolution can be either slow or fast. The most rapid change has been measured in guppies in Trinidad and finches in the Galápagos Islands, in both of which the rate of change per generation of some traits caused the mean value to continue to shift so rapidly over many generations that at least half of the offspring in each generation attained a value that ranked them above 70% of the previous generation. Microevolution can be very fast, much faster than one would infer from fossils, and it is fastest in large populations for traits with a great deal of genetic variability experiencing strong selection, as happens when predatory fish eat small guppies.

The context of selection depends on the thing selected

The type of selection experienced by a gene in a genome, or a trait in an organism, or an organism in a group, results from an interaction between the environment, the focal gene, trait or organism, and the other genes, traits, and organisms contributing to survival and reproduction. Selection thus has both

> ● **KEY CONCEPT**
> The puzzle is not how evolution could have produced so much, but why it has sometimes been so slow.

> ● **KEY CONCEPT**
> Selection acts at many levels; each level has its own context.

external and internal causes. It always involves interactions between the thing selected and its environment.

- For a gene, the other genes in the genome are important.
- For a trait in an organism, the other traits in the organism are important.
- For an organism in a population, the other organisms, especially potential mates and competitors, are important.

The external environment is not an absolute but a relative agent of selection, for even the selective impact of a physical factor like temperature can depend on such interactions—the impact of cold or heat on animals may depend on competition for places protected from extreme temperatures.

Whether a species is sexual or asexual is the most important feature of the context in which genes evolve. In an asexual clone the entire genome is transmitted like a single gene, and any advantages or disadvantages created by a new mutation in one gene are transmitted reliably into future generations. For example, a gene that alters the mutation rate of the whole genome produces a reliable consequence in an asexual organism. In contrast, in a sexually reproducing organism, recombination and the production of an offspring in which half the genome comes from another organism ensure that a gene will find itself with many new partners in every generation. Altering their mutation rates will not produce reliable consequences.

The chromosomal neighborhood is also an important part of the context of selection on genes. When a gene comes under selection and starts to change in frequency, it will cause changes in the frequencies of all genes close to it on the chromosome. The action of **indirect selection** is described as a selective sweep, by analogy to a broom that sweeps up whatever it encounters, including much besides the object of attention. In asexual organisms, all genes are subject to selective sweeps caused by selection acting anywhere in the genome. In sexual organisms, genes usually only encounter selective sweeps caused by selection acting on genes that sit nearby on the same chromosome.

> **Indirect selection:** *When a gene increases in frequency not because it is being selected but because it is on a chromosome near a gene that is being selected.*

Some organisms choose the environment in which they are selected

Organisms can sometimes choose the environmental context in which they will live and thus the kinds of selection pressures that they will encounter. For example, phoronids are marine worms that live in sand. Their planktonic larvae actively choose sand with phoronids living in it when they settle. By selecting an environment in which other phoronids have already survived, they greatly increase their chances of surviving. Phoronids are so-called living fossils that have not changed much for hundreds of millions of years. One reason for the lack of change may be that their larvae have always chosen the same environment in which the adults live, an environment that has always been available and has not changed very much. Another example is provided by house sparrows, which prefer to live in and around human settlements. It is hard to find one deep in a forest, where selection pressures are quite different. The presence of humans protects house sparrows from many predators, and house sparrows choose a safe habitat. If performance in human settlements trades off with performance in wild habitats, then the habitat choice of house sparrows is leading to a type of adaptive evolution that will cause them to suffer if human settlements disappear. They are domesticating themselves.

Selection to benefit groups at the expense of individuals is unlikely but not impossible

One can still occasionally hear the claim that something evolved 'for the good of the species.' We now discuss why that is unlikely.

We introduce the subject by asking why red grouse in Scotland cut back on reproduction at high density. Wynne-Edwards (1962) stimulated an illuminating controversy by claiming that the reason that red grouse reduce reproduction at high density was to keep the population from over-cropping their food supply. There are, however, good reasons for individuals *not* to reduce reproduction to preserve the food supply. Such an action would produce benefits for the entire population, not only for their own offspring. If there were variation in the population in the tendency to reduce reproduction at high density, those individuals who reduced it as little as possible would have the greatest reproductive success. They would benefit from the reproductive modesty of others and increase in numbers (Williams 1966).

To generalize this example, we return to the definition of natural selection. The ability of any selection process to produce evolutionary change is determined by three things (see p. 2): the variation in the reproductive success of the units being selected, the strength of the correlation between the trait under selection and the variation in the reproductive success of the unit (individual or group), and the genetic variation of the trait among the units. When we compare selection operating on groups such as populations or species with selection operating on genes or the individuals that carry them, we notice striking differences, as follows.

- The correlation between traits and reproductive success is often much weaker for groups than for individuals.

- Variation in reproductive success among individuals is often greater than variation in reproductive success among groups. The potential *strength* of individual selection is therefore normally much greater than the potential strength of group selection.

- The amount of genetic variation in a trait that can be accounted for by differences among individuals is normally much larger than the amount that can

● KEY CONCEPT

The relative strength of selection at each level of a hierarchy depends on the amount of genetic variation and variation in reproductive success at each level.

be accounted for by differences among groups. The potential *response* to individual selection is therefore normally much larger than the potential response to group selection.

- The generation time of individuals is often much shorter than the generation time of groups, and the number of individuals is much larger than the number of groups. Therefore the number of incidents of selection on individuals in a unit of time is *very* much larger than the number of incidents of selection on groups.

- Individuals are discrete things with measurable reproductive success; the boundaries of groups in space and time can be diffuse.

The combination of these conditions usually makes it easy for types that emphasize their own reproductive interests to take over populations of organisms that are not reproducing as fast as possible. The exceptions, discussed in Chapter 9, are fascinating.

So why do red grouse in Scotland cut back on reproduction at high population density? Not for the good of the entire population, but because there are so many of them that each one has less to eat. When they starve, they cannot make as many offspring. It is as simple as that.

Four factors can limit adaptation

KEY CONCEPT

Adaptation is not an inevitable outcome of evolution; it only happens when conditions permit.

Natural selection operates whenever there is variation in reproductive success and the variation in reproductive success is correlated with heritable variation in the trait. Because there is always some variation in reproductive success, some trait is usually correlated with reproductive success, and many traits display heritable variation, natural selection is usually acting in all populations, including our own. Because natural selection acts on all variable traits that contribute to survival and reproduction, if such a trait is not in the state best for survival and reproduction, then something must be limiting its evolution. Four limiting factors are particularly important: gene flow, sufficient time for adaptation to occur, tradeoffs, and constraints. We now discuss each in turn.

Gene flow: *Genes flow from one place or situation to another when organisms born in one place or situation move to another where they have offspring that survive to reproduce there.*

Gene flow can cause maladaptation

Maladaptation: *The state of a trait that leads to demonstrably lower reproductive success than an alternative existing state. Maladaptations can arise when local populations are swamped by gene flow.*

Genes 'flow' from one place to another when organisms born in one place or situation reproduce successfully in another. When natural selection favors different things in different places, the movement of organisms between habitats transports genes that have been successful in one place to places where they may not be so successful. Like mutation, gene flow introduces new genetic variants into local populations, and it can produce local **maladaptations**. For example,

Figure 2.11 Blue tits nesting in downy and holm oaks in southern France have greater success on downy oaks and time their reproduction on both trees to the peak of insect abundance on downy oaks, even though they would do better by delaying reproduction on holm oaks. (By Dafila K. Scott.)

blue tits in southern France breed both in deciduous downy oaks and in evergreen holm oaks (Figure 2.11).

These birds should start to breed later in holm oaks where the peak of insect abundance comes later in the season, but in fact they breed at the same time on both oak species. Gene flow is strong, for adult birds move freely between downy and holm oaks, and the limitation on nesting territories on downy oaks forces some birds to nest on holm oaks, where they are maladapted and where clutches laid later in the season would result in more surviving offspring (Blondel et al. 1992). Blue tits cannot adapt to holm oaks because reproductive success is so much greater on downy oaks that most genes in the population only have a history of having experienced reproduction on downy oaks.

Gene flow not only causes maladaptation. It can also introduce advantageous alleles to small, isolated populations where they are unlikely to arise through mutation, and it can reduce the effects of inbreeding in small populations.

Strong selection can produce local adaptation despite gene flow

Gene flow does not always lead to maladaptation. Despite gene flow, adaptations can evolve when selection is strong. A classic example is heavy metal tolerance in plants. Plants on mine tailings grow on toxic soil and rapidly evolve adaptations to deal with it. Antonovics and Bradshaw (1970) sampled plants from a transect across a zinc mine tailing and into an uncontaminated pasture. They found extremely rapid change in zinc tolerance at the boundary between the mine tailing and the uncontaminated pasture. The index of tolerance changed from 75% to 5% in less than 10 m. Plants that were zinc tolerant did survive but flowered later and were smaller, suggesting that tolerance had costs. They also suffered less from inbreeding than those that were not zinc tolerant, which suggests an evolved response to local mating with relatives. Even more

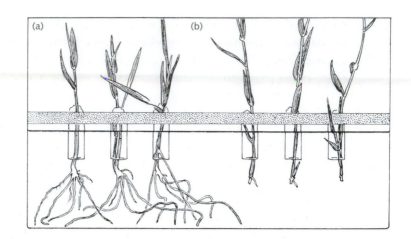

Figure 2.12 Grass that has adapted to zinc from power poles (a) grows much more impressive roots in zinc solution than does grass grown from seed collected just a few meters away that is not adapted to zinc (b). (By Dafila K. Scott from a photograph by A.D. Bradshaw.)

rapid evolution and stronger selection were suggested by another study of grasses growing near galvanized-steel power poles, which appeared much more recently than the mine. Plants growing within 10 cm of a power pole had significantly higher zinc tolerance than those just 20 cm from it (Figure 2.12). Grasses are wind pollinated; their pollen can fertilize seeds many meters away. Thus gene flow is occurring on a much larger scale in this example than are the effects of local selection pressures.

Species often consist of a patchwork of genetically different populations each displaying different adaptations. For gene flow to prevent local adaptation, natural selection must be weak and the mean distance that genes move in each generation must be large (Endler 1977).

It takes time to respond to selection

Even without gene flow, it takes time for a population to adapt to environmental change. Consider the absorption of milk sugar, lactose, by adult humans. Like other mammals, human children are equipped with the enzymes needed to digest milk, and most children lose that ability because those enzymes are no longer expressed after the age at which they used to be weaned—at about 4 years. However, some humans retain the ability to digest fresh milk into adulthood, including people living in northern Europe and some in western India and sub-Saharan Africa (Simoons 1978). This genetic difference explains why dairy products are more prominent in French than in Chinese cooking. The ancestral condition was the inability to digest fresh milk after the age of 4 years, and the recently evolved condition is that ability.

How long would it take that ability to evolve? The domestication of sheep and goats occurred about 10–12 000 years ago, the subsequent origin of dairying can be traced to 6–9000 years ago, and the ability to digest fresh milk after the age of 4 has a simple genetic basis: it behaves as a single dominant **allele** that

Allele: *One of the different homologous forms of a single gene.*

is on one of the normal chromosomes rather than on a sex chromosome. Dominant alleles increase in frequency under selection more rapidly than do recessive alleles (Chapter 4, p. 78), and knowing that the gene is not on a sex chromosome simplifies the prediction of how its frequency will change under selection (see the Genetic appendix for details).

Imagine a Stone-Age human population in which goats and sheep have been domesticated and milk production has begun. In that population, a new mutation arose that allowed people to utilize fresh milk after the age of 4. (The mutation is not in the lactose-digestion gene itself but in an enhancer element; Enattah et al. 2002). It had an advantage over the ancestral state, for those who could absorb lactose benefited from a high-quality food source, rich in calcium and phosphorus, especially at times when other food was scarce, and especially for nursing mothers and growing children. As the culture of milk-drinking as adults began to spread, those who drank milk but could not absorb lactose suffered from flatulence, intestinal cramps, diarrhea, nausea, and vomiting, which affected their physical state and reduced their reproductive success. Suppose that the ability to absorb lactose conferred a selective advantage of 5%, so that for every 100 surviving and reproducing children of non-absorber parents, the same number of absorber parents produced 105.

At the beginning, the lactose-tolerant allele was rare and simply because it was rare it could only increase slowly, for very few people carried it, enjoyed its advantages, and produced a few more surviving children than did those who did not carry it. As the allele increased in frequency, and more people carried it, it began to spread more rapidly through the dairying culture. However, when it became common, its rate of spread decreased, for then most people carried it, and there were very few who suffered from the disadvantage of not having it. How long did it take to increase from a single new mutation to a frequency of 90%? The answer is about 350–400 generations or 7–8000 years. It would take longer if we assumed a weaker selective advantage. Thus if the age of its origin is even approximately accurate, adult milk-drinking increased fitness substantially.

That argument assumes a causal connection between the ability to absorb lactose and individual fitness, an ability that arose with the origin of agriculture. Agriculturists had an advantage over hunter-gatherers for several reasons, not just dairying and the ability to absorb lactose. They spread both their culture and their genes, including the allele for lactose absorption, from the Near East to all of western Europe. Thus the rapid rise in frequency of the lactose-tolerant allele in Europe did not depend just on its digestive advantages: it also hitchhiked with a culture that expanded and dominated.

Note that selection in nature can be much stronger and evolution much more rapid than it appears to have been in the lactose example (see above), and selection on domesticated plants and animals has often yielded dramatic changes quite quickly, in just a few generations.

Dominant: *An allele is dominant if it is expressed in the phenotype in the heterozygous diploid state.*

Recessive: *An allele is recessive if it is not expressed in the phenotype in the heterozygous diploid state.*

Tradeoffs constrain changes in traits connected to other traits with effects on fitness

Tradeoff: *A tradeoff occurs when a change in one trait that increases fitness causes a change in the other trait that decreases fitness.*

Another reason why individual traits may not become well adapted lies in tradeoffs. A tradeoff exists when a change in one trait that increases reproductive success is linked to changes in other traits that decrease reproductive success. The reasons for such linkages are not always well understood, but we know that tradeoffs occur frequently. If there were no tradeoffs, then natural selection would drive all traits correlated with reproductive success to limits imposed by constraints. Because we find many traits that are clearly correlated with reproductive success varying well within such limits, tradeoffs must exist. A common tradeoff is that between reproduction and survival. For example, fruit flies selected to lay many eggs early in life have shorter lifespans (Rose 1991). Because they cannot both reproduce a lot early in life and have a long life, this tradeoff shapes the evolution of their lifespan.

Other important tradeoffs occur between the ability to eat one thing and the ability to eat many things, especially important in the ecology of plant-eating insects, and between mating success and survival, a key feature of the secondary sexual characteristics produced by sexual selection through mate choice. There are many others. Whenever one analyzes the costs and benefits of changes in traits, tradeoffs are usually found. They limit how much reproductive success can be improved by changing traits, for when traits cannot be changed independently of one another, the benefits gained by changes in one trait are often rapidly balanced by costs incurred in others, causing the response to selection to stop.

Constraints can limit the response to selection

Organisms are not soft clay out of which adaptive evolution can sculpt arbitrary forms. Natural selection can only select variants currently present in the population, variation that is often strongly constrained by history, development, physiology, physics, and chemistry. Natural selection cannot anticipate future problems, nor can it redesign existing mechanisms and structures from the ground up. Evolution proceeds by tinkering with what is currently available, not by proactively designing ideal solutions, and the variation currently available is often limited by constraints.

Some constraints are imposed by physics and chemistry. The diameter of a spherical organism without a circulatory or respiratory system cannot be much greater than 1 mm, a limit set by the rate at which oxygen diffuses through water. Water-breathers face a major obstacle to becoming warm-blooded because they have to pass large volumes of water across large gill surfaces to extract oxygen, and moving water rapidly strips heat from warm bodies. (Some water-breathers, including tuna and sharks, have overcome this constraint with special adaptations to limit heat loss.) The limbs of terrestrial animals must be

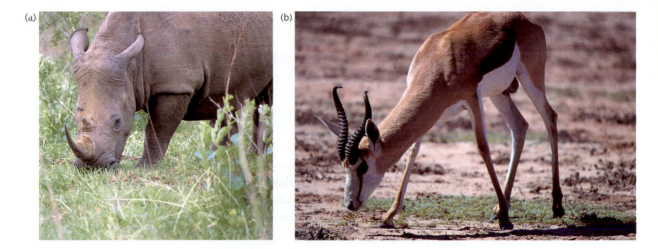

(a) (b)

thicker in heavier organisms, for the strength of a limb is determined by its cross-sectional area, whereas weight is determined by volume, which grows more rapidly (x^3) with the length of an organism (x) than does the cross-sectional area (x^2) of one of its limbs. That is why the legs of rhinoceros are thicker, in proportion to their body lengths, than the legs of antelopes. Physics cannot be avoided, and physics does not evolve (Figure 2.13).

Also interesting are evolved constraints. Past adaptations can become future constraints, placing the imprint of history on a lineage, as we saw in the plethodontid salamanders discussed in Chapter 1 (p. 11). Here are four examples.

First, asexuality has many advantages as a mode of reproduction (Chapter 8, p. 177): it is straightforward, efficient, and does not require a partner. If that is so, then why are there no asexual mammals? Early development in mammals requires one egg-derived and one sperm-derived haploid nucleus. The two types of nucleus are marked differently in the germ line of the parents by the attachment of methyl groups to the DNA molecules, a pattern known as genetic imprinting. Early development requires the expression of some genes derived from the father and some genes derived from the mother, which is determined by the sex-specific imprinting that occurs in the germ line of the parent. The parental patterns are erased later in development, allowing the offspring to imprint the genes that are appropriate to their own sex and making it possible to clone some mammals from adult cells. If all the genes came from the mother, then some normally paternal genes would not be turned on at the right time, and early development would fail.

Second, why are there no asexual frogs? In a freshly inseminated frog egg, the sperm donates the organelle that replicates to form the two devices that pull the chromosomes apart when the cell divides. Activated eggs that lack the paternally contributed device divide abortively (Elinson 1989). As we will see in Chapter 9 (p. 197), this pattern may have resulted from genomic conflict between the paternal and maternal genes over the issue of asexuality—the pattern is

Figure 2.13 The legs of a rhinoceros (a) must be thicker in proportion to body length than the legs of an antelope (b). ((a) Photo courtesy of Melinda Smith; (b) Photo courtesy of Beverly Steams.)

Genetic imprinting: *The silencing of specific genes in the germ lines of the parents. The effect is to turn those genes off during the early development of the offspring.*

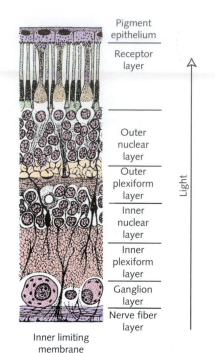

Pigment
epithelium

Receptor
layer

Outer
nuclear
layer

Outer
plexiform
layer

Inner
nuclear
layer

Inner
plexiform
layer

Ganglion
layer

Nerve fiber
layer

Light

Inner limiting
membrane

Figure 2.14 The pigment epithelium of the vertebrate eye containing the light-sensitive cells develops inside the nerves and blood vessels through which light must pass, a design no engineer would approve. (From Romer 1962.)

consistent with the interpretation that the paternal genes have been selected for a pattern of development that prevents the spread of asexuality and the elimination of males.

Third, the vertebrate eye, admired for its precision and complexity, contains a basic flaw. The nerves and blood vessels of vertebrate eyes lie between the photosensitive cells and the light source (Goldsmith 1990). No engineer would recommend a design that obscures the passage of photons into the photosensitive cells (Figure 2.14). Long ago, vertebrate ancestors had simple, cup-shaped eyes that were probably originally used only to detect light, not to resolve fine images. Those simple eyes developed as an out-pocketing of the brain, and the position of their tissue layers determined where the nerves and blood vessels lay in relation to the photosensitive cells. If the layers had not maintained their correct positions, relative to one another, then the mechanisms that control differentiation, in which an inducing substance produced in one layer diffuses into the neighboring layer, would not work. Once such a developmental mechanism evolved, it could not be changed without destroying sight in the intermediate forms that would have to be passed through on the way to a more 'rationally designed' eye.

The fourth example concerns the length and location of the tubes connecting the testicles to the penis in mammals. In the adult cold-blooded ancestors of mammals, and in present-day mammalian embryos, the testicles are located in the body cavity, near the kidneys, as are ovaries in adult females. Because mammalian sperm develop better at temperatures lower than those found in the

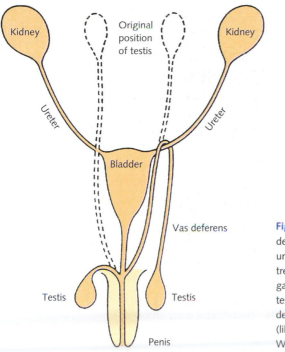

Figure 2.15 The mammalian vas deferens is wrapped around the ureter like a hose wrapped around a tree by someone watering a garden. If the development of the testis were not constrained, the vas deferens could be much shorter (like the one on the left). (From Williams 1992.)

body core, there was selection, during the evolutionary transition from cold- to warm-bloodedness, to move the testicles out of the high-temperature body core into the lower-temperature periphery and eventually into the scrotum. This evolutionary progression in adults is replayed in the developmental progression of the testes from the embryo to adult (Figure 2.15). As they descend from the body cavity into the scrotum, the testes wrap the vas deferens around the ureters like a person watering the lawn who gets the hose caught on a tree. If it were not for the constraints of history and development, a much shorter vas deferens would have evolved that cost less to produce and might have done a better job.

Conclusion

Natural selection is a powerful process not included in physics or chemistry. The only mechanism known that can increase the complexity of organisms, it can rapidly produce states that would otherwise appear to be highly improbable. We call such states adaptations. To detect selection of organisms, we can measure variation in lifetime reproductive success. To detect selection of traits, we can manipulate the trait and note the consequences for lifetime reproductive success. To detect selection of genes, we can measure changes in gene frequencies and rule out other causes, such as drift. Such measurements suggest that the strength of selection in nature varies from strong to zero.

Natural selection is not limited to organisms. When things vary, reproduce, have some form of inheritance, and some non-zero correlation between heritable variation and reproductive success, adaptive evolution occurs. Such things do not have to be organisms.

If heritable variation is not correlated with reproductive success, then the traits or genes involved will drift at random within boundaries set by constraints. Much of the genome has been shaped not by adaptive evolution driven by selection but by neutral evolution driven by genetic drift. That is the topic of the next chapter.

● SUMMARY

- Natural selection occurs when variation in a trait is correlated with variation in reproductive success. There are many reasons for such correlations; whatever the reason, their consequence is that selection occurs.

- Lifetime reproductive success is always relative to the other variants currently existing in the population; it is not absolute. Therefore natural selection is not guaranteed to find the best version; it simply finds the best one available.

- Random drift occurs when reproductive success is not correlated with variation in a gene or trait.

- A trait will only respond to selection when some of its variation is heritable.

- Variation in lifetime reproductive success is always present in natural populations.

- Heritable variation is common; it is the rule rather than the exception.

- Natural selection can rapidly produce extraordinary consequences when variation is heritable because the information about what led to greater reproductive success in the parents is preserved in the offspring and then, in the next generation, improved further.

- Adaptations produced by natural selection consist of changes that increase reproductive success. Adaptations can accumulate to produce structures and processes that are intricate and precise.

- Natural selection has been demonstrated experimentally in populations of short-lived organisms, including viruses, bacteria, algae, insects, and fish.

- Selection in nature can be strong and can rapidly produce impressive responses in large populations containing a great deal of genetic variation.

- But responses to selection can also be limited by gene flow, time, tradeoffs, and constraints.

- Constraints can be physical, chemical, or biological. Biological constraints have evolved; adaptation in one trait can produce constraint as a byproduct in another.

● RECOMMENDED READING

Clutton-Brock, T.H. (ed.) (1988) *Reproductive success*. University of Chicago Press, Chicago.

Dawkins, R. (1999) *The extended phenotype. The long reach of the gene*, revised edition. Oxford University Press, Oxford.

Grant, B.R. and Grant, P.R. (1989) *Evolutionary dynamics of a natural population. The Large Cactus Finch of the Galápagos*. University of Chicago Press, Chicago.

● QUESTIONS

2.1. What conditions are necessary for natural selection? Which are sufficient?

2.2. Which of the conditions necessary for natural selection must change for neutral evolution to occur?

2.3. If a trait is not well adapted to current local conditions, what might be the reasons why it is not, and how might one check to see which of those reason(s) is correct?

2.4. You buy a home with several prior owners, all of whom liked to tinker and repair things themselves. Your friend buys a new home built from the ground up to her specifications. Shortly thereafter, you both decide to make changes to the kitchen and the main bathroom. When you compare notes, you discover that it was much more difficult and expensive for you to make the changes in your previously owned home than it was for your friend to make similar changes in her new home. What does this example have to do with evolved constraints and with evolution as a process that tinkers with the variation that is available?

2.5. Do ideas, agricultural practices, or computers evolve in response to natural selection? What then corresponds to reproductive success and to inheritance? How does cultural change connect to biological evolution? Neither reproduction nor inheritance can be as precisely defined for cultural change as they can for biological evolution, but the analogy is worth exploring.

CHAPTER 3

Neutral evolution

How do gene frequencies change when there is no selection?

● **KEY CONCEPT**

Selection is not the only cause of changes in gene frequencies.

Chapter 2 pointed out that the correlation between traits and reproductive success determines whether evolution is adaptive or neutral. If the correlation is close to zero, evolution is neutral. Otherwise, when that correlation is non-zero and some of the variation in reproductive success is heritable, evolution is adaptive and changes in the genetic composition of the population improve adaptation. From this follow important questions. What is the quantitative relationship between variation in reproductive success and genetic variation? How much genetic variation is there? How often do genetic differences imply differences in reproductive success? The answers to these questions determine whether adaptive evolution can occur; we address them in Chapters 4 and 5 (pp. 70, 99). Before we consider adaptive genetic change, it helps first to know what to expect when evolution is not adaptive, when it is neutral. Here are some important questions about neutral evolution.

- How do gene frequencies change when there is *no* relationship between genetic variation and reproductive success?
- What is the evolutionary significance of *non*-heritable variation in reproductive success?
- Will variation in reproductive success that is *not* correlated with genetic variation *not* change the genetic composition of the population?
- Is such variation therefore *un*important in evolution?

Fitness: *As a first approximation, fitness is synonymous with reproductive success.*

In addressing these questions we will now use **fitness** as a synonym for reproductive success. Fitness is a key concept in evolutionary biology, and equating it with reproductive success is not very precise, but for present purposes this will do. A more precise definition is given in Chapter 4, pp. 85 ff. To answer the questions posed above, we need to understand how genes, in principle, connect to fitness.

Neutrality arises in the connections between genotypes, phenotypes, and fitness

There are at least three levels between which the connection between genetic variation and variation in fitness could be uncoupled, and variation relevant for evolution occurs at all three: the genotypic, the phenotypic, and fitness. In Figure 3.1, which sketches the connections between the three levels, we represent the different genotypes in a population by points on the genotype plane (a realistic genotypic space would have many dimensions; this one has just two for convenience). Similarly, points on the phenotypic plane represent different phenotypes, and points on the fitness plane represent different fitness values. Because the connections between genotype, phenotype, and fitness depend on the environment, one picture of this sort only holds in one environment.

Figure 3.1 illustrates both genotypic redundancy with respect to phenotype and phenotypic redundancy with respect to fitness. This means that many genotypes can produce the same phenotype (G_1 and G_2 both correspond to P_1), and many phenotypes can yield the same fitness (P_1 and P_2 both have fitness F_1). Because the three genotypes G_1, G_2, and G_3 all have the same fitness, their differences are *neutral* with respect to fitness: the variation among them is not correlated with variation in reproductive success. On the other hand, the difference between any one of them and G_4 causes a difference in fitness and evokes a response in the population leading to adaptive evolutionary change.

The example in Figure 3.1 is theoretical. Does selectively neutral genetic variation occur in reality? We describe next an experiment illustrating neutral evolution.

> **Phenotype:** *The material organism, or some part of it, as contrasted with the information in the* **genotype** *that provides the blueprint for the organism.*

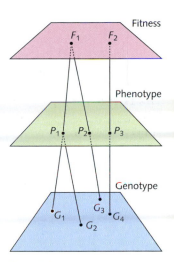

Figure 3.1 Schematic relations between variation at the levels of the genotype, the phenotype, and fitness. G_i, where $i = 1$–4, denotes different genotypes; P_i, $i = 1$–3, denotes different phenotypes; and F_i, $i = 1, 2$, denotes different fitnesses. The variation between the genotypes G_1, G_2, and G_3 is neutral with respect to fitness.

Experimental evolution in a bacterium confirms some reasons for neutrality

Conclusions from an experiment are considerably strengthened if the same outcome is observed every time the experiment is repeated. At first sight you might think that this would make experimental evolution problematic. Evolution is, after all, an historical process, and an evolving population changes. Repeating the process requires that conditions be recreated precisely as they were in the past, which is not possible. However, experiments done by Lenski and his colleagues come close to replicating evolution—not by repeating an evolutionary process exactly but by studying evolution in replicated populations evolving simultaneously in identical environments. They work with the gut bacterium *Escherichia coli*, whose short generation time allows experiments to extend over thousands of generations. The replicated populations were initially identical, consisting of a single genotype, and can be revived after freezing at $-80\,°C$. Thus derived and ancestral populations can be compared directly, for example by measuring fitness differences in competition. We consider one experiment that reveals much about the repeatability of evolution.

At the phenotypic level evolution was repeatable

Twelve initially identical populations were allowed to evolve independently in identical glucose-limited environments (Travisano et al. 1995). After 2000 generations the fitness of the bacteria in the derived populations had improved relative to their common ancestor by 35%. The 12 replicate populations differed in fitness from one another by only a few per cent and resembled each other closely in several respects: they all had higher maximal growth rates, larger individual cells, and fewer cells at stationary phase than their common ancestor. Thus their phenotypes had changed in similar ways. So far evolution appears to have been repeatable and largely deterministic: identical starting conditions and identical environmental conditions produced almost identical evolutionary outcomes in 12 replicate populations.

The result, however, could mean either of two things, as the authors realized. The 12 populations might have achieved the observed phenotypic adaptations by an identical evolutionary process involving the same genetic and physiological changes, or they might have reached the same end result through different changes at the genetic and physiological levels. These two possibilities are illustrated in Figure 3.2.

But evolution repeated in phenotypes was not repeated in genotypes

To discriminate between the two possibilities, the authors introduced all 12 evolved populations to novel environments, substituting other sugars (maltose or lactose) for glucose in the culture medium. When they measured the fitness of the bacteria relative to the ancestral genotype in the novel environments, they observed 100-fold greater genetic variation for fitness among the 12 populations than they had in the glucose environment in which the populations had evolved. Some populations were able to grow much better in a novel environment

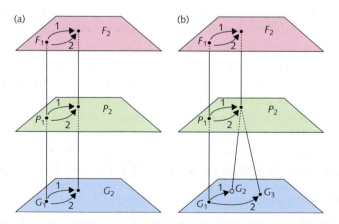

Figure 3.2 Experimental evolution in two replicate populations, 1 and 2. In case (a) the starting genotype G_1 has evolved to the same genotype G_2 in both populations. In case (b) the starting genotype G_1 has evolved to different genotypes, G_2 and G_3, in the two populations, but both these genotypes correspond to the same phenotype P_2 (at least in the environment in which selection occurred).

than others. Although the replicate populations had evolved identical changes at the *phenotypic* level during the 2000 generations of evolution, they nevertheless differed at the *genotypic* level.

Chance mutations led to an outcome of nearly neutral genetic differences

The authors' interpretation was that the mutations that distinguish the different lines produce the same phenotype in the glucose environment but differ in their side effects in the novel environments, suggesting that Figure 3.2(b) was more correct than Figure 3.2(a). Whatever the interpretation, these experiments demonstrate that different mutations in each of the 12 independently evolving populations were an essential ingredient of the evolution observed.

In these experiments both adaptive and neutral evolution occurred simultaneously. The adaptive evolution in the 12 populations resulted from a response to heritable variation in fitness in the glucose environment. The 12 populations diverged genetically in part because they experienced different mutations that were selected. However, the differences in their eventual genetic composition were close to *neutral* in the glucose environment because they all had about the same fitness. A mixture of the 12 evolved populations tested in the glucose environment would have contained considerable genetic variation but it would be uncorrelated, or very weakly correlated, with fitness.

Note that the term neutral always refers to a comparison. In this experiment, neutrality refers to the absence of differences in fitness, measured by competition with the ancestor, among the 12 evolved populations. When a new mutation occurs that has the same fitness effects as the allele from which it mutated, the mutation is called neutral. A neutral mutation may affect the reproductive success of its carrier. The characteristic that makes it neutral is that its effect does not differ from that of the allele from which it originated.

The rest of this chapter explores the causes and consequences of neutral evolution. Next we consider what types of mutation are likely to experience neutral evolution. Because neutral alleles are, by definition, not correlated with variation in fitness, their frequencies in populations experience random change caused by chance effects. We then discuss the processes that cause such change and the role of neutral genetic change in molecular evolution. We end with some applications and analogies to neutral evolution at levels higher than the gene.

Why variation in genes may not produce variation in fitness

Some genetic variation is not expressed in proteins

Some nucleotide changes in DNA, called *synonymous substitutions*, do not change the coded amino acid because the genetic code is redundant, mostly in the third position of the nucleotide triplets that code for amino acids. For example, AAG (adenine-adenine-guanine) and AAA both code for the amino acid

> ● **KEY CONCEPT**
>
> Neutrality can arise at several levels: the genetic code, proteins, development, and interactions with the environment.

phenylalanine. The redundancy for the amino acid leucine is even more impressive: six codons—AAT, AAC, GAA, GAG, GAT, and GAC—code for leucine. If all possible nucleotide substitutions in all triplet codons are equally likely, then about 25% of all substitutions are synonymous; at the third position, up to 70% (Li and Graur 2000). Because synonymous substitutions do not change the phenotype, they are neutral or nearly neutral.

The reason we say nearly neutral is **codon bias**, which means that synonymous codons do not always occur in equal frequency. In many species usage of synonymous codons is non-random, indicating selection favoring particular codons, possibly caused by differences in abundance of the corresponding transfer RNAs. Because the selection involved is believed to be very weak, the assumption of neutrality is approximately true.

Other mutations not likely to have phenotypic effects occur in **introns** and in **pseudogenes**. Introns are sequences within eukaryotic genes that are removed after transcription into RNA. Because introns do not code for proteins and are not expressed in the phenotype, mutations in introns are likely to be neutral. (Some caution is appropriate, for in an increasing number of cases a mutation has been found in an intron that affects the expression of a gene.) Pseudogenes are DNA sequences derived from functional genes by gene duplication; they are no longer expressed and therefore nonfunctional. Mutations in pseudogenes have no phenotypic effect and are neutral. Selective neutrality of mutations in introns and pseudogenes is supported by molecular evidence that introns evolve faster than the translated sequences (*exons*) and that pseudogenes evolve faster than functional genes.

Codon bias: *Synonymous codons do not occur with equal frequency.*

Intron: *A sequence within a gene that is removed after transcription and before translation by gene splicing; its DNA sequence is not represented in the RNA sequence of the spliced mRNA or the amino acid sequence of the resulting protein; introns occur in eukaryotes but not prokaryotes.*

Pseudogene: *A nonfunctional copy of a gene; it is not expressed.*

Changes in amino acids that do not affect protein function are nearly neutral

Some changes in the sequence of amino acids of a protein do not affect its function; such regions show higher rates of amino acid substitution when proteins of related species are compared. For example, apolipoprotein molecules carry lipids in the blood of vertebrates and have a lipid-binding site consisting of hydrophobic amino acids. Comparisons of apolipoprotein sequences from several groups of mammals suggest that in these domains one hydrophobic amino acid can be replaced by another without affecting function (Luo et al. 1989). If such amino acid substitutions have an effect on reproductive success it is probably very small: they are approximately neutral.

Genes affecting a canalized trait are neutral when not expressed

Some characters show no, or very little, phenotypic variation, despite considerable environmental and genetic variation. These are called canalized characters because the final phenotypic outcome is kept constant, as though development were confined within a canal that does not allow deviations from its course.

When environmental or genetic variation is extreme, the canalization may break down, revealing genetic variation for the trait that had been hidden and demonstrating that the normal state was canalized. For example, *Drosophila melanogaster* normally have exactly four scutellar bristles, but in flies homozygous for the mutation *scute* the number of bristles is reduced to an average of two, with some variation. The mutation thus does two things: it reduces the average number of bristles, and it allows hidden genetic variation for bristle number to be expressed. This variation can be used to select for lower or higher bristle numbers; thus it is genetic. Extreme-temperature treatments also reveal the underlying genetic variation that is normally invisible due to canalization. Because of developmental buffering, the phenotypic effect of mutations in genes affecting a canalized trait is usually suppressed; such mutations are neutral so long as they are not expressed. In this example, genes causing variation in bristle number would be neutral in normal flies and potentially non-neutral in flies homozygous for the mutation *scute*.

Variation adaptive in one environment may be neutral in another

Genetic variation may result in fitness differences in one environment, but not in another. Such is the case with antibiotic-resistance mutations in bacteria and fungi. When these organisms are exposed to an antibiotic, mutations are selected that confer resistance to the antibiotic. If these resistance mutations had negative side effects (costs of resistance), they would be deleterious in the absence of the antibiotic and would disappear fairly rapidly. But negative side effects are often absent or very small, which means that resistance mutations are effectively neutral in an antibiotic-free environment and will take a long time to disappear. Moreover, when a cost of resistance initially exists, natural selection tends to reduce the cost by favoring secondary mutations that remove the negative side effects, and the resistance mutation gradually becomes neutral, or nearly so (Anderson and Levin 1999). This is reason for concern, as mutations for antibiotic resistance that increase in frequency when antibiotics are applied may persist for a long time in the population if they become neutral after the antibiotic has been discontinued.

Neutral genetic variation experiences random processes

Can neutral genetic variation be responsible for evolutionary change? Because neutral variation by definition does not lead to differences in fitness, natural selection will not favor one variant over another. Therefore, you might think that neutral genetic variation, when it occurs, will be maintained stably and remain without evolutionary consequences as long as it remains neutral—very much like the stable coexistence of two alleles at a locus in Hardy–Weinberg

● **KEY CONCEPT**

When alleles do not experience selection, their frequencies drift randomly.

equilibrium (see the Genetic appendix, p. 517). But actually this expectation only holds in populations that are infinitely large, where random increases will balance random decreases with fine precision. In reality any population is finite, the balance of random increases and decreases is never exact, and this implies that chance inevitably affects the relative frequencies of genetic variants.

Here we use the term chance to refer to all causes—often unknown and unpredictable in their effects—that are unrelated to the genotypes we are considering. Chance affects which alleles an individual transmits to its offspring; for example, a man with three daughters did not transmit any Y chromosomes to the next generation, although the probability of transmitting a Y or an X chromosome is the same: 50% per offspring. Chance also affects the number of offspring; for example, identical twins may have different numbers of children, despite the fact that they have the same genotype. Several mechanisms cause random changes in the genetic composition of a population. They affect both neutral genetic variations and genetic variation that is responding to selection—adaptive genetic variation is not immune to random change.

Random change in the genetic composition of a population caused by chance events is called *genetic drift*. Neutral alleles drift aimlessly unless they are located close to a gene undergoing selection, in which case they 'hitchhike' with the selected gene. Genetic variation correlated with fitness is subject to change directed both by natural selection and by genetic drift. Which of the two forces determines the outcome depends on their relative strengths. Thus the two types of gene, neutral and selected, actually lie along a continuum. Change in neutral genes is dominated by drift but influenced by selection. Change in selected genes is dominated by selection but influenced by drift.

The strength of selection on a trait decreases with the correlation of the trait with reproductive success and with the amount of variation in reproductive success, and genetic drift is stronger in small than in large populations. Next we discuss the processes that contribute to genetic drift, and then we model the drift of gene frequencies as a statistical sampling process.

The Mendelian lottery causes chance variation in alleles transmitted to offspring

Mendelian lottery: *A particular allele will or will not be represented in the offspring because of the segregation of alleles at meiosis and the random chance that any particular gamete will form a zygote.*

That chance plays an essential role in the sexual transmission of genes from parents to their offspring is reflected in the phrase the **Mendelian lottery**. Consider an individual that is heterozygous at some locus. Although it produces equal numbers of gametes that carry one or the other allele at this locus, how many copies of each allele it transmits to its offspring is subject to chance. In a large population individual variations in transmission tend to cancel each other out, but in small populations this usually does not occur, and allele frequencies change as a consequence. The Mendelian lottery can be seen easily in the distribution of sons and daughters within families. When many families are taken together, roughly equal numbers of sons and daughters are observed, as expected, but within individual families striking deviations from the 1:1 expectation are common. What holds for sex chromosomes holds for all chromosomes: small samples of a random process often deviate strikingly from average expectation.

Variation in family size also causes random changes in allele frequencies

Differences in family size (the number of offspring per parent) also contribute to genetic drift. Individuals may have different numbers of offspring due to factors that have nothing to do with the particular alleles that they carry at some genetic locus. Family sizes vary both because individuals encounter different environments and because they carry different genes. Thus an allele may be disproportionately represented in the next generation because it happened to occur in parents that had many offspring for reasons other than having that allele. Such accidental changes in allele frequencies are greater in small than in large populations, for in the latter the positive and negative deviations tend to cancel each other out.

The founder effect and genetic bottlenecks produce sampling errors

New populations are sometimes founded by a few individuals forming a small group in which gene frequencies differ considerably from the frequencies in the parent population simply by chance. This is called the **founder effect**. Some alleles common in the original population may be completely absent; others rare in the parental population can reach high frequency in the new population simply because they happened to be present in a founder. For example, in the 16th and 17th centuries small groups of European colonists gave rise to the Afrikaans-speaking population in South Africa, derived from Dutch founders, and to the French-speaking population of Quebec, derived from French founders. Some genetic diseases rare in Europe occur at relatively high frequency in these populations. The disease porphyria variegata, an autosomally inherited, dominant disorder of heme metabolism, is very rare in most populations but occurs in about 1 in 300 Afrikaners. Most of the estimated 10 000–20 000 carriers in South Africa are descendants of a single Dutch couple, Gerrit Jansz, a Dutch settler in the Cape, and Ariaantje Jacobs, who was one of eight women sent in 1688 from an orphanage in Rotterdam to provide wives for Dutch settlers in the Cape (Dean 1972, Jenkins 1996). It was probably Ariaantje who carried the mutation, for a son of her half-sister (another of the eight orphans) also developed the disease. At the time the Cape settlement numbered a few hundred people.

Similarly, if because of some catastrophe only a few individuals survive to breed, the genetic composition of the population changes dramatically as it passes through a **genetic bottleneck**. Many alleles are lost and others rise to high frequency. Even if the few surviving individuals do so because of a

Founder effect: *Major changes in gene frequencies can occur in a population founded as a small sample of a larger population.*

Genetic bottleneck: *A reduction in population size to a low-enough level for long enough that many alleles are lost and others are fixed (risen to a frequency of 1.0).*

selectively superior genotype, adaptive evolution is likely to affect the allele frequencies directly only at a few loci. Changes at most loci will be random. On Pingelap, which consists of a few coral atolls in Micronesia, an abnormally high fraction of the population suffers from achromatopsia, total colorblindness and intense sensitivity to sunlight due to the complete absence of cones in the retina. The condition is caused by a recessive mutation that is very rare elsewhere with an incidence of 1 in 50 000–100 000. Among the 3000 people on Pingelap, about 1 in 12 are achromatopes. In 1775 Pingelap had a population of 1000; then a typhoon and subsequent starvation cut that number to 20. Of the survivors, one male passed along the recessive mutation for achromatopsia, and four generations later the condition began to appear. For details see Sacks (1998).

Genetic drift: the gene-pool model

● **KEY CONCEPT**

Genetic drift *reduces* genetic variation *within* populations but *increases* genetic variation *among* populations.

The Mendelian lottery, variation in family size, founder effects, and genetic bottlenecks all have one thing in common: they all involve statistical sampling. In the Mendelian lottery, the gametes (and with them the alleles they contain) that form the actual offspring are a random sample from all the gametes available for fertilization. Variation in family size unrelated to the genetic variation under consideration can be seen as the result of sampling offspring from the population of potential parents. A founder effect and a genetic bottleneck can both be viewed as drawing a small sample from a much larger population to obtain the next generation.

The gene-pool model represents genetic drift as a sampling process

Therefore we now consider a simple model of genetic drift as a statistical sampling process. This model is called the *gene-pool model*. In it the formation of the next generation is represented by sampling gametes from a large gamete population, very much like drawing red or blue balls from an urn. The red and blue balls correspond to *A* and *a*, respectively, two alleles at one genetic locus. This model approximates the genetics of a population in which individuals mate randomly. We characterize the individuals in the population by their genotype at the *A* locus. Thus individuals are either *AA*, *Aa*, or *aa*. A new generation is obtained as follows. Each parent produces an equal and large number of gametes; the gametes from all parents collectively form a gene pool from which an offspring is formed by drawing two gametes (genes) at random. This is repeated n times to produce an offspring generation of n individuals (Figure 3.3). The frequency of the *A* allele in the offspring is given by a binomial distribution because the repeated drawing of a gene from the gene pool is a repeated chance event with two possible outcomes, like drawing a red or a blue

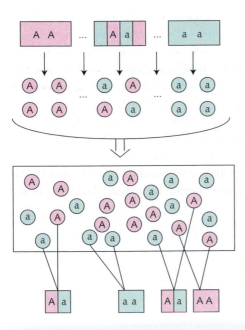

Figure 3.3 The gene-pool model of genetic drift. Diploid parents contribute haploid gametes to a gene pool, from which *n* diploid offspring are drawn at random.

ball. Strictly speaking the binomial distribution only applies if the sampling is with replacement, which means that the frequencies of the *A* and *a* gametes in the gene pool remain the same irrespective of which offspring already have been formed. In reality this is not the case, but if the total number of gametes in the gamete pool is large, the binomial distribution is a very good approximation.

Genetic drift disperses the frequency of a neutral gene among replicate populations

What can we conclude from this model? Imagine we start with many identical parental populations, and for each an offspring generation of n individuals is formed as described above, independently. Then we expect that among the populations the gene frequency *p'* of *A* in the new generation will vary according to a binomial distribution. We do not expect a change in gene frequency when we average over all populations: upward and downward deviations from the original gene frequency *p* will be equally likely. If we repeat the same procedure over many generations, results like those in Figure 3.4 are expected: increasing dispersion of the gene frequency among the replicate populations.

Sooner (when n is small) or later (when n is larger), either the *A* or the *a* allele will be lost (see Table 1.1 in Chapter 1, p. 4).

The dispersion of gene frequencies due to chance is called genetic drift. In statistics it would be called the propagation of sampling error. It is stronger in small populations than in large populations, because a small random sample from a population is likely to deviate more from the population composition than is a large sample. Another consequence of genetic drift is that it tends to make populations homozygous, because it leads to loss or fixation of alleles.

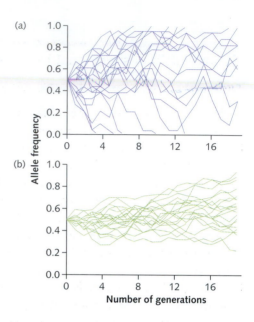

Figure 3.4 Genetic drift at a locus with two alleles with initial allele frequencies of 0.5 has been simulated in (a) 20 small populations (nine diploid individuals) and (b) 20 larger populations (50 diploid individuals). In the absence of mutation, fixation of one of the alleles occurs inevitably, but more rapidly (on average) in the smaller populations. Fixation occurs when the frequency goes either to 1.0 or to 0.0. (From Ridley 1996.)

Thus, genetic drift *reduces* genetic variability *within* a population but *increases* genetic differentiation *among* populations.

Cavalli-Sforza and his colleagues (Cavalli-Sforza 1969) tested the predictions of the genetic-drift model in humans. From villages and towns near Parma, Italy, they collected blood samples from which they estimated gene frequencies for the ABO, MN and Rh blood-group loci. The neutral model predicts greater divergence of the gene frequencies among small, relatively isolated mountain villages than among the larger populations of the towns from the plain. The data clearly confirm this prediction (see Figure 3.5).

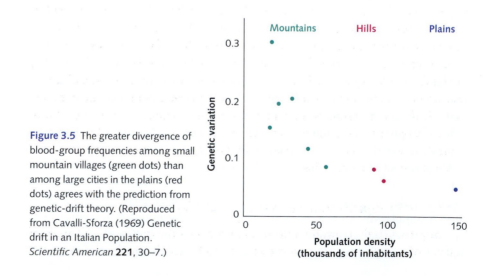

Figure 3.5 The greater divergence of blood-group frequencies among small mountain villages (green dots) than among large cities in the plains (red dots) agrees with the prediction from genetic-drift theory. (Reproduced from Cavalli-Sforza (1969) Genetic drift in an Italian Population. *Scientific American* **221**, 30–7.)

Genetic drift is significant in molecular evolution

Different proteins have different, roughly constant rates of change

In the 1960s new technology revealed the amino acid sequences of proteins. By comparing the sequences of proteins such as hemoglobin and cytochrome c from different species, and using fossil estimates of times to last common ancestors, biologists could estimate the rate of evolutionary change in the protein sequences. For example, dogs and humans, which split from a common ancestor about 100 million years ago, are separated by 200 million years of independent evolution (100 million years down each branch of the tree), and the protein sequences of hemoglobin α subunit in dogs and humans differ at 23 of the 141 amino acids, or 16.3%. This suggests that the mean number of substitutions per amino acid is about one every 1.23 billion years (Li and Graur 2000). A strikingly similar figure is obtained from comparisons of α-subunit hemoglobin evolution between other pairs of vertebrates. Other proteins also have characteristic and roughly constant rates of change, rates that differ considerably among proteins. For example, histone proteins evolve very slowly, while fibrinopeptides change about 80 times faster. The differences in rates are caused by chemical and functional constraints. Histones interact very closely with the DNA, where almost every amino acid has a precise role and is hard to replace without loss of function. In fibrinopeptides, involved in blood clotting, many amino acid changes have little effect on function.

The rate of change in DNA sequences is also roughly constant

The first data on molecular evolution were on amino acid sequences, but since further technological developments enabled us to look directly at nucleotide sequences in DNA, molecular evolution has usually been described as nucleotide substitutions in DNA sequences. Just as there is a fairly constant and characteristic rate of change per amino acid in a protein or class of proteins, there is also a roughly constant and typical rate of nucleotide substitution in a DNA sequence or class of DNA sequences. The rate of synonymous substitutions, substitutions that do not change the coded amino acid, has a large random component and varies little among proteins. In contrast, some non-synonymous substitutions, which change the coded amino acid, experience strong selection, while some are neutral. It is thus the rate of change of non-synonymous substitutions that varies greatly among proteins.

Each protein appears to have its own molecular clock

The approximately constant rate of amino acid substitutions in all lines of descent of protein sequences suggests that each protein has its characteristic

● **KEY CONCEPT**

Drift is more characteristic of DNA sequences than of phenotypes.

Molecular clock: *The approximately constant rate of nucleotide substitution for particular genes and classes of genes within particular lineages. The constancy of the rate depends on the randomness with which particular nucleotides mutate and then drift to fixation.*

pace of evolutionary change, its own molecular clock. This can be used as a method to estimate the time elapsed since the divergence of independent evolutionary lines. Molecular clocks are further discussed in Chapter 13 (pp. 325 f.). There has been much discussion and controversy about their constancy, and several instances have been found where the clock appears to tick faster in one lineage than in another. Nevertheless, evolution in the parts of the DNA that we can expect to be neutral appears to proceed at a fairly constant rate. The constancy of the rate indicates the randomness of the mechanism driving the change.

This is in sharp contrast to morphological evolution. Many cases have been found in which evolutionary rates in genes and morphology seem to be uncoupled. Near constancy of morphology over hundreds of millions of years may go together with large changes at the genetic level. Conversely, large changes in morphology may occur in short time periods during which most genes change little, probably because such morphological change is driven by changes in a few genes with large effects.

Kimura proposed that random drift causes most evolutionary change in DNA

In an influential paper, Kimura (1968) proposed that most of the evolutionary change at the molecular level occurs as a consequence of random genetic drift of mutant alleles that are selectively neutral or nearly neutral. A heated controversy followed. At the time, many evolutionary biologists could not accept the idea that evolutionary change could be a random process; they maintained that changes in allelic frequencies in populations are adaptive and largely determined by natural selection. Note, however, that Kimura did not suggest that all evolutionary change is driven by genetic drift, only most of the nucleotide changes observed at the molecular level. There was never disagreement on the adaptive significance of much morphological, life-history, and behavioral evolution.

Despite recent progress, how much of the genetic variability measured by molecular methods is produced by random genetic drift and how much by adaptive evolution is still not clear. On the one hand, it is clear that the selection forces driving the evolution of DNA sequences that are not expressed and have no direct function must be weak. On the other hand, it also has become clear that several suspected neutral genes have been subjected to weak selection. The controversy over genetic drift versus natural selection in molecular evolution is further discussed in Chapter 5.

Species dynamics resemble genetic dynamics

Species drift affects community and ecosystem ecology and biodiversity

If two or more species are very similar to each other, they may be virtually neutral—virtually substitutable—with respect to the ecological forces that affect their abundance, such as competition, predation, pathogens, and abiotic factors. Just as mutation is a source of new alleles, speciation is a source of new species. If the speciation rate is reasonably high, enough new, essentially similar species may be injected into a continental or oceanic species pool to maintain local biodiversity at a fairly high level as they drift slowly through to extinction. A group of nearly equivalent species is called a guild.

This process has been suggested as the mechanism maintaining the pool of species in several guilds of coral reef fishes (Sale 1977) and in the *Enallagma* damselfly guild of eastern North America (Turgeon & McPeek 2002) at several times the numbers that could be maintained at equilibrium if the species were not nearly ecologically neutral. If this is generally true, then many species could go extinct without loss of ecological function, for as long as at least one member of a guild survived and its population expanded to compensate for losses due to another species' extinction, the function performed by that guild would be maintained.

Species founder effects introduce noise into biogeography and paleontology

Just as genetic founder effects can create gene pools with high frequencies of genes that were rare in the founder population, so, after a mass extinction or colonization of an isolated island, can a few species founders that were rare in the previous geological era or in the continental source pool suddenly achieve high frequency in the new geological era or on an oceanic island. If the reasons for their survival or colonization are fairly arbitrary, such that they represent a nearly random sample of the species that were available, then they introduce noise into the historical and biogeographic process. Isolated islands and new geological eras then fill up with groups of organisms that elsewhere or previously were rare or inconspicuous. Geographic examples include the radiations of land snails, lobelias, and fruit flies in Hawaii. Paleontological examples include the radiation of the mammals in the Cenozoic era after the extinction of the dinosaurs at the end of the Mesozoic era. Which of the available groups actually comes to fill islands or eras with species may, to a certain extent, be an arbitrary accident of history (see Chapter 17, p. 403).

The theory of neutral evolution has generated much controversy and stimulated intensive research on the relationship between molecular variation

● **KEY CONCEPT**
Genetic drift is mirrored at a higher level by the drift of ecologically equivalent species.

and fitness variation. In the next chapter, we examine the contrasting consequences of different types of inheritance—sexual versus asexual, Mendelian versus quantitative—for directed genetic change under selection.

● SUMMARY

- Different alleles with the same effects on fitness are neutral with respect to each other.

- Genetic variation not expressed in the phenotype is likely to be neutral, as is variation that causes functionally unimportant phenotypic variation.

- Neutral genetic variation is subject to random evolutionary change.

- Several processes cause random changes in allele frequencies. The most important are sexual transmission (the Mendelian lottery), random variation in family size, founder effects, and genetic bottlenecks.

- The random change in allele frequencies caused by these processes is called genetic drift; it has stronger effects in small populations than in large populations.

- The influential neutral theory claims that much of the evolutionary change measured with molecular methods is determined by genetic drift.

- One argument for the neutral theory is the relatively constant rate of change of the amino acid composition and the DNA sequence of a protein in all lines of descent from a common ancestor. Each protein is characterized by a characteristic rate of change: it evolves according to its own molecular clock.

- Just as neutral alleles drift in populations, neutral species drift in communities.

- Just as founder effects change the frequencies of genes sampled from the source population, so too do founder effects change the balance of biotas on isolated islands and in eras following mass extinctions.

● RECOMMENDED READING

Kimura, M. (1983) *The neutral theory of molecular evolution*. Cambridge University Press, Cambridge.

Li, W.-H. and Graur, D. (2000) *Fundamentals of molecular evolution*, 2nd edition. Sinauer Associates, Sunderland, MA.

Sacks, O. (1998) *Island of the colorblind and Cycad Island*. Vintage Books, New York.

● QUESTIONS

3.1. Some genetic differences are subject to natural selection in some environments but are selectively neutral in others. Was some human genetic variation that is now neutral formerly under selection? Which traits were involved? What consequences do you expect from the change in selection regime?

3.2. A population of dairy cows may consist of millions of animals, while at the same time due to artificial insemination practices only few males (perhaps a hundred) are used for breeding the next generation. Would you expect genetic drift to play a significant role in such a population?

3.3. Does genetic drift only affect the frequencies of neutral alleles, or does it also affect the frequencies of alleles that are subject to natural selection?

3.4. Genetic drift and genetic founder effects have analogs at the species level. Do they also have analogs in the cultural evolution of language?

3.5. The reproductive success—the fitness—of genes is not immediately clear because genes do not reproduce. Organisms do. And the inheritance of traits is also not immediately clear, for in sexual species the gene combinations that determined that traits were successful or unsuccessful in one generation are broken up and *recombined* in the next. Before reading the next chapter, discuss: how might we conceive of the influence that a gene has on the reproductive success of the organisms in which it occurs? And how might we conceive of the inheritance of traits that are determined by many genes?

CHAPTER 4

The genetic impact of selection on populations

Genetic change is a key to understanding evolution

KEY CONCEPT

All evolutionary change is based on genetic change.

Chapters 1 and 2 discussed several examples of evolutionary change produced by natural selection acting on traits. Under natural selection trait values change from generation to generation when they are correlated with reproductive success, but only if the trait values are inherited. Thus when traits change as a result of selection, genes must also have changed. Knowing how genetic change occurs is essential for understanding and predicting an evolutionary process, for the direction and rate at which traits change under selection depends on the relation between genotype and phenotype and on the genetic composition of the population. This chapter discusses the genetic response to selection.

Population and quantitative genetics are two approaches to genetic change under selection

Population genetics is used when the genotype–phenotype relationship is fairly simple, when genetic differences between two alleles at one locus have phenotypic effects large enough to be detected unambiguously. Many traits whose variants differ qualitatively have a simple genotype–phenotype relationship, including the garden pea (*Pisum sativum*) used by Mendel in the experiments from which he deduced the rules of genetic transmission. His plants differed in several traits including flower color (purple and white), seed color (green and yellow), and seed shape (round and wrinkled). In each of these traits the phenotypic differences correspond to simple differences between alleles at one locus, with each locus on a different chromosome. For example, at the locus determining seed color two different alleles were present in Mendel's experimental plants; the allele *Y* determining a yellow seed color and the allele *y* determining green seeds. *Y* is dominant over *y*, causing heterozygous *Yy* plants to produce yellow seeds. Other examples include some coat-color differences in mammals and many rare genetic diseases in humans. When the genotype can be inferred from the phenotype, population genetics can describe a population in terms of genotype frequencies and can model evolution as changes in these frequencies.

Quantitative genetics is used in cases of a continuous distribution of phenotypic variation instead of discrete phenotypic classes. For example, height, weight, longevity, milk yield in cows, and oil content in seeds vary continuously and quantitatively among individuals. The continuous distribution of the trait values may be due both to the sensitivity of the trait to environmental variation and to the number of genes affecting the trait. Quantitative traits are often affected by many genes, no one of which by itself has an effect large enough to create a recognizable phenotype that can be assigned unambiguously to a specific gene. Because quantitative traits do not segregate into a few recognizable phenotypic classes in genetic crosses as do qualitative traits, the genotype cannot be inferred from the phenotype, and quantitative genetics cannot describe populations in terms of genotype frequencies. Instead it focuses on the variances of traits, estimates how much of the phenotypic variation is due to genetic differences between individuals, and uses that estimate to predict how fast a trait will change under selection.

We now consider simple models of evolutionary change in populations, taking first the population and then the quantitative genetic approach. The population genetic approach requires that we specify the genetic system of the species under consideration. The genetic system consists of the processes involved in transmitting genetic information from parent(s) to offspring; it includes the type of reproduction (sexual or asexual) and the ploidy level of the reproducing individuals (haploid or diploid). Therefore, we first consider a simple classification of the genetic systems in the main groups of organisms.

Sexual: *A mode of reproduction in which the genetic material from two parents is mixed to produce offspring that differ genetically from both parents.*

Asexual: *A type of reproduction in which a parent produces an offspring either by dividing in two or by producing an egg that develops without fertilization. In most cases offspring produced asexually are genetically identical to the parent.*

Haploid: *An organism or cell having a single set of chromosomes. In humans the haploid set contains 23 chromosomes and the diploid set contains 46.*

Diploid: *Diploid cells have two copies of each chromosome, usually one from the father and one from the mother.*

Genetic systems are sexual or asexual, haploid or diploid

The impact of selection on a population depends strongly on the genetic system of the organisms. The two key features of a genetic system are whether reproduction is sexual or asexual and whether the adult organisms are haploid or diploid (Figure 4.1).

We next describe these four genetic systems, then analyze the population genetics of a few cases to see how selection produces genetic change and what difference the genetic system makes.

● **KEY CONCEPT**

The genetic system determines how genes respond to selection.

(a) Sexual haploids include some algae, most fungi, and the mosses

Sexual organisms with predominantly haploid life cycles include some red algae, some conjugating green algae, most fungi, including the ascomycetes (e.g. *Neurospora*), and the mosses (Figure 4.2). In these organisms meiosis occurs

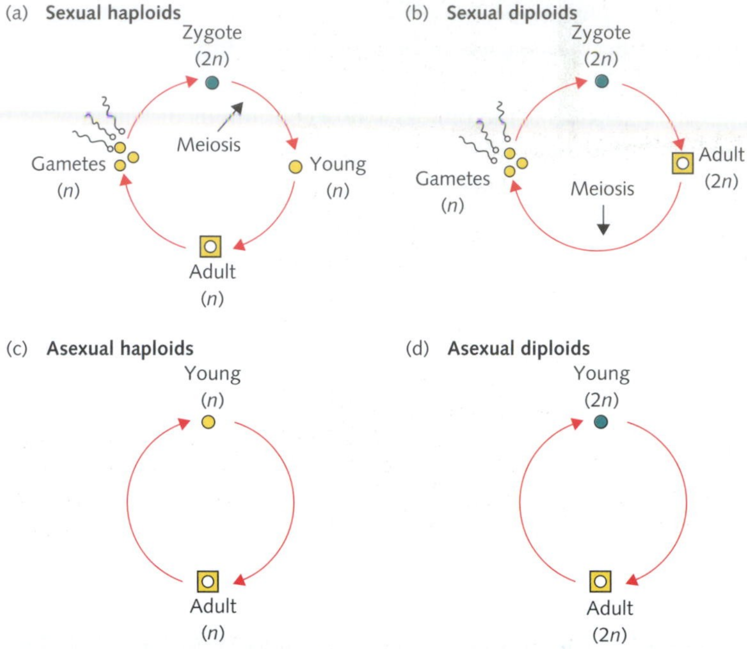

(a) **Sexual haploids**

Zygote (2n)

Meiosis

Gametes (n)

Young (n)

Adult (n)

(b) **Sexual diploids**

Zygote (2n)

Adult (2n)

Gametes (n)

Meiosis

(c) **Asexual haploids**

Young (n)

Adult (n)

(d) **Asexual diploids**

Young (2n)

Adult (2n)

Figure 4.1 The four major genetic systems represented as life cycles. Selection can occur at every stage in the life cycle, but is assumed here to act only as differential survival during growth from zygote or young to adult. The genetic systems differ in whether the adults are diploid (green) or haploid (yellow).

Figure 4.2 A moss, a sexually reproducing organism in which the dominant phenotype in the life cycle is haploid. (Photo courtesy of Robert Thomas, Bates College.)

immediately after zygote formation, and the body of the resulting adult individual is built from haploid cells.

There are fewer than 10 major groups with this genetic system, but the system is not rare—the algae, fungi, and mosses are represented by many species that can occur in large populations.

Figure 4.3 A sexual diploid animal and plant: a hawkmoth pollinating a flower. (Photo courtesy of David Baum, University of Wisconsin.)

(b) Sexual diploids include most animals and plants and some algae, protozoa, and fungi

Sexual organisms with predominantly diploid life cycles include about 20 animal phyla, the multicellular plants with alternating haploid and diploid stages whose diploid stage is larger and longer lived (ferns, cycads, conifers, flowering plants), and several groups of algae, protozoa, and fungi. Sexual diploids tend to be large and long-lived (Figure 4.3). In these organisms the diploid zygotes do not undergo an immediate reduction division. The adult body is built from diploid cells, and the reduction division occurs later in the life cycle just before the haploid gametes are formed.

(c) Asexual haploids include prokaryotes, some fungi, and cellular slime molds

Asexual organisms with predominantly haploid life cycles are mostly prokaryotes, i.e. bacteria and Archaea (see Figure 4.4). This genetic system also occurs in some fungi, such as *Penicillium*, and in the cellular slime molds. With the bacteria, these amount to about 20 major groups of organisms, including most of those that cause infectious diseases. Asexual haploids outnumber by far all the other organisms on the planet.

> **Prokaryotes:** *Single-celled organisms lacking a nucleus and organelles; they include Eubacteria and Archaea.*

(d) Asexual diploids include dinoflagellates and some protoctists

Asexual organisms with predominantly diploid life cycles include the dinoflagellates, more than 10 groups of protoctists (unicellular algae, protozoa, and

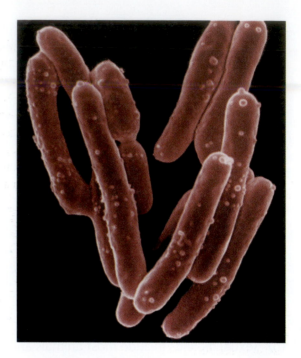

Figure 4.4 A haploid asexual bacterium, *Mycobacterium tuberculosis*, the pathogen that causes tuberculosis. (Photo courtesy of Elizabeth Fischer and Clifton E Barry 3rd, National Institute of Allergy and Infectious Disease, National Institutes of Health.)

Figure 4.5 A diploid asexual bdelloid rotifer, one of the scandalous ancient asexuals that is not supposed to have survived (see Chapter 8, p. 177). (By Dafila K. Scott.)

unicellular groups resembling fungi), and several groups of multicellular animals (Figure 4.5). Although most of the species in these groups are diploid, some may be haploid.

The two commonest types of life cycle are asexual haploids and sexual diploids. The other two life cycles are less common but still important.

Many organisms alternate haploidy with diploidy or sexual with asexual reproduction

As often happens in biology, many groups cannot be so simply classified. Among these are about six groups of sexually reproducing organisms that alternate between haploid and diploid phases with neither one dominating. These include the Foraminifera, important marine protozoa with a long fossil record, the Basidiomycetes, of which mushrooms are well-known representatives, the Microsporidians, an important group of parasites, and the Apicomplexa group of protozoans, which includes *Plasmodium*, the agent of malaria. In addition, there are eight predominantly diploid animal phyla containing species that alternate sexual and asexual reproduction, among them rotifers, cnidarians, annelids, and arthropods, including aphids and water fleas (*Daphnia*).

Thus each of the four major systems has numerous and important representatives, and some organisms do not fit into this binary classification. In the next section we consider simple models of population genetic change under selection, in particular paying attention to the effects of ploidy and of sex. These models allow us to predict how fast an allele will spread through (or disappear from) the population depending on the strength of selection and the genetic system.

A brief comment on the role of models in science

In the next section you will encounter abstract models of genetic change. If you are not yet used to theory in science, you may wonder why we use these models. Here we would like to list briefly several reasons why theoretical models are useful in science.

● **KEY CONCEPT**

To understand, we must simplify, focus only on essential features, and aim for quantitative, not just qualitative, predictions.

1. The natural world is very complex, so complex that we could not think about it clearly if we tried to hold all of its parts and their dynamic interactions in our heads at once. We must simplify it in order to understand it. This necessary simplification could be done in many different ways. Science has tried many of them out. Those that remain are the ones that so far have proven to be the most useful.

2. By modeling a process as a set of equations expressed in parameters and functions, we can represent the process quantitatively as well as qualitatively. Quantitative statements are much more useful than qualitative statements when we are trying to test ideas with experiments, for they are much riskier and easier to falsify. If we say, for example, that we expect an allele to increase in frequency, then any increase will confirm the prediction. But if we say that we expect the allele to increase at a rate of no more than 1% per generation, and we observe an increase of 5% per generation, then it is clear that we have not yet understood something important about the process.

3. Once we have succeeded in understanding a process so well that it can be represented in simple models that have been confirmed repeatedly and independently in risky experiments, we can then connect it meaningfully to other processes. Without a model that makes a few points clear, we would not be able to place the process in context and connect it to other processes, for we would not know which of many things made the essential differences.

With that as background, we now introduce models of genetic change under selection.

Genetic change in populations under selection

● KEY CONCEPT

The rate of adaptive genetic change depends on the genetic system as well as the strength of selection.

In the cases analyzed here we start with the same standard situation, a population into which a new allele that confers greater fitness has entered by mutation. Then we work out the fate of this new allele in different genetic systems. To this end we develop a population genetic model that is built on the following simplifying assumptions.

1. The change in the genetic composition of the population is only caused by the selective differences between the genotypes specified in the model; other possible influences such as genetic drift (treated in Chapter 3, pp. 62 ff.) are neglected.

2. In sexual populations we assume random mating: the probability of a mating does not depend on the genotype of the male and female involved. The random-mating assumption greatly simplifies the analysis, as explained in the Genetic appendix (p. 517), and is a reasonable approximation in many cases.

3. We assume that the selective differences between the genotypes are only manifested as differential survival during development from zygote (or young) to adult. This difference in survival is expressed in the model by a parameter, assigned to each genotype, called *relative fitness*, a number that represents the fraction of the individuals of a given genotype that survive to adulthood and reproduce. By convention, one of the relative fitnesses (usually the highest) is set equal to 1, and the fitnesses of the other genotypes are scaled relative to this one. For example, in the haploid model developed in Box 4.1 (see below), the two genotypes A_R and A_S are assigned relative fitnesses of 1 and of $1 - s$. The parameter s is called the *selection coefficient*. A value of $s = 0.1$ means in this case that A_S individuals survive on average 10% less well than A_R individuals.

The fitness concept in our population genetic models is limited and neglects many aspects that matter in real life. However, it helps to keep the models simple. After we have treated the models we will return to the concept of fitness.

We start with a population of haploid organisms, for example *Chlamydomonas*, a genus of unicellular haploid green algae, some of which live in the soil. Under conditions that are favorable for growth, they reproduce asexually, but when the food gets scarce they differentiate into gametes and reproduce sexually. Let us suppose that an agricultural field is treated with a herbicide that kills weeds and also affects the *Chlamydomonas* cells. A mutation causing resistance to the herbicide confers a selective advantage and is expected to increase in frequency. We assume that a gene at locus *A* is involved in making the *Chlamydomonas* cells resistant to the herbicide, and that two alleles occur in the population at

BOX 4.1

Population genetic model of selection at a single locus in a haploid population

	Genotype	A_R	A_S
	Relative frequency	p	q
	Relative fitness	1	$1-s$

Selection →

		A_R	A_S
	Relative frequency after selection	$p' = \dfrac{p}{1-sq}$	$q' = \dfrac{q(1-s)}{1-sq}$

Reproduction →

Asexual

	A_R	A_S
Relative frequency in in next generation	$p' = \dfrac{p}{1-sq}$	$q' = \dfrac{q(1-s)}{1-sq}$

Sexual

	A_R		A_S
Relative frequency among gametes	$p' = \dfrac{p}{1-sq}$		$q' = \dfrac{q(1-s)}{1-sq}$
Genotype of zygotes	$A_R A_R$	$A_R A_S$	$A_S A_S$
Relative frequency	p'^2	$2p'q'$	q'^2
Genotype of next generation	A_R		A_S
Relative frequency in next generation	$p' = \dfrac{p}{1-sq}$		$q' = \dfrac{q(1-s)}{1-sq}$

this locus, A_S (causing cells to be sensitive to the herbicide) and A_R (conferring resistance). We will now analyze how fast the resistant allele is expected to spread in this population, assuming that initially it is present at a low frequency.

The haploid selection model is specified in Box 4.1 both for the case of asexual reproduction and for sexual reproduction. The models produce a formula that allows calculation of the relative frequency p' of the resistance allele in the next generation, given its relative frequency p in the present generation and a value of s, the fraction by which the sensitive genotype survives less well than the resistant genotype. By instructing a computer to repeat this calculation for many consecutive generations we obtain the change in allele frequency over many generations; it is shown in Figure 4.6 (curve a).

In developing this population genetic model we follow the life cycles shown in Figure 4.1 (a) and (c). Starting at the zygote (or young) stage, we specify the genotypes and the relative frequencies at which they occur in the population. Selection acts in the form of differential survival to the adult stage and is represented by the relative fitness values. The population composition after selection (among the adults) is calculated by dividing the surviving fractions (p and $q(1 - s)$) by their sum (note that $p + q(1 - s) = 1 - sq$, because $p + q = 1$).

Then the adults reproduce, and here we have to consider asexual and sexual reproduction separately. The case of asexual reproduction, where every adult produces its own type, is straightforward: because we assume that there are no

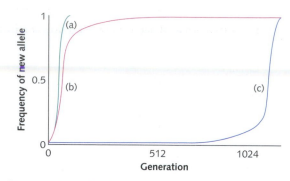

Figure 4.6 Allele-frequency change by selection. The curves show the increase of an initially rare allele with a selective advantage of 10% ($s = 0.1$): (a) in a haploid population (both asexual and sexual); (b) in a diploid sexual population when the allele is dominant, and (c) in a diploid sexual population when the allele is recessive. The degree of recessiveness has a large effect on the rate of spread of a favorable allele in a sexual population. Completely recessive alleles spread very slowly.

fertility differences among the genotypes, the genotype frequencies among the newborn individuals are the same as among their parents. Sexual reproduction is more complicated, for it consists of gamete formation, fusion of pairs of gametes to produce zygotes, then meiosis in the diploid zygotes to make a new generation of haploid individuals. Since again no fertility differences are assumed, the genotypes among the gametes will reflect those among their parents. The assumption of random mating results in the Hardy–Weinberg distribution among the zygotes, which does not affect the allele frequencies (see Genetic appendix, p. 517). Finally, the process of meiosis that produces the haploid offspring also does not change the allele frequencies (see Genetic appendix, p. 512). Thus, although the sexual model is more complicated than the asexual model, the final outcome is identical here. Genetic change at a single locus under selection in a haploid population is the same in sexual and asexual populations. We will see later that this conclusion is not general, for it does not apply to diploid populations or to populations described by the genotypes at more than one locus.

Next we develop a similar model for selection at a single locus in a diploid population. Elaborating on the example we used before, we can now consider a diploid weed species growing in the agricultural field that is treated with herbicide, and model the spread of an initially rare resistance mutation in a population of these plants. Again assuming that sensitivity to the herbicide is controlled at a single locus and that two alleles occur in the population, A_S causing plants to be sensitive and A_R conferring resistance, we can specify relative fitness values for the three diploid genotypes as is detailed in Box 4.2. Because in diploids there can be heterozygotes, we need a new parameter, h, to describe the fitness of heterozygotes, which is defined in our model as $1 - hs$. The value of h determines the dominance relationship between the two alleles. Complete dominance of A_R over A_S can be indicated by setting $h = 0$, while complete recessiveness of

BOX 4.2

Population genetic model of selection at a single locus in a diploid population

Genotype	A_RA_R	A_RA_S	A_SA_S	A_R	A_S	
Relative frequency	x	y	z	$p = x + \frac{1}{2}y$	$q = z + \frac{1}{2}y$	
Relative fitness	1	$1 - hs$	$1 - s$			Asexual model
Selection→						
Relative frequency after selection	$\dfrac{x}{V}$	$\dfrac{y(1-hs)}{V}$	$\dfrac{z(1-s)}{V}$			
		$V = x + y(1 - hs) + z(1 - s)$				
Reproduction→						
Relative frequency in next generation	$x' = \dfrac{x}{V}$	$y' = \dfrac{y(1-hs)}{V}$	$z' = \dfrac{z(1-s)}{V}$	$p' = x' + \frac{1}{2}y'$	$q' = z' + \frac{1}{2}y'$	

Genotype	A_RA_R	A_RA_S	A_SA_S	A_R	A_S	
Relative frequency	p^2	$2pq$	q^2	p	q	
Relative fitness	1	$1 - hs$	$1 - s$			Sexual model
Selection→						
Relative frequency after selection	$\dfrac{p^2}{W}$	$\dfrac{2pq(1-hs)}{W}$	$\dfrac{q^2(1-s)}{W}$	Relative frequency in gametes	$p' = \dfrac{p^2 + pq(1-hs)}{W}$	
		$W = p^2 + 2pq(1 - hs) + q^2(1 - s)$			$q' = \dfrac{q^2(1-s) + pq(1-hs)}{W}$	
Reproduction→						
Relative frequency in next generation	p'^2	$2p'q'$	q'^2	p'	q'	

A_R corresponds to $h = 1$. Remember that a recessive allele is masked in a heterozygote, which explains the identical fitnesses of the dominant homozygote and the heterozygote. Often dominance or recessiveness of an allele is not absolute, and both alleles have some effect in heterozygotes. In those cases h will take other values; for example when the heterozygotes are exactly intermediate in fitness between the homozygotes, $h = 0.5$. The model is worked out in detail in Box 4.2, both for an asexual and for a sexual population.

In developing this population genetic model we follow the life cycles shown in Figure 4.1 (d) and (b). The relative frequencies of the diploid genotypes are specified (Box 4.2, left-hand side) as well as the relative allele frequencies (Box 4.2, right-hand side). Relative frequencies after selection are calculated by dividing the surviving fractions by their sum, which is designated by the symbol V in the asexual model and by W in the sexual model. The way the model is worked out follows the same sequence of steps as we used in the haploid model (see the explanation in Box 4.1).

In contrast to the results from the haploid model, in which genetic change due to selection at a single locus is the same in sexual and asexual populations, these two reproductive systems do make a difference in the diploid model. The calculation of genetic change requires specification of the diploid genotype frequencies, because selection acts in the diploid phase of the life cycle. In a sexual population random mating produces Hardy–Weinberg proportions among the offspring, which implies that knowledge of the relative allele frequencies is sufficient to describe the population in terms of the diploid genotype frequencies (in the two-alleles case the genotype frequencies p^2, $2pq$, and q^2 follow directly from the allele frequencies p and $q = 1 - p$). In an asexual diploid population the allele frequencies do not determine the diploid genotype frequencies. For example, $p = 0.6$ and $q = 0.4$ when the genotypes $A_R A_R$, $A_R A_S$, and $A_S A_S$ have relative frequencies of 0.3, 0.6, and 0.1, but also when the population composition is 0.4, 0.4, and 0.2, and in many others. This explains why in the sexual model we can calculate the genetic change due to selection with a formula that contains the allele frequency as the only variable (for fixed values of the parameters s and h), while in the asexual model such a convenient calculation is not possible. There the outcome will be different for each different initial distribution of the diploid genotypes. For this reason we only consider the dynamics of genetic change in the sexual diploid model. In Figure 4.6 (curves b and c) we contrast the outcomes for the two different extreme cases: complete dominance ($h = 0$) and complete recessiveness ($h = 1$) of the allele A_R. The degree of recessiveness determines how rapidly a favorable allele will spread; with complete recessiveness the process occurs very slowly.

What population genetics implies for evolutionary biology

Allele substitution by selection in natural populations is slow-fast-slow

● KEY CONCEPT

The spread of advantageous alleles is slow-fast-slow in all systems; it is slowed by recessiveness in diploid sexuals, and is faster in haploids than in diploids.

In all cases the curve describing the genetic change under selection is S-shaped (Figure 4.6). The models predict slow change at low and high allele frequencies and fast change at intermediate allele frequencies. Therefore rare phenotypes are expected to increase or decrease very slowly (depending on whether they are selected for or against). When alleles are at intermediate frequencies, evolutionary change can be much faster. The basic reason is that the increase or decrease per generation of an allele with frequency p is proportional to $p(1 - p)$, as can be shown after some manipulation of the formulae developed in Boxes 4.1 and 4.2. This implies that for values of p close to 0 and close to 1, the change is very slow.

There are not many well-documented examples of allele substitutions in populations. One reason is that in most cases where an advantageous allele is replacing an existing prevalent allele, the complete process is too slow to be witnessed within a human lifetime. It is therefore not surprising that the examples we have are either from microorganisms with very short generation times or for species confronted with drastic environmental changes that have produced strong selection for alleles that could cope with the new conditions. Figure 4.7

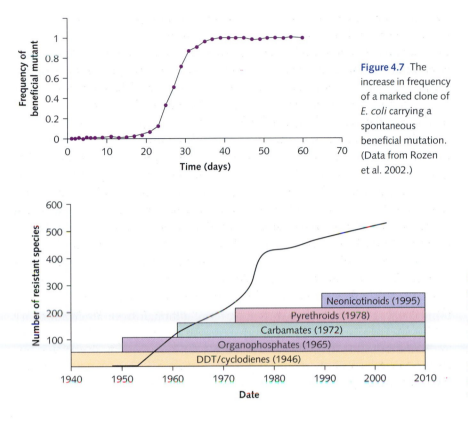

Figure 4.7 The increase in frequency of a marked clone of *E. coli* carrying a spontaneous beneficial mutation. (Data from Rozen et al. 2002.)

Figure 4.8 The number of insect species resistant to pesticides has increased rapidly to over 500; the colored bars show when particular insecticide groups have been used, and dates in parentheses are the year in which resistance was first documented. (From Denholme et al. 2002.)

shows the spread of a favorable mutant in a culture of *E. coli* bacteria over a period of 60 days, which amounts to over 4000 generations at 20 min per generation. The pattern is clearly slow–fast–slow.

The spread of resistance to pesticides has been rapid

A dramatic example of recent, rapid genetic change is the spread of resistance to pesticides (Figure 4.8). Since World War II large amounts of insecticides, particularly DDT (dichlorodiphenyltrichloroethane) and malathion, have been used to control insects that attack crops and spread human diseases. Initially most insects were sensitive to the pesticides, chemical control was effective, and by the early 1960s the incidence of diseases transmitted by insects was much reduced. But by 1970 many insects had evolved genetic resistance to the principle insecticides. Resistance is based on several mechanisms. Some resistance mutations change the properties of the cell membrane, preventing the toxic molecules from entering the cell, while other mutations allow the cell to break these substances down into harmless molecules. The spread of resistance is exemplified by estimates of the incidence of malaria. Malaria declined to about 70 million cases per year in 1960 but increased again to 200 million by 1970 due to the rapid spread of resistance to DDT among the mosquitoes that transmit the disease. By 2000 over 500 species of insects were resistant to the common pesticides.

In hundreds of species similar evolutionary processes occurred: rare alleles conferring resistance increased in frequency after the massive applications of pesticides started. During the first 10–15 years this increase had little impact at the population level because of the slow initial increase in frequency of the resistance alleles. Then came a period of fast change during which resistance spread in most species. This pattern, consistent with the models of gene substitution considered above, suggested that genetic resistance to pesticides is often based on single alleles. We now know that resistance is often determined by a single gene interacting with several modifying genes. Our genetic model predicts that even if pesticide application is continued, the non-resistant alleles will remain in populations at low frequencies for a long time because the final phase of allele substitution is slow.

Note that decision makers would not have perceived any problem for 10–15 years, a characteristic of evolutionary processes that makes it difficult for human institutions to deal with them. Most evolutionary processes occur on a timescale that humans perceive as long-term, although from an evolutionary perspective they are rapid and short-term. An exception is the evolution of antibiotic resistance in the bacteria that cause infections and disease in humans. That happens so rapidly that the medical profession recognizes it as a major problem and has developed countermeasures, with some success.

In diploids the degree of dominance strongly affects rate of change

In diploids, a recessive allele cannot be selected when it occurs in heterozygotes. This effect is most pronounced in asexual diploids. A newly recessive mutation will initially be present in a heterozygous individual, and if the new allele is completely recessive, selection cannot change its frequency, for it is not expressed in the phenotype of heterozygotes. In sexual diploids, a recessive mutation is not completely shielded from selection, for although it initially occurs only in heterozygotes, matings between heterozygotes will produce some homozygous recessive offspring that are exposed to selection. Random mating distributes the alleles at a particular locus over the diploid genotypes in Hardy–Weinberg proportions. When A_R is recessive and rare (p is small, for example 0.001), the selectively favored $A_R A_R$ genotypes occur with frequency p^2, which means they are extremely rare (in this example, 0.000001). In a sexual population these rare $A_R A_R$ genotypes will have mostly $A_R A_S$ offspring, because almost all $A_R A_R$ individuals will mate with the common $A_S A_S$ genotype. Thus there is little scope for selection of A_R because the recessive allele is not expressed and is 'invisible' to natural selection in heterozygotes. This explains the very slow spread of a rare favorable recessive mutation (Figure 4.6c).

These arguments apply both to the spread of favorable mutations and to the removal from diploid populations of deleterious recessive alleles by selection.

Particularly at low allele frequencies, most of these alleles are expected to reside in heterozygotes in which they are shielded from selection because they are not expressed in the phenotype. This conclusion is supported by data on the frequency of recessive lethal alleles in wild populations. In haploid populations such alleles cannot be maintained, but in diploid populations, both of *Drosophila* and of a few vertebrate species, the average number of recessive lethal alleles is between one and two per individual (McCune *et al.* 2002). They are distributed over many loci.

For dominant genes the situation is different. They are expressed in heterozygotes and selection can 'see' the allele in both homozygous and heterozygous genotypes. This explains why a dominant favorable mutation will spread much faster (Figure 4.6b).

Because recessive favorable mutations spread much more slowly than dominant ones, it is not surprising that many of the mutations conferring pesticide resistance are dominant, for we only detect the ones that responded quickly. Recessive resistance mutations may well be present in insect populations, but they have not yet had time to reach high frequencies.

Genetic change is faster in haploid than in diploid sexual populations

The models suggest that if the same mutation conferring a fitness increase occurred in a haploid and in a sexual diploid population, it would spread faster in the haploids, particularly if it were largely recessive. Because most microorganisms are haploid, while most animals and seed plants are sexual diploids, populations of microorganisms are expected to respond to the same selection pressure more rapidly than populations of higher eukaryotes. Because the timescale used in comparing the two genetic systems is in units of generation, the difference is even greater when measured in absolute time, for microorganisms can have generation times as short as 20 min. Therefore pathogenic microorganisms should have the upper hand in coevolutionary interactions with sexual diploid hosts, particularly multicellular hosts with longer generation times. The hosts have evolved several solutions, including immune responses and endosymbiotic partner microorganisms that respond on the same timescale as the pathogens.

Eukaryote: *An organism with a cell nucleus surrounded by a nuclear membrane, usually with organelles, such as mitochondria and chloroplasts, that have their own circular DNA genome. Eukaryotes include the protists, fungi, plants, and animals.*

Single-locus models do not capture the important consequences of sexual recombination

Our models analyze genetic change at a single locus at which two alleles differ in their effects on fitness. Because real organisms contain thousands of loci, this is a gross simplification. To what extent do the conclusions we have drawn remain valid for models that consider selection involving several loci? We have seen that in haploid populations allele frequency change at one locus is identical under asexual and sexual reproduction. But the one-locus haploid model is exceptional in producing identical formulae for allele-frequency change in

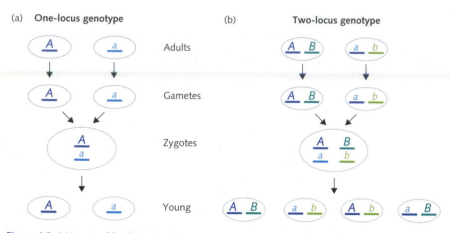

Figure 4.9 (a) In sexual haploids single-locus genotypes of the offspring are the same as the parental genotypes, as under asexual reproduction. (b) When the genotypes are characterized at two loci, offspring genotypes may differ from the parents due to sexual recombination.

asexual and in sexual populations, and it would be wrong to conclude that sex has no effect on the dynamics of genetic change in haploids, for that conclusion changes as soon as we consider two loci.

When we consider two or more loci, we see that sex has large and important consequences for genetic change, both in haploids and in diploids. Whereas asexual organisms have offspring identical to their parent, sexual organisms have offspring that differ genetically from the parents. At a single locus, this effect is not observable in haploids, but it is in diploids, where two heterozygous parents can produce a homozygous offspring. When we follow two loci, we can observe offspring with a genotype different from that of either parent both in diploids and in haploids. For example, if the two loci A and B are on different chromosomes, sexual reproduction between haploid parents *AB* and *ab* would generate 50% offspring with the parental genotypes and 50% recombinant *Ab* and *aB* offspring (see Figure 4.9).

Sex affects genetic change by 'liberating' an allele from its genetic background

Thus sex can 'liberate' an allele from its current genetic background (the rest of the genome) and put it in a different background, in which it may be more (or less) exposed to selection. A favorable mutation that finds itself in a genotype carrying many deleterious genes has little chance to spread in an asexual system where its fate is determined by the low overall fitness of the genotype. In contrast, in a sexual system this mutation can escape from its low-fitness background by recombination and will have a better chance of spreading in the population. Similarly, selection in an asexual population against a deleterious mutation in a high-fitness background will be less effective than in a sexual population.

For a fuller discussion of the important effects of sexual recombination on the genetics of populations see Chapter 8 (p. 184).

The fitness concept in population genetics

The fitness of a genotype is abstract and convenient but hard to measure

The term fitness is shorthand for reproductive success. This is not a very precise definition, but sufficient for our purposes. A heritable trait is favored by natural selection if individuals carrying this trait have on average higher fitness (greater reproductive success) than individuals without this trait. In evolutionary biology fitness is a very important concept, for it provides a way to quantify natural selection. In the population genetic models above, fitness appears as a parameter associated with a single-locus genotype. For example, in a haploid model genotype A_S is said to have fitness $1 - s$ relative to the fitness 1 of genotype A_R. What do we mean when we assign a fitness value to an allele? It seems odd to assign a measure of reproductive success to an allele rather than to an organism. What is meant is the *average* reproductive success (fitness) of many individuals carrying allele A_S relative to the average fitness of many individuals carrying A_R. Ideally, the estimation of the fitness of a genotype should involve contrasting groups of individuals that *only* differ at the locus under consideration and have identical genotypes otherwise. This is practically impossible in the sexual diploid organisms that include most flowering plants and animals. It is in principle possible to approximate such conditions with genetic techniques in bacteria like *E. coli* and some fungi like yeast and *Aspergillus*.

> ● **KEY CONCEPT**
>
> Fitness in population genetics represents the impact of single alleles on the reproductive success of organisms. Such impacts can extend to relatives.

Fitness: *Relative lifetime reproductive success. In some situations, measures that account for differences in the timing of reproduction are more appropriate.*

The concept of inclusive fitness generalizes individual fitness and helps to explain altruism

Assigning a fitness value to a genotype allows an important generalization that would not be possible if the fitness concept were restricted to the reproductive success of individuals. This generalization leads to the concept of inclusive fitness, which consists of two components:

- personal reproductive success of a focal individual;

- the increase in the reproductive success of other individuals of the same genotype that results from the actions of the focal individual.

Inclusive fitness has important applications to the evolution of social behavior developed by Hamilton (1964a,b) in two important papers that form the basis for an evolutionary explanation of altruistic traits.

The existence of altruistic traits poses a difficult problem for evolutionary theory as long as fitness is defined as individual reproductive success. How can self-sacrificing behavior evolve if it lowers fitness? Yet self-sacrificing behavior does occur. Striking examples are found among the social insects; for example,

Inclusive fitness: *Genetic contribution to the next generation through both one's own reproduction and the increase in the reproductive success of relatives caused by aid given to them.*

Altruism: *Behavior that increases the reproductive success of others while reducing one's own reproductive success.*

sterile worker bees that have abandoned reproduction to help the queen rear her offspring. Hamilton's theory explains such behavior. Before introducing his main result, we need to define relatedness, r. The relatedness of parent and offspring is $r = 0.5$, as it is for full siblings, while the relatedness between cousins is $r = 0.125$. Hamilton's result, now known as Hamilton's Rule, states that a heritable altruistic behavior may evolve if $c/b < r$, where c is the cost of the altruistic behavior to the individual displaying the behavior (the actor), measured as reduced fitness, b is the benefit to the recipient individual, measured as increased fitness, and r is a measure of the genetic relatedness between the actor and the recipient, which can take values from 0 (completely unrelated) to 1 (genetically identical). Hamilton's Rule shows that altruistic behavior is expected to evolve between close relatives that have high values of r.

We can restate Hamilton's Rule as a directive: 'Help your relatives if the effects of your aid, as measured by the *increase* in the number of copies of your genes present in the next generation coming from all relatives affected by your action, more than compensate for the *decrease* in the number of copies of your genes that you are able to pass on yourself that results from your action.' In this light a famous statement made by Haldane becomes understandable: 'Would I lay down my life to save my brother? No, but I would to save two brothers or eight cousins.'

Selection operating on differences in inclusive fitness is called **kin selection**, because such interactions are usually between relatives. As early as 1930 Fisher had suggested a similar mechanism for the evolution of distastefulness in some butterfly larvae. A bird that tries to eat a distasteful larva will learn to avoid others of the same type although this is of no help to the one eaten. But if the eggs have been laid in a cluster, the other larvae close by are probably full siblings of the victim. The distastefulness can then be seen as an altruistic trait: the one eaten is sacrificed, but this may save a few brothers and sisters. The role of kin selection in transforming potential conflict into cooperation is further discussed in Chapter 15 (p. 373).

Kin selection: *Adaptive evolution of genes caused by relatedness; an allele causing an individual to act to benefit relatives will increase in frequency if that allele is also found in the relatives and if the benefit to the relatives more than compensates the cost to the individual.*

Quantitative genetic change under selection

Quantitative genetics deals with traits influenced by many genes

● **KEY CONCEPT**

Quantitative genetics explains the inheritance of traits influenced by many genes; it focuses on the causes of changes in *traits* rather than on the causes of changes in *genes*.

Unlike the examples in the previous section, many traits vary continuously and quantitatively among individuals in a population, including traits with considerable ecological and evolutionary significance, such as body size, learning ability, running speed, fecundity, and survival. Continuously distributed traits result both from the sensitivity of traits to environmental change and from development influenced by many genes, such that no single allele has a large-enough effect to create an easily recognizable phenotype. Usually both types of cause play a role. Because quantitative traits do not segregate into clear phenotypic classes in genetic crosses, we cannot use Mendelian genetics to infer

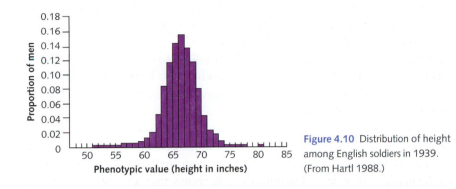

Figure 4.10 Distribution of height among English soldiers in 1939. (From Hartl 1988.)

genotype–phenotype relations. The frequency diagrams of such traits measured on many individuals are often bell-shaped curves (Figure 4.10).

Continuous phenotypic variation thus usually has two causes:

- genetic variation among the individuals;
- variation in the environments they experience.

That genetic variation is at least partially responsible for phenotypic differences follows from plant and animal breeding, where artificial selection almost always results in heritable change (Figure 4.11a). We know that the environment also affects phenotypic variability because individuals from inbred lines or from the same clone still vary phenotypically despite being genetically similar or identical (Figure 4.11b).

We cannot use population genetics to analyze genetic change caused by selection on quantitative traits, for we have no idea what genotypes are present and cannot describe genetic change in terms of allele frequencies. Instead we take an

Artificial selection: *Selection carried out by humans with the aim of changing particular trait values, widely used in plant and animal breeding to improve desired traits and in evolutionary research to test hypotheses.*

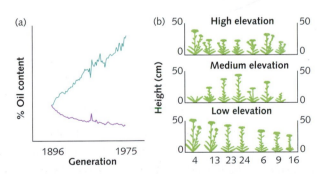

Figure 4.11 Both genetic and environmental variation can cause phenotypic differences. (a) Since 1896 a population of corn (*Zea mays*) has been selected for high and low oil content. The response to selection, particularly in the high line, points to the continued presence of genetic variation within this population. (From Dudley 1977.) (b) Cuttings of seven different genotypes of *Achillea* were grown at three different elevations. The resulting phenotypes show that variation within a single environment is caused by genetic differences and that variation within a single genotype is caused by environmental differences. The numbers on the *x* axis are labels for the different genotypes. (From Griffiths et al. 1996.)

approach that analyzes phenotypic variation to predict evolutionary change. It makes two simplifying assumptions, as follows, each of which is discussed below in detail.

1. Traits are affected by many loci of similar effect.
2. Phenotypic variation combines genetic and environmental effects.

Quantitative genetics assumes traits affected by many loci of similar effect

The first important assumption is that a quantitative trait is affected by alleles at many loci and that most of these alleles have a small, similar effect on the trait. For example, suppose that variation at 10 loci, each on a different chromosome, affects differences in height among human males. At each locus the population contains two alleles, denoted + and −. The first has a positive effect and the second a negative effect on height. The allele frequencies may differ between the loci but are assumed to be intermediate, say 0.2 to 0.8. At some loci the + allele may be more frequent in a population, at others the − allele. Under these assumptions most individuals will have a genotype with roughly equal numbers of + and − alleles, and only rarely will genotypes occur that contain mostly + or mostly − alleles. Thus, this model is consistent with the approximately bell-shaped phenotypic distribution of height (Figure 4.10). Note that we cannot infer that bell-shaped phenotype distributions must be caused by variation at many loci, for they can also result from variation in the environment. For example, the distribution of Figure 4.10 could, in the absence of genetic variation, also occur when adult height is sensitive to the quality and quantity of nourishment during childhood, with most people having grown up under average conditions, while minorities experienced favorable or unfavorable conditions.

Phenotypic variation results from genetic variation and environmental effects

How can we find out how much phenotypic variation is caused by genetic differences? This is an important question, for selection on parents will only result in phenotypic change in offspring if some of the phenotypic variation is heritable. Only then will the greater reproductive success of particular phenotypes result in genetic change (see Figure 4.12).

There are straightforward experimental procedures to estimate what part of the total phenotypic variance in a population is due to genetic and what part to environmental variation. The underlying assumption is that the phenotype P is the sum of genetic effects G and environmental effects E:

$$P = G + E \qquad\qquad\qquad [4.1]$$

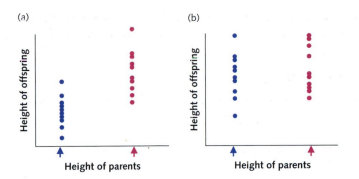

Figure 4.12 Experimental procedure to test the heritability of a trait. From a population containing plants of different sizes, two of the smallest plants are crossed with each other; the same is done with two of the largest plants. The arrows along the *x* axis indicate the average sizes of the pairs of parents. From both crosses 10 seeds are harvested and sown in a greenhouse, to make sure that all offspring will grow up under the same conditions. The diagrams show the offspring sizes from the crosses. If the offspring from the small × small cross are smaller than those from the large × large cross (as in part (a)), we conclude that the size differences in the population are in part due to genetic factors. If the offspring from both crosses do not differ significantly in size (as in part (b)), we conclude that the size differences observed in the original population must have been caused by environmental factors.

P, G, and E are measured in phenotypic units, for example height measured in centimeters or inches. When G and E are not independent, for example because different genotypes respond differently to environmental variation, the additive assumption (eqn 4.1) may not hold. If genes and environment act independently, we can write for the variances of the phenotypic, genetic, and environmental values:

$$Var(P) = Var(G) + Var(E) \qquad [4.2]$$

This formula suggests how to estimate what part of the total phenotypic variance is due to genetic variation in the population and what part can be ascribed to environmental variation. In controlled experiments, either $Var(G)$ or $Var(E)$ can be made practically zero. For example, when we cultivate many cuttings from a single plant, the genetic variation is absent and the phenotypic variance $Var(P)$ estimates the environmental variance $Var(E)$; when we grow offspring from several crosses under strictly controlled conditions in a greenhouse, the environmental variation is very low, and the observed phenotypic variance $Var(P)$ estimates the genetic variance $Var(G)$.

Generally genetic variation is easier to control than environmental variation, for one often does not know which environmental variables should be controlled. Genetic variation can be reduced to near zero in some plants by using cuttings from a single plant so that all experimental plants have the same genotype. In microorganisms and fungi genotypes can easily be replicated through asexual reproduction. In animals the most practical procedure is to generate inbred lines by continued inbreeding for many generations. Among individuals of an inbred line there is little genetic variation, and crosses between two highly inbred lines yield heterozygotes that are virtually the same.

You might think that once the genetic and environmental variances as fractions of the total phenotypic variance have been estimated, we are in a position to judge how effective selection can be in causing genetic change in the population. If most phenotypic variability is due to environmental variation, selection is not expected to result in genetic change, while a high genetic variance should give much opportunity for selection to bring about genetic changes in consecutive generations. This reasoning is roughly correct, but a complication arises from the fact that not all genetic variance allows a response to selection because several interactions among different genetic factors may influence the trait in ways that inhibit a response to selection.

The following example illustrates such an interaction, in this case between the alleles at a single locus. Suppose that selection favors cold-resistant plants and that a population shows genetic variation for this trait. In particular, suppose that at a locus that contributes to cold resistance the A_1A_2 heterozygotes are more cold resistant than either homozygote. The plants favored by selection and producing most of the offspring will be the A_1A_2 genotypes. But only half of the offspring from these heterozygotes will be heterozygous themselves, while the other half will consist of less well-adapted homozygotes. Even if only heterozygotes breed to form the next generation, 50% of the offspring will be homozygous, preventing a further response to selection. Interactions between alleles at different loci may similarly prevent a selection response, despite the genetic variability they represent.

Therefore, we need to estimate the part of the genetic variance that is actually available for a response to selection. This part of the genetic variance is called the *additive genetic variance Var(A)*, because it results from genotypic differences caused by additive allelic effects—that is, effects that are independent of the complicating interactions. In our example, the allelic effects are additive if the heterozygotes A_1A_2 are exactly intermediate in cold resistance between the homozygotes A_1A_1 and A_2A_2. Then the alleles A_1 and A_2 are acting additively—neither is dominant, and their combined impact on the phenotype is the numeric average of their independent effects. When this is the case, a response to selection for cold resistance is expected for as long as there is genetic variation at the locus. For this reason, quantitative geneticists focus on the portion of total phenotypic variance in a trait that can be attributed to additive genetic effects.

Heritability: *The fraction of total phenotypic variance in a trait that is accounted for by additive genetic variance; measures the potential response to selection.*

The heritability determines the potential response to selection

The additive genetic variance expressed as a fraction of the total phenotypic variance is called the heritability.

$$h^2 = Var(A) / Var(P)$$

[4.3]

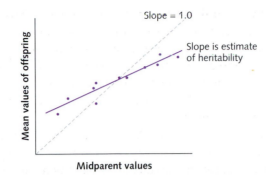

Figure 4.13 One procedure to estimate h^2. Several crosses are made, and from each cross the combination of the midparent value (the mean of the trait values of the two parents) and the mean of the offspring values is plotted. If the points obtained lay on the dotted line with slope 1.0, heritability would be perfect: in every cross the mean value of the offspring would be fully predictable from and exactly equal to the mean of the parental values. In practice environmental factors almost always affect the trait values. Some parents will have very low trait values because they happened to have experienced unfavorable environmental conditions, while other parents will have very high values because of a favorable environment. Because this environmental component is not transmitted to the offspring, the offspring lie closer to the population mean. The slope of the regression line through the data points estimates the heritability. This line fits the data best in the sense that it is the one out of many possible lines that minimizes the sum of the squared deviations of the points from the line.

The squared symbol, which may look a bit strange, derives from the 1920s when h was used as the corresponding ratio of standard deviations. The heritability is a useful measure of the expected response to selection. Estimation of the heritability of a trait requires that we divide the estimated genetic variance $Var(G)$ into an additive component $Var(A)$ and non-additive components. The additive variance $Var(A)$ can be estimated from comparisons of measurements among various types of relative. Details of the statistical procedures involved are not discussed here but can be looked up in any text on quantitative genetics. An intuitive understanding of how we can estimate heritability by comparing relatives can be obtained from a graph in which trait values of offspring and parents are plotted (Figure 4.13).

Traits more strongly correlated with fitness have lower heritabilities

Table 4.1 summarizes 1120 experimental estimates of narrow-sense heritabilities for natural, outbred animal populations (Mousseau and Roff 1987). The traits are grouped into four categories: life-history traits like fecundity, viability, survival, and development rate; physiological traits like oxygen consumption and resistance to heat stress; behavioral traits like alarm reaction and activity level; and morphological traits like body size and wing length. The average heritabilities differed significantly between all pairs of categories except physiology and behavior.

Table 4.1 Summary of 1120 heritability estimates. (From Mousseau and Roff 1987.)

Trait category	Life history	Physiology	Behavior	Morphology
Mean heritability	0.262	0.330	0.302	0.461

In general, traits directly connected to reproductive success have low heritabilities and traits less strongly correlated with fitness have higher heritabilities. Part of the reason is that continued selection on a trait tends to exhaust additive genetic variation by fixing advantageous alleles, causing the heritability to decrease, so long as beneficial mutations do not occur frequently enough to replace the variation fixed by selection. Because fitness is under continuous selection in natural populations, we expect low heritabilities for traits that are strongly correlated with fitness—and we find them.

The concept of heritability must be treated with caution

Heritability tells us about the contribution of genetic *variation* to the phenotypic *variation* in a population. It does *not* tell us to 'what extent a trait is genetic or environmental'. Taken literally, that phrase is meaningless. Try for example to explain the meaning of the following statement: adult body size in humans is determined 70% by genes and 30% by environmental conditions like nutrition and physical exercise. It is obviously ridiculous to assume that in a person of 180 cm, 126 cm are due to genes and 54 cm to environment. Clearly, both genes and suitable environmental conditions are necessary for a human to exist and to possess any trait. It is also not correct to apply an estimate of heritability from a particular population to the whole species or to the same trait in other species. When environmental conditions are well controlled in an experimental population, there will be little environmental variance, and consequently heritability estimates will be higher than when environmental conditions are allowed to vary between individuals. And when a population is inbred, as is often the case in the laboratory or in domestic plants and animals, there will be relatively little genetic variation, and heritability estimates will be lower than in an outbred population. Estimates of heritabilities are only reliable for the population and environment in which they are measured.

Evolutionary implications of quantitative genetics

High heritabilities combine with strong selection to produce rapid change

● **KEY CONCEPT**

Quantitative genetics provides us with a theoretical standard by which to judge claims about the rate of evolution, such as those concerning stasis and punctuation (Chapter 17).

The model developed above predicts that under directional selection the rate of phenotypic change depends on the narrow-sense heritability. When much of the total variability in the population is due to additive genetic effects, the response

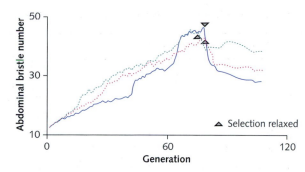

Figure 4.14 Artificial selection on a trait with a high heritability can produce an impressive response. Here the result is shown from a long-term selection experiment on abdominal bristle number in *Drosophila*. The tendency to return to lower values after relaxation of selection indicates that the continued upward selection produced some negative side effects. (From Yoo 1980.)

to selection can be rapid and strong. Little additive genetic variance implies a poor response to selection. Artificial selection experiments support the validity of the model.

For example, Yoo (1980) got an impressive response to selection on abdominal bristle number in *Drosophila*, a trait with a high heritability (Figure 4.14). After 90 generations of intense selection the increase in bristle number was 16 times the standard deviation in the original population (recall that less than 1% of a normally distributed population is more than three standard deviations from the mean). On the other hand, selection on rate of egg production in chickens, a trait with low heritability, has been largely unsuccessful (Nordskog 1977).

Many examples from plant and animal breeding testify that strong directional selection can produce large phenotypic changes. Just think of the different breeds of dogs and the many cereal crops that are so different from their wild ancestors. The body mass of domestic dogs can range from chihuahuas weighing less than 1 kg to newfoundlands of 80 kg. This range of body sizes is far greater than can be found among the wild species in the genus *Canis*. Such cases of fast change under strong directional artificial selection contrast sharply with what we know of long-term evolutionary change under natural selection, where rates of change estimated from the fossil record over very long periods of time are many orders of magnitudes slower (Gingerich 1983).

Continued strong responses to directional selection may be rare in nature

How can we explain these different rates of evolutionary change? One possibility is that continued strong directional selection, as applied in artificial selection, is rare in nature. A well-documented example comes from the Grants' study of the seed-eating Galapagos finch *Geospiza fortis* (Grant and Grant 1989). During a severe drought in 1977, large, hard seeds were more frequent, individuals with larger beaks survived better, and the strong selection for large beaks resulted in an increase of some 4%, or about half a standard deviation. Seven years later, weather conditions had changed, small seeds were more frequent, and after this season beak size decreased by about 2.5%, or about 0.2 standard deviations. Thus strong natural selection produced rapid phenotypic change, but the direction reversed within 7 years. Occasional strong selection of variable direction may look like very slow change when averaged over long time periods.

Another reason for the slower rate of evolution in natural systems may be that the extreme trait values produced by directional selection reduce fitness because of negative side effects of the alleles selected. A common experience in artificial selection is that after relaxation of selection the selected trait tends to return towards its original value. This indicates that natural selection is opposing artificial selection. Yoo's experiments (Figure 4.14) are a good example. This expresses in genetic language what tradeoffs express in physiological language. Traits involved in tradeoffs cannot be selected very far in any direction without causing fitness to be lost through side effects on the other traits to which they are connected.

Population and quantitative genetics are being integrated

● **KEY CONCEPT**

Studies that apply both views have revealed that Mendelian traits have a quantitative dimension, and quantitative traits have a Mendelian dimension.

Population genetic models assume one locus or a few loci with genotypic differences visible in phenotypic differences. Quantitative genetic models assume many loci with small additive effects on the phenotype. So presented, the two approaches have little in common. Population genetics is framed in genetic language and focuses on genotypes, quantitative genetics uses statistical terms and focuses on phenotypes. The two approaches arose and were developed independently, but it has been known since Fisher's work in 1918 that the principles of quantitative genetics can be derived from and are consistent with population genetics (Fisher 1918). Recent developments in molecular techniques are yielding practical applications of Fisher's theoretical insights.

Molecular genetics finds many genes affecting traits previously thought to be affected by only one

For two reasons, the two approaches will be better integrated in the future. First, molecular analysis of loci affecting qualitative, Mendelian traits has uncovered many alleles with variability in phenotypic effects. This tends to obscure the precise genotype–phenotype relationships whose recognition is essential for population genetics. Because of this, quantitative genetics becomes relevant for the analysis of qualitative characters. Second, DNA sequences mapped to particular positions on chromosomes can be used to detect hitherto unknown loci that affect quantitative traits. When some of the genes affecting quantitative traits are known, the population genetic approach becomes relevant. Thus, population genetics is becoming more quantitative-genetic and quantitative genetics is becoming more population-genetic. The best examples are from medical genetics and plant breeding.

Phenylketonuria is a recessive genetic disease with >50 different alleles

Phenylketonuria (PKU) is a well-known genetic disease in humans in which the conversion of the amino acid phenylalanine (Phe) to tyrosine (Tyr) is blocked, causing severe mental retardation in untreated patients. It results from homozygosity for a non-functional allele at the locus coding for the enzyme phenylalanine hydroxylase (PAH). Thus it is a recessive genetic disease usually modeled as one locus with two alleles—a population genetic model.

However, molecular analysis of the PAH locus has revealed that it has at least 50 different alleles (reviewed by Weiss 1993). Most of these seem to be normal: they do not cause the disease. Among 206 European PKU patients, eight PAH alleles accounted for 64% of all PKU chromosomes. The effects of these eight alleles on PAH activity varied. In the homozygous state four of them caused less than 1% of the normal PAH activity, but the others caused enzyme activities of 3, 10, 30, and 50%. Since in randomly mating populations with so many alleles at a locus the heterozygosity is very high, most PKU patients are actually heterozygotes. The severity of the disease in heterozygous patients can be predicted reasonably well from knowledge of the diploid genotype and the independently measured allelic effects of the alleles that reduce PAH activity by assuming additive gene action.

Thus, when the phenotype is measured precisely we find variation in the disease that we can try to explain from genetic and non-genetic variation. This is basically a quantitative genetic approach but, unlike most quantitative traits, here we already know many of the alleles responsible for the disease.

Quantitative traits are determined by a few major and many minor genes

We can now characterize individuals with respect to many molecular markers. These are DNA sequences, such as restriction sites (sites where an enzyme that recognizes a specific DNA sequence will cut the sequence into two parts), which have a known chromosomal position whose presence or absence in an individual can be determined. If we call these markers A, B, . . . and use a capital letter to denote presence and a lower case letter for absence, chromosomes can be classified as being A or a, B or b, and so forth. These marker sequences are not genes, just positions on the chromosome where variation in nucleotide sequence occurs, but they segregate as normal Mendelian alleles. The basic principle of quantitative trait locus (QTL) detection is to find associations between the marker genotype and extreme phenotypic trait values, indicating that a marker is located close to a QTL affecting the trait. If the association between the marker and the QTL is sufficiently strong, it may be possible to find the QTL and to infer from its DNA sequence a possible function of the gene product. With such methods, we can take a trait whose genetic determination initially appeared to be quantitative, locate the genes that are affecting it, and then understand their evolution as a problem in population genetics.

Throughout this chapter we have assumed that genetic variation is present in the populations and models discussed, and we have seen that genetic change can only occur if there is genetic variation. We have not yet explained the origin and maintenance of genetic variation, which are discussed in the next chapter.

● SUMMARY

- This chapter describes how to understand and predict evolutionary change by modeling the genetic changes that underlie changes in traits.

- The two approaches to analyzing genetic change are population genetics, applied to genes whose phenotypes are visible and discrete, and quantitative genetics, applied to continuously varying traits whose genetic determination is obscure.

- The progress of a rare, advantageous mutation towards fixation is slow-fast-slow: slow when it is rare or common, and fast at intermediate frequencies.

- The evolution of resistance of insects to pesticides and of bacteria to antibiotics are good examples of slow-fast-slow change.

- Genetic change is faster in haploid sexual than in diploid sexual populations.

- In a diploid sexual population advantageous mutations will spread much more slowly when they are recessive than when they are dominant.

- Sex liberates genes from their genetic backgrounds; asex embeds genes in one genetic background, where selection is on the genotype as a whole.

- The variation in a population is caused in part by genes and in part by the environment. The contribution of each can be estimated.

- Additive genetic variation is that portion of genetic variation that is not influenced by genetic interactions between alleles.

- Additive genetic variation responds directly to selection; the proportion of the total variation for a trait in a population that is additive and genetic is defined as the heritability (h^2) of that trait.

- Traits correlated strongly with reproductive success have lower heritabilities than traits correlated weakly with reproductive success.

- Quantitative genetics successfully predicts the results of artificial selection experiments, the responses to selection in domestic plants and animals, and helps us to interpret evolutionary change in quantitative traits in wild populations.

- Population and quantitative genetics are becoming integrated.

- Diseases previously thought to be caused by simple Mendelian loci are now being revealed by molecular genetics to be caused by many alleles that combine in ways reminiscent of quantitative traits.

- The analysis of quantitative traits with molecular markers is revealing quantitative trait loci (QTLs) that behave like Mendelian genes and allow us to use population genetics to understand evolutionary change in continuous traits.

● RECOMMENDED READING

Fisher, R.A. (1930) *The genetical theory of natural selection*. Oxford University Press, Oxford.

Griffiths, A.J.F., Miller, J.H., Suzuki, D.T., Lewontin, R.C. and Gelbart, W.M. (2000) *An introduction to genetic analysis,* 7th edn. W.H. Freeman, New York.

Haldane, J.B.S. (1990) *The causes of evolution*, reprint edn. Princeton University Press, Princeton.

Hartl, D.L. and Clark, A.G. (1989) *Principles of population genetics*, 2nd edn. Sinauer Associates, Sunderland, MA.

Roff, D.A. 1997. *Evolutionary quantitative genetics*. Chapman and Hall, London.

● QUESTIONS

4.1. Suppose in a diploid, random-mating population two alleles occur at a locus A. A_1 is very common (frequency, $p = 0.95$). Due to recent environmental change A_1 has become disadvantageous and is selected against. Consider two cases:

		A_1A_1	A_1A_2	A_2A_2
(a) A_1 is recessive;	fitness	$1 - s$	1	1
(b) A_1 is dominant;	fitness	$1 - s$	$1 - s$	1

In which case will A_1 reach a frequency of 0.5 faster? Why?

4.2. A botanist measured seed weight and estimated its narrow-sense heritability in three widely separated *Phlox* populations.

Population	Mean seed weight (Mg)	Narrow-sense heritability
A	15	0.60
B	12	0.65
C	17	0.58

Because the heritabilities were high and approximately equal, he concluded that the differences in seed weight between the populations were largely due to genetic differences. Do you agree? Explain.

4.3. Suppose in a human population a mutation occurs that increases fitness by 10%. Do you think that this mutation will reach a frequency of 0.95 in less than 1000 years? If not, what is a more realistic timespan needed for this change? What factors influence the length of this period?

4.4. What implicit assumptions about his own and his relative's future reproductive success was Haldane making when he made his famous statement about being willing to sacrifice himself to save two brothers or eight cousins?

4.5. For those of you who are comfortable with calculus, show that if the rate of change of gene frequency is equal to $p(1 - p)$, then the rate of change is maximal when $p = 0.5$.

CHAPTER 5

The origin and maintenance of genetic variation

Without genetic variation, there can be no evolution

Chapter 4 discussed the genetic response to selection, and it assumed that genetic variation was present. If all individuals in a population were genetically identical and produced offspring identical to themselves, evolutionary change would be impossible. Both adaptive and neutral evolution require heritable differences among individuals to change the genetic composition of a population. Genetic variation is essential for evolutionary change.

It is not just the presence but the amount of genetic variation that influences the rate of evolutionary change. If there were very little genetic variation, the rate of evolutionary change would be limited by rare favorable variants. Most individuals would have a standard set of genes (termed *wild type* by the classical geneticists). Natural selection would remove deleterious variants, and occasionally a favorable variant would spread through the population. If, in contrast, there is a great deal of genetic variation, as we now know is usually the case, then individuals will differ genetically in many traits. Classification of most of the population as wild type is not possible, and the rate of evolutionary change is not limited by the occurrence of new favorable mutations. In an extreme case, with great genetic variability associated with great variability in fitness, selection could be too strong, removing such a large fraction of the population each generation due to low survival or fertility that it might go extinct.

Clearly, how much genetic variation is present, how it is maintained, and how much of it is correlated with fitness are crucial issues. In this chapter we first consider the origin of genetic variation, then its maintenance, and finally its relevance for adaptive evolution.

● **KEY CONCEPT**
The amount of genetic variation affecting fitness limits the response to selection.

Mutations are the origin of genetic variation

● **KEY CONCEPT**

The ultimate origin of all genetic variation is mutation.

Mutation: A hereditary change in the DNA sequence or in chromosome number, form, or structure.

All genetic variation originates through mutation. The different types of mutation are explained in more detail in the Genetic appendix (pp. 515 ff.). Most mutations arise from errors during DNA replication. Mutations can occur in somatic cells (body cells) as well as in the germ line (cells that end up as eggs and sperm). Somatic mutations can affect the function of individual organisms, both positively and negatively. For example, somatic mutation—together with other mechanisms—helps to generate antibody diversity in the immune system and thus contributes to the defense against pathogens; in contrast, some somatic mutations cause cancer. Only very rarely do mutations enhance reproductive success. An overwhelming majority of both somatic and germ-line mutations are deleterious or neutral. Germ-line mutations are more important for evolution, for unlike somatic mutations they are transmitted to future generations. In plants and fungi, which do not have a germ line, reproductive tissue can develop from somatic cells. Thus in these organisms some somatic mutations are inherited.

While mutations are necessary for evolution, too frequent mutation can prevent evolution, for with a very high mutation rate, not enough of the well-adapted genes would be transmitted unchanged to the next generation. Their loss would prevent the evolution and maintenance of adaptations. Thus there is likely to be an optimal mutation rate: not too few and not too many mutations. This optimal rate need not be the same for all species and all genes. That a population sometimes cannot survive a high mutation rate was shown in an experiment (Zeyl et al. 2001) in which replicate populations of two different yeast strains were propagated. In the strain with a 200-fold enhanced mutation rate, extinction was observed in two out of 12 replicate populations, whereas no extinctions were observed in the 12 populations with a normal mutation rate.

We have good reason to think that the mutation rate is to some extent under genetic control. In several species (mainly microorganisms) genetic variation for the mutation rate has been observed. For example, in bacteria so-called mutator strains are known, which have an enhanced mutation rate due to less-efficient repair of DNA damage. Thus mutation rates can be changed by natural selection.

Optimal mutation rates are easier to achieve in asexual organisms

Sexual and asexual species differ in the ease with which mutation rates can be adjusted by selection. In asexual organisms, where the whole genome is trans-mitted intact to the offspring, evolution of the mutation rate is easy in principle, for the genes that affect the mutation rate stay together with the genes whose mutation rate they adjust. If conditions favor a higher mutation rate, a mutation that enhances the mutation rate at all loci enjoys a selective advantage and will increase in frequency because it stays associated with the genotype that is benefiting from the higher mutation rate. In contrast, in sexual organisms a gene affecting mutation rate does not remain associated with the genome on which it has its effect because recombination can separate the gene determining the mutation-rate gene from the genes that mutate. Therefore evolution of the mutation rate to a value that maximizes the rate of adaptive evolution is expected to occur more readily in asexual species than in sexual species.

Rates of mutation

The average mutation rate per nucleotide pair per replication is about 10^{-10}—1 in 10 billion—in organisms with DNA genomes (Drake et al. 1998). Some viruses (for example those causing influenza and HIV) have a genome coded in RNA instead of DNA; they have much higher mutation rates, because repair of damage is less efficient in RNA genomes than in DNA genomes. The figure given is only a rough generalization, for mutation rates per nucleotide pair can vary by orders of magnitude among loci and among species. In microorganisms with DNA genomes the mutation rate appears to be strikingly constant when calculated per genome instead of per nucleotide pair: despite huge variations in genome size, all microbes have a mutation rate of approximately 1 in 300 per genome per replication. This implies that the mutation rates per nucleotide pair must also vary considerably. In higher eukaryotes estimates of the mutation rate vary between 0.1 and 100 per genome per generation. This figure is substantially higher than in microbes and probably arises because higher eukaryotes have both much larger genome sizes and many cell divisions per generation. However, the mutation rate in higher eukaryotes per cell division per *effective* genome (the part of the genome that codes for functional genes) may be of the same order of magnitude as in microbes (1 in 300). These numbers are derived from measurements on DNA sequences and do not tell us about the phenotypic consequences of the mutations for the fitness of individuals.

● **KEY CONCEPT**

Mutations rates differ strikingly in DNA and RNA genomes, for mutations of small versus large effects, and in males and females.

Mutations with large effects are much less frequent than those with small effects

A classical way to determine the likelihood of a mutation is to observe the spontaneous occurrence of abnormal phenotypes known to result from single allele changes. Since in this approach the number of mutations is given as a function of the number of individuals or gametes measured, it is best to call the resulting estimate *mutation frequency* to avoid confusion with the earlier-mentioned *mutation rates*, which are based on numbers of mutations per unit of time (replication or generation). Through the study of spontaneous mutations that cause human diseases mutations have been estimated to occur with a frequency of about 10^{-5}—1 per 100 000—per gamete. Similar figures have been obtained from studies in mice and *Drosophila*.

Using another approach, *Drosophila* geneticists (Mukai 1964, Mukai et al. 1972, Houle et al. 1992) have accumulated recessive (or partially recessive) mutations over many generations on a chromosome that is kept heterozygous and prevented from recombining. At regular intervals the fitness effect of the chromosome is measured in homozygotes, where the recessive mutations are expressed. The results suggest that the mutation frequency in *Drosophila* is about one mutation with a small deleterious effect per zygote and that mildly deleterious mutations greatly outnumber lethal ones.

Because humans have much more DNA than *Drosophila* and mutation rates per locus per generation are similar in humans and *Drosophila*, an average human might carry tens of new mutations, but many of them would be in DNA that did not code for proteins.

More mutations occur in males than in females

Recent molecular data on human genetic diseases suggests higher point-mutation rates in males than in females in some genes. Extreme examples are achondroplasia and Apert Syndrome, two dominantly inherited disorders. In both, all new mutations occurred in the father in more than 50 cases. The higher male point-mutation rate may be related to the much higher number of cell divisions in the male than in the female germ line (Crow 1997). That point mutations are associated with cell division makes this explanation plausible.

How random are mutations?

● **KEY CONCEPT**

The effects of mutations have no systematic relationship to the needs of the organism.

It is often stated that natural selection produces adaptations by acting on variation resulting from random mutations. What does the word random mean in this context? Because some parts of a genome experience much higher rates of mutations than other parts, mutation is not random with respect to where it occurs. Mutations can also be triggered by a specific signal, for example, in the fungus *Neurospora crassa*, where newly duplicated sequences trigger a specific mutational response (called RIP) that deactivates the repeated sequence (Selker 1990). RIP is an adaptive mutation, for it prevents the harmful accumulation of non-functional repeated sequences. Enhanced mutation rates at places in the genome where a high level of genetic variability is advantageous are also adaptive. Examples include the high level of somatic mutation in immune receptor genes in the vertebrate immune system and the highly mutable bacterial genes involved in the interactions of pathogenic bacteria with their hosts (Moxon et al. 1994). Mutations do not occur at random with respect to their location in the genome. Some genes mutate more frequently than others.

The critical question, however, is this: do mutations with a specific phenotypic effect occur more often when they are advantageous than when they are not? If so, adaptations could be produced by mutation alone, and natural selection would be less important. Such a directed mutational process is called Lamarckian because it resembles Lamarck's idea that an adaptation acquired by an organism during its lifetime can be transmitted to its offspring (see Chapter 1, pp. 13, 15 ff. for Lamarck's role in the history of evolutionary thinking). This would be the case, for example, if an animal could transmit to its offspring the immunity to a disease that it had developed through an immune response—but it cannot. We cannot at present rule out Lamarckian mutations entirely, but there is no evidence for them at the level of genetic mutations (changes in DNA sequence), and there is no evidence that they are very important.

Mutations, on the other hand, are certainly random in the sense that there is no systematic relationship between their phenotypic effect and the actual needs of the organism in which they occur. Note that it is the *specific* phenotypic effect of a mutation that matters here. Vertebrates require antibody diversity to produce an effective immune response, and a mutational process helps generate this diversity. However, this is not a Lamarckian process because the presence, for example, of influenza virus does not affect the probability that a somatic mutation yields resistance to influenza virus.

DNA duplication events increase the number of genes, providing a substrate for the evolution of new functions

Mutations that change the amount of DNA or the number of genes are key events in evolution. They include **polyploidization** and **duplication** of genes or gene clusters. Polyploidization—together with other mutations affecting the number and structure of chromosomes—is the main process responsible for differences in the number of chromosomes and in the total amount of DNA among species. Polyploidy is found in many plants and is thought to have played a very important role in the origin of new species (see Chapter 12, pp. 293 ff.). It is also common in certain vertebrate groups, such as the salmonid fishes and frogs. DNA duplication increases the total amount of DNA in the genome, providing material for the evolution of new functions. Its role has been substantial in evolution. For example, recent analyses of the human genome have revealed that over 15% of human genes are duplicates (Li et al. 2001). Duplicate copies of genes will accumulate mutations independently and may diverge to acquire a new function. Several mechanisms have been proposed to explain how such functional divergence between duplicated genes may work (Prince and Pickett 2002). In addition to DNA duplication, other mechanisms have been discovered that increase the size of the genome.

Polyploidization: *A doubling of the complete chromosome set.*

Duplication: *Copying of a DNA sequence without loss of the original, increasing the size of the genome by the size of the sequence copied.*

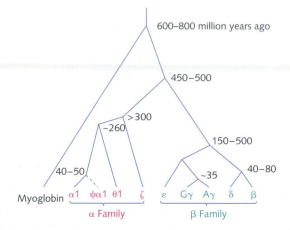

Figure 5.1 Phylogenetic tree of human globin genes, illustrating a series of gene duplications. The human globin gene family consists of three groups, the myoglobin gene on chromosome 22, the α-globin genes on chromosome 16, and the β-globin genes on chromosome 11. Several pseudogenes, remnants of duplicated genes that have become nonfunctional due to mutations, are in this gene family. Hemoglobin has two protein chains, one coded by a gene from the α and one by a gene from the β group. The various combinations differ in oxygen-binding affinities and appear at different developmental stages (embryo, fetus, adult). (From Li and Graur 1991.)

Repeated gene duplication produces multigene families

There are many *multigene families*, consisting of genes that have arisen by duplication from a common ancestral gene and have retained similar function. Examples in mammals include genes coding for heat-shock proteins (involved in protection of cells against environmental stress), globin proteins (involved in oxygen transport), apolipoproteins (involved in lipid metabolism), oncogenes (implicated in cancer), *HOX* genes (very important in development; discussed in Chapter 6, pp. 137 ff.), and genes involved in the immune system. Figure 5.1 depicts the evolutionary history of the human globin genes. An ancient duplication allowed divergence into two types of functional globin protein: myoglobin, for oxygen storage in muscles, and hemoglobin, for oxygen transport in blood.

Further duplications and divergence have produced the α and β families of hemoglobin, which consist of functional genes, like α1, θ1, and ζ in the α family and ε, γ, δ, and β in the β family, and pseudogenes, nonfunctional remnants of once functional genes, such as ψα1 in the α family.

● **KEY CONCEPT**

Recombination during meiosis creates great genetic diversity among offspring.

The effect of recombination on genetic variability

Because of recombination during meiosis, sexual individuals produce haploid gametes that differ genetically from the gametes that formed them. Thus an *AaBb* individual, grown from a zygote that resulted from the fusion of an *AB*

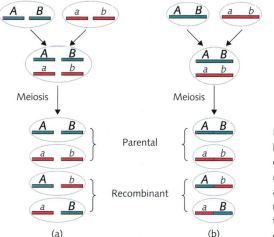

Figure 5.2 Recombination between loci on different chromosomes may result from Mendelian independent assortment (a) and recombination between loci on the same chromosome from crossing over (b).

and an *ab* gamete, will produce gametes both with the parental **haplotypes** *AB* and *ab* and with the recombinant haplotypes *Ab* and *aB* (Figure 5.2).

> **Haplotypes:** *Groups of closely linked genes that tend to be inherited together.*

When two DNA sequences are located on different chromosomes, they segregate independently at meiosis. When they are on the same chromosome, they segregate together unless a crossover occurs between them. In either case recombination is between genes and may result in new combinations of genes on a chromosome or in an individual. Thus recombination affects genetic variation among individuals when combinations of many loci are considered. Only rarely does it affect variation at a single locus.

Recombination produces phenotypes well outside the starting range

The effectiveness of recombination at converting potential into actual variability is spectacular in animal and plant breeding, where, starting from a uniform population established by crossing two inbred lines, individuals can be selected in a few generations with traits well outside the range of the original population. The examples mentioned in Chapter 4 (p. 93) of large and rapid phenotypic change under directional selection in *Drosophila* and in dogs illustrate this point. Selection can be so effective because the traits are affected by many genes whose recombination generates many combinations of alleles across multiple loci.

The amount of genetic variation in natural populations

The amount of genetic variation affecting fitness is important to know but hard to measure

To understand adaptive evolution, we must know how much genetic variation there is in natural populations and how much of it affects individual reproductive success. This has been one of the major questions of population

> ● **KEY CONCEPT**
>
> Molecular methods have revealed tremendous genetic variation in natural populations.

genetics since about 1920. Attempts to answer this question have been hampered by two related problems. Both stem from our ignorance of the relationships between genotypes and phenotypes. First, except where large phenotypic differences show Mendelian segregation patterns in crosses, we do not know the genetic variation that underlies the observable phenotypic variability. Second, we can only measure fitness effects of individual genetic variations when they are fairly large. The first problem—not knowing the genetic variation underlying phenotypic differences—was the main obstacle to estimating the amount of natural genetic variation until molecular methods were introduced in the 1960s. Then the problem of measuring fitness effects took priority.

In the mid-1960s biologists started to apply the biochemical technique of enzyme electrophoresis to samples of individuals from natural populations of animals and plants. Electrophoresis separates proteins on the basis of their mobility through a gel under the influence of an electric current. Proteins that differ in their net electrical charge move at different speeds. This can be observed by staining the proteins after they have moved through the gel for some time. The technique greatly improved estimates of genetic variation, for it made visible the variation at loci that could until then not be inferred from the phenotypes. The amount of genetic variability in populations is usually measured by the *genetic diversity*, h, defined as the probability that two alleles chosen at random from all alleles at that locus in the population are different. The easiest way to compute this probability is by seeing that it equals 1 minus the probability that two randomly chosen alleles are identical. If the relative frequency of allele A_1 is x_1, the probability that two randomly chosen alleles are both A_1 equals x_1^2. This applies similarly for all other alleles in the population. Therefore the probablity that two randomly chosen alleles are the same is the sum of the probabilities for each allele separately. Denoting the frequency of allele i by x_i, we get

$$h = 1 - \sum_i x_i^2 \qquad [5.1]$$

So, when a population has little genetic variation, the probability of two alleles being identical is high, and h will be close to 0. If, on the other hand, there is much genetic variability, the probability of two alleles being identical is low, and h will be high. When the population mates randomly, the two alleles at a locus in an individual form a random pair. Therefore under random mating the genetic diversity h equals the actual *heterozygosity*, H, the proportion of the population that is heterozygous at a locus. You can check with eqn 5.1 that for the case of two alleles $h = 2x_1x_2$, which is the familiar Hardy–Weinberg frequency of heterozygotes. If we average over loci, H can also be interpreted as the average proportion of loci that are heterozygous per individual.

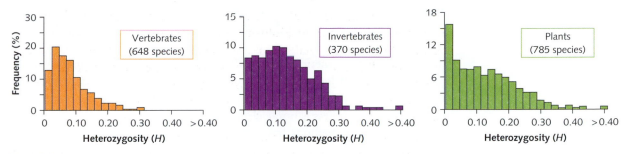

Figure 5.3 Estimates of heterozygosity, H, based on protein-electrophoretic surveys in many different species. (From Avise 1994.)

Electrophoretic heterozygosity is about 10% and varies among populations and species

Protein electrophoresis has been applied to many samples from populations of many species. The results of many studies suggest that H is about 10% and varies between populations and species (Figure 5.3). Such high levels of genetic variability were unexpected and, as we shall discuss later in this chapter, stimulated the development of theories that could explain the stable maintenance of all this variation.

Since 1980 more-refined molecular techniques have yielded measurements of genetic variability at higher resolution. One approach is to isolate DNA and cut it with restriction enzymes that recognize particular short nucleotide sequences. The resulting DNA fragments can be separated by gel electrophoresis according to molecular weight and visualized as stained bands. Differences between homologous chromosomes in the location of restriction sites (the short nucleotide sequences recognized by the restriction enzymes) can thus be measured. Another approach is to sequence the DNA to get the nucleotide sequence itself (e.g. . . . . AATGCTTCGA . . .). This became practical with the development of the polymerase chain reaction (PCR), which amplifies small amounts of DNA, even the DNA from a single cell.

Nucleotide diversities are about 0.0001–0.01 within populations

Both restriction analysis and sequencing allow us to estimate genetic variability at the level of nucleotides. The genetic diversity h is not a good measure of the variability of DNA sequences, for when long homologous sequences are compared all nucleotide sequences differ from each other, and h is close to 1. A better measure is the *nucleotide diversity*, the average number of nucleotide differences per site between randomly chosen pairs of sequences. Nucleotide diversities are typically in the range 0.0001–0.01.

Molecular methods do not *solve* the problem of deducing the underlying genetic variation from observed phenotypic variability: they *circumvent* it.

Molecular methods give direct access to genomic information without using phenotypic variation to draw conclusions about the genotype. They tell us how much genetic variation is present in a particular part of the genome, but they do not tell us how this genetic variation affects phenotypic variation. In a sense, the patterns of stained bands on a gel representing protein variants or pieces of DNA are phenotypes made visible by molecular techniques, but the relationship of those bands to fitness is rarely clear.

Thus molecular methods have revealed enormous genetic variation. How much of this variation causes fitness variation and serves as a substrate for adaptive evolution? Or, to put the same question the other way round, how much molecular genetic variation is selectively neutral? This is an empirical question. In Chapter 3 we discussed Kimura's neutral theory (p. 66), which claims that most variation at the molecular level is neutral. His theory caused considerable controversy about the relative importance of genetic drift and adaptive evolution in molecular evolution. We next discuss some attempts to measure fitness consequences of molecular genetic variation. Then we consider some models that aim to understand how mutation, genetic drift, and natural selection affect the level of genetic variation in a population.

Evidence of natural selection from DNA sequence evolution

Functionally important sites in DNA molecules experience natural selection

● **KEY CONCEPT**

Both selection and drift have played important roles in the evolution of DNA sequences.

No one believes that all genetic variation is selectively neutral. The abundant evidence of adaptation through natural selection (Chapter 2, pp. 35 ff.) must be reflected in DNA sequences. The question is how much of the variation in DNA sequences can be considered neutral. In a few cases strong indirect evidence of adaptive evolution has been obtained from comparisons of homologous DNA sequences. For example, Hughes and Nei (1989) compared the DNA sequences of the antigen-recognition sites of major histocompatibility (MHC) genes of humans and mice, genes involved in immune responses (their name refers to the role these genes play in the rejection of organ transplants or tissue grafts). Rates of substitution were estimated by counting the number of nucleotide differences between homologous stretches of DNA, and synonymous and non-synonymous substitution rates were distinguished. Non-synonymous substitutions change the amino acid coded; synonymous substitutions do not. Synonymous changes are usually more frequent than non-synonymous ones because amino acid replacements often reduce protein function and are selected against. Based on 36 protein-coding genes, the mean rate of synonymous substitution had been estimated to be five times higher than for non-synonymous substitution (Li and Graur 1991). Hughes and

Nei (1989), however, found that in the antigen-recognition region more non-synonymous than synonymous substitutions had occurred, indicating natural selection for fast change in the antigen-recognition properties of histocompatibility genes. This suggests that the higher level of variation in histocompatibility genes is adaptive and not selectively neutral.

Genome-wide sequence data suggest important roles for both selection and drift

Statistical tests have been developed that discriminate between natural selection and neutrality as the most likely cause for observed differences in homologous DNA sequences within and between species. They include the often-used McDonald–Kreitman test (McDonald and Kreitman 1991). Several studies based on these methods now agree that a large fraction (perhaps 30–45%) of protein divergence is adaptive and driven by selection (Fay et al. 2001, 2002, Smith and Eyre-Walker 2002). It thus appears that both natural selection and genetic drift play a significant role in evolutionary change at the molecular level.

Genetic variation is maintained by a balance of forces

We will now discuss some of the theoretical models that try to predict the quantitative effects of genetic drift and selection on the genetic variability of a population. They all share the important assumption that populations are in genetic equilibrium, which implies a balance between the forces that increase and decrease genetic variation. They therefore concentrate on the equilibrium states caused by mutation, selection, migration, and drift. There are two reasons for this approach. One is mathematical convenience. The other is that the periods during which the system is in a state of change, for example when an advantageous allele is spreading through a population, are thought to be short compared to the periods when the allele is fixed in the population. Whether this assumption is valid remains to be seen, and we will return to it.

● **KEY CONCEPT**

Genetic variation is held at equilibrium when mutational increases are balanced by loss to selection or drift.

Neutral genetic diversity is balanced between origin through mutation and loss through drift

In 1968 Kimura postulated that most of the evolutionary changes at the molecular level are neutral or nearly neutral (Kimura 1968, 1989). This idea has some *a priori* plausibility, as discussed in Chapter 3 (pp. 57 ff.). Many nucleotide substitutions do not cause phenotypic change because they do not change an amino acid and some amino acid changes do not affect protein function.

Mutations that do not change the fitness of their carriers are solely subjected to genetic drift.

We now consider the fate of such mutations as predicted by the unique mutation model, also known as the *infinite-allele model*. This model assumes that every mutation is unique in the sense that it results in a new allele, a reasonable assumption if we characterize the mutations not by their phenotypic effect but by their DNA sequence. Because the probability of a specific nucleotide change in DNA is extremely small (10^{-10} per replication), the same nucleotide change in a different individual would probably represent a new allele, for some of the 1000 or so surrounding nucleotides forming the complete gene would also differ. Thus the assumption is plausible.

Every new unique mutation will eventually either disappear from the population or become fixed. Fixation of a mutation means that all homologous chromosomes in the population carry a copy of the mutation. The process by which a new mutation eventually becomes fixed is called *gene substitution*. The probability that a new mutant allele will reach fixation is the *fixation probability*, and the time it takes to become fixed is the *fixation time*, measured in generations. Using the infinite-allele model several predictions have been made for these quantities.

The substitution rate for unique neutral mutations equals the mutation rate

One important and remarkably simple result is that the substitution rate for neutral mutations equals the mutation rate. This can be shown as follows. The probability that an allele becomes fixed by genetic drift is equal to its current relative frequency. A new mutation in a diploid population of N individuals has an allele frequency of $\frac{1}{2N}$, since the population contains $2N$ alleles at the locus at which this mutation occurs. Thus, in a diploid population of N individuals a new neutral mutation has a probability of $\frac{1}{2N}$ to be fixed and $1 - \frac{1}{2N}$ to be lost. It follows that most neutral mutations will be lost. If a new mutation occurs with probability u per locus per generation, there will be $2Nu$ new mutations in the population each generation. Since each of these mutations will be fixed with probability $\frac{1}{2N}$, the substitution rate for neutral mutations is $2Nu \cdot \frac{1}{2N} = u$. Thus one neutral mutation will reach fixation on average per locus every $\frac{1}{u}$ generations, which in many species is a very long time because u is of the order of 10^{-5} or 10^{-6}. The substitution rate is independent of population size, for the higher fixation probability of a mutation in a small population precisely compensates for the smaller total number of mutations in smaller populations.

That neutral mutations are fixed at a constant rate equal to the mutation rate supports the concept of a molecular clock (Chapter 3, pp. 65 ff.), which proposes that the evolution of nucleic acid and protein sequences occurs at a constant rate in all lines of descent. Molecular clocks play an important role in the estimation of divergence times and the construction of phylogenetic trees (Chapter 13, pp. 317 ff.).

It takes much longer to fix a neutral mutation than to lose it; many more are lost than fixed

In discussing the probability that a neutral mutation will be fixed, it is helpful to define the concept of **effective population size**. That definition varies with the context, depending on the process being considered that causes the effective size to differ from the actual size. In this context it is the size of an abstract population consisting exclusively of individuals with equal reproductive success that would experience the same amount of genetic drift as a real population of size N. In real populations individuals differ in reproductive success. Real populations may also be subdivided, vary in size, and have different sex ratios, factors that also influence the rate of genetic drift. The effective size is therefore often much smaller than the actual size.

> **Effective population size:** *The size N_e of an abstract population that would experience the same amount of genetic drift as a real population of size N.*

For the mutations that do eventually become fixed, the expected time to reach fixation is $4N_e$ generations (we do not derive this result here). There is a big difference between the fixation time of the mutations that get fixed and the survival time of the mutations that get lost. In a population averaging 10^6 individuals the fixation time is in the order of 10^6 generations whereas mutations that are lost survive on average less than 10 generations. Thus the theory of unique neutral mutations yields a picture of a continual flux of genetic variants through a population. There is a steady input of new mutations, most of which are quickly lost, while a few drift slowly to fixation, their number representing a balance between mutation and drift (Figure 5.4).

Adaptive genetic diversity is balanced between origin through mutation and loss through selection

Turning now to mutations that—unlike neutral mutations—do affect reproductive success, we consider a different modeling approach. If we characterize

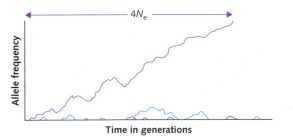

Figure 5.4 The fate of neutral mutations. N_e = effective population size. Most neutral mutations are lost quickly; a few drift slowly to fixation (the expected time to fixation is $4N_e$ generations). (From Nei 1987.)

mutations not by their DNA sequence but by their phenotypic effect, we can exchange the model of unique mutations for one of recurrent mutations. In some cases several alleles that differ in their nucleotide sequence all produce the same phenotype, e.g. phenylketonuria (Chapter 4, p. 95). When these alleles are lumped into one category, a recurrent-mutation model is justified. This is the classical population genetic approach to calculate the expected frequency of deleterious mutations.

We now calculate the expected frequency of a recurrent deleterious mutation at mutation–selection balance. The calculation is simple for the case of a haploid population and is shown in Box 5.1. The result is:

$$q = \frac{u}{s}$$

This implies that we can expect deleterious mutations in a population to occur at frequencies that are roughly in the order of the mutation rate. If the selective disadvantage of the mutation is very small (that is, for very low values of s) the frequency will be higher, which makes sense because then selection is less effective in removing such mutations from the population. For example, if a mutation occurs with a probability of 10^{-6} and it causes a reduction in reproductive success of only 1%, so $s = 0.01$, the expected frequency of the mutation in the population will be 1 in 10 000.

We suppose that in a haploid population two alleles at locus A occur. The common allele A_1 with relative frequency p (p is close to 1) is assumed to mutate

---

BOX 5.1

Population genetic model of the mutation–selection balance at a single locus in a haploid population

Genotype	A_1	A_2
Relative frequency	p	$q = 1-p$
Relative fitness	1	$1 - s$

$\leftarrow$ Selection

Relative frequency after selection	$\dfrac{p}{1-sq}$	$\dfrac{q(1-s)}{1-sq}$

$\leftarrow$ Mutation

Relative frequency after mutation	$\dfrac{p-up}{1-sq}$	$\dfrac{q(1-s)+up}{1-sq}$

$\leftarrow$ Random mating

Relative frequency in next generation	$p' = \dfrac{p(1-u)}{1-sq}$	$q' = \dfrac{q(1-s)+up}{1-sq}$

In equilibrium: $p' = p \rightarrow 1 - sq = 1 - u \rightarrow \boxed{q = \dfrac{u}{s}}$

to A_2 with probability u per generation. Each generation the sequence of events is: selection, mutation, reproduction (by random mating). The relative allele frequencies after selection are calculated in the usual way, by dividing the surviving fractions by their sum. The relative frequencies after mutation are obtained by subtracting a fraction u from the A_1 frequency and adding it to the A_2 frequency. Random mating does not affect the allele frequencies (see Box 4.1), so the allele frequencies in the next generation are easily obtained. The equilibrium assumption implies that at mutation–selection balance the allele frequency remains constant in consecutive generations. Equating $p' = p$ yields after some algebraic manipulation the desired expression for the expected equilibrium frequency of the mutant allele A_2.

For a diploid population a similar calculation yields

$$q = \sqrt{\frac{u}{s}}$$

for recessive mutations and

$$q = \frac{u}{s}$$

for dominant mutations.

Because the selective disadvantage s of a mutation is hard to estimate accurately, it is not easy to test this theory in real populations. The most reliable estimates are for mutations where affected individuals do not reproduce ($s = 1$), as in several human hereditary diseases. Unfortunately, a small fitness effect in heterozygotes causing a deviation from strict recessiveness or dominance can change the expected equilibrium frequency considerably. Mildly deleterious mutations seem very often to be partially recessive (Mukai et al. 1972; Peters et al. 2003); thus there is often some effect in heterozygotes. Moreover some mutant phenotypes may be caused by mutations at several loci. For example the chlorate-resistance phenotype in the fungus *Aspergillus nidulans* can be caused by mutations at any of about 10 different loci. Similarly, several human genetic diseases that until recently were thought to be caused by a mutation at one locus are now known to be caused by mutations at several loci.

The assumption of equilibrium between mutation and selection may also be questionable for many populations. Migration and genetic bottlenecks can move a population well away from equilibrium, and the return to equilibrium can take a long time. Many human genetic diseases caused by single-gene mutations, such as enzyme deficiencies leading to metabolic disorders, show strikingly different frequencies in different populations. This suggests that many populations are not in mutation–selection equilibrium for these loci, probably due to historic population events. For example, phenylketonuria (also called PKU; see also Chapter 4, p. 95) has a higher incidence in eastern than in western Europe. The Scandinavian countries, especially Finland, and Japan have a very low frequency of phenylketonuria, whereas the disease is found at high frequency in Ireland. Another striking example of a deviation from mutation–selection equilibrium due to population history is the so-called South African malady, porphyria (see Chapter 3, p. 61).

Whether or not it leads to equilibrium, mutation–selection balance is a powerful mechanism for maintaining genetic variation. It guarantees a certain amount of genetic variability at all loci. As we will see below, this is particularly important for quantitative traits, which are affected by many loci.

In overdominance heterozygotes have higher fitness

The possibility of a stable allelic polymorphism due to superior fitness of heterozygotes has been recognized since the 1920s. Intuitively, this seems to make sense, because if the heterozygotes have superior fitness, the genotype favored by selection contains both alleles, so that both are 'protected.' We can analyze this in a formal population genetic model, using the same approach as in the models for genetic change in Chapter 4. We calculate the expected allele frequency change over one generation and then assume an equilibrium at which the allele frequencies remain constant. The model is the same as the diploid sexual model in Box 4.2 (p. 79):

Genotype	A_1A_1	A_1A_2	A_2A_2
Relative frequency	p^2	$2pq$	q^2
Relative fitness	1	$1 - hs$	$1 - s$

The analysis, worked out in Box 5.2, yields the following formula for the allele frequency at equilibrium:

$$p = \frac{1-h}{1-2h} \tag{5.2}$$

We consider the diploid sexual model analyzed in Box 4.2. There we obtained the formulae that predict how the allele frequencies change from one generation to the next. We now will find out under which conditions (in terms of the parameters s and h) an allele frequency equilibrium can exist. At such an equilibrium the allele frequencies should remain constant. By equating the allele frequencies in two successive generations we obtain as a solution the expression for p shown in the rectangular box in Box 5.2. Next we must examine under which condition this formula can represent an allele frequency; that is, will take values in the interval [0,1]. For positive values of the selection coefficient s this condition turns out to be that the dominance parameter h should be negative, implying superior fitness of heterozygotes.

An equilibrium can be stable or unstable. When the allele frequency equilibrium is stable, the frequency is expected to move towards the equilibrium value; when it is unstable, the frequency is expected to move away from it. It appears that in overdominance the equilibrium is stable, but if the heterozygotes have a *lower* fitness than both homozygotes (underdominance), the allele frequency

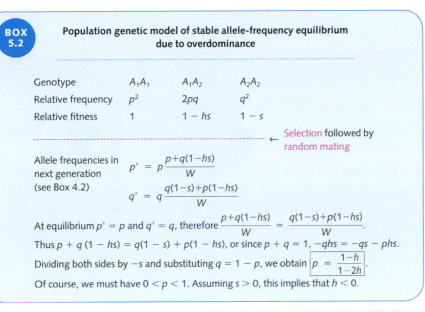

BOX 5.2

Population genetic model of stable allele-frequency equilibrium due to overdominance

Genotype	A_1A_1	A_1A_2	A_2A_2
Relative frequency	p^2	$2pq$	q^2
Relative fitness	1	$1 - hs$	$1 - s$

← Selection followed by random mating

Allele frequencies in next generation (see Box 4.2)

$$p' = p\frac{p+q(1-hs)}{W}$$

$$q' = q\frac{q(1-s)+p(1-hs)}{W}$$

At equilibrium $p' = p$ and $q' = q$, therefore $\dfrac{p+q(1-hs)}{W} = \dfrac{q(1-s)+p(1-hs)}{W}$.

Thus $p + q(1 - hs) = q(1 - s) + p(1 - hs)$, or since $p + q = 1$, $-qhs = -qs - phs$.

Dividing both sides by $-s$ and substituting $q = 1 - p$, we obtain $\boxed{p = \dfrac{1-h}{1-2h}}$.

Of course, we must have $0 < p < 1$. Assuming $s > 0$, this implies that $h < 0$.

equilibrium is unstable. Question 5.5 at the end of this chapter considers how to test the stability of the equilibrium given by eqn 5.2.

Sickle-cell hemoglobin is a classical example of overdominance

The classical example of overdominance is provided by hemoglobin variants associated with malaria in humans. The protein components of hemoglobin molecules are coded by the α-globin cluster on chromosome 16 and the β-globin cluster on chromosome 11. Several mutant hemoglobins produce varying degrees of anemia as homozygotes. The best-studied variant is sickle-cell hemoglobin (HbS), which takes its name from the altered shape of the red blood cells that contain it. Where malaria is common, the homozygote for normal hemoglobin is susceptible to malaria, the homozygote for the HbS allele suffers from sickle-cell anemia, but the heterozygote has little anemia and is protected from malaria. This overdominance explains the high frequencies of the HbS allele in parts of Africa and India where malaria is common and the low frequencies of the allele in non-malarial environments. Several hemoglobin alleles have been found that also protect from malaria in the heterozygous condition (see Figure 5.5). The main ones are HbS (mainly occurring in Africa and India), HbC (West Africa), HbE (mainly south-east Asia), and HbD (northern parts of India and Pakistan).

The dynamics of competition between the hemoglobin alleles are complex and much depends on the poorly known fitness effects of the various homozygotes and heterozygotes. Moreover, because populations containing the different alleles are not well mixed, the long-term outcome is difficult to predict. One of

Figure 5.5 The darker-shaded areas indicate where particular alleles for abnormal hemoglobins are more common. (From Vogel and Motulsky 1997.)

these mutant alleles may ultimately out-compete the others; a stable polymorphism involving two or more alleles is also possible. As with mutation–selection balance, the equilibrium assumption is not fulfilled.

When selection favors rare types, they increase until they become common

● **KEY CONCEPT**

Fitness is often frequency-dependent, depending on whether a variant is common or rare. When rare variants are favored, variation will be maintained.

Many traits are involved in interactions between individuals, whether these belong to the same or to different species. They include traits involved in sexual, social, and antagonistic behavior and in predator–prey, host–parasite, or mutualistic interactions. Natural selection on these traits is frequency-dependent when the success of a particular behavior or interaction depends on what others are doing. Selection favoring rare types may then be common. Consider, for example, two genotypes that differ in the type of prey they eat. With a constant supply of the two types of prey, whenever one of the genotypes is rare, its fitness is high because plenty of food is available, whereas the fitness of the common genotype is low because of strong competition for food. Thus rare genotypes become more frequent, common genotypes become less frequent, and at some point the two balance.

Hori (1993) found a beautiful example of negative frequency-dependence in scale-eating fish in Lake Tanganyika. Several species of the scale-eating cichlid fish genus *Perissodus* have asymmetrical mouth openings (Figure 5.6), some individuals being 'left-handed' (sinistral) and others 'right-handed' (dextral). This asymmetry of the mouth is an adaptation for efficiently tearing off the scales of prey. *Perissodus microlepis* approaches its prey from behind to snatch scales from the flank. Dextral individuals attack the victim's left flank, and sinistral ones the right flank. Handedness appears to be genetically determined by a

Figure 5.6 Asymmetrical mouth opening of a scale-eating cichlid fish, *Persissodus microlepis*. A right-handed individual is attacking its prey from the left side. (From Hori 1993.) (By Dafila K. Scott.)

simple one-locus/two-allele system with dextrality dominant over sinistrality. Hori examined the ratio of handedness in *P. microlepis* over an 11-year period. He found that both types were on average equally frequent with small oscillations in the ratio with a period of 5 years. During periods when sinistral individuals were more numerous, the prey suffered scale-eating from dextral individuals more often than from sinistral individuals, and vice versa. When the prey became more alert to the common type of predator, they let predators of the rare type be more successful than those of the common type: a clear case of negative frequency-dependent selection.

Negative frequency-dependent selection is readily modeled

That negative frequency-dependent fitnesses can maintain genetic variability is readily demonstrated in a population genetic model that can be analyzed graphically (Figure 5.7). Assuming haploid organisms (diploids require a slightly more complicated model), we can let the relative fitnesses of the two alleles decrease linearly with frequency to model the simplest type of negative frequency-dependence. Thus:

Genotype	A_1	A_2
Relative fitness	$1 - ap$	$1 - bq$

where p and q represent the allele frequencies and a and b are parameters specifying how strongly the relative fitnesses depend on the allele frequencies. Figure 5.7 shows that when A_2 is rare, it has a higher fitness than A_1 and increases in frequency. Similarly, when it is common it has lower fitness and decreases, so that when both fitnesses are equal the frequency equilibrium is stable.

How much evidence is there for negative frequency-dependent selection? The botanical and agricultural literature report that performance often depends on

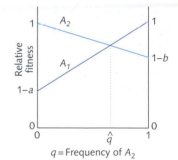

Figure 5.7 The relative fitnesses of genotypes A_1 and A_2 in a haploid population are shown as linearly decreasing functions of their relative frequency (respectively $1 - ap$ and $1 - bq$). Such negative frequency-dependent fitnesses generate a stable allele frequency equilibrium at $\hat{q} = a/(a+b)$. In the example shown, $a = 0.6$ and $b = 0.3$, giving an equilibrium value $\hat{q}$ of 0.67.

neighboring types. If a plant surrounded by plants of its own type performs more poorly than when it is surrounded by plants of different type, there is an advantage to rarity because rare plants mostly have neighbors of another type. At least two mechanisms are plausible and could both be operating. If different types have different resource requirements, resource competition will be strongest between plants of the same type. And if diseases are communicated more easily among plants of the same type, then rare types get sick less often.

Selection for rare types occurs in grass and bacteria

For example, Antonovics and Ellstrand (1984) planted clonal tillers of the grass *Anthoxanthum odoratum* into natural sites so that some test plants were surrounded by plants of their own genotype and some by plants of a different genotype. Clones in minority situations had approximately twice the fitness of those in majority situations. The causes of minority advantage remain unknown.

A study by Rozen and Lenski (2000) demonstrated coexistence due to frequency-dependent selection of two clones of the bacterium *E. coli* during 14 000 generations of serial propagation in a common glucose-limited medium. The two clones—called *L* (large cells) and *S* (small cells)—differed in morphology and ecological properties. *L* had a 20% higher maximal growth rate than *S* in fresh medium, but in conditioned medium *L* and *S* secreted metabolites that promoted the growth of *S* but not of *L*. Thus at high *L* frequency *S* cells were fitter, while at low *L* frequency *L* cells were fitter (Figure 5.8). The negative frequency-dependence was confirmed by the rapid convergence of the two types on the same relative abundance, starting from different initial frequencies (Figure 5.9).

This study is especially interesting because *S* arose in an evolving population of *E. coli* that was founded from a single asexual cell and had been kept

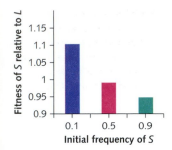

Figure 5.8 Frequency-dependent relative fitnesses of the *L* and *S* strains. At low frequency strain *S* has higher fitness than strain *L*, at high frequency the opposite is the case.

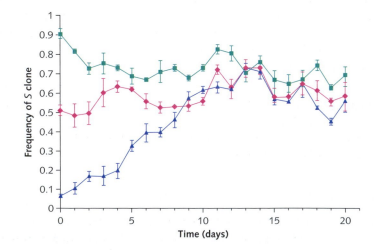

Figure 5.9 Convergence on a stable equilibrium composition of the population. Type *S* increases in relative frequency when rare and decreases when common. (From Rozen and Lenski 2000.)

continuously on the same culture medium with glucose as the sole carbon and energy input. Thus in this experiment the whole process could be followed starting from the occurrence of a novel mutation until the establishment of a stable allele frequency equilibrium. Rozen and Lenski showed that S had originated by mutation around generation 4000.

Genetic diversity of complex quantitative traits

Stabilizing selection is common in quantitative traits

Many quantitative characters are under stabilizing selection: some intermediate trait value is the best, smaller and larger values reduce function. The general reason is probably tradeoffs between different functions of a trait: a higher metabolic rate provides more potential for growth and activity, but a lower rate requires less food and resources; stronger bones give better support for the body, but lighter bones are cheaper to make and require less energy to carry; higher blood pressure promotes rapid transport of substances via the blood-stream, but lower blood pressure is better for the vascular system; laying more eggs means more potential offspring, but laying fewer eggs costs less and allows better care of each offspring.

If artificial selection of wild species can produce big changes both up and down, then natural selection must have been stabilizing. Artificial selection has often demonstrated that genetic variability exists to change body size in both directions: consider the sizes of the different breeds of dogs. Much of the genetic potential for variation in body size in dogs must have been present in wolves, from which dogs descend. Yet wolves have had a constant size for millions of years, which must mean that larger and smaller wolves are selected against.

Thus natural selection often favors some intermediate trait value at which the net benefit of the different functional aspects of the trait is highest. This means that often selection is stabilizing: deviations from the optimum phenotype are selected against. This is even likely to be true when paleontological evidence indicates a long-term directional change. For example, horses have evolved from ancestors that were about the size of a very small pony 50 million years ago, but the mean rate of change was so slow that it can be explained by very weak directional selection or even by genetic drift. Throughout that long history the body size of horses could have often been under stabilizing selection. One of many well-documented examples of stabilizing selection is the birth weight of human babies (Figure 5.10).

Genetic variation for quantitative traits is abundant

There is genetic variation for almost all quantitative traits in natural populations. Plant and animal breeders know that artificial selection on practically any trait

> ● **KEY CONCEPT**
>
> Quantitative traits are usually under stabilizing selection, and genetic variation for such traits is probably maintained by the mutation–selection balance.

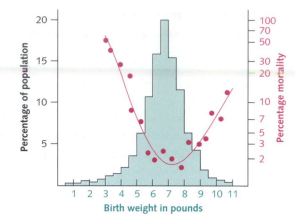

Figure 5.10 The distributions of birth weight (bars) and early mortality (circles) among 13 730 babies. These data (from Karn and Penrose 1951) indicate stabilizing selection on birth weight, since the optimum birth weight is associated with the lowest mortality. (From Cavalli-Sforza and Bodmer 1971.)

will produce a selection response, confirming the presence of genetic variability for the trait. As explained in Chapter 4 (pp. 86 ff.), the extent of this genetic variation cannot be expressed in terms of the genetic diversity or heterozygosity at the loci involved, because those loci are almost always unknown. Instead, we can estimate what fraction of the phenotypic variation is due to genetic variation. So although we cannot point to the genes that vary, we know that populations do contain much genetic variation for quantitative traits.

Overdominance and frequency-dependence are not plausible ways to maintain quantitative variation

How is genetic variation maintained under stabilizing selection? In theory, over-dominance or negative frequency-dependent selection could be responsible. It is hard to tell as long as the genes involved are not known. On *a priori* grounds it does not seem likely that overdominance predominates at the many loci involved in quantitative traits under stabilizing selection. Because firm evidence for single-locus overdominance is scarce (the sickle-cell polymorphism in malarial areas is one of the few well-documented cases), assuming that it occurs at many loci affecting quantitative traits is unjustified. The *a priori* case for frequency-dependent selection is perhaps stronger. If the optimal phenotype under stabilizing selection is created by roughly equal numbers of + and − alleles, a shortage of, for example, + alleles at the population level will move the average phenotype away from the optimum towards the − side and selection will favor the rare + alleles. But selection pressures on individual alleles are then likely to be very weak (see below), and empirical evidence of the involvement of wide-spread frequency-dependent selection is lacking.

Mutation–selection balance can maintain quantitative variation

A plausible explanation is the mutation–selection balance. A mutation that slightly increases or decreases an optimal phenotype will experience very weak selection because fitness declines away from the optimal value (Figure 5.11) and because mutations are moved by recombination among many different genetic backgrounds.

A mutation that slightly increases the phenotypic value will be advantageous when it is present in an individual with a phenotype lower than the optimum just as often as it is disadvantageous in an individual with a phenotype higher than the optimum. Similar reasoning applies to a mutation that slightly decreases the phenotypic value. Thus the average effect of a mutation on fitness will be very small, and such mutations may reach relatively high equilibrium values at the mutation–selection balance. Although the mutation rate per locus is low, there may be quite a bit of mutational input of genetic variability because most quantitative traits are influenced by many loci.

The mutation–selection balance explanation does not exclude other explanations for the maintenance of genetic variability. There may be overdominance at some loci and some frequency-dependent selection. Changes in the direction of selection may also play a role. The optimal phenotypic value may also shift with environmental conditions (remember selection on beak size in Darwin's finches; chapter 4, pp. 93 f.), contributing to the maintenance of alleles by increasing the frequency at which they shift between being advantageous and disadvantageous. But when we consider all factors together, it seems most likely that much of the genetic variation for quantitative traits under stabilizing selection is maintained by mutation.

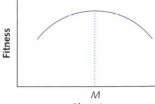

Figure 5.11 A model of stabilizing selection. Fitness decreases in proportion to the square of the distance from the optimum phenotypic value, M. (From Crow 1986.)

● SUMMARY

This chapter considers the origin and maintenance of genetic variation, the existence of which is necessary for evolution to occur.

- All genetic variation ultimately derives from mutation, the rate of which appears to be adjusted to an optimal level. Both too little and too much mutation would limit the rate of evolution, the former by a shortage of genetic variability and the latter by destroying well-adapted genotypes.

- Adjustment of the mutation rate to optimize adaptive evolution should occur more easily in asexual than in sexual organisms.

- Mutation rates may differ among DNA nucleotide sequences, genes, and species. In *Drosophila* the frequency of slightly deleterious mutations is about one new mutation per zygote, in humans it is even higher.

- Recombination increases genetic variability between individuals in a population by creating many multi-locus genotypes.

- The amount of genetic variation in natural populations was long debated. Molecular techniques have uncovered great variation in DNA sequences, most of which is consistent with both natural selection and neutrality. Recently genome-wide DNA sequence data suggest that both selection and drift are significant in driving evolution at the molecular level.

- The most important processes affecting the maintenance of genetic variation are drift and selection. Theories that explain the maintenance of genetic variation all assume that populations are in genetic equilibrium, and that the forces increasing and decreasing variation balance each other. If this assumption is false, the theories are at best only approximately valid.

- Neutral theory depicts a continual flux of genetic variants through a population. Most are lost quickly; a few drift slowly to fixation.

- Selection can maintain genetic variation if the effects of different selection forces balance. The two main possibilities are overdominance (selection favoring heterozygotes) and negative frequency-dependent selection (selection favoring rare alleles). The evidence suggests that the latter is more common.

- Many quantitative characters are under stabilizing selection favoring intermediate trait values. Artificial selection provides ample evidence of the presence of genetic variation for almost any quantitative trait. Much of this variation is probably maintained by mutation.

It is not enough that genetic variation be present and be maintained in populations for evolution to occur. Genetic variation must also be expressed in the phenotype, where development can produce a pattern of variation quite different from a simple view based only on genetic variation. How development can do that is the subject of Chapters 6 and 7.

● RECOMMENDED READING

Hartl, D.L. and Clark, A.G. (1997) *Principles of population genetics*, 3rd edn. Sinauer Associates, Sunderland, MA.

Li, W.-H. and Graur, D. (1991) *Fundamentals of molecular evolution*. Sinauer Associates, Sunderland, MA.

● QUESTIONS

5.1. Meiotic recombination can result in the formation of new haplotypes (haploid multi-locus genotypes). Consider two populations: one with a high and one with a low genetic diversity. In which will recombination result in more novel haplotypes?

5.2. Many rare human genetic diseases are caused by an abnormal recessive allele. Usually between 1 in 10 000 and 1 in 100 000 persons have such a disease, while typical mutation rates are in the range $10^{-4} - 10^{-6}$. Do you think these diseases can be explained as an equilibrium between mutation and selection?

5.3. Cystic fibrosis is a serious recessive genetic disease caused by a single gene, characterized by malfunction of the pancreas and lung, and found in about 1 in 4000 newborn babies. Do you think mutation–selection equilibrium is a likely explanation of its frequency? What are some alternative explanations?

5.4. If cystic fibrosis patients do not reproduce, due to early death or infertility, how much overdominance (superior fitness of heterozygotes) would be needed to explain the observed disease frequency?

5.5. The stability of an allele frequency equilibrium can be inferred from an analysis of the formula for the change Δp in allele frequency p over one generation. If $\Delta p > 0$, the change is positive and p is expected to increase; if $\Delta p < 0$, p will decrease. Of course, $\Delta p = 0$ exactly in the equilibrium value $\hat{p}$. By sketching the graph of Δp as function of p, the stability of the equilibrium can be judged: if p is expected to move towards $\hat{p}$ both from the left and from the right, $\hat{p}$ is stable; if the frequency is expected to move away from $\hat{p}$, there will be an unstable equilibrium.

(a) Derive an expression for Δp in the model used for analyzing overdominance. *Hint*: we derived an expression for p' (the expected frequency after one generation of selection); remember that $\Delta p = p' - p$.

(b) Sketch the graph of Δp as a function of p. Analyze the stability of the equilibrium given by eqn 5.2.

(c) For what values of s and h will the equilibrium in eqn 5.2 be unstable? Do you think that cases of unstable allele frequency equilibrium often occur in natural populations? Why?

CHAPTER 6

The importance of development in evolution

● **KEY CONCEPT**

Development links genes to organisms.

Development: *The process that builds complex multicellular organisms from single cells.*

Developmental control genes: *Genes that control the expression of other genes, often in combination, through the production of proteins called transcription factors that bind to the control regions of the genes whose expression is being regulated. Also called REGULATORY GENES, which refers to a broader class including genes that regulate anything, including development.*

In Chapters 3, 4, and 5 we looked at evolution in a fairly abstract way, asking how do alleles change in frequency under drift and selection, what is the origin of genetic variation, and how is genetic variation maintained? We did so without specifying what genes actually did. In this chapter we provide some details. Many genes control how organisms get built. When they change, organisms change in structure, function, and appearance. The link between genes and organisms is made by development.

The construction of complicated organisms from single cells is astonishing, one of the most precise and complicated processes in biology. Imagine what is involved when development turns a single-celled egg into a multicellular organism (Figure 6.1).

In recent years, we have learned a great deal about how genes control development. Some developmental control genes are broadly shared among distantly related groups, and some of the processes they control are surprisingly similar in organisms as superficially different as flies and mice. That is a macroevolutionary message. But this is a chapter in a section on microevolutionary principles, so it is logical to ask, why put this chapter at this point in the book?

This chapter is here because ancient, conserved developmental control genes, to which you are about to be introduced, construct the framework within which microevolution occurs—which traits exist in any particular organism, how they interact, and how they vary. Thus organisms do not simply transmit genetic information passively. Through development they actively shape the variation in traits that is presented to selection. This complicates the connection between genotype and phenotype in important and interesting ways.

We begin this chapter by explaining why evolutionary biologists are interested in development. After a brief description of development in which we introduce the process and concepts relevant to evolution, we then summarize the major messages of recent work on the role of development in macroevolution, a role that creates the context within which microevolution occurs. In the next chapter we describe how organisms are fine-tuned by development to the particular set of environmental circumstances that they encounter while they

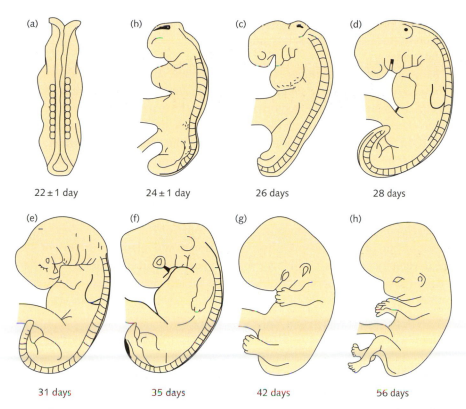

Figure 6.1 Human development resembles that of many vertebrates. First the basic features are laid down, then the finer ones. The anterior–posterior axis is determined early, as is the basic bilateral left/right symmetry. After that the major parts of the body with their characteristic organs—head, thorax, abdomen—are determined. Only then do the limbs begin to form.

are growing up, and how that fine-tuning produces reactions that can change genetic and phenotypic variation.

The study of development answers important evolutionary questions

How do diverse descendants arise from similar ancestors?

The major groups of organisms differ strikingly in their body plans, different body plans result from differences in development, and the differences in development have evolved. For example, sea anemones, jellyfish, and corals—the cnidarians—have radial symmetry, lack segments, and have no hard skeleton. Arthropods and vertebrates have bilateral symmetry and are segmented, arthropods have an exoskeleton and vertebrates have an endoskeleton, and arthropods and vertebrates share ancestors with cnidarians. Between those

● **KEY CONCEPT**

Development has important roles in both micro- and macroevolution.

ancestors and these descendants quite different body plans evolved. Thus one reason to study development is to understand the evolution of major *differences* in body plans.

Moreover, when we compare some major groups, such as arthropods and vertebrates, we find as many *similarities* in body plans as we do differences. Both groups are bilateral, segmented organisms. Both have a body axis running from the front (anterior) to the back (posterior), dividing the body into two parts (left and right), each of which is a mirror image of the other. In both the body is divided front to back into parts—head, thorax, abdomen—with characteristic organs and functions. In organisms as different as flies and mice the genetic control of the development of body plans is strikingly similar, genes of similar sequence and organization acting in similar parts of the body at similar stages of development to elicit similar structures. Some of the genes involved in the development of flies and mice are even present and play important roles in the development of jellyfish. How evolution has maintained some basic similarities while creating other major differences is a fundamental macroevolutionary question.

How is genetic information transformed into material organisms?

A second reason to study development follows from viewing a life cycle as a repeated sequence of three processes; one ecological, one genetic, and one developmental:

1. Ecology reduces the cohort of newborn organisms to those that have managed to survive and reproduce, producing the variation among individuals in lifetime reproductive success that causes natural selection (Chapter 2, p. 27).

2. Genetics and mating transform the genotypes of the surviving and reproducing parents into the genotypes of the offspring (Chapters 3–5, pp. 54–122).

3. Development maps the information in the genotypes of the offspring into material phenotypes (this chapter, p. 124, and Chapter 7, p. 152).

Thus each life cycle contains an element—ecology—in which material interactions predominate, an element—genetics—in which transmission of information predominates, and an element—development—that converts information into matter. We cannot understand evolution without understanding what difference each of the three processes makes to the outcome.

What is the role of development in trait evolution?

We also study development because of interest in the evolution of traits. True enough, neutral changes in DNA and protein sequences can occur without much consequence for traits. But other than that, all evolutionary change in the appearance, function, organization, and performance of the organism involves development. None of those things are present in the egg, and all of them are present in the adult.

Developmental mechanisms are needed to explain innovations and parallelisms

A fourth reason to study development is that only by incorporating development in our thinking can we understand two major features of evolution: innovations and parallelisms.

Innovations are structures in descendent lineages that were not present in ancestral lineages. Examples include the cephalopod eye, the tetrapod limb, the angiosperm flower, the bird feather, and the insect wing. Without evolutionary innovation, living things would not be morphologically diverse. Explaining innovations requires us to imagine how structures that appear to be completely new could have evolved from pre-existing elements. One pattern suggests that innovations originate when a developmental control gene is duplicated, allowing the new copy to be used to control a process in a new context while the old copy continues to function in the old context. For example (see details below), the developmental control genes that determine the major divisions of the vertebrate body—head, thorax, abdomen, tail—were duplicated twice in the lineage leading to tetrapods, and one copy now functions to control the development of the tetrapod limb. Another pattern suggests that a developmental control gene with one function early in development can be used to initiate different structures in different lineages later in development. Thus a gene involved in early segmentation in all insects is used later in development in butterflies to initiate eyespot development in the middle of wings and in flies to determine what types of structure form on the ends of legs (see Chapter 7, pp. 165–70, for details).

Parallelisms are the patterns produced when lineages descended from a common ancestor evolve in similar ways because they share developmental mechanisms rather than because they encounter similar selection pressures. A good example concerns patterns of digit loss in the feet of frogs and salamanders (Figure 6.2). Frogs and salamanders are both amphibians, but they diverged in the Jurassic more than 135 million years ago (Ma) and have evolved independently longer than any two lineages of eutherian mammals. Thus they are less closely related to each other than whales are to shrews.

> **Innovations:** *Traits in descendants not present in ancestors.*

> **Parallelisms:** *Parallel patterns of evolution in related species caused by shared developmental constraints, not by shared experience of selection.*

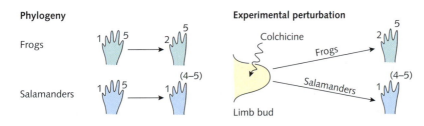

Figure 6.2 Parallelism in digit loss in amphibians. Both in phylogeny and in experimental manipulations of development (using colchicine), frogs lose the first digit, salamanders the fifth. In both cases the digit lost is the last one to form. (From Alberch & Gale 1985.)

Frogs and salamanders can lose one or more of their original five digits, producing a foot that has four toes. In evolution this has happened repeatedly in species whose closest relatives have not lost digits. The pattern can also be elicited by experimental manipulations during development, and within a lineage the same pattern is elicited in experiments as is found in evolution. The important point is that frogs share one pattern, whereas salamanders share another. Frogs lose the first digit; salamanders lose the fifth. In both cases, the digit lost is the last one to form in development (Alberch and Gale 1985). Thus frogs have one developmental program for digit development; salamanders have another. Digit loss, when selected, occurs at the same stage in the development process in both groups, but because the order of digit development is different in the two groups, different digits are lost. There is one parallelism within the many species of frogs, where first digits are lost, another within the many species of salamanders, where fifth digits are lost.

We could not understand either innovations or parallelisms if our evolutionary explanations consisted only of natural selection acting on genetic variation without paying attention to how traits are produced by development. Natural selection acts on organisms produced by developmental mechanisms that have a long history and that differ among lineages. Those differences in developmental mechanisms make important differences in how organisms from different lineages will respond to the same selection pressures. Some will do it one way, some another, and we can only understand why if we include the evolution of development in our thinking.

What happens during development from egg to adult?

Lives are lived in cycles that develop in characteristic stages

● KEY CONCEPT

Development transforms genetic information into material organisms by controlling the properties of cells.

Lives are lived in cycles. A sexual multicellular organism starts the cycle as an egg fertilized by a sperm or pollen cell. It develops into an embryo that passes through several stages before becoming an adult. The study of the development of eggs into adults, called developmental biology, provides a perspective on development that concentrates on detailed mechanisms. But as seen by evolutionary biology, development does not stop with the construction of the adult phenotype; it continues after maturation through reproduction to death, including the entire life cycle from start to finish. That provides a perspective on development that concentrates on broad consequences. Combining the two views of development is helpful, for the nature and breadth of the consequences depend on the details of the mechanisms.

Development occurs in stages that characterize groups of organisms. The developmental stages of many frogs include eggs, tadpoles, several stages during which the tail disappears and limbs develop, and adults. The developmental stages of mammals include zygotes, many intrauterine embryonic stages

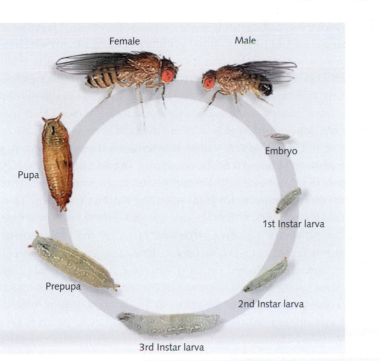

Figure 6.3 The life cycle of *Drosophila melanogaster*. Flies cycle from eggs through fertilized embryos, larval stages and pupae to adults, then back to eggs and sperm. During the pupal stage the body plan is completely remodeled. All these transitions are under the control of genes. (*Source*: With kind permission from Christian Klambt, University of Müenster.)

(Figure 6.1), infants, juveniles, and adults. Genes control the progression through stages of development in all organisms. How they do so is particularly well understood in the fruit fly, *Drosophila melanogaster*, whose life cycle is typical of insects like butterflies, bees, and beetles that undergo complete metamorphosis (Figure 6.3). The major features of life cycles, the striking patterns of development, vary among the major groups of organisms. Why should that be the case, and what difference does it make? And how can the DNA in the nucleus of a single cell possibly contain enough information to build an organism consisting of hundreds of cell types and billions of cells with all parts in the right place? That we now know many of the details should not diminish our amazement that the process exists. Development is among the most complex and precise processes that have evolved. Complex as it is, we can understand it in simpler terms.

Development happens in cells under the control of genes

Development consists of cell fusion, cell division, cell growth, and cell differentiation. The egg itself contains the first elements of the body plan: it has a front and a back. This is called **positional information**: information on where one thing is relative to another in space. The haploid egg fuses with a sperm or pollen cell to form a single diploid cell, called the zygote in animals, the embryo in flowering plants. That first diploid cell then divides repeatedly, forming an

Positional information: *Information used by cells during development that establishes where they are relative to other cell types.*

embryo with many cells. During those divisions the original positional informa-tion is preserved and in animals the body axis is laid down, which means that each cell is given the information it needs to know whether it is at the front or at the back, on the top or on the bottom, on the left or on the right.

Positional information is established by concentration gradients of molecules diffusing away from a source. High concentration indicates that the source is nearby, low concentration that it is far away. Those signaling molecules are often protein gene products called **transcription factors**; they may also be smaller non-protein hormones. Transcription factors and hormones are regula-tory signals that, alone or in combination, turn genes on or off. Transcription factors are more important for local signaling within the cell and from a cell to its neighbors; hormones, which are smaller and diffuse faster, are more import-ant for long-distance signaling, from one part of the organism to another. Transcription factors are produced by developmental control genes. Those genes are themselves turned on or off by responding to other transcription factors. Thus networks of developmental control genes are built up that control gene expression over much of the genome. The control of gene expression is the cause of cell differentiation, for the nuclei of all cells contain all the genes needed to build the organism, and the differences among the cells forming different tissues are not differences in which genes they contain, but in which genes are expressed and which remain silent.

Cells interact during development through cell–cell signaling, maintaining a picture of what stage of development the whole organism and their immediate neighboring cells are in. That information determines the next step that they take.

In animals, development proceeds as a process of sequential subdivision: first the larger parts are formed, then they are subdivided into smaller parts, and then those are further subdivided until all the body parts and tissue types are formed. Each step in the subdivision is signaled by the expression of transcrip-tion factors specific to the structure that will start to emerge. In plants, devel-opment is dominated by hormones produced in the growing tip. A plant can consist of many such growing tips, each controlling the growth of a module that it organizes. A similar mechanism exists in colonial animals with modular construction, such as colonial hydroids. As in animals, local signaling among neighboring plant cells is often exercised by protein transcription factors. In both plants and animals gene regulation precisely determines which genes are switched on and which are switched off in every cell at every step in development.

Developmental control genes can have other functions. A gene that controls the development of a large part of the organism early in develop-ment may affect a much smaller and different part of the organism late in development. Thus developmental control genes associated with basic body plans and broad phylogenetic patterns can have microevolutionary functions as well.

Transcription factors: *The protein products of developmental control genes. They bind to the regulatory sequences of genes, where they interact with other transcription factors to determine whether the genes will be switched on or off.*

Eukaryotic gene expression is controlled by transcription factors acting in combinations

Transcription factors bind to DNA in a regulatory region, or promoter, upstream from the coding sequence of the gene whose expression they control. If a gene has several such binding sites, then the same transcription factors can act in various combinations to elicit different gene expressions in different tissues and at different stages of development. Such **combinatorial control** has great power and precision to determine many different types of gene expression when and where they are needed. It solves one central problem of development, which is how to take cells that were originally identical and turn them into cells that are appropriately different: brain cells in brains, liver cells in livers, skin cells in skin, leaf cells in leaves, root cells in roots, and so forth.

The cell is like a symphony orchestra. The genes are instruments; the notes that can be played on the instruments correspond to levels of gene expression; the control of the instruments—the keys on a piano, the fingering and bowing of a violin, the finger-holes of a flute—correspond to the transcription factor binding sites. For example, in the fruit fly, *Drosophila*, there are 10–20 binding sites in the control region of an average gene, a transcription factor may bind to the control region of anywhere from one to several hundred genes, and there are roughly 13 000 genes. Those numbers give some idea of the number of different kinds of cells and different functions within a single cell type that can be produced through combinatorial control of gene expression. Just as an orchestra can play Mozart or Wagner by varying the instruments used and the notes expressed in a precise temporal sequence, a cell can become a brain cell or a liver cell by varying which genes are expressed and orchestrating the interactions of the expressed products. Within the organism, which consists of thousands to billions of cells, the precise temporal sequence and spatial location of cell types and states are orchestrated to integrate function.

When discussing transcription factors, one often hears of **boxes**, for example the **homeobox** or the MADS box. Boxes are highly conserved sequence motifs found in the DNA coding for a particular family of transcription factors. They are conserved because they have important functions that require a particular structure. First, their DNA sequence determines the part of the protein sequence that binds the transcription factor to the DNA in the control region of the gene (Figure 6.4). Second, they contain part of the molecule that interacts with the nuclear pore through which the transcription factor, which is produced in the cytoplasm, must move to get into the nucleus to regulate a gene. Third, they contain part of the coding region for amino acids involved in interactions with other transcription factors at the binding site. Because they are highly conserved for these reasons, boxes are a signal in the DNA sequence indicating that the gene functions as a developmental switch. The function of that developmental switch is quite sensitive to changes in the DNA sequence inside the box. If there were a mutation within the box that changed an amino acid in the binding site

Combinatorial control: *The effects elicited by developmental control genes depend upon the combination of such genes expressed in a given tissue.*

Boxes: *DNA sequences that specify protein structures that bind to the DNA double helix. They characterize families of transcription factor genes, each family having its own 'box'. Shared boxes suggest shared origins.*

Homeobox: *A 180-base-pair sequence in important regulatory genes that codes for a highly conserved protein segment that is a key part of a transcription factor.*

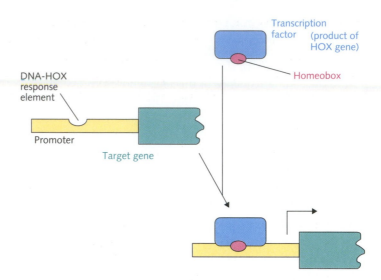

Figure 6.4 The **homeobox gene** codes for the part of the protein transcription factor that binds with the control region of the target gene. The arrow indicates that in this case the target gene is then expressed. (Reproduced with kind permission of Bill Daniel, University of Illinois.)

Homeobox gene: These include the HOX genes as well as other genes that share the DNA sequence motif that translates into a protein sequence, part of which binds the transcription factor to the control region of the genes whose expression is being regulated.

of the protein transcription factor, then the binding site would be the wrong shape, it would not bind to the DNA, and it would not be able to control the gene expression necessary for proper development. Thus a part of developmental control genes has been conserved, and it has been conserved for several reasons. The remainder of the sequence of these genes, the part that is not in the box, has not been conserved; it can be quite variable.

Transcription factors are categorized into families by sequence motifs, not by functions, which may be diverse within a family of transcription factors. For example the genes in the MADS family perform many different functions at many stages of development in organisms as diverse as fungi, animals, and plants, but they all contain a DNA sequence recognizable as the MADS box. Shared sequence motifs indicate a history of gene duplication. The diverse functions of control genes with similar motifs suggest that following gene duplication the new copies of the genes have acquired new functions. They were free to do so because the old copies continued to control the part of development that had existed up to that point.

How do cells sense where they are and what genes to activate?

Cells sense their position in the organism and keep track of the progress of development so that they can activate genes appropriately. How do they do this? The details can be summarized in five general concepts (See Carroll et al. 2001):

- The initial stages of development often involve a *concentration-dependent response* by control genes. For example, in fruit flies the genes that divide the embryo into a front and a rear half are sensitive to the concentration of maternal proteins diffusing from the front and rear ends of the embryo.

- Genes that activate transcription interact with genes that repress transcription to define areas in which only a very specific subset of genes are expressed.

- Many genes are regulated by two or more activators and repressors in combination. Combinatorial control is a general feature of developmental logic that makes possible the production of many cell types under specific conditions.

- Control genes producing transcription factors are activated by control genes producing transcription factors. This hierarchy allows the same developmental switches used early in development to specify patterns in the embryo to be used later in development in a different place to specify the formation of a different tissue, for example, in the brain.

- The genes are activated in a sequence in which first the broad pattern is laid down, then the parts of the pattern, then the details of the parts. In this regulatory hierarchy the activity of one control gene triggers the activity of the next, ensuring the correct temporal sequence of events.

With that brief sketch of the mechanics of development in hand, we now discuss how developmental patterns are associated with phylogeny, then how to explain those patterns in terms of the evolution of developmental control genes.

Developmental patterns are associated with phylogeny

Development evolved independently in plants and animals

Living organisms form natural groups whose relationships are defined by shared ancestry (Chapter 13, p. 303). Members of a group share more recent common ancestors with each other than they do with members of other groups. They also share developmental patterns, some of the most basic of which are controlled by ancient genes that have retained their function across groups that are now only distantly related. Recently much has been learned about the evolution of the genetic control of major patterns of development in two groups: the multicellular animals (the metazoans) and the seed plants. In each of those two large groups the coordination of cell–cell signaling and the production of patterns of differentiated tissues had to evolve independently. The significance of these patterns can only be appreciated when they are placed in a phylogenetic context. Therefore, before proceeding to the patterns, we first look at the phylogenies.

The major metazoan groups, taken in the rough order in which they evolved (Figure 6.5), are the Porifera (sponges), which lack symmetry and cell layers in their body plan, then two radially symmetrical groups, the Ctenophores (comb jellies) and the Cnidarians (corals, jellyfish, hydroids, and their relatives) with two clearly defined cell layers, and the bilaterally symmetrical animals (the Bilateria, all the rest of the animals) with three embryonic cell layers—endoderm, mesoderm, and ectoderm. In addition to three cell layers the Bilateria all have,

● KEY CONCEPT

Major differences in the architecture of organisms are associated with ancient events in the evolution of development.

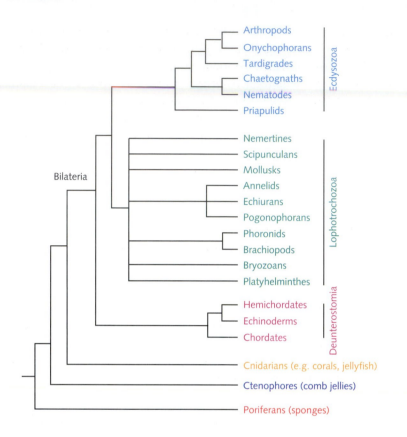

Figure 6.5 A phylogenetic tree of the major metazoan groups. We belong to the chordates, which include the vertebrates, the lancelets (cephalochordates), and the tunicates (hemichordates). (After Aguinaldo et al. 1997.)

at some stage of their development, a front and a back end as well as a left and a right, and a top and a bottom. Each group has its own body plan formed during development, and the groups that originated more recently have body plans that are more complex.

Major developmental changes in animals include symmetry, body axes, skeletons, and larvae

In keeping with this pattern, the Bilateria are more complex than the sponges and jellyfish. Among their additional features are segments, a hard skeleton, a tube-within-a-tube design with a mouth and gut forming a digestive tract, a central nervous system with ganglia and often a brain, a circulatory system with a heart or hearts, an excretory system with nephridia or kidneys, a respiratory system with gills or lungs, a reproductive system with gonads, sensory organs appropriate to the environment lived in, and an endocrine system that coordinates the internal state of the organism.

The Bilateria are subdivided into the Deuterostoma, the Ecdysozoa, and the Lophotrochozoa. The Deuterostoma have an endoskeleton; they include vertebrates and their chordate relatives as well as the starfish, sea urchins, and their relatives (Echinodermata) and the acorn worms and their relatives (Hemichordata). The Ecdysozoa, to which the insects, crustacea, velvet worms, round worms, and their relatives belong, have an external cuticular body covering that serves as a rigid exoskeleton or as the container for a hydrostatic skeleton. The Ecdysozoa get their name from the way they grow: by molting, or ecdysis. The Lophotrochozoa, to which the annelids, several other worm-like groups, and the mollusks belong, share, at least primitively, a characteristic larval stage, the trochophore larva, and other features.

Thus the broadest features of metazoan development are the ability to produce:

- symmetry—none, radial, or bilateral;
- cell layers—one, two, or three;
- body axes—front–back, left–right, top–bottom;
- a skeleton—internal or external;
- organ systems—digestive, circulatory, respiratory, nervous, excretory, reproductive, sensory, and endocrine.

If we want to understand how development maps genetic information into phenotypic matter in multicellular animals, we must at least understand the evolutionary and mechanical origins of symmetry, cell layers, body axes, skeletons, and organ systems.

Next we look at the evolution of land plants to see whether there might be comparable developmental issues there.

The major developmental changes in land plants involve apical growth, roots and shoots, leaves, and flowers

To colonize the land and compete with other evolving plants, plants had to develop a method of growing into large structures capable of extracting water and nutrients from the soil, of growing vertically to shade competitors, of capturing light energy efficiently with leaves, and of transporting water, nutrients, and the products of photosynthesis upwards from the roots and outward from the leaves (Friedman et al. 2004). Those functional and morphological elements were assembled in pieces over roughly 150 million years from the mid-Ordovician (450 Ma) to the Carboniferous (355–295 Ma).

The first land plants, the embryophytes, were liverworts derived from green algae (Charales, Figure 6.6). Extant liverworts lack apical growth and roots; some have leaves.

The origin of apical growth driven by a meristem capable of indefinite division appears to have come in the mosses, which lack roots but have shoots at the tip of each of which is an apical meristem.

Roots first appear in the lycophytes (club mosses and their relatives). Both their phylogenetic position and their developmental genetics suggest that roots are derived from shoots. Roots evolved again, apparently independently, in the euphyllophytes (ferns and their relatives plus the seed plants—Figure 6.6).

Leaves have a more complex history than roots, and to understand it we have to recall that land plants alternate gametophyte and sporophyte generations.

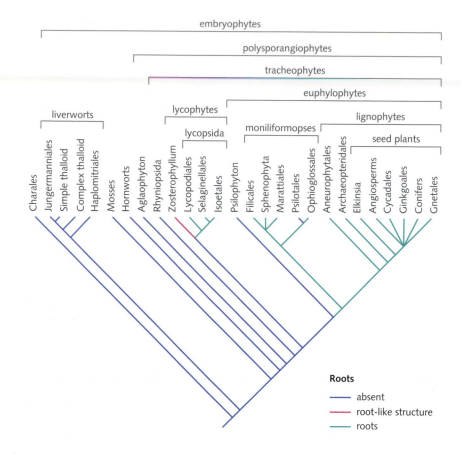

Figure 6.6 In the phylogeny of land plants, roots evolved twice, once in the lycopsids (club mosses and their relatives), and once in the euphyllophytes (ferns and their relatives plus seed plants). (From Friedman et al. 2004).

Leaves apparently evolved a minimum of three times in the gametophyte generation: twice in the liverworts, and once in the ancestor of extant mosses. They also appear to have evolved three times in the sporophyte generation: once in the lycopsids (club mosses), once in the monoliiformopses (ferns and their relatives), and once in the lignophytes (seed plants and their woody relatives).

True flowers originated in the late Jurassic or early Cretaceous with the flowering plants (angiosperms), about 150–120 Ma. Their arrival set off an explosion of plant-insect coevolution as pollinators and herbivores specialized on new species in the rapidly radiating clade.

Thus if we want to understand how development maps genetic information into phenotypic matter in land plants, we must at least understand the evolutionary and developmental origins of apical meristem, shoots and roots, leaves, and flowers.

At first sight there is much that is different in the development of plants and animals. Animal cells can move during development; plant cells cannot. In both there is a body axis, but the body axis in plants is a growth axis, whereas that of bilateral animals is an axis of sense and movement. Because of the different organization of their growth, plants have evolved their own way to construct an inside and an outside, a top and a bottom, a supportive structure, and a transport system.

At a deeper level, however, both plants and animals need a method to specify into what state a cell will differentiate depending upon its position within the organism. We will see that both groups have independently evolved a similar solution to that problem by using the general principle of combinatorial control implemented in plants through MADS genes and in animals through HOX genes.

Developmental control genes are lineage-specific toolkits for constructing organisms

The HOX genes are ancient and, in Bilateria, colinear

The HOX genes were discovered as mutants in silk moths and fruit flies. Mutations in these transcription factor genes result in homeotic mutations, mutations that cause the development of appendages in inappropriate places, such as a leg where an antenna would normally develop. After the HOX genes were sequenced, it was discovered that they all coded for transcription factors and contained a DNA motif, a so-called box, that coded for a binding site that was given the name homeobox. Later other genes were found to contain the same conserved DNA sequence; they are also called homeobox genes. The HOX genes are thus a subset of the homeobox genes, the ones that were found first. They also form a cluster, sitting next to each other on one chromosome. The other homeobox genes occur outside this cluster. In early development the HOX genes of most animals determine the fate of the segments along the body axis; later in development they play many different roles.

The HOX genes have both a remarkable evolutionary history and a remarkable genetic configuration (Figure 6.7). They are ancient, with DNA sequences so well conserved that the orthologous nature can be established of genes that last shared common ancestors towards the end of the Precambrian era about 600 Ma. The same HOX genes that determine where the head, thorax, and abdomen of a fly develop do the same job in the embryo of a mouse. We can recognize that the same genes are doing the same job because the DNA sequence in the homeobox has been so well conserved.

The HOX genes also occur on the chromosome in a linear sequence that corresponds to the part of the body axis whose development they control (Figure 6.8). This property of linear position on chromosome corresponding to position of effect along body axis is called colinearity.

The phylogeny, function, and chromosomal organization of the HOX genes suggest a certain evolutionary history. First single genes were duplicated in tandem to form a cluster of paralogous genes with related functions sitting next to each other on the same chromosome. Then the entire cluster was duplicated. With each duplication event, old functions were preserved under the control of

MADS genes: *A family of genes coding for transcription factors present in plants, animals, and fungi.*

HOX genes: *These produce transcription factors that control the expression of genes that determine the major parts of the body axis; similar genes determine similar parts of the bodies of flies and mice.*

● **KEY CONCEPT**
Some of the largest patterns in the Tree of Life are being illuminated by studying the genes that control development.

Orthologous: *Similarity in DNA sequence because of descent from a shared ancestor; usually applied to genes in different species.*

Colinearity: *A property of HOX genes: they have the same relative position on the chromosome as the part of the body that the gene affects.*

Paralogous: *Similar in DNA sequence because one gene is a duplicate of the other. Usually applied to genes in the same organism.*

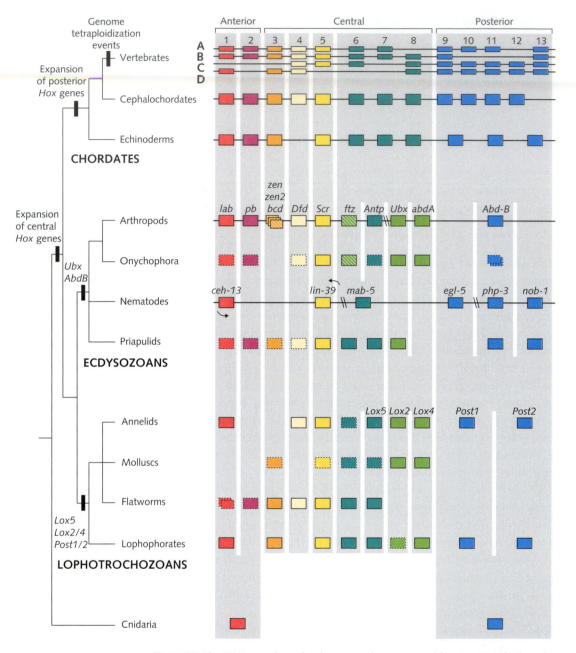

Figure 6.7 The HOX genes have deeply conserved sequence and function. Note both tandem duplication (increase in the number of genes in the cluster) and cluster duplication (increase in the number of copies of the whole cluster). Duplication is associated with increased capacity to control the development of new structures. (Reproduced from Carroll et al. (2001) *From DNA to diversity*, with permission from Blackwell Publishing.)

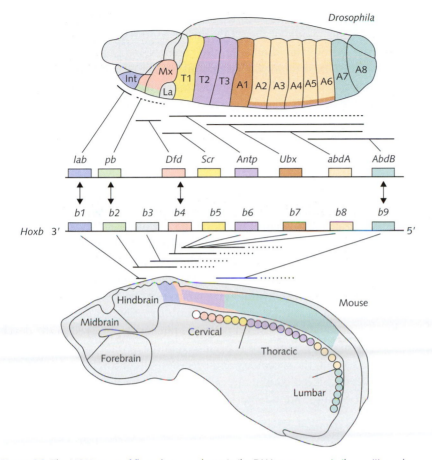

Figure 6.8 The HOX genes of fly and mouse share similar DNA sequences, similar positions along the chromosome, and control the development of similar parts of the body. They have such different names in fly and mouse because of the historical sequence in which they were discovered and the different naming traditions in fly and mouse research. The development of human embryos up to this stage (see Figure 6.1) is very similar to that of mice and is under the control of the same genes. (From Wilkins 2002.)

the original copies of the genes, and new functions could be added under the control of the new copies. Thus as genes were duplicated it became possible to build more-complicated organisms. Supporting this interpretation are the following observations (see Figure 6.7): some HOX genes are found in Cnidaria, but by no means all of them. In the Lophotrochozoa (annelids, mollusks, and their relatives), there are from six to eight HOX genes; in the Ecdysozoa (arthropods, nematodes, and their relatives), from six to 10; in the Deuterostoma (chordates, echinoderms, and their relatives), from 10 to 14. In the branch leading to the vertebrates the entire HOX cluster was duplicated twice; vertebrates have four copies of most of the cluster.

Both the deep conservation and the colinearity of the HOX genes can be seen in Figure 6.8. In both fly and mouse, the HOX genes located to the left control anterior development, those located to the right control posterior development, and the genes in the middle of the HOX cluster control the development of the middle of the body. The double-headed arrows indicate strong sequence similarity, establishing orthology. Flies and mice most recently shared ancestors 550–650 Ma. This pattern of developmental control has persisted since then, over a period so long that it is hard to imagine, in a form that can still be recognized.

Duplicating control genes makes possible new structures; colinearity and combinatorial control are preserved

When the HOX cluster was duplicated twice in the ancestors of the vertebrates, not all of the copies were needed to control the ancestral developmental processes. Some became available to control the development of novel structures. One of those structures is the vertebrate limb, and most of the HOX genes used to control its formation are the last five members of the first (the A) and the fourth (the D) copy. Two striking facts about how these five transcription factors control limb development emerge from knockout experiments in mice (Figure 6.9).

First, their impact along the axis of the limb corresponds to their position along the chromosome. Thus the property of colinearity observed in the developmental control of the body axis is preserved in the developmental control of the limb axis. Second, the specificity and precision of control over the different limb elements is achieved combinatorially, by assigning meanings to different combinations of transcription factors, as follows (Wilkins 2002).

- Attachment to trunk: if A9 and D9, then make a pectoral or pelvic girdle.

- Upper leg or arm: if A10, then make a femur; if A10 and D11 (and possibly D10), then make a humerus.

- Lower leg or forearm: if A11 and D11, then make a radius and ulna (and possibly C11 make a tibia and fibula).

- Wrist, hand, and fingers; ankle, foot, and toes: if D11, D12, and D13, then make carpals, metacarpals, and phalanges; if A13, then make tarsals, metatarsals, and phalanges.

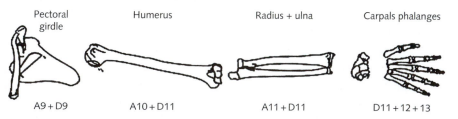

Pectoral girdle	Humerus	Radius + ulna	Carpals phalanges
A9 + D9	A10 + D11	A11 + D11	D11 + 12 + 13

Proximal: *Located towards the body.*

Distal: *Located away from the body.*

Figure 6.9 HOX gene expression in the vertebrate forelimb is colinear and combinatorial. The genes are arranged on the chromosome from left to right and from low to high numbers; they control portions of the limb that are arranged from **proximal** on the left to **distal** on the right, and they do so in specific combinations.

Bilateral animals share a basic body plan, and arthropods are upside-down vertebrates

One of the most charming features of our expanding understanding of the role of development is its ability to answer basic questions posed by biologists in the 19th century. The great debate between Cuvier and Geoffroy St. Hilaire concerned the extent to which the body plans of animals could be homologized. The correspondence between the evolution of the HOX genes and the evolution of the body plans of bilateral animals clearly suggests that Geoffroy St. Hilaire was right: all bilateral animals do share a body plan, and the reason that they share that body plan is that they all use the same set of developmental control processes, inherited from a common ancestor, to construct their bodies (Slack et al. 1993).

Geoffroy St. Hilaire also suggested that arthropods are upside-down vertebrates. We now know that the dorsoventral patterning of the embryos of vertebrates and flies is controlled by genes with shared ancestry. The expression of some of these genes occurs in a striking pattern. Those that are expressed *dorsally* in the fly embryo are related to genes expressed *ventrally* in vertebrate embryos. This supports the idea that after arthropods and chordates diverged there was a reversal of the dorsoventral axis in one of the two lines. In other words, vertebrates do appear to be upside-down arthropods, and arthropods do appear to be upside-down vertebrates (Nublerjung and Arendt 1994).

Downstream genes evolve; many are needed to build a structure, few to initiate its development

The fact that a sequence in a developmental control gene has been conserved for a long period of time does not mean that the structure whose development that gene controls has likewise been conserved. Genes coding for transcription factors control the spatial and temporal expression of **downstream genes**. The genes included in the set of downstream genes can evolve. This makes it difficult to infer the morphological phenotype of the last common ancestor from the fact that certain organisms currently living share developmental control genes with impressive DNA sequence homologies. For example, the gene for the transcription factor that initiates the development of eyes in mice is orthologous to the gene that initiates the development of eyes in flies (Halder et al. 1995), but the mouse eye differs so dramatically from the fly eye that all we can infer about the eye of the ancestor is that it was an epidermal light-sensitive organ connected to the central nervous system in a multicellular organism.

Nor should we let the discoveries of deeply conserved developmental control genes cause us to forget the number of genes involved in building a complex organ. Hundreds to thousands of genes are involved in the development of an eye, a limb, or a brain. Many more genes are involved in building a structure than in controlling when and where it will appear in development, and if the

Downstream gene: *A gene under the control of a regulatory gene; the genes downstream from a regulatory gene constitute a regulatory pathway.*

degree to which they are conserved did not vary considerably, the diversity of life would not be possible.

In echinoderms old genes found new functions

No group better displays new uses for Hox genes than the echinoderms, which change symmetry patterns during development. In echinoderms Hox genes are used to control the patterning of structures with a geometry unlike anything found in other phyla, including 5-fold radial symmetry in adults (Figure 6.10).

The way these genes are used, which differs from class to class among the echinoderms, suggests that the role of some transcription factors was changed in evolution to organize new developmental functions. Transcription factors changed roles both in the divergence of radial adult echinoderms from bilateral

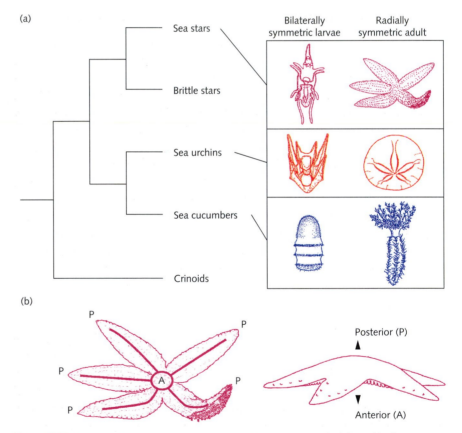

Figure 6.10 Although Hox genes are associated with the patterning of the bilateral body axis in arthropods and chordates, the echinoderms have put them to new uses in different ways in several classes. (a) The bilateral symmetry of the larvae is transformed into the radial symmetry of the adults. (b) The anterior–posterior axis of the larvae may have become either the axis of the radial nerve cords (left) or the oral–aboral axis (right). (Reproduced from Carroll et al. (2001) *From DNA to diversity*, with permission from Blackwell Publishing.)

ancestors and in the divergence of the extant echinoderm classes from each other. This required transcription of the genes at new times and places to change the determination of basic symmetry patterns and the developmental control over very different larval and adult morphologies (Lowe and Wray 1997).

Molecular genetics thus sheds light on the origin of the developmental control genes used to build the diverse morphology found in much of the animal kingdom. Many of these control genes originated between 1000 and 500 Ma and a critical piece of them—the sequence that defines them as transcription factors—has been conserved in sequence since then. The cells in animals know where they are in the body because they have positional information, and they can turn genes on or off appropriately by controlling gene expression with transcription factors. Similar mechanisms control development in plants. In particular, we examine the development of flowers, which evolved 150–120 Ma.

The MADS genes control the development of angiosperm flowers

The MADS box is a highly conserved sequence motif found in a large family of transcription factors that occurs in species from all eukaryotic kingdoms. After the conserved domain was recognized in the sequences of the first four members of the family—MCM1, AGAMOUS, DEFICIENS, and SRF—the name MADS was then constructed from their initials. Most MADS transcription factors play important roles in developmental processes, and their most prominent role is in the angiosperms, the flowering plants, where they are the molecular architects of flower morphogenesis. Their role in helping to structure the angiosperm flower put flesh on the bones of ideas that Goethe had 200 years ago (Coen 2001, Coen et al. 2004). Goethe suggested that the parts of a flower are all derived from leaves, not only the sepals and petals, which look a bit like leaves, but even the anthers and pistil, which have been transformed into cylinders. The architectural role of MADS genes is best understood in *Arabidopsis thaliana*, a cress that is a member of the cabbage and mustard family, valued by scientists for its rapid development and well-understood genetics. It has a small but beautiful flower with all of the parts normally found in angiosperm flowers (Figure 6.11).

Unlike the HOX genes, the MADS genes are scattered throughout the genome. In *Arabidopsis*, they occur on all five chromosomes, and there is nothing resembling the colinearity of position and function of HOX genes. Instead of considering individual genes, it is useful to arrange the MADS genes involved in flower development into three groups, the A, B, and C groups. Within each group, the MADS genes share phylogenetic relationship, suggesting that each group formed from its own common ancestor through gene duplications. The A, B, and C groups combine to determine what part of the flower will develop depending on which set of genes is expressed (Weigel and Meyerowitz 1994), and thus this view of angiosperm flower development is called the ABC model (Figure 6.12).

Figure 6.11 The flower of *Arabidopsis thaliana* has sepals, petals, anthers, a pistil, and an ovary (at the base of the pistil): all the standard parts of an angiosperm flower. (Photograph courtesy of Jürgen Berger, Tübingen University.)

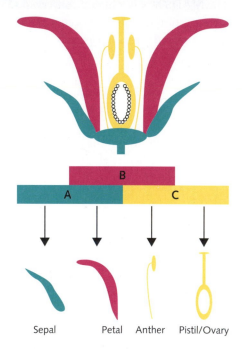

Figure 6.12 The ABC model of angiosperm flower development. (Becker)

Here we see the underlying similarity of plant and animal development: the principle of combinatorial control. In flowers, it works like this:

- if only genes from the A group are expressed, then cells make a sepal.
- if genes from the A and B groups are expressed, then cells make a petal.
- if genes from the B and C groups are expressed, then cells make an anther.
- if only genes from the C group are expressed, then cells make a pistil and ovary.

The same combinatorial principle has evolved both in MADS and HOX genes through completely independent evolutionary paths—a beautiful example of convergence, and strong evidence for the claim that this method of control must be an extremely good one.

Given that this is how the MADS genes function in the angiosperm flower, and that these genes, or their relatives, are found in all plants (as well as in animals and fungi), it is natural to ask, what function did they have before flowers evolved? We do not yet know what functions the MADS genes that control angiosperm flower formation have in gymnosperms. Related MADS genes help to control leaf development in ferns. Thus in the plants we do not yet have a clear history associating changes in the function of developmental control genes with major morphological changes. In the Ecdysozoa (the animals that grow by molting) we have pieces of that history, and it suggests that changes in body plans were brought about through changes in the parts of the body controlled by particular HOX genes.

Turning the ancestor of a velvet worm into a fly involved changing the domains of expression of HOX genes

The onychophorans, or velvet worms, are the basal group in the Ecdysozoa; they resemble the ancestors of the group. The most recently evolved groups are the fleas and the flies. The onychophorans evolved about 550 Ma; the flies evolved about 150 Ma; there was thus about 400 million years available to turn an onychophoran into a fly. What was involved?

The velvet worms (Figure 6.13) have features that place them between worms and arthropods. With worms they share soft bodies, hydrostatic skeletons, ciliated excretory ducts, and smooth muscle layers in the body wall. With arthropods they share legs, respiratory tracheae, a heart with valves, longitudinally partitioned blood sinuses, and jaws derived from appendages. They are not internally segmented like annelids, and they do not look much like a fly (Figure 6.13).

To turn a velvet worm into a fly, at least this much must have happened: the production of legs must be restricted to head and thorax, the legs on the head must be turned into antennae, wings must be invented and then restricted to thorax, and the head must be reshaped into more-complex eating, sensory, and central nervous systems.

Hydrostatic skeleton: *A cavity containing an incompressible fluid surrounded by a flexible covering to which muscles passing through the fluid attach.*

Skeleton: *A structure that permits muscles to be stretched back to their original length following contraction.*

(a)   (b)

Figure 6.13 (a) *Aysheaia pedunculata*, a fossil marine onychophoran from the Burgess Shale of British Columbia, is a basal Ecdysozoan. (Photo courtesy of Doug Erwin, Smithsonian Institute.) (b) *Drosophila melanogaster*, a contemporary fly, is a recently evolved Ecdysozoan. (Photo courtesy Eye of Science/Science Photo Library.)

The conversion of velvet worm to fly may have involved the following steps (this is just one of several possible scenarios). (1) The leg-bearing segments were progressively restricted to the thorax of flies by shifting the expression domains of the HOX genes Ultrabithorax (Ubx = Hox 7) and Abdominal-A (abd-A = Hox 8) forward from the tail towards the head and controlling other genes determining segment identity so that the expression of legs was eventually turned off behind the thorax in all insects, leaving them with three pairs of legs. The myriapoda (millipedes and centipedes), chelicerata (spiders and their relatives), and crustacea (crabs, shrimp, water fleas, and their relatives), represent intermediate steps in this process (Figure 6.14). (2) In early insects wings may

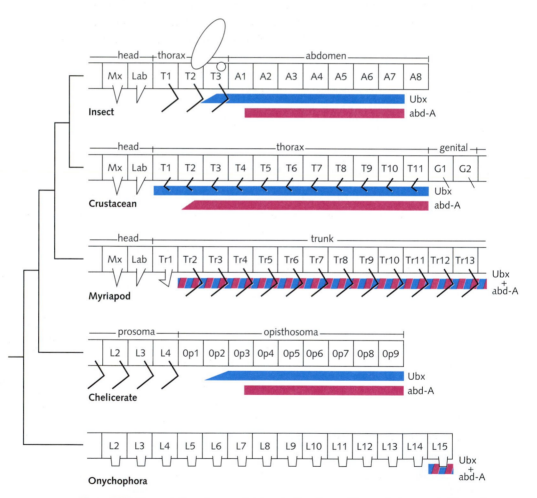

Figure 6.14 The evolution of segment determination changed from velvet worms to flies as the range over which two HOX genes, Ultrabithorax (Ubx = Hox 7) and Abdominal-A (abd-A = Hox 8) expanded from back to front, progressively restricting the segments in which legs were expressed. (Reproduced from Carroll et al. (2001) *From DNA to diversity*, with permission from Blackwell Publishing.)

have been expressed in all segments behind the head. The wing-bearing segments were progressively restricted to the central thorax. While this was happening in insects, the expression domains of the HOX genes along the body axis did not change; they continued to provide the basic patterning of the thorax and abdomen. Changes in the expression domains of wing-forming genes were involved, probably dependent on combinatorial control involving HOX genes as well as other as-yet-unknown control genes. (3) The fly head was sculpted by restricting the expression domains of other HOX genes, including Antennipedia (= Hox 6), Ultrabithorax, and Labial (= Hox 1), from several segments in spider-like ancestors to one or two segments in insects.

Thus the general process was as follows. First make a generalized segment with legs and wings and repeat it many times, forming a long segmented animal with appendages on most segments. Then progressively restrict the expression of appendages to specific segments by altering the expression domains of control genes, and use the specificity gained to alter the appendages so that a different front and back leg and a different fore- and hindwing (which in flies is a balancing organ, the haltere) could be produced.

We get further insight into the great age and functional nature of developmental control genes from their role in eye development.

The control of eye development is ancient and deeply conserved

The transcription factor gene that controls the initiation of eye development, *pax-6* or *eyeless*, is one of many developmental control genes in animals that is not a HOX gene. An extra copy of the gene can be inserted into the fly genome and turned on in various parts of the body that do not normally produce eyes: antennae, thoracic segments, and legs (Figure 6.15).

Remarkably, such ectopic eyes can also be elicited by expressing the transplanted mouse version of the same gene. This example illustrates several points, as follows:

- The genes that initiate eye development in flies and mice have remained so similar for more than 600 million years that the mouse gene still functions to initiate eyes in the fly.

- The common ancestor of the mouse and the fly, a much simpler worm-like animal, had at least a light-receptive organ, perhaps a simple eye.

- Many more genes are required to build the eye, for there are huge differences in the types of eyes that can be built after initiation by the same single gene—a vertebrate camera-like eye with a single lens or an arthropod compound eye with hundreds of lenses.

- The fact that eye development has been initiated by *pax-6* for 600 million years has not constrained the type of eye that is built, simple or compound. The developmental constraints on eye development (see Chapter 2, p. 50) arose later.

Ectopic: *In an abnormal position or place.*

Figure 6.15 Ectopic eyes on thorax and antenna produced in a genetically transformed fly by targeted expression of the *eyeless* gene. (Photograph courtesy of W. Gehring, from Halder et al. (1995) Induction of ectopic eyes by targeted expression of the eyeless gene in Drosophila. *Science* **267**.)

All evolutionary change involves changes in development

The evolution of development must be done without disrupting function

● **KEY CONCEPT**

Development both constrains and enables evolution.

Evolutionary change in organisms is like remodeling a car while it is moving. There is no opportunity to take the organism into the shop to reconstruct it. Function must be preserved while changes are made. One way to do that is to duplicate genes, leave the originals with their prior functions, and involve only the copies in the evolution of new functions. Whatever changes are made, they must result in the development of a functional organism that can survive and reproduce. All other changes will disappear.

The genes that control development, genes coding for transcription factors, are very old, hundreds of millions of years old. During that time, some of them, such as *pax-6*, have retained an ancient function. Others have acquired new functions, such as HOX D9–D13, the control genes used in the evolution of the tetrapod limb from the fish fin, and the HOX genes in echinoderms.

Development constrains evolution more in the short than in the long term

Do developmental mechanisms constrain evolution in the sense that they make possible only certain phenotypes and not others? The answer is often yes in the short term and often no in the long term. Development is complex, highly integrated, and involves some mechanisms that have not changed for a very long time. It can only be changed while retaining organismal function. This all suggests that the range of functioning phenotypes that can be produced by evolutionary changes in development must be restricted, and there are examples that suggest that it is. However, in the long term, transcription factors can be duplicated and the copies can acquire new functions. The echinoderms—the starfish, sea urchins, and sea cucumbers—have evolved radial symmetry from bilaterally symmetrical ancestors and deploy HOX genes in entirely different contexts than their bilaterally symmetrical relatives. Their example shows us that developmental systems, such as the HOX determination of the body axis that appears so deeply conserved in the comparison of mice and flies, can evolve to be applied to shape fundamentally different outcomes. Genes that regulate the expression of other genes have retained their function to switch some process on or off, but the processes they regulate can, over a long period of time, change completely.

Development is an interaction of genes, cells, biological materials, and the environment

Do genes cause development? That is a metaphor, not an explanation. True enough, there is variation in the number and the function of transcription factor genes, and that variation is associated with differences in development. But those observations do not imply that the genetic element of control is a sufficient cause of development, that when we have understood the role of the genes, we have understood all that is necessary to explain development. Genetic effects are useful pointers to processes that are larger and more complex than genetic analysis alone can reveal. Those processes are fundamentally cellular. Genes control the duplication, differentiation, and, in animals, the movement of cells. They cannot make changes that cannot be implemented by a cell.

Moreover, genes have to build organisms out of materials with specific physical and chemical properties. In so doing they can only work through the properties of biological materials. If they commit to building an exoskeleton out of keratin, they will never be able to produce a terrestrial organism heavier than a few hundred grams, for any organism heavier than that would require an exoskeleton too thick to function both as an epidermis and as a support structure. If they commit to building a water-transport system out of xylem embedded in wood, they will never be able to produce a tree more than about 130 m tall, for any tree taller than that would have to produce such strong evaporative suction to get water to its highest leaves that the connection to the roots would fail. If they commit to building an animal body using the HOX cluster in a colinear fashion to produce a bilateral organism, they will only be able to build some kinds of animals and not others.

Development integrates genetic and environmental information. During development genetic information is expressed in signals transmitted within and between cells. The environment influences the course of development by intervening in the internal signaling pathways used by the genes. In this way *all* aspects of the phenotype become the products of gene–environment interactions to a greater or lesser extent.

Development is a key to understanding the evolution of organisms

We began this chapter with a transition from the population genetics presented in Chapters 3–5. We hope that we have now convinced you that evolution involves much more than changes in genes. Changes in genes provide us with a convenient, compact history of evolution, a method of keeping track of changes, but they are a very indirect record of changes in traits. Genes are linked to organisms by development, and development gives to organisms a role in evolution missed by a strictly gene-centered approach. Not only the genes' method of making more genes, organisms determine through development and through interactions with the environment how genes are expressed and which genes will survive.

A gene not embedded in a functioning life cycle has no future, and a functioning life cycle is one in which development succeeds in producing an organism that can deal successfully with ecological challenges to grow, differentiate, survive, and reproduce. In the next chapter we examine how genes interact with the environment to influence performance within the lifespan of a single organism.

● SUMMARY

- Development turns single cells into adults with hundreds of cell types and billions of cells.

- Development is one of three essential transformations that occur in every life cycle—the others involve ecology and genetics.

- Development maps the information in the genotype into the material of the phenotype as a function of the environments encountered.

- All evolutionary change in phenotypes involves changes in development.

- Cells are the units of development. All cells contain all the genes needed to build an organism; cell differentiation is thus regulation of which genes are expressed, not which genes are there.

- Many regulatory genes produce transcription factors, proteins that bind to DNA in the control regions of other genes whose expression is thereby regulated.

- Regulatory genes occur in families, including HOX genes in animals and MADS genes in eukaryotes. Such gene families are recognized by highly conserved DNA sequences called boxes, which form only a small portion of each gene. The other portions vary.

- Boxes are conserved for at least three reasons: they code for the amino acid sequences in the region of the protein that must bind precisely to DNA; they code for the region of the protein that must interact precisely with the nuclear pore through which it travels to get from the cytoplasm, where it is produced, to the gene that it regulates; they code for the region of the protein that participates in the protein–protein interactions required for combinatorial control of gene expression.

- The control of cell–cell communication, the production of positional information, and the control of cell fate by transcription factors that activate or repress the transcription of other genes, all evolved independently in plants and animals. That is what makes the next point so striking.

- Both the HOX genes, which control the differentiation of the body axis in most animals and the limbs of tetrapods, and the MADS genes, which control the differentiation of angiosperm flowers, achieve their precise and efficient control through combinatorial logic. Several transcription factors bind in specific combinations to the control regions of the genes they regulate. That is how genes can be turned on or off in several different contexts—as many contexts as there are combinations of transcription factors.

- The evolutionary history of morphological change in some major groups can be partially understood as changes in the expression domains of developmental control genes. For example, the ancestors of velvet worms—onychophorans—were transformed into fruit flies through changes in the expression of genes controlling the number and position of the segments that bear legs or wings.

- A developmental gene, *pax-6*, that initiates the development of light receptors in the epidermis, has retained its function for more than 600 million years.

- In many cases, ancestors had most or all of the genes involved in control over development. Much of the evolution of development consisted of changing the specificity of expression of those genes in time and space and the specificity of the receptors that reacted to their products.

- One key process in the evolution of development that makes possible the evolution of new structures while retaining old ones is gene duplication. The HOX genes are a good example. They duplicated first in tandem, on the same chromosome, to form a cluster of paralogous genes. Later, in some lineages, the entire cluster was duplicated. With each duplication event, the old copies retained their original function, and the new copies were released to acquire new functions. That is one important source of innovations in evolution.

This chapter sketched what is known about the mechanisms that produce an organism from an egg. The next chapter discusses how organisms interact with their environment while they develop to produce a range of outcomes.

RECOMMENDED READING

Akam, M. (1998) Hox genes: from master genes to micromanagers. *Current Biology* **8**, R676–8.

Carroll, S.B., Grenier, J.K. and Weatherbee, S.D. (2001) *From DNA to diversity: molecular genetics and the evolution of animal design*. Blackwell Science, Oxford.

Coen, E., Rolland-Lagan, A.G., Mathhews, M., Bangham, J.A. and Prusinkiewicz, P. (2004) The genetics of geometry. *Proceedings of the National Academy of Sciences, USA* **101**, 4728–35.

Slack, J.M.W., Holland, P.W.H. and Graham, C.F. (1993) The zootype and the phylotypic stage. *Nature* **361**, 490–2.

QUESTIONS

6.1. From the point of view of development, how does the life cycle of a bacterium resemble the life cycle of an elephant? How does it differ?

6.2. How does the deployment of HOX genes in the development of the tetrapod limb illustrate the opportunism of evolution?

6.3. How much of the information needed to build the organism is contained in DNA sequences and how much is contained in the properties of cells and biological materials? To take one example, look up the properties of hydoxyapatite, one of the constituents of vertebrate bone (e.g. Currey 1984), and discuss them with respect to the information they contain that might not have to be coded in a DNA sequence. Also discuss how the genes can shape the properties of bone without being able to change the properties of hydroxyapatite.

CHAPTER 7

The expression of variation

● **KEY CONCEPT**

Developmental mechanisms specific to lineages limit phenotypic variation despite genetic variation.

We often discuss evolutionary genetics (Chapters 3–5, pp. 54–123) as though genetic variation were directly expressed as phenotypic variation. This is only occasionally true, for the links that development makes between genotypes and phenotypes are often indirect and complex (Chapter 6, p. 124). This chapter explores some complexities in how development links genotype to phenotype. Particularly important are effects that cause phenotypic variation to differ from genetic variation, for natural selection operates with phenotypic variation (Chapter 2, p. 27), whereas the response to selection depends on genetic variation (Chapters 4 and 5, pp. 70–123). Such developmental effects can have at least two causes: ancient mechanisms shared by many species within a lineage, and the current interactions of developing organisms with their environments. Ancient developmental mechanisms shared by an entire lineage may only permit the expression of certain phenotypes, as we saw in Chapter 6 with the different ways that frogs and salamanders respond to selection for fewer digits. Such lineage-specific constraints combine with the laws of physics and chemistry to influence the interactions of developing organisms with their environments to change the relationship between genetic and phenotypic variation. To illustrate such effects we begin this chapter with two examples that show how lineage-specific developmental mechanisms interact with variation in the environment to structure phenotypic variation.

Butterfly wing patterns are produced by a conserved developmental mechanism specific to the lineage

Butterfly wings are among the most beautiful products of evolution (Figure 7.1). They are also the product of a well-understood, phylogenetically conserved, developmental mechanism (Nijhout 1991). Butterfly wings develop during pupation from wing discs that are easily accessible for manipulation through the walls of the cocoon. They are also nearly two-dimensional sheets with an upper and a lower side subdivided into compartments or cells by wing veins and the wing margin, giving them a standard geometry. This allows us to specify patterns

Figure 7.1 The butterfly families Nymphalidae (N), Papilonidae (P), and Satyridae (S) illustrate the huge diversity of wing patterns that can be elicited from evolutionary modification of a shared developmental system (Figure 7.2). Top row (left to right): *Heteronympha merope* (N), *Junonia coenia* (N), *Papilio demodocus* (P); second row: *Papilio dardanus meriones* (P), *Teinopalpus imperialis bhumipholi* (P), *Inachis io* (N); third row: *Pierella nereis* (S), *Paranassius smithii* (P), *Cethosia bibles* (N); fourth row: *Cithaerias menander* (N), *Morpho catenarius* (N), *Precisa octavis* (N); fifth row: *Papilio chikae* (P), *Xois sesara* (N), *Papilio bianor kotoensis* (P); sixth row: *Eurytides thyastes* (P), *Eurytides philolaus* (P). (Butterfly photos credit to Terry Dagradi; specimens courtesy of the Peabody Museum, Yale University, arranged by Raymond Pupedis, Curatorial Assistant in Entomology.)

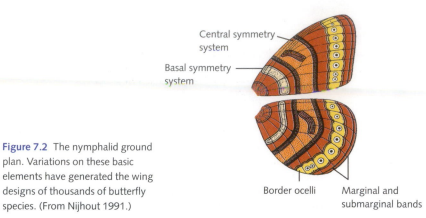

Figure 7.2 The nymphalid ground plan. Variations on these basic elements have generated the wing designs of thousands of butterfly species. (From Nijhout 1991.)

at precise positions, and makes butterfly wings an ideal model system for the study of pattern formation. Many butterfly wing patterns are variations on a single basic plan, which is called the nymphalid ground plan after the family of butterflies in which it is most clearly expressed, the Nymphalidae.

Mutations affecting wing patterns only produce changes within this framework, not something totally new

The units of the plan are the wing cells. The formation of a pattern within each wing cell, on the upper and lower surfaces of that cell, and in the fore- and hindwings, can be independently controlled. The pattern of the whole wing consists of repetition with variation of the patterns in the cells. Working with the basic plan of the butterfly wing (Figure 7.2), evolution has been able to produce tremendous diversity by varying the expression of each element.

This developmental pattern has been conserved in the radiation of thousands of species of butterflies, functioning as a filter that determines what phenotypes are possible. Mutations affecting wing patterns only produce changes within this framework, not something totally new. Thus the developmental mechanisms that control this pattern shape the expression of genetic variation in the phenotype.

Developmental mechanisms not only produce patterns in morphology. They also, very importantly, determine how organisms grow. Differences in growth patterns among major lineages have strongly affected how the species in those lineages have evolved.

Discontinuous growth combines with maturation thresholds to produce surprising consequences

Arthropods must grow discontinuously by molting because they have hard exoskeletons. In the water flea *Daphnia*, discontinuous growth interacts with a threshold size at which maturation is initiated. Individuals vary in their size at

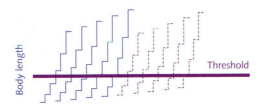

Figure 7.3 The growth curves for 10 hypothetical *Daphnia* individuals from the first instar until the first reproductive instar. Each horizontal step represents 1 instar, and the length of the next instar is 1.3 times the length of the previous instar. Maturation is initiated when the threshold length is reached. Because maturation takes two instars, reproduction occurs at the beginning of the third instar after the threshold is passed. Individuals that were smaller at birth (dark blue) need more instars to mature than individuals that were larger at birth (pink). The group of five individuals on the left needed six instars to reach maturation; the group of five individuals on the right needed only five because they were larger at birth. (From Ebert 1994.)

birth for genetic, environmental, and maternal reasons, and they vary in their growth rates because of variation in food and temperature. The lineage-specific growth pattern interacts with variation in size at birth and growth rate to produce surprising consequences (Ebert 1994).

Consider a series of newly hatched *Daphnia* of increasing size (Figure 7.3). The growth line for the smallest animal crosses the size threshold for maturation during the growth following the third molt; it matures after two more molts, having had five juvenile instars.

The next four *Daphnia* also cross the threshold during their third molt, have five juvenile instars, and define, with the first individual, an instar group. Within this group, size at maturity increases with size at birth, an unremarkable result. The sixth individual, however, is sufficiently larger at birth to cross the threshold in the *second* molt and matures after only four juvenile instars. With the next four individuals, it forms another instar group. The sixth individual is larger at birth than the first, but smaller at maturity. Thus discontinuous growth coupled with a size threshold for maturation can reverse the normal relationship between size at birth and size at maturity between instar groups.

Discontinuous growth, found in most arthropods, generates phenotypic variation that interacts with size-specific selection in surprising ways. Consider Figure 7.4. In its upper portion are plotted the sizes at birth and sizes at maturity of seven individuals in a six-instar group, which were smaller at birth, and seven individuals in a five-instar group, which were larger at birth. Within each group, size at maturity increases with size at birth, but there is no such relationship between instar groups. So, the six-instar individuals, although smaller at birth than the five-instar individuals, do not remain smaller when they reach maturity, for then the two groups overlap completely in size with mean sizes of about 3.3 mm.

Now think about what happens when there is size-specific mortality on adults and a range of sizes at birth, as indicated by the shaded bars in the lower portion of

Instar: *The period between molts, usually used for arthropods.*

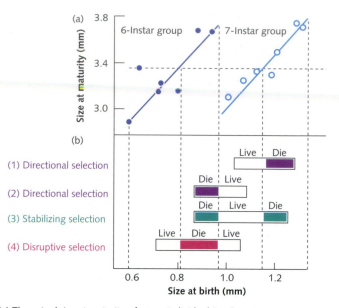

Figure 7.4 (a) The actual size at maturity of seven individual *Daphnia magna* in a six-instar group (dark blue) and seven individuals in a seven-instar group (light blue). Within an instar group, size at maturity increases with size at birth. Between groups, there is no clear relation. (b) The type of selection that would be experienced on size at birth given threshold selection against adult individuals larger than 3.3 mm at maturity, for four different ranges of size at birth. (1) **Directional selection** against larger adults selects against larger newborn (individuals in the dark size range die, those in the white size range survive); (2) selects for larger newborn; (3) selects for intermediate-sized newborn (stabilizing selection); (4) selects for very small and very large newborn (disruptive selection). (From Jones et al. 1992.)

Directional selection: *Selection that always acts in a given direction, for example, always to increase the value of a trait.*

Stabilizing selection: *Selection that eliminates the extremes of a distribution and favors the center.*

Disruptive selection: *Selection that favors the extremes and eliminates the middle of a frequency distribution of trait values, for example, increasing the frequency of small and large individuals and reducing the frequency of medium-sized individuals.*

Figure 7.4. If a predator were preferentially to eat *Daphnia* larger than approximately 3.3 mm, it would select for small size at birth if there were only one instar group generated by a narrow range of birth sizes, for within each instar group, there is a consistent relationship between size at birth and size at maturity. However, if the range of birth sizes were large enough to generate two or more instar groups, as is usually the case, then selection against large adults could produce stabilizing selection on birth sizes for one range of birth sizes (approx. 0.9–1.2 mm), and disruptive selection on birth sizes for another range (approx. 0.7–1.1 mm). Thus the kind of selection operating on size at birth in *Daphnia* depends in a complex way on the range of offspring sizes at birth and on growth rates.

Size at birth in *Daphnia* varies genetically among females, increases with age within each female, and is affected by the food supply. Because offspring of various sizes are always entering *Daphnia* populations, there can never be a single, stable type of selection on offspring size driven by selection on adult size. Here a pattern of growth and maturation shared by many arthropods acts as a filter, introducing into the distributions of phenotypes patterns that can only be partially and indirectly affected by changes in genes.

The butterfly wing plan and discontinuous growth in arthropods exemplify lineage-specific developmental mechanisms that determine the kinds of variation presented to natural selection. Other mechanisms have evolved primarily to produce phenotypes appropriate to the environment where development occurs, mechanisms that rely on receiving information from the environment to determine the kind of phenotype that will be produced.

The environmentally induced responses of one genotype produce several phenotypes

Induced responses are adaptations that illustrate the capacity of one genotype to produce quite different phenotypes depending on the environment encountered either by its mother or during its own development. According to Williams (1966) and Curio (1973), who sought clear definitions of adaptations, an adaptation is the ability of a genotype to produce a change in a phenotype that occurs in response to a specific environmental signal and improves reproductive success. Otherwise, the change does not take place. Many induced responses are adaptations by this definition.

Some *Daphnia* develop helmets, spines, and neck teeth that protect them against predators, but only when they detect predators

In water fleas of the genus *Daphnia* a signal produced by predators induces a defensive response. *Daphnia* respond to molecules dissolved in water that indicate the presence of invertebrate and fish predators by developing tail spines, helmets, and neck teeth (Figure 7.5). Predators feed less effectively on spiny, helmeted *Daphnia*. Helmets and spines are costly; individuals that do not produce them have higher reproductive rates than individuals that do produce them; and therefore when predators are not present the spines and helmets are not produced (Dodson 1989).

In the presence of predators, barnacles develop bent shells and birch trees develop resistant leaves

Barnacles of the genus *Chthamalus* react to the presence of a predatory snail, *Acanthina*, by altering their development. If the snail is present, the barnacles grow into a bent-over form that suffers less from predation but pays for it with a lower reproductive rate. If the snail is not present, the barnacles develop into a typical form with normal reproduction (Lively 1986).

Birch trees subjected to natural defoliation by caterpillars and artificial defoliation by scientists develop resistant leaves. Resistance is measured by bioassay: caterpillars reared on trees previously subjected to defoliation pupate

> ● **KEY CONCEPT**
>
> Just as many genotypes can produce one phenotype, one genotype can produce many phenotypes in response to environmental variation.

> **Induced response:**
> *Developmental response to a specific environmental signal that has a functional relationship to that signal, resulting in improved growth, survival, or reproduction.*

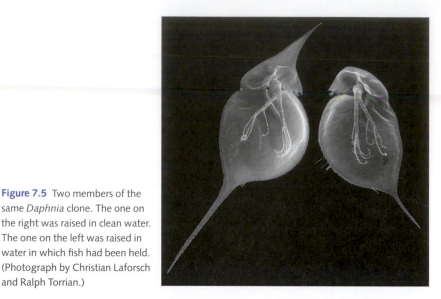

Figure 7.5 Two members of the same *Daphnia* clone. The one on the right was raised in clean water. The one on the left was raised in water in which fish had been held. (Photograph by Christian Laforsch and Ralph Torrian.)

at smaller sizes than caterpillars reared on trees with no defoliation. The reaction persists in trees treated with fertilizer, suggesting that the leaves are not simply of lower quality because the trees have been damaged, but contain chemicals harmful to caterpillars (Haukioja and Neuvonen 1985).

Thus induced responses demonstrate that one genotype can produce several phenotypes in reaction to environmental signals.

Reaction norms help us to analyze patterns of gene expression

A population of genotypes is a bundle of reaction norms

● **KEY CONCEPT**

Reaction norms describe how genotypes react to the environment to produce a range of phenotypes.

A naive view of the relation of genes and traits is that variation in genes has a direct relation to variation in traits, and that a given amount of genetic change would result in an equivalent amount of phenotypic change. However, the examples just given show that this view is incorrect. We need conceptual tools that enable us to think clearly about the roles of genetic variation, developmental mechanisms, and environmental variation in producing patterns of phenotypic variation. This section discusses some concepts that help.

Reaction norm: *A property of a genotype that describes how development maps the genotype into the phenotype as a function of the environment.*

The first tool is the concept of the reaction norm. The reaction norm of a trait is a property of one trait, one genotype, and one environmental factor. It is measured by raising individuals from one clone at different levels of the environmental factor, measuring the trait at each level, and plotting it as a function of the environmental factor. The resulting line (Figure 7.6a) describes how development maps the genotype into the phenotype as a function of the environment. A population of genotypes can be described as a bundle of

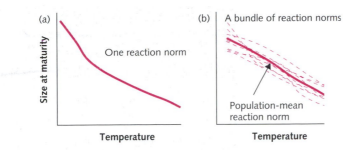

Figure 7.6 (a) An example of a reaction norm, which is a property of a single genotype: individuals that all belong to a single clone mature at smaller sizes when reared at higher temperatures. (b) An example of a bundle of reaction norms. Each dashed line represents the sensitivity of a single genotype to temperature; the solid line represents the population-mean reaction norm.

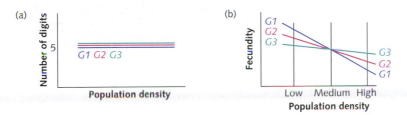

Figure 7.7 (a) The number of digits in the hand of a lizard is not sensitive to population density and does not vary genetically; it has a flat, canalized reaction norm. (b) The fecundity of the same three genotypes would be sensitive to population density; here sensitivity to density varies with genotype.

reaction norms (Figure 7.6b), and the average reaction of the population to the environmental factor can be described as the population-mean reaction norm.

Measurements of genetic variation can depend on the environment in which they are made

Depicting trait variation as a bundle of reaction norms is an effective way to see at a glance how genes and environments interact to determine the trait. Consider an artificial example of three genotypes (*G1*, *G2*, and *G3*) sampled from a population of **parthenogenetic** lizards, reared as clones, and raised at three population densities, low, medium, and high (Figure 7.7). We measure two traits, number of digits per foot and fecundity. Figure 7.7(a) depicts the reaction norms of the three genotypes for number of digits per foot. In fact, they would lie on top of one another, for every individual in the entire population has five digits per foot at all population densities, but in the figure they are separated to show that three genotypes were measured. These are perfectly flat reaction norms. The trait is insensitive to environmental variation, expresses no genetic variation, and cannot respond to selection.

For fecundity, the situation is quite different (Figure 7.7b). All three genotypes reduce their fecundity at higher population densities, but the sensitivity of

Parthenogenesis: *Asexual reproduction from an egg cell that usually does not involve recombination. In most cases the daughters are exact genetic copies of the mothers.*

the three genotypes to changes in density differs. *G1* is quite sensitive. It has the highest fecundity at low population density and the lowest fecundity at high population density. *G3* is not very sensitive. It has the lowest fecundity at low population densities and the highest fecundity at high population densities. *G2* is intermediate.

If we were to measure the genotypes only at low or only at high population densities, we would say that the heritability of fecundity was high. If we were to measure them only at medium population densities, we would say that the heritability of the trait was near zero and that there was little genetic variation for fecundity in the population. Thus plotting reaction norms is an easy way to see how measurements of genetic variation depend on the environment in which they are made. A trait like fecundity in Figure 7.7(b) can respond to selection at low and high population densities, where it expresses genetic variation, but not at medium density.

The plasticity of a trait can respond to selection when slopes of reaction norms are heritable

> ● **KEY CONCEPT**
>
> The extent to which a trait responds developmentally to environmental variation is called plasticity and is itself under genetic control. Adaptive phenotypic plasticity 'fits' organisms to the variation in their environments. Not all plasticity is adaptive.

Phenotypic plasticity:
Sensitivity of the phenotype to differences in the environment. This term is less precise than reaction norm.

The sensitivity of a trait to change in an environmental factor—the slope of the reaction norm—measures its phenotypic plasticity. In Figure 7.7(b) there is genetic variation for the sensitivity of fecundity to changes in density. Selection to increase the sensitivity of fecundity to density would increase the frequency of *G1*. Selection to decrease the sensitivity of fecundity to density would increase the frequency of *G3*. When survival of offspring born at high density is extremely poor, and survival of offspring born at low density is quite good, then selection will increase the sensitivity of fecundity to density. When there is little difference in offspring survival at low and high densities, selection may decrease the sensitivity of fecundity to density. Thus natural selection can shape phenotypic plasticity to react appropriately to ecological problems, but only if there is genetic variation for the slopes of reaction norms. Sensitivity to environmental variation raises the question of the adaptive value of such sensitivity.

Some plasticity can be adaptive, but not necessarily all

Deciding whether a plastic response is an adaptation is not straightforward, for some responses may be natural consequences of physical, chemical, and ancestral biological processes that would have occurred whether or not evolution had adapted the population to its current circumstances. For example, if a plant is deprived of water, it must grow more slowly—it has no other option. We know that some plastic responses are not adapted (they might be fortuitously adaptive) because many organisms react to environmental variation never encountered in evolution. It is therefore important to be able to decide how much of a plastic response, a reaction norm with a non-zero slope, is the product of adaptive evolution and how much is just a byproduct of other processes.

Variation in plasticity can change genetic correlations across environments

The response to selection of a single quantitative trait depends on its heritability, the amount of phenotypic variation that can be ascribed to additive genetic effects (Chapter 4, p. 90). When one trait is selected, but is genetically correlated to another trait because some genes influence both traits, the second trait will respond to selection on the first trait whether it is directly under selection or not. Thus the response of a trait to selection depends both on its heritability and on its genetic correlations to other traits also under selection.

Variation in slopes of reaction norms that is large enough for reaction norms to cross can change heritabilities and genetic correlations when environmental factors change. Consider Figure 7.8. In figure 7.8(a), reaction norms for three genotypes are plotted from 20 to 28°C for an arbitrary trait, Trait 1. Let the same three genotypes also affect another trait, Trait 2; their reaction norms for Trait 2 are given in Figure 7.8(b). Because the same three genotypes are affecting two traits, those traits are genetically correlated. When we plot Trait 1 against Trait 2, in Figure 7.8(c), we see that the traits have a negative genetic correlation at 20°C that changes to a positive genetic correlation at 28°C. Selecting Trait 1 upwards at 28°C will produce a correlated *upward* response in Trait 2, but selecting Trait 1 upwards at 20°C will produce a correlated *downward* response in Trait 2.

This could have serious consequences. For example, consider two traits on butterfly wings (see Figure 7.12, below). Let Trait 1 be eyespot size (*y* axis) and Trait 2 the color (light, or dark) of the ring around the eyespot (*x* axis; light to the left, dark to the right) on butterfly wings. Suppose the two environments are a cool, dry season (upper-right group of points in Figure 7.8c) and a warm, wet season (lower-left on Figure 7.8c). Large, conspicuous eyespots with a light ring around them are adaptive in the wet season; small, inconspicuous eyespots with a dark ring around them are adaptive in the dry season. But if the reaction norms for the two traits behave like those in Figure 7.8, then selection to increase eyespot size in the warm, wet season would cause the color of the ring around the eyespot to become lighter, not darker, in the next dry season. That

Genetic correlation: *The portion of a phenotypic correlation between two traits that can be attributed to additive genetic effects.*

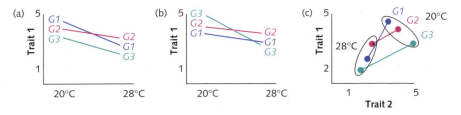

Figure 7.8 We combine the reaction norms for two traits to visualize changes in genetic correlations. (a) The reaction norms of three genotypes for Trait 1 as a function of temperature. (b) The same for Trait 2. (c) Trait 1 plotted against Trait 2 for each of the three genotypes and each of the two temperatures. The genetic correlation shifts from positive at 28°C to negative at 20°C.

would make the butterfly more conspicuous to predators and easier to catch. For the sake of these imaginary butterflies, let us hope that is not the case. In the next section, we will see that the pattern in Figure 7.8(c) actually occurs.

Spadefoot toads illustrate how genetic correlations can change across environments

Whereas we can only measure reaction norms as properties of *genotypes* in organisms that can be cloned, they can be estimated in sexually reproducing organisms as the reaction of the mean value of a *family* of siblings to changes in an environmental factor. Newman (1988) made such estimates by taking eggs from five families of spadefoot toad tadpoles, rearing half of each family in ponds of short duration and half of each family in ponds of long duration (two levels of an environmental factor), and measuring body length and age at metamorphosis (two traits). Overall, individuals reared in ponds of small duration tended to metamorphose earlier than those reared in ponds of long duration. In ponds of short duration, individuals with longer bodies tended to metamorphose earlier than those with shorter bodies, whereas in ponds of long duration, individuals with longer bodies tended to metamorphose later than those with shorter bodies. Thus in ponds of short duration, the means of the two traits were negatively correlated among families, whereas in ponds of long duration, they were positively correlated (Figure 7.9).

Given the correlations between the two traits described in Figure 7.9, selection for more rapid development strictly within ponds of short duration would select for larger body size at metamorphosis. Selection for more rapid development strictly within ponds of long duration would select for smaller body size at metamorphosis. This illustrates the points made in Figure 7.8: because the

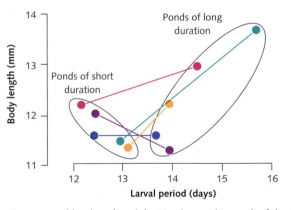

Figure 7.9 The reaction norms of families of spadefoot toads reared in ponds of short and long duration behave like the model depicted in Figure 7.9: a negative genetic correlation in one environment shifts to a positive genetic correlation in the other. The lines joining the means of siblings from split broods reared separately show how both traits react to the environmental difference for each family taken separately.

genetic correlations of pairs of traits can change with the environment in which they are expressed, the potential response to selection of a set of correlated traits depends on the environment.

How important is the influence of the environment on the expression of genetic variation? It depends on the trait examined. Weigensberg and Roff (1996) compared the heritabilities of morphological, life history, and behavioral traits measured in the field with heritabilities measured in the laboratory for organisms of many kinds. The environment in the laboratory is usually assumed to differ from the environment in the field, but in this study heritabilities in the two environments were quite similar, and they were more similar for morphological traits than for life-history traits. Thus although the effects depicted in Figure 7.9 do happen, they do not always happen.

Next we will see that changes in the genes do not always cause changes in the phenotype even when the genes are expressed.

The genotype–phenotype map has some important general features

Organisms that are similar phenotypically can differ genetically

The ciliates are sexual, heterotrophic, single-celled organisms covered with cilia and having a micronucleus for inheritance and a macronucleus for RNA production. *Paramecium* is a well-known example. Ciliates have existed at least since the Cretaceous period (135–65 Ma), and they now live in both fresh and salt water. Their closest relatives are the dinoflagellates and the plasmodial slime molds (see Chapter 17, p. 420). The ciliates in the *Tetrahymena pyriformis* complex (Figure 7.10) are indistinguishable morphologically but differ both in their genes and in the proteins that build the indistinguishable structures. The genetic differences are large and indicate that the many members of the complex have been evolving independently from one another for over 100 million years

● **KEY CONCEPT**

The genotype–phenotype map as determined by development plays a key role in determining both the strength of selection and the response to it.

Figure 7.10 The ciliate *Tetrahymena pyriformis* consists of a group of species so similar that they are virtually impossible to discriminate by eye although they have been separate species for over 100 million years. (Photo courtesy of Louis de Vos, Free University of Brussels.)

(Nanney 1982), despite which they now look exactly the same. Here the DNA sequences of the different species have diverged considerably while the phenotypic structures have remained the same. We do not know whether this is because there is strong stabilizing selection for a particular structure, or whether developmental mechanisms only permit certain structures to be produced.

Organisms that are similar genetically can differ phenotypically

The reverse of this also occurs. In cases of induced defensive responses, the genome contains the information necessary to produce several phenotypes depending on the environment. Precisely the same genotype could produce, for example, a plant that produced high concentrations of poisonous chemicals in its leaves and was therefore resistant to caterpillar predation or a plant that produced fewer chemicals and had better reproductive performance as a result. To that we can add metamorphosis: from tadpoles to frogs in amphibians, from caterpillars to pupae to butterflies in insects, from underwater to aerial leaves in aquatic buttercups, and other developmental transformations in the life of a single individual. In all such cases the genome contains the information needed to produce several strikingly different phenotypes.

An extreme example of such genetic potential can be found in cases of environmental sex determination associated with marked sex dimorphism. If a larva of the marine worm *Bonellia* settles on normal substrate, it metamorphoses into a female and grows into an adult that resembles a 10–20-cm-long sausage with a long feeding tube ending in a pair of palps. If a larva settles, however, on a female, it metamorphoses into a male that becomes a parasite embedded in the body of the female, reduced to little more than testes making sperm. Thus the same genotype can make either a large, complex, independent female or a tiny, simple, parasitic male.

There are at least two reasons for loose coupling of genotype and phenotype

Genotype and phenotype are loosely coupled in eukaryotes for two basic reasons. First, there is a great deal of DNA in the genome that is either never transcribed or whose transcription products have little or no effect on the phenotype. Some of it is parasitic or symbiotic, including transposons and selfish genetic elements, some of it appears to be accumulated junk with no clear role, some may play an adaptive role that has not yet been determined, and some of it codes for synonymous amino acid substitutions. Whatever the reasons for its existence, the fact that much of the eukaryotic genome is not transcribed means that whatever genetic changes occur in that portion of the genome will have little effect on the phenotype.

The second reason for loose coupling was suggested in Chapter 6 (p. 124) and above in the examples of induced defenses and environmentally determined sexual dimorphism: **regulatory genes** that turn entire developmental pathways on or off. If the same sets of developmental pathways exist in all individuals of a species, and all that is needed to produce different phenotypes is to flip switches at a few branch points, then a single gene on a sex chromosome could switch on male rather than female development, a signal from a predator could switch on one gene that controlled development of a defensive phenotype, and a signal of sun or shade could induce the formation of different types of leaves. When regulatory genes turn developmental pathways on or off as a function of the environment, many relations between genotype and phenotype become possible. This has happened in many multicellular organisms whose developmental mechanisms are becoming better known. Why it has not happened in single-celled ciliates is not clear.

> **Regulatory gene:** *A gene that turns another gene, or group of genes, on or off. Small changes in regulatory genes can cause large changes in phenotypes.*

Macro- and microevolution meet in the butterfly wing

In the course of evolution, the four-winged condition of butterflies evolved before the two-winged condition of flies. The genes controlling wing development in butterflies and flies still have sufficiently similar DNA sequences that one can take well-studied genes from *Drosophila*, use their sequences to locate the orthologous genes in butterflies, then study the patterns of expression of the butterfly genes. Such studies have revealed that the spatial pattern of the *Drosophila* wing is determined by the expression of genes orthologous to those that determine the spatial pattern of the wing of the buckeye butterfly, *Precis coenia* (Carroll et al. 1994). Moreover, some of the genes found in butterflies by 'fishing' with *Drosophila* genes known to be involved in wing patterns turned out to be old genes with new roles, roles that involve them in the construction of elements—midline stripes, chevrons, and venous stripes—in the wing cells of the nymphalid butterfly family ground plan (Nijhout 1991).

> ● **KEY CONCEPT**
> The butterfly wing paradigm reveals the complex nature of a genotype–phenotype map in which macro- and micro-evolution both influence the same developmental pathways.

Particularly interesting was the discovery of the role played in butterfly wings by a gene called *distalless* in *Drosophila*. *Distalless* was discovered as a mutant that transforms distal antennal structures into distal leg structures, probably by producing a protein that acts as a transcription factor, activating the genes needed for distal structures (Wilkins 1993). Thus in *Drosophila* the gene helps control the parts of appendages that are furthest from the body. In butterflies, it has a different role: it helps to control the formation of eyespots on wings. Wings are appendages, but eyespots are formed in the middle of the wing, not at its distal edge, and eyespots do not look at all like antennae. Thus the homology in the function of *distalless* in *Drosophila* and butterflies concerns not what will be produced, nor where it will be produced in an appendage, only that it will be produced in an appendage and not somewhere else in the body.

The study of the developmental mechanisms that control gene expression is becoming much easier because of the continuing development of new tools in molecular genetics. Although thousands of genes may be involved in the production of a trait, it now seems likely that only a few regulatory genes that produce transcription factors are involved in determining when and where that trait is produced within the body plan and how the pattern of the trait is determined (see Chapter 6, pp. 129 ff.). Because regulatory genes are highly conserved, we can take a gene out of *Drosophila*, where its function is well known, clone it, use it to find orthologous genes in some other organism, study their expression, and have a good chance of uncovering a role for that orthologous gene that sheds as much light on the development of that organism as it did in *Drosophila*.

Next we consider an example where the developmental control of gene expression can be followed from genetic and developmental studies in the laboratory to the fitness consequences of changes in expression in organisms released in the field.

Butterfly wing spots may change seasonally across generations

We began this chapter by discussing the plan that determines the expression of patterns in many butterfly wings. There it served as an example of a conserved developmental pattern that influenced how genetic variation could be expressed in a large group with many species. We then discussed reaction norms as important patterns in the expression of phenotypic variation. After noting that organisms that are similar phenotypically can differ genetically and those similar genetically can differ phenotypically, we discussed reasons for that loose coupling: the structure of the eukaryotic genome and the genetics of development, particularly the role of regulatory genes, small changes in which can cause big changes in phenotypes. Some regulatory genes are highly conserved, playing similar roles in organisms as distantly related as mammals and insects and allowing us to use genes whose role is well understood in fruit flies to search for genes with similar DNA sequences and potentially similar roles in other organisms. Such a search of butterfly genes revealed that the gene *distalless* that controls structures expressed on the distal end of *Drosophila* appendages has a homolog in butterflies that controls the expression of eyespots on the distal portion of wings. That observation brought us back to the nymphalid ground plan in the search for the conserved molecular mechanisms responsible for the conserved pattern of butterfly wings.

At the same time that some of the regulatory genes responsible for laying down the insect body plan were being discovered, a project on the ecological and evolutionary significance of seasonal variation in the patterns of butterfly wings was being carried out in the field, in Malawi, and in the laboratory, in the Netherlands and Scotland. This rapidly growing body of work can be accessed through Carroll et al. (1994), Brakefield (2001), Brunetti et al. (2001), Beldade and Brakefield (2002), McMillan et al. (2002), and Monteiro et al. (2003). They study tropical butterflies in the genus *Bicyclus*, which display striking seasonal variation, or **polyphenism**.

Polyphenism: *In contrast to continuous reaction norms, distinctly different, discrete phenotypes expressed by the same genotype in reaction to an environmental signal.*

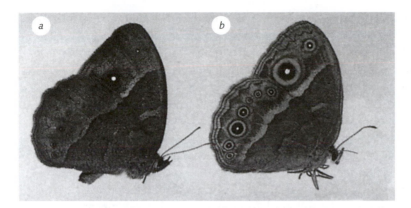

Figure 7.11 (a) Dry- and (b) wet-season forms of the African butterfly *Bicyclus anyana*, a polyphenism in the field that is a reaction norm when tested across a set of intermediate conditions in the laboratory. (Courtesy of Paul Brakefield.)

Bicyclus anynana, for example, has a wet-season form with large, striking, yellow-black eyespots, and a dry-season form with greatly reduced eyespots (Figure 7.11). This is a remarkably useful system in which to analyze the causation of wing patterns with molecular genetics, quantitative genetics, transplant experiments in developing pupae, and field trials of the survival of manipulated phenotypes to test hypotheses about adaptation. Briefly, here is what is known so far.

First, natural selection favors cryptic forms with small eyespots to blend into a background of dry leaves in the dry season and striking forms with large eyespots to deflect predator attacks in the wet season (Brakefield and Larsen 1984). Second, the wet- and dry-season forms are not discrete phenotypes; they represent the end points of continuous reaction norms that react to the temperature of the environment during development (Windig 1993). The discrete appearance of the wet- and dry-season forms is thus caused by the difference of temperatures in the field. Seasonal variation in this butterfly's eyespots is adaptive and is a continuous reaction norm. Third, the size of the eyespots responds to artificial selection; when one eyespot is selected, others respond as well; and selection of eyespots on the ventral surface (the more striking surface) does not produce very strong correlated responses on the dorsal surface (Holloway et al. 1993). This suggests that the developmental regulation of eyespot production extends to all the eyespots on a wing surface, resulting in an eyespot pattern that behaves as an integrated unit; that there is partially independent control over the development of the ventral and dorsal wing surfaces; and that there are many genes that modify the intensity of the basic pattern.

Fourth, the developmental foci that produce eyespots can be grafted into new positions on the wings of the same or different individuals. Such grafts have been done both from lines selected for large eyespots into lines selected for small eyespots, and in the other direction. The grafts make clear that both the eyespot foci and the cells surrounding the eyespot respond to selection for eyespot size. When an eyespot focus from a line selected for large eyespots is grafted into a

Cryptic: *Disguised to look like the background.*

line selected for either large or small eyespots, it produces an eyespot with more than twice the area than does a similarly grafted focus from a line selected for small eyespots. The cells around the eyespot focus also influence the size of the eyespots, for either type of focus produces an eyespot in a line selected for large eyespots that is at least 50% larger than it does in a line selected for small eyespots. Thus in the line selected for large eyespots, the cells surrounding the focus cause it to grow bigger, whereas the same focus transplanted into a line selected for small eyespots will not grow as large because the surrounding cells are either not having this influence or are causing the eyespot to remain small. Thus the genetic changes involved in a response to selection for larger or smaller eyespots influence both the characteristics of the eyespot focus and the characteristics of the surrounding cells (Monteiro et al. 1994).

Does one set of genes lay down the pattern and another set of genes fine-tune its interactions with the environment?

This system appears to give a straightforward answer to the question: how does plasticity interact with conserved developmental patterns? The shared ancestors of butterflies and flies had developmental control genes that shaped appendage structure. In the lineage leading to flies they evolved to control the development of legs and antennae; in the lineage leading to butterflies they evolved to control the development of wing spots. In both lineages the places in which the developmental control genes are now expressed are constrained. In butterflies they can only induce the production of spots in certain parts of the wing and not in others. Once, however, it is determined that an eyespot will develop in a certain place, its size and color can be altered during the development of a single individual by temperature. Thus some temperature-transduction system had to interact with the developing eyespot. This yields a picture of a phylogenetically conserved developmental pattern expressed early in development that lays down structures that are then subject to fine-tuning by plasticity later in development (Figure 7.12).

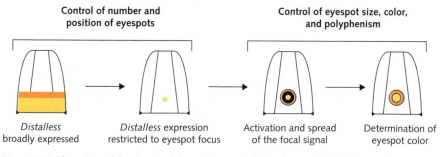

Control of number and position of eyespots		Control of eyespot size, color, and polyphenism	
Distalless broadly expressed	*Distalless* expression restricted to eyespot focus	Activation and spread of the focal signal	Determination of eyespot color

Figure 7.12 The action of developmental control genes in butterfly eyespot development. Broad early expression of *distalless* in the distal part of the wing narrows to the eyespot focus with involvement of other genes that assist in the control of eyespot size, color, and polyphenism. (Courtesy of Jeffrey Marcus.)

One could imagine that there were developmental control genes that produced the conserved structure and plasticity genes that adjusted the interactions with the environment. The developmental control genes created a vase, the eyespot, that held a bundle of reaction norms for size and color like a spray of flowers, representing the phenotypic plasticity of the butterfly population in seasonal Africa. Just as a vase holds a bouquet of flowers on a table, so might the phylogenetically conserved developmental control over the eyespot hold the reaction norms in position on the wing.

The causes of developmental patterns lie in systems of interactions among gene products and cells, not in artificial categories of gene types

This analogy may not be the best way to think about the system. It is at least incomplete and may mislead through the assumptions it makes about the nature of genetic causation (Sultan and Stearns 2005). Consider the following more recent experimental results.

- Artificial selection can uncouple the size of the anterior and posterior eyespots, essentially removing one while retaining the other (Beldade et al. 2002b). Thus it is easy to eliminate one eyespot module while retaining another.

- One of the genes responding to selection for eyespot size is *distalless* itself (Beldade et al. 2002a). Thus the gene thought to represent a deeply conserved phylogenetic constraint itself responds rapidly to microevolutionary selection pressures with allelic change.

- In *B. anynana* there are normally seven eyespots. X-ray mutagenesis yields mutants that remove some eyespots but not others. These results suggest that the formation of eyespot foci in each wing cell is controlled by focus-regulator genes that are under the control of regional regulator genes. Mutations to the local regulators have local effects; mutations to the regional regulators have regional effects. Some mutants yield wing patterns not found in any existing species of *Bicyclus*, including some that were thought to be forbidden by the nymphalid ground plan (Monteiro et al. 2003).

> **Mutagenesis:** *The production of mutations in an experimental population using either chemicals or radiation.*

Thus not all conceivable changes in butterfly wing patterns can be selected; nor does mutagenesis produce all conceivable variants. The concept of a basic butterfly wing pattern remains intact, as does the idea that an eyespot is a module. But many more modifications of the basic pattern can be selected than had at first been thought. And questions are raised about how one should think about genetic determination, for the same gene is involved in both macroevolutionary patterns and microevolutionary change. There may well be a cascade of regional and local effects, but they are effects felt in a network of hundreds of genes, and some of the genes thought to be triggering the cascade, such as *distalless*, are themselves part of the fine-tuning of the module. The causes of the nymphalid ground plan are thus not readily assigned to different categories of genes, some of them control genes, others structural genes. They are rather to be sought in the system of interactions that arises among gene products and cells in particular geometric configurations. Developmental control genes are not hands-off managers that initiate development and then leave the scene; they are micromanagers that remain involved in the details of the process until it is complete (Akam 1998).

Plasticity and constraints are determined in part by the same genes

The example of the butterfly wing shows that a phylogenetically fixed developmental mechanism widespread in butterflies is controlled by identified genes of known effect. That suggests the notion of a phylogenetic constraint holding a bundle of reaction norms like an old and durable vase holding an ephemeral bouquet of flowers. In some general sense that may be an appropriate analogy, but it may mislead us as much as it guides us. We may think we see plasticity fine-tuning the phenotype within a long-established framework of phylogenetic and developmental constraint, when in fact one network of interactions may be causing the entire pattern, interactions of effects that cannot be cleanly assigned on the one hand to phylogeny and constraint and on the other hand to plasticity, interactions that produce the appearance of constraint in one part and the appearance of plasticity in another part of the same system, both as byproducts. And as we have seen, some things that appear to be constraints in this system can be modified by selection and mutation.

Next we consider an example where the causal chain that produces the phenotypic response has been studied in detail in a plant.

Adaptive plasticity in leaf development is mediated by phytochromes

● **KEY CONCEPT**

The molecular mechanisms underlying phenotypic plasticity are particularly well understood for plant phytochromes.

Phytochromes: *Light-detecting molecules that function as the 'eyes' of plants. The information they receive influences germination and growth.*

The phytochromes, light-detecting molecules found in most plants, are an important link in the chain of mechanisms that produce adaptive responses to light conditions, including seasonal timing, germination, and shade avoidance (Smith 1995). They can be used to monitor the intensity, quality, direction, and duration of light and to modify development appropriately.

There are several **phytochromes**, labeled A, B, and so on, each of which can exist in two states. For example, phytochrome B is tuned to respond to red and to far-red wavelengths. Its two forms are called Pr (for phytochrome-red) and Pfr (for phytochrome-far-red). When the Pr form is exposed to red light (600–700 nm), it switches to the Pfr form, and when the Pfr form is exposed to far-red light (700–800 nm), it switches to the Pr form. Because there are many copies of these molecules per cell, the ratio of the number of molecules in each state measures the red/far-red ratio (R:FR ratio) in the environment. That ratio is informative, for chlorophyll preferentially absorbs red light. Beneath a leaf, or beneath a forest canopy, where red light is scarce and far-red light is relatively abundant, the R:FR ratio is low. It is also low in light reflected off a leaf. In the open—for an isolated plant, or a plant growing in a gap in the forest—the R:FR ratio is high. Thus the R:FR ratio provides information on shading and on the nearby presence of other plants competing for light. In this sense, the phytochromes are the 'eyes' of plants.

Phytochrome B helps to mediate the germination response. Imagine you are the seed of a forest tree lying in the soil, waiting to germinate. You ask yourself, am I in a light gap in which I might survive to adulthood if I germinated now, or should I wait here in the soil until a nearby tree falls and opens up some space? For an answer, you can monitor the R:FR ratio in your tissues and, when it passes a certain threshold, start germinating.

Phytochromes allow plants to respond adaptively to encroachment by competitors and alter morphology from sun to shade

Phytochrome B also helps plants to detect light reflected from neighbors (Schmitt and Wulff 1993). Imagine that you are a plant growing in full sunlight, and one day you start to detect light reflected off a rapidly growing neighbor. If you can use that information to change the direction of your growth before you are directly shaded, you will respond much more effectively, for it takes time to grow, and it is harder to grow out from under a neighbor that is shading you directly than it is to grow away from an encroaching neighbor while you are still in full sunlight. The phytochrome system enables plants to do this.

Plants can also use the phytochrome system to change their morphology from sun-adapted to shade-adapted forms. For example, the ribwort plantain (*Plantago lanceolata*) forms heavier rosettes with more, larger, broader leaves in the shade than it does in the sun. The response appears to be mediated by several hormones, each of which mimics some elements of the phytochrome response (van Hinsberg 1997).

The phytochrome system reveals surprising sensory capacities in plants. The different phytochromes have discrete properties, are differentially expressed, and mediate the perception of different light signals. In *Arabidopsis*, for example, mutations deficient in four of the five phytochrome genes have been isolated; their physiological functions overlap considerably, but each appears to trigger a different downstream signalling pathway (Whitelam et al. 1998). Such insights into the molecular and physiological genetics of phytochromes show how the phytochrome system can, like the butterfly wing, integrate genetics, biochemistry, and physiology with the ecology and evolution of induced responses. Here the mechanisms underpinning adaptive plasticity can also be laid bare.

● **SUMMARY**

This and the previous chapter summarize how to think about the developmental connections between genotype and phenotype in an evolutionary context. The critical question is this: how does development structure the phenotypic variation that causes natural selection, and how does it filter the genetic variation that determines the response to selection?

- Butterfly wing patterns and discontinuous growth in arthropods exemplify structured phenotypic variation and filtered genetic variation. They are controlled by lineage-specific developmental mechanisms.

- Induced responses, many of which are adaptive, illustrate the capacity of a single genotype to produce several phenotypes in response to a specific environmental signal and thereby improve reproductive success.

- Reaction norms help us to think clearly about connections between genotype and phenotype. A reaction norm, a property of a genotype, describes how development maps the genotype into the phenotype as a function of the environment. A population of genotypes is a bundle of reaction norms.

- The sensitivity of a trait to change as a function of an environmental factor—the slope of its reaction norm—measures its phenotypic plasticity. Natural selection can shape phenotypic plasticity when there is genetic variation for the slopes of reaction norms. Not all plasticity is adaptive. By comparing an ancestral with a derived population, one can decide how much of the plasticity is adaptive by performing a balanced, common garden experiment in two or more environments.

- Because both the heritabilities of single traits and the genetic correlations of pairs of traits can change with the environment in which they are expressed, the potential response to selection of a set of correlated traits depends on the environment and can be visualized by plotting reaction norms.

- Organisms that are similar genetically can differ phenotypically, and organisms that are similar phenotypically can differ genetically. Genotype and phenotype are partially uncoupled, and only loosely connected. Because some genes are not always expressed, and because a few developmental control genes have large effects, there is huge variation from gene to gene in the impact of genetic changes on phenotypes.

- Seasonal polyphenism in butterfly wings and the phytochromes in plants both exemplify systems in which the molecular genetic mechanisms underlying adaptive plasticity can be studied in detail. Here specialties are integrated—evolution, development, genetics, and physiology—that traditionally have been separated.

- The production of structures and the fine-tuning of those structures to environmental conditions are not done by separate sets of genes, for some of the same genes are involved in both functions.

● RECOMMENDED READING

Schlichting, C.D. and Pigliucci, M. (1998) *Phenotypic evolution. A reaction norm perspective*. Sinauer Associates, Sunderland, MA.

www.acsu.buffalo.edu/~monteiro/ancientwings/ancient-wings.swf

● QUESTIONS

7.1. In Figure 7.7(b) replace fecundity by IQ on the *y* axis, population density by per-capita income on the *x* axis, and the parthenogenetic lizards by three human mothers who bear identical triplets separated at birth and adopted into different families, with one of the triplets raised at each of the income levels. Does the question, is intelligence determined by genes *or* environment?, have any meaning? Is it clearly formulated? Does it make appropriate assumptions?

7.2. If organisms with different body plans—such as arthropods and vertebrates—share orthologous developmental control genes, how might developmental mechanisms cause body plans to be shared within each of those large groups but not between them?

7.3. Discuss the roles of external and internal factors in evolution in the light of this chapter. Are both necessary for adaptive or neutral change? Is either sufficient for either type of change?

7.4. Use the concept of genotype-specific reaction norms to critique the notion that some plastic phenotypic trait is caused either by nature (the genes) or by nurture (the environment).

PART 2

Design by selection for reproductive success

Part 1 of this book dealt with the basic mechanisms of microevolution: selection, drift, inheritance, and development. Selection acts on material organisms. The strength of selection on a trait is determined by the correlation between variation in the trait and variation in reproductive success. If the trait is heritable, the response to selection is recorded in genetic information. Development transforms genetic information into material organisms in interaction with the environment. It has two key features: conserved mechanisms that lay down the major features of body plans, and plasticity determined by interactions of genes with environment.

We now consider how natural selection has designed organisms for reproductive success. Consider an organism that is about to start life and imagine its options. Should it reproduce sexually or asexually? What conflicts does it face with other evolving genomes? At what age and size should it start to reproduce? How many times in its life should it attempt reproduction? Should its offspring be few in number but high in quality and large in size, or should they be small and numerous but less likely to survive? Should it concentrate its reproduction early in life, neglect maintenance, and have a short life as a consequence, or should it have fewer offspring, invest more in maintenance, and live longer? Should it be born as a male or as a female? If it is a sequential hermaphrodite, at what age and size should it change sex? If it belongs to the limiting sex—usually the females—how should it choose among its suitors? Should it choose a mate that controls resources and would be a good parent or a mate that advertises genes for pathogen resistance or sexual attractiveness?

When these questions have been answered, many key features of the organism have been set. How and why those questions get answered in particular ways is the subject of Part 2.

CHAPTER 8

The evolution of sex

Genes that increase reproductive success will spread through a population, whereas genes that lower reproductive success will disappear. Thus most—if not all—traits have proven their usefulness for reproductive success and have been shaped by natural selection to contribute directly or indirectly to reproduction. Viewed this way, reproduction is the center of biology, and everything else—development, physiology, behavior, and genetics—is at its service.

To be sexual or asexual—that is the question

In this chapter we examine the major feature of genetic systems: whether organisms are sexual or asexual. Sex often, but not always, coincides with reproduction, and there are many types of reproductive system. We begin with reproduction that does not involve sex. Single-celled organisms like bacteria or yeasts reproduce by division into two daughter cells (Figure 8.1a); many multicellular algae and fungi

● **KEY CONCEPT**

Sex is complex, inefficient, and costly—but very common, and thus a paradox that evolutionary biologists are trying to resolve.

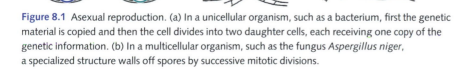

Figure 8.1 Asexual reproduction. (a) In a unicellular organism, such as a bacterium, first the genetic material is copied and then the cell divides into two daughter cells, each receiving one copy of the genetic information. (b) In a multicellular organism, such as the fungus *Aspergillus niger*, a specialized structure walls off spores by successive mitotic divisions.

bud off single cells (spores) that develop into new multicellular individuals (Figure 8.1b).

Both exemplify asexual reproduction, the kind of efficient reproduction that an engineer might design. It is basically a copying process: copy the genetic instructions on how to build an organism and package them in a suitable carrier from which the new organism can start to grow.

Sex produces genetically diverse offspring; asex does not

Sex contrasts sharply with asex, for in sexual reproduction the genetic instructions from two lines of descent fuse to create a new individual (Figure 8.2). This new individual contains genetic information from both parents. Because the parents will mostly be genetically different, the offspring will be different from either parent. This is a fundamental difference from asexual reproduction, where the offspring is identical to the parent.

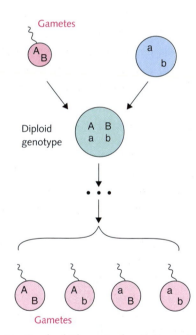

Figure 8.2 Sexual reproduction. Two haploid gametes fuse to form a diploid zygote. In multicellular organisms the zygote develops into an adult which then produces gametes by meiosis. In many unicellular organisms where the zygote directly undergoes meiosis, the meiotic products are vegetative offspring, which may later form gametes by mitotic divisions. At meiosis two important processes occur, recombination and segregation (for details see the Genetic appendix). Recombination ensures that the gametes an individual produces contain genetic information derived from both its parents. Segregation ensures that every gamete contains a complete haploid set of chromosomes and that the paternally and maternally derived alleles are equally represented among the gametes.

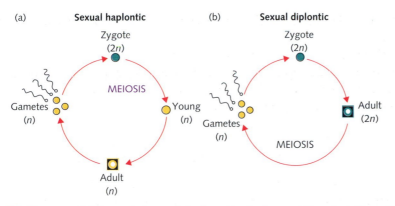

Figure 8.3 The essential difference between diplontic and haplontic sexual life cycles is the position in the cycle where meiosis occurs. If zygotes directly undergo meiosis, the resulting cells function as haploid offspring that produce gametes by mitosis. If zygotes do not undergo meiosis, they develop into diploid offspring that produce gametes by meiosis.

Sex occurs in two types of life cycle: diplontic and haplontic

There are two basic types of sexual life cycle, as discussed in Chapter 4 (pp. 71 ff.). For clarity we repeat part of Figure 4.1 in Figure 8.3. In a **diplontic life cycle** the adults are diploid and the haploid gametes are produced by meiosis. In a **haplontic life cycle** haploid adults produce gametes by mitosis and the newly formed diploid zygotes immediately undergo meiosis to produce haploid individuals.

Diplontic life cycle: *A life cycle in which diploid somatic adults produce haploid gametes by meiosis that fuse to form diploid zygotes that develop into somatic adults.*

Sex is complex, inefficient, and costly

No sensible engineer would ever propose such a process as sex when asked to design a reproduction machine. In fact, sex probably did not evolve for reproduction. The intimate association of reproduction and sex in many organisms should not distract us, for it is a derived state, the product of long evolution after the origin of sex. In its essence sex—where two genomes merge, recombine, and segregate—has nothing to do with reproduction, for it does not increase the number of individuals.

Sex is less efficient and more complicated than asex, for it takes longer than mitotic division and involves more complicated mechanisms. Several difficulties with sex are discussed below; we mention here an obvious one—logistics. An asexual organism can reproduce by itself, but a sexual organism must find a partner to mate. This can be a big problem, especially for sessile organisms and for small, short-lived organisms that live at low population densities. Many elaborate adaptations have evolved to help mating organisms find each other, including the precisely timed swarming of palolo worms (Chapter 2, p. 34) and the sex pheromones produced by female butterflies and detected with extraordinary sensitivity by males. Other organisms can even be used to unite the gametes of sessile organisms, as in the many plants that depend on insects for pollination.

Haplontic life cycle: *A life cycle in which haploid adults produce haploid gametes by mitosis; the diploid zygotes immediately undergo meiosis to produce haploid individuals.*

So why is there sex?

This incomplete list of complex, inefficient, and costly aspects of sex explains why evolutionary biologists are fascinated by the widespread occurrence of sexual reproduction. Despite its costs sex is very common, especially among large multicellular organisms, and it affects their biology deeply. In a world without sex there would be no males and females with their many differences, no showy flowers and no insects, birds, or bats specialized in pollinating or feeding on them, no extravagant color and form like the peacock's tail, and no behaviors aiming at finding, selecting, and defending mates. The great riddle of the evolution of sex is why sex should be better than asex. Something so striking and prevalent with such large effects on almost every aspect of biology must surely have an obvious adaptive explanation, but so far we have had a very hard time finding it. As Bell (1982) remarked, 'Sex is the queen of problems in evolutionary biology.'

Variation in sexual life cycles

> ● **KEY CONCEPT**
>
> Sex and reproduction occur in many forms, both alone and in combination.

Most eukaryotic species share the basic features of sex, but they differ greatly in the extent to which reproduction depends on it and in how frequently it occurs. In prokaryotes, processes similar to eukaryotic sex also result in genetic recombination. To illustrate these differences we describe sex in several organisms.

Bacterial conjugation transfers DNA from donor to recombinant recipient

> **Conjugation:** *A form of bacterial sex involving DNA transfer via a plasmid from a donor to a recipient cell, which gets a recombinant genome. The cells do not reproduce.*

The bacterium *Escherichia coli* reproduces by asexual division (Figure 8.1a). However, in a process called conjugation (Figure 8.4) DNA from a donor cell is transferred into a recipient cell that ends up with a recombinant genome. To that extent bacterial conjugation is analogous to eukaryotic sex, in that the

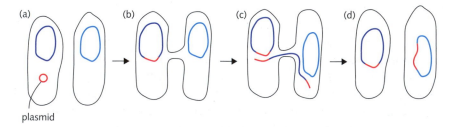

plasmid

Figure 8.4 Bacterial conjugation. (a) A small circular DNA molecule called a plasmid (in *E. coli* the so-called F factor) integrates into a copy of the bacterial chromosome. (b) During cell-to-cell contact with a bacterium lacking the F factor, the plasmid transfers part or all of this bacterial chromosome into the recipient cell (c); (d) The transferred segment recombines with homologous parts of the recipient cell's chromosome.

daughter cells differ genetically from the parents. However, in bacteria the DNA that recombines almost never consists of two complete single genomes. Instead, an intact genome makes contact with small 'foreign' DNA fragments that recombine at homologous places by crossing over, breakage, and reunion (see Figure 8.4). As a result, the recipient cell ends up containing a recombined chromosome, while the remaining part of the transferred DNA is lost during subsequent cell growth. Thus in bacterial sex, unlike meiosis, there is no segregation and no reproduction. Donor cells that can transfer part of the bacterial chromosome into a recipient cell are rare under natural conditions.

In an alga mating only occurs between different mating types

The unicellular green alga *Chlamydomonas eugametos* lives in small freshwater ponds and in wet soils. Asexual reproduction occurs by fission under conditions favoring growth. Under other conditions, including nitrogen-limitation, vegetative cells may turn into gametes. The gametes are not morphologically differentiated into eggs and sperm, but there are two genetically determined mating types, called + and −. All cells of a clone have the same mating type, and sexual fusion to yield zygotes is only possible between gametes of different mating type. Thus no sex is possible in a local population that derives clonally from a single cell. Zygotes develop a thick protective wall, and germination of zygotes requires special conditions not easily achieved in the lab. *Chlamydomonas* has a haplontic life cycle; when the zygote germinates, it undergoes meiosis and cell division, producing two recombinant haploid offspring.

> **Mating types:** *Sets of potential mating partners. Matings can occur between partners of different type but not with partners of the same type.*

In a fungus an individual can form sexual spores without recombination

The ascomycete fungus *Aspergillus nidulans* lives on organic matter in dry soils and reproduces by forming either asexual conidiospores or sexual ascospores. Ascospores differ from conidiospores in having a thick protective wall that enables them to survive temporarily unfavorable conditions. The fungal soma consists of hyphae—long, multi-nucleate, cylindrical cells that grow at one end. By branching and intra-individual fusion the hyphae form a network called a mycelium. A mycelium contains millions of haploid nuclei, which are—except for mutations—all identical. The species is self-fertile: a single individual can form sexual spores. From a genetic point of view selfing in a haploid organism differs essentially from selfing in diploids, for the fusion of identical haploid nuclei and subsequent meiotic segregation produce ascospores that are genetically identical to the parental nuclei and to the asexually produced conidiospores. Thus sex in this self-fertile haploid organism does not involve genetic recombination. In contrast, selfing in diploids, as in many plants, usually involves recombination, for fusion can occur between genetically different gametes (provided the plants are not completely homozygous). When different

A. *nidulans* individuals are grown together, some outcrossing occurs and results in recombinant ascospores, but how often such outcrossing occurs under natural conditions is not known.

In dandelions asexual clones arise in sexual populations repeatedly but infrequently

Taraxacum officinale, the common dandelion, is geographically widespread in Europe, occurring from the Mediterranean to the Arctic. Near the Mediterranean most plants are diploid and sexual, whereas asexual triploid forms predominate to the north. Only asexual forms occur in North America. Like the sexual ones, the asexual plants produce bright yellow flowers, but reproduction, by a process called **apomixis**, produces seeds with genes identical to those in the parent. Asexual plants are thought to arise repeatedly but infrequently in a sexual population, producing asexual lineages (clones).

Apomixis: *Asexual formation of seeds in plants without a genetic contribution from a male gamete. Apomictic seeds are genetically identical to the mother plant.*

Thale cress (*Arabidopsis*) reproduces mostly via self-fertilized sexual seed formation

Arabidopsis thaliana, thale cress, is a tiny annual flowering plant much used in genetic studies. Unlike many other flowering plants that have the option of both sexual reproduction and asexual propagation by vegetative runners, it reproduces exclusively via sexual seeds. Most seeds are produced by self-fertilization; the offspring are thus less genetically diverse than would be offspring produced exclusively by outcrossing.

Aphids are viviparous and cyclically parthenogenetic

Cryptomyzus is a genus of aphids, a group of plant-sucking insects that cause much damage in agriculture. Aphids, in contrast to most other insects, are viviparous: instead of producing eggs females give birth to small larval aphids. In the ovarioles of these larvae development of new offspring is already taking place; the daughter starts reproduction before her mother has finished giving birth to her. Their life cycle is also marked by **cyclical parthenogenesis**. From spring to fall females, without being fertilized, produce daughters that are genetically identical to their mother. In the autumn, when males and sexual females are produced, mating and sexual reproduction take place, and females lay eggs that can survive the winter.

Cyclical parthenogenesis: *A life cycle typical of aphids, rotifers, cladocerans and some beetles in which a series of asexual generations is interrupted by a sexual generation. The offspring of the sexual generation are often adapted to resist extreme conditions and to disperse.*

Mammals are exclusively sexual because of genetic imprinting

Homo sapiens is a mammal with a worldwide distribution and, like all mammals, only reproduces sexually. Asexual reproduction does not seem to be possible in

H. sapiens, or in any other mammal, because of genetic imprinting. The expression of some genes important during early embryonic development depends on whether they spent the previous generation in a male or in a female germ line. At critical stages of development only the allele that came from the mother is expressed and the paternal allele is inactivated, while at other stages the paternal allele is active and the maternal allele is silent. Imprinting is usually removed after early development in the offspring. A mutation for parthenogenesis, causing the formation in a female of a diploid egg that could start development on its own, would not survive, for development would cease as soon as an essential paternal gene was required to carry out an essential function.

Evolutionary aspects of genomic imprinting are further discussed in Chapter 9 (pp. 210 ff.).

> **Genetic imprinting:** *The silencing of certain genes in the germ lines of parents prior to gamete production, accomplished by methylating the DNA sequence.*

Distribution patterns of sexual reproduction

Comparative studies have yielded several generalizations about the phylogenetic and ecological distributions of sex and asex (e.g. Glesener and Tilman 1978, Bell 1982, and Bierzychudek 1987).

> ● **KEY CONCEPT**
> Asex is roughly correlated with simple, disturbed habitats with extreme environmental conditions. Sex is roughly correlated with biotically complex and undisturbed environments.

Exclusively asexual species often have a relatively recent evolutionary origin

Many organisms can reproduce both asexually and sexually: only rarely is sex wholly absent. Purely asexual species often originated relatively recently; they appear to be the short-lived offshoots of sexual ancestors (Bell 1982, Barraclough et al. 2003). There are a few striking exceptions, the so-called 'scandalously ancient asexuals' (Judson and Normark 1996). The most famous scandalous asexuals are the bdelloid rotifers, in which males have never been found, nor any evidence of meiosis, and for which molecular evidence suggests that they have been asexual for millions of years (Welch and Meselson 2000). Exclusively sexual reproduction also occurs in a few taxa, most notably mammals. Among the majority with both options, asexual reproduction predominates among small organisms with very large populations, whereas sex predominates in large organisms with smaller populations.

Asexuals are more frequent at high latitudes and in disturbed habitats and tend to have broad distributions

In the many species with both options, and therefore within which comparisons are well controlled for other factors, asexual lineages tend to be found at higher latitudes and in more disturbed habitats than sexuals. In general, asexuals have a wider distribution than the sexuals, and sexuals are found in habitats where selection appears to be driven more by interactions with other organisms than by interactions with the physical environment.

Sex has important consequences

Recombination: it occurs both within and among chromosomes

● KEY CONCEPT

Sex produces offspring that differ genetically from their parents.

In eukaryotes sex combines two different (normally haploid) single genomes into a double (normally diploid) genome, which then reduces again to single genomes (Figure 8.3). Meiosis regulates the genetic reduction from diploid to haploid cells. It differs from mitosis in allowing crossing-over and in halving the number of chromosomes per cell. In the diploid nucleus of a cell undergoing meiosis the chromosomes are arranged in homologous pairs whose two chromosomes are placed into each of the two daughter cells. Because of meiotic segregation, the haploid cell products of meiosis each contain a complete set of genetic instructions. In most eukaryotes, for meiotic segregation to be accurate, chromosomal crossing over cannot be avoided. During chromosomal crossing-over non-sister chromatids of homologous chromosomes come into physical contact at homologous sites and exchange chromosome parts by breaking and rejoining (Figure 8.5).

Since the two homologous chromosomes involved in crossing-over often have different alleles at many loci, crossing-over produces chromosomes containing alleles from both parents of the individual in which the meiosis occurs. Crossing-over causes *intra-chromosomal* recombination, but is not the only reason for recombination. The independent segregation of chromosomes causes *inter-chromosomal* recombination of genes on different chromosomes (see the Genetic appendix). Recombination is thought to explain the evolutionary success of sexual reproduction, for reasons to be discussed below on pp. 188–91. Because recombination is the key feature of sex, processes causing recombination in prokaryotes like bacterial conjugation or transformation (Figure 8.4) are also called 'sexual'.

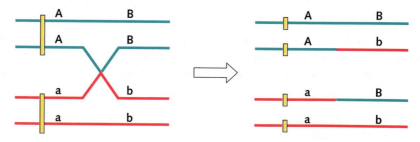

Figure 8.5 Crossing-over of non-sister chromatids during meiosis, resulting in exchange of homologous chromosome parts by breakage and reunion, can produce chromosomes with a new allelic composition. Recombination produces new genetic variants.

Anisogamy: isogamous sex preceded anisogamous sex

Recall that eukaryotic sex is essentially the fusion of two haploid nuclei to form one diploid nucleus, which sooner or later produces haploid nuclei by meiosis. There is nothing inherently asymmetric about this process. It is quite conceivable that sexual fusion could occur between similar haploid cells that differ only in their genotype. However, with few exceptions, sexual fusions are asymmetric, where two cases can be distinguished. The most common situation is anisogamy, in which the fusing gametes have different sizes and shapes. The big, scarce gametes are female (egg cells or ovules), and the small, numerous gametes are male (sperm cells or pollen). The less-common situation is isogamy, where the fusing gametes do not differ in shape and show no male/female distinction, but differ in some other respects. Isogamy occurs in ciliates, unicellular algae, and many fungi. It was the ancestral state; anisogamy is the derived state. Therefore we need to understand how anisogamy evolved from isogamy; how sexual species with gametes of two different sizes evolved from sexual species in which all gametes were the same size.

In many of the isogamous species, the two gametes that form a zygote, although of equal size, do differ in behavioral or in physiological details, each playing a different role in the fusion process. Thus isogamous species can have two or more different gamete types. Because the words male and female or sperm and eggs are not applicable here, these are called mating types and are designated as + and −, or as A and a. Organisms with more than two mating types include numerous mushroom species, which have many mating types; in these species mating is only possible between different types.

The models that have been studied to understand what might select for anisogamy assume that individuals may either produce a few big gametes or many small ones. More precisely, they assume that the product of gamete size multiplied by gamete number is the same for every individual. When larger zygotes are much fitter than smaller ones, gamete specialization becomes advantageous: one type (the egg) specializes in provisioning the offspring during its early development, and the other type (the sperm)—which does not need to invest in provisioning because its partner has already done so—specializes in finding the egg (Parker et al. 1972, Hoekstra 1987).

The explanation of the evolution of mating types from symmetric gamete fusion is more complicated and is further discussed in Chapter 9 (pp. 208 f.). Once mating types exist, there will often be selection to increase their number, for the greater the number of mating types present in the population, the more likely it is than an individual encountered will be of different mating type and hence a potential mate.

Sexual selection: because eggs are 'expensive' and sperm is 'cheap,' females became choosy

In anisogamy the large female and small male gamete differ in 'value.' A female gamete represents a large investment of resources, the male gamete a small

● **KEY CONCEPT**
Some sexual organisms have evolved gametes of different sizes; they are *anisogamous*.

Anisogamy: *Having gametes of different sizes; large eggs and small sperm.*

Isogamy: *Mating partners have gametes of the same size.*

Ancestral state: *A state characteristic of an ancestor shared with related groups.*

Derived state: *A state that evolved after the ancestral state in this lineage.*

Sexual selection: *The component of natural selection that is associated with success in mating.*

investment. This difference in investment drives sexual selection, which involves competition for and choice of mates. Because a female gamete is expensive and a male gamete is cheap, one expects females to be choosy, selecting sperm of high quality for fertilization, and males to compete with each other for access to females. Anisogamy is thus a necessary prior condition for sexual selection, which has then caused the evolution of many differences between males and females (Chapter 11, pp. 251 ff.).

Because all isogamous gametes require the same investment, no further differences have evolved between isogamous individuals of different mating types, for in these species sexual selection is non-existent or very weak.

Sex allocation: how much effort to allocate to each sex?

Another indirect consequence of sex follows from the male/female difference. It is caused by natural selection acting on the distribution of the male and female functions among individuals, on the allocation of reproductive effort to male versus female offspring, and on the relative frequency of males and females. These and related issues, collectively termed sex allocation, are treated in Chapter 10 (pp. 236 ff.).

Sex allocation: *The allocation of reproductive effort to male versus female function in hermaphrodites and to male versus female offspring in species with separate sexes.*

To summarize, a plausible scenario for the evolution of male/female differences starts with selection for better-provisioned zygotes, leading to anisogamy in which two types of gamete are produced: big, expensive female gametes and small, cheap male gametes. The difference in the cost of the two types of gamete led to sexual selection, and sexual selection (Chapter 11, p. 251) then shaped the often striking differences between male and female organisms. Sex allocation then shaped investment in the two sexual functions.

The evolutionary maintenance of sex is a puzzle

When and why is sex favored over asex?

● **KEY CONCEPT**

Sex has many advantages and disadvantages. Their balance is obscure.

We now turn to one of the most intriguing questions in evolutionary biology: when and why is sexual reproduction favored over asexual reproduction? The question is obvious, and from Darwin onwards biologists have tried repeatedly to answer it. Since about 1970 attempts to reach a conclusive understanding of the significance of sex have intensified, and many books and articles have been devoted to a problem that seems to be very difficult. A definitive answer is still not within reach. It is surprising that it has been so hard to understand the evolution of such a conspicuous and widespread trait.

How should one answer the question in principle? Understanding the maintenance of sex requires us to list and quantify the selective advantages and disadvantages of sex and asex, then judge the balance between the two. That list is now so long that a simplifying classification of the hypotheses has become necessary (Kondrashov 1993). Here we describe some of the main ideas.

The disadvantages of sex

The two-fold reproductive disadvantage

Suppose that in a sexual population a mutant female arises that reproduces asexually instead of sexually, for example by a mutation that suppresses meiosis and allows eggs to develop by mitotic division into offspring genetically identical to the mother. If the asexual female produces the same number of offspring as sexual females, and if all offspring have the same average fitness, the asexual mutation would initially have a two-fold fitness advantage, for asexual females would produce twice as many daughters as sexual females. The proportion of asexual females in the population would rapidly increase, and the sexual females would soon be completely replaced by asexual females (Figure 8.6).

This argument assumes that males contribute only genes to their offspring. Then sexual females 'waste' half their reproductive potential on sons. If males enhanced the fitness of the offspring by providing parental care, the advantage of an asexual mutation would be less, for asexual females would have to provide all parental care themselves.

Sex opens the door to nuclear meiotic drivers and selfish cytoplasmic genes

Strict asexual reproduction implies co-transmission of all genes, nuclear and cytoplasmic. A mutation that lowers fitness will not spread, for its fate is completely linked to that of the organism and its other genes. Sexual reproduction implies that chromosomes and alleles segregate and recombine in every

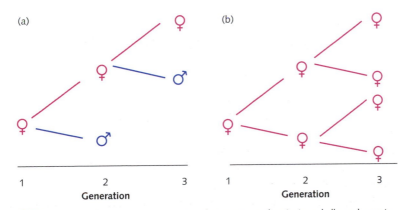

Figure 8.6 The two-fold cost of sex. A mutation that suppressed meiosis and allowed eggs to develop without fertilization would enjoy a two-fold increase relative to its standard sexual allele, if sexual and asexual females produce on average the same number of (equally viable) offspring. Here we assume that (a) a sexual female produces one son and one daughter and that (b) an asexual female produces two daughters.

generation; not all genes are transmitted together to the offspring. This opens the door to the possible spread of mutants that cause unfair transmission at the expense of their non-mutant colleagues. Such mutations are called selfish because they promote their own spread at the expense of alternative alleles or the host organism. They include nuclear meiotic drivers (genes distorting meiosis to their own advantage) and selfish cytoplasmic genes. Several mechanisms are treated in Chapter 9 (pp. 203 ff).

> **Meiotic drive:** *Genes distort meiosis to produce gametes containing themselves more than half the time.*

> **Selfish cytoplasmic gene:** *A gene located in an organelle, plasmid, or intracellular parasite that modifies reproduction to cause its own increase at the expense of the cell or organism that carries it.*

Recombination both creates and destroys favorable gene combinations

Another important effect of sex is that recombination both creates new combinations of alleles and destroys existing combinations. Suppose that in a haploid population the allele combinations *AB* and *ab* have positive effects on fitness, while *Ab* and *aB* have negative effects. For example, the loci A and B might code for enzymes that function in the same metabolic pathway, and the enzymes specified by the different alleles might differ slightly in chemical characteristics. Since the two enzymes have to cooperate, it may happen that some combinations (*AB* and *ab*) function better than other combinations (*Ab* and *aB*). The fitness consequences of recombination then depend on the population composition. If *AB* and *ab* are common and *Ab* and *aB* are rare, then when the alleles are combined independently and mating is at random, recombination destroys more *AB* and *ab* genotypes than it creates, while more *Ab* and *aB* genotypes are created than destroyed. This is easiest to see in the extreme case of a population that initially consists only of *AB* and *ab* genotypes. If the two genes are on different chromosomes, then any mating between *AB* and *ab* will produce up to 50% maladapted *Ab* and *aB* offspring by recombination. Here recombination has an unfavorable effect. On the other hand, when the population initially consists only of *Ab* and *aB*, recombination will have a positive effect. Thus natural selection may favor recombination or not, depending on the distribution of the two-locus genotypes in the population.

The advantages of sex

Recombination accelerates the rate of evolution

In a sexual population, recombination can rapidly bring together in the same genome two (or more) beneficial mutations that originated in different organisms, whereas in an asexual population this genotype can only be created when the second mutation occurs in a genotype already carrying the first mutation (Figure 8.7). In large populations, in which different beneficial mutations occur in different organisms within a short period of time, evolution will often be faster in sexual than in asexual populations. Sexual recombination can also liberate an advantageous mutation from a genome with several deleterious mutations in

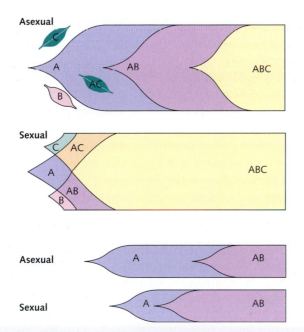

Figure 8.7 Sex can combine beneficial mutations that arise independently in different individuals; with asexual reproduction the mutations must be accumulated sequentially in the descendants of a single mutant individual. This advantage of sex is much greater in large than in small populations, for the probability of two beneficial mutations arising at nearly the same time is much greater in a large population. The upper graph shows clonal interference (Gerrish and Lenski 1998): a single asexual clone carrying a beneficial mutation A comes to dominate the population for some time because of its competitive superiority, thereby preventing the establishment of other beneficial mutations (B and C). (From Crow and Kimura 1965.)

Clonal interference: *The inhibiting effect on the spread of advantageous mutations through an asexual population caused by the presence of clones that lack the mutations but still have a temporary fitness advantage from genes at other loci.*

it. Under asexual reproduction the advantageous mutation would be likely to disappear if deleterious mutations reduced the fitness of its carrier.

Sex can cleanse lineages of mutational damage

Sexual recombination enables selective elimination of deleterious mutations and thus repairs genetic lineages with mutational damage. Muller (1964) pointed out one way in which this can happen. An asexual individual can only produce offspring with a number of mutations that is either the same or higher than its own if reversions (mutations fully restoring the original function) are so rare that we can safely ignore them. Consequently, the number of deleterious mutations in the genome of an asexual lineage will tend to increase over time. At best it could stay constant if the population size is so large that selection maintains genomes in which no new mutations have occurred, but even in large populations mutations with very small deleterious effects will accumulate. Small asexual populations accumulate mutations faster than large populations. This process is called Muller's Ratchet. In a sexual population, in contrast, an optimal genotype lost by drift can be reconstituted by recombination.

Muller's Ratchet: *A mechanism operating in finite asexual populations whose effect is that the number of deleterious mutations can only increase over time.*

Recombination can eliminate several mutations at once

Sexual recombination also contributes to the elimination of deleterious mutations by promoting the elimination of several mutations simultaneously. This principle was also suggested by Muller and has been worked out by several population geneticists (Crow and Kimura 1979, Kondrashov 1982, 1988). The basic idea is that a deleterious mutation can only be removed from the population by failure of the individual carrying the mutation to reproduce (a 'genetic death'). This implies that a high rate of deleterious mutations imposes a heavy burden on a population because many genetic deaths are then required to prevent accumulation of the mutations in the population.

Calculations suggest that if the mutations reduced fitness independently (and were therefore eliminated independently, as in asexual reproduction), severe problems would arise if the deleterious mutation rate per zygote were about 1.0 or higher. In fact, a recent estimate of the rate of deleterious mutation in humans suggests that it is at least 1.6 per zygote (Eyre-Walker and Keightley 1999). So how do we cope with deleterious mutations? Sexual reproduction may help: sexual recombination creates variation in the number of mutations per individual every generation. Each pair of parents produces some zygotes with fewer and some zygotes with more mutations than either of them carries. Theory suggests that if different mutations tend to enhance each other's negative effects, then sexual recombination eliminates deleterious mutations quite effectively, for individuals carrying many mutations have very low fitness. The genetic deaths occur more often among individuals carrying several mutations each, so that the elimination of relatively few individuals removes many mutations from the population.

Sex creates rare genetic combinations that flourish when common types suffer

Almost every organism is host to parasites and pathogens that enhance their fitness by exploiting the host and lowering its fitness. Host–parasite interactions are often frequency-dependent, for parasites adapt to the commonest host type, giving rare host types an advantage. In fact, most interactions between species where the partners affect each other's fitness are likely to result in frequency-dependent selection on the traits that mediate the interaction.

Consider the level of virulence of a pathogen and the level of resistance of its host. There is evidence of genetic variation for both. Natural selection in the pathogen population will favor virulence mutants that have the most reproductive success in interaction with the commonest resistance type in the host population. Similarly, natural selection in the host population will favor resistance types that minimize damage in interaction with the most common virulence type among the pathogens. This is simply because a pathogen most often encounters a common host genotype, and a host most often encounters a common pathogen genotype. Rare resistance types are favored because selection on pathogens for specific virulence against them is weak, and rare virulence variants are favored because selection on hosts for specific resistance against them is weak—until their success makes them common and natural selection against them intensifies.

Thus negative frequency-dependent selection favors rare genotypes in both sexual and asexual species of hosts and pathogens. Such biotic interactions lead to frequent changes of direction of selection. Because a rare genotype is selectively favored, it increases in frequency. As it becomes common, it is selected against. This again causes it to become rarer and thus again selectively favored, and so on. In such a system it is important that a genotype that is temporarily selected against does not completely disappear before it is favored again—selection should not change genotype frequencies too quickly at low frequencies.

Here sex enters the story. When selection regularly changes direction, sexual populations, which recreate genotypes by recombination, retain temporarily bad genotypes longer than asexual populations under most conditions (Jaenike 1978, Hamilton 1980). A simple model of an extreme situation illustrates this for one locus with two alleles. Suppose that selection often changes direction. In one period the heterozygotes A_1A_2 die, but in the next both homozygotes A_1A_1 and A_2A_2 die. An asexual population would go extinct in the second period, but a sexual population could persist because it recreates the genotypes that are now favored but were previously lethal. Similar models can be devised for multi-locus genotypes.

Evidence on the function of sex is scarce but increasing

In contrast to the abundant theory on the maintenance of sex, relatively few empirical tests have been done, for they are difficult. Several types of test are conceivable: examining whether existing distribution patterns of sex and asex obey theoretical predictions, performing experiments in species that have both a sexual and an asexual life cycle to see which conditions promote which mode of reproduction, and testing whether the assumptions of a theory are met.

> ● **KEY CONCEPT**
> Several explanations for sex have been proposed, but as yet no definitive answer is agreed upon.

Distribution patterns of sex and asex can be explained by several hypotheses

Using the observed distribution patterns is problematic, for they cannot discriminate between alternative explanations. It is true that asex is roughly correlated with relatively simple, abiotically extreme and disturbed habitats and that sex is correlated with biotically complex and undisturbed environments. However, this can be explained both by the relative absence of parasites from simple and extreme habitats and by arguing that some correlate of asexual reproduction is responsible for the better performance of asexuals in extreme habitats. For example, the asexual dandelions that predominate at higher latitudes are triploid, sexual dandelions are diploid, and triploid plants can often endure lower temperatures than diploids. Is the distribution caused by sex or ploidy? Who knows?

Sex accelerates evolution in a unicellular alga

Recall (Figure 8.7) that in a large population, clonal interference slows the adaptation of asexual clones, but it has less effect in small populations. Studies on bacteria and viruses have demonstrated that clonal interference indeed does occur in asexual populations (Miralles et al. 1999, Rozen et al. 2002). Colegrave (2002) then tested the idea that sex can increase the rate of adaptation in a large population in which several beneficial mutations can spread at the same time. He allowed 20 populations of the unicellular alga *Chlamydomonas reinhardtii* to evolve for 250 generations in a simple laboratory environment containing a novel food source. He then varied the effective population size by regularly bottlenecking the populations to different degrees. To examine the effect of sex, he took samples from the populations at generation 150 and passed each of these populations through eight rounds of sexual reproduction followed again by asexual reproduction. The relative fitness of the sexual populations was estimated by comparing the growth rate of each sexual population with that of its asexual control (Figure 8.8). Sex had little advantage in small populations, but in large populations it had a marked beneficial effect. Colegrave ascribed this difference to the role of clonal interference.

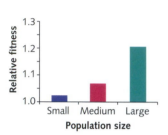

Figure 8.8 The bars show the mean relative fitness of the sexual lines compared to their asexual controls in unicellular algae. The advantage of sex in speed of adaptation is greatest in large populations where the adaptation of asexual populations is slowed by clonal interference. (From Colegrave 2002.)

Populations of New Zealand snails with more parasites have more sex

Lively (1992) tested the idea that host–parasite interactions favor sex. He sampled many populations of *Potamopyrgus antipodarum*, a dioecious freshwater snail native to New Zealand. In this species both diploid sexual and triploid asexual females occur, reminding us of the dandelions discussed earlier. The snails host several species of parasitic worms. Lively found a positive correlation between the number of males per sample (indicative of the frequency of sexual reproduction) and the level of worm infection, as expected if a high level of parasitism favors sexual over asexual reproduction. However, the observed correlation may have other causes. Sexual individuals may be more easily infected than asexual females, for sex may help to transmit the trematodes. Lively argues that these alternative explanations are unlikely, that the simplest explanation of his findings is that asexual clones have displaced sexuals from habitats with low parasite risk, and that sexual individuals can persist in habitats where parasites are common.

Variation in resistance and virulence favor sex in *Daphnia* and *Pasteuria*

Carius et al. (2001) used the freshwater crustacean *Daphnia magna* and its bacterial pathogen *Pasteuria ramosa* to investigate genetic variation for susceptibility among host clones and for infectivity among parasite isolates. The hypothesis that host–parasite interactions may favor sex assumes:

- that there is genetic variation in susceptibility and infectivity, and
- that genotype-specific host–parasite interactions must occur, whereby the resistance and infectivity depend on the genotype of both opponents, thus enabling the operation of negative frequency-dependent selection.

They found precisely the type of genetic host–parasite interactions that would favor sex. No single host clone is more resistant than any other host clone, and no single parasite clone is more infective than any other parasite clone (Table 8.1).

We need more data to evaluate mutational effects

Theory predicts that if genomic mutation rates are higher than one deleterious mutation per genome per generation, sex can be maintained due to its greater efficiency of eliminating deleterious mutations (Kondrashov 1982, 1988). This effect of sex would work particularly well if deleterious mutations enhanced each other's negative effects on fitness. As mentioned in Chapter 5 (pp. 101 ff.), both in *Drosophila* and in humans the genomic deleterious mutation rate is likely to be greater than 1. Several studies have tried to measure whether deleterious mutations enhance each other's effect on fitness, but the results have been inconclusive.

Table 8.1 Variation in host–parasite interactions between nine different isolates of the bacterial parasite *Pasteuria ramosa* and nine different clones of its host *Daphnia magna* within the same population. The figures shown are percentages of hosts that can successfully be infected. (Modified from Carius et al. 2001.)

Pasteuria isolate	Percentage of hosts sucessfully infected									
Daphnia clone . . . 1	3	4	5	7	8	13	14	15	Average
1	78	89	94	11	0	11	17	0	83	42
3	83	89	89	61	11	56	50	16	89	60
4	0	78	61	94	55	0	0	50	33	41
5	0	89	67	94	50	0	0	61	28	43
7	0	78	55	83	39	0	0	39	28	36
8	0	78	0	0	0	56	33	0	0	19
13	33	78	44	11	0	67	44	0	44	36
14	0	67	39	83	22	0	0	16	22	28
15	89	89	100	0	0	0	0	0	100	41
Average	31	81	61	49	20	21	16	20	48	

Some studies found supporting evidence; others failed to show an effect. Thus, although some information on genomic mutation rates and on mutation interaction is available, it is too early to draw firm conclusions. Data from many more species are needed to decide whether deleterious mutations could be a general force maintaining sex.

A pluralistic explanation of sex may be correct

A pluralistic explanation of sex is not attractive but cannot be excluded

The main theoretical ideas about the maintenance of sex do not exclude each other. They should all enter the book-keeping scheme that quantifies the advantages and disadvantages of sex and asex. But how should we quantify them? Are they all valid for all species? If not, can there be one sufficient evolutionary explanation for sex? Should we take the pluralistic view that in many species sex is maintained by a combination of factors, as has been argued by West et al. (1999)? Or is some intermediate between these two extremes the best we can hope for: identifying a few factors that dominate in the maintenance of sex in most organisms, with many special cases of species where some particular factor plays a dominant role? These questions are not easily answered, and evolutionary biologists are still struggling with them. A pluralistic explanation of sex is not attractive, for it would imply that one of the most conspicuous and influential processes in biology does not exist because it has one or two important general functions, but because it affects many processes, with history largely determining which functions are relevant in which species. Nevertheless, the pluralistic view cannot be excluded, and if it is correct, then the search for a general explanation of sex may have taken so long because it is an illusion.

Sexual reproduction often occurs in life cycles with confounding elements

There are certainly good arguments for a form of weak pluralism. For example, in mammals genomic imprinting appears to preclude any success of parthenogenetic mutants. Because this seems to be an absolute constraint, it is meaningless to discuss the potential optimality of sexual or asexual reproduction in mammals. Perhaps natural selection would favor asexual reproduction in humans if it could occur. We are forced to conclude that sex occurs among mammals for historic reasons that fixed genomic imprinting in the lineage, making the asexual alternative impossible, not because natural selection favors sex for some reason.

Another example is provided by organisms such as algae, fungi, and aphids that have both an asexual and a sexual life cycle, in which sex is necessary to produce spores or eggs that can resist frost, drought, or other harsh conditions. This ecological factor must enter the book-keeping for these species, for it may be among the most important factors maintaining sex, but it is hard to see that producing resistant survival structures is an essential general function of sex. It may be another example of a historical contingency: for whatever reason, once established the association between sex and resistance of spores may be difficult to break and prevents the loss of sex.

SUMMARY

This chapter discusses why evolutionary biologists are puzzled by the existence of sexual reproduction and the solutions they have proposed for that problem.

- Asexual copying is an efficient and simple way to reproduce; sex is more complicated, takes more time and energy, and requires finding and selecting a good partner. Despite that fact most organisms are sexual.

- Evolutionary biologists have failed to find an obvious general explanation of why natural selection has produced and maintains sex. Many theories have been suggested. Most concentrate on the effects of genetic recombination at meiosis.

- Recombination may speed up adaptive evolution, may help in purging the genome of deleterious mutations, and may be important in the coevolutionary struggle between hosts and pathogens.

- It is not yet clear if a single hypothesis can provide a sufficient and general explanation of sex. There is some reason to think that it cannot.

RECOMMENDED READING

Hurst, L.D. and Peck, J.R. (1996) Recent advances in understanding of the evolution of sex and maintenance of sex. *Trends in Ecology and Evolution* **11**, 46–52.

Rice, W.R. (2002) Experimental tests of the adaptive significance of sexual recombination. *Nature Reviews Genetics* **3**, 241–51.

West, S.A., Lively, C.M., and Read, A.F. (1999) A pluralist approach to sex and recombination. *Journal of Evolutionary Biology* **12**, 1003–12; plus commentaries in the same issue, pp. 1013–53.

QUESTIONS

8.1. Many organisms that can reproduce both sexually and asexually share a reproductive pattern. When the environment is favorable for growth, reproduction is asexual. When the environment deteriorates, sex occurs.

Sexually produced offspring remain dormant until better conditions arrive. Which of the theoretical ideas about the function of sex is supported by this observation? And which of the disadvantages of sex are less of a problem?

8.2. In many old texts and in some recent popular articles sex is seen as necessary to provide the genetic variability that will keep species from going extinct when the environment changes significantly. What do you think of this explanation?

8.3. Williams argued that in species with both sexual and asexual reproduction a long-term advantage of sex to the species would not explain its maintenance, and sexually derived offspring must have an immediate advantage over asexually produced offspring. Do you agree with his argument? Give reasons.

8.4. Sheep and mice have been cloned from adult somatic cells. Does this invalidate the idea that there can be no parthenogenetic mammals because genes essential for early development are imprinted differently in the male and female germ lines? What evidence would you need to decide?

CHAPTER 9
Genomic conflict

In previous chapters adaptive evolution was described as the product of the differential reproductive success of organisms that differ genetically. Some organisms contribute more offspring to the next generation than others, and traits correlated with reproductive success become more common if they are heritable. Thus adaptive evolution requires variation, reproduction, and heredity. Note that this description lacks any genetic detail except the very general statement that the variation must be in part heritable. That is why Darwin could describe adaptive evolution in essentially this way, without knowing about genes, how they are transmitted, the nature of genetic information, or where it is located. In this chapter we concentrate on some consequences of genetic details for adaptive evolution. Those consequences are sometimes very important, for properties of the genetic system can make the response to natural selection surprisingly subtle and complex. We emphasize the two features of genetic organization that produce the conditions under which conflict among different parts of the genome can become serious: *multilevel selection* and *asymmetric transmission*.

Multilevel selection occurs in a nested hierarchy of replicators

Natural selection works at several levels within one organism

A multicellular eukaryotic organism (see Figure 9.1) contains cells; a cell contains a nucleus and mitochondria; a nucleus contains chromosomes, and a chromosome contains coding DNA sequences (genes) and non-coding sequences. Each mitochondrion contains several genomes in the form of circular DNA molecules; these DNA molecules contain coding and (sometimes) non-coding sequences. All these structures satisfy the three conditions for adaptive evolution: variation, heredity, and reproduction. Individual organisms vary, part of

> ● **KEY CONCEPT**
> When independent genomes are nested within levels, the divergence in short-term reproductive interests of genomes at different levels leads to conflicts.

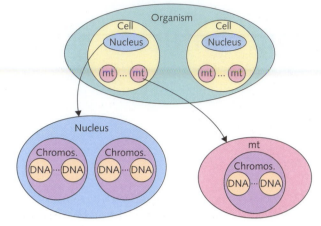

Figure 9.1 An organism as a nested hierarchy of replication levels: mt, mitochondrion; Chromos., chromosome; DNA, DNA sequence on chromosome.

their variation is heritable, and they can reproduce: thus natural selection operates on individual organisms. But cells also show variation: there are different types like epithelial cells, lymphocytes, liver cells, and cancer cells; they reproduce by mitotic division, and they show heredity: liver cells, for example, normally divide to produce liver cells. Thus cells are also subject to natural selection. So are mitochondrial genomes and repeated DNA sequences on nuclear chromosomes.

Thus a multicellular organism is a hierarchy of replicating units, several of which may be undergoing adaptive evolution simultaneously. This is multilevel evolution. One important consequence of **multilevel evolution**, **genomic conflict**, occurs when a trait is favored at one level but selected against at another, or when different genes affecting the same trait experience contradictory selection pressures because they follow different transmission rules. Male sterility in flowering plants is briefly discussed here to introduce the essential characteristics of genomic conflict. It is treated in greater detail later on in this chapter (p. 209). The key feature to notice is the difference in transmission of nuclear and mitochondrial genes.

Multilevel evolution: *Adaptive evolution occurring simultaneously at several levels of a biological hierarchy, e.g. nuclear and cytoplasmic genes.*

Genomic conflict: *Occurs when genes affecting the same trait experience different selection pressures because they obey different transmission rules or experience opposing selection at different levels of a nested hierarchy.*

Selection favors hermaphroditic flowers for nuclei but male-sterile flowers for mitochondria

Several hermaphroditic plant species contain some individuals that produce no viable pollen. These male-sterile plants are effectively females, in contrast to normal individuals that have flowers with both male and female reproductive structures (see Figure 9.6 later in the chapter). That the mutation for male sterility is in the mitochondrial genome gives us a clue to its evolutionary success, which otherwise would be a puzzle. Because mitochondria are transmitted only in the female line, mitochondrial mutations that enhance female fitness are favored by natural selection, irrespective of their impact on male fitness. A mitochondrial mutation that reduces male function and improves female function

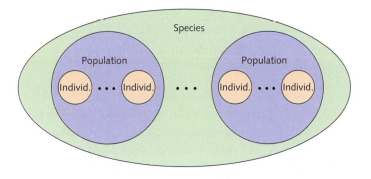

Figure 9.2 A nested hierarchy of replication levels upwards from individual to species. See Chapter 2 (pp. 41 ff.) for a discussion of the conditions that set the strength of selection at each level.

will be selected. Nuclear genes, in contrast, are transmitted with equal probability through pollen and ovules, and nuclear genes affecting reproductive organs are selected for equal allocation to male and to female function (see Chapter 10, pp. 236 ff.). Thus there is genomic conflict: natural selection favors hermaphroditic flowers coded by genes in the nucleus and male-sterile flowers coded by genes in the mitochondria.

The concept of multilevel selection can be extended to higher levels. Individuals occur in groups and groups make up a species. Groups and species also satisfy the criteria for adaptive evolution in some sense: they vary, they reproduce by division, and they show heredity (Figure 9.2).

Multilevel selection shapes virulence in myxoma virus in rabbits

Just as conflicts can occur between genetic levels within an individual organism, so do conflicts arise between the effects of natural selection operating at higher levels. Consider, for example the evolution of the virulence of a pathogen infecting a host population (May and Anderson 1983). The myxoma virus infects rabbits. Its replication rate within a rabbit determines the virulence of the virus. Fast replication leads to more severe disease and more rapid death of the rabbit than does slow replication. When there is genetic variation among viruses within a rabbit, selection within rabbits increases virulence. However, selection on the virus for transmission between rabbits favors viruses that have the greatest probability of being transmitted to new hosts, i.e. those living in rabbits that survive for a long time. Evolution at this level—among rabbits considered as transmission vehicles—reduces the virulence of the virus. Thus there is a conflict between two levels: selection among viruses within hosts increases virulence, and selection among hosts reduces virulence. The outcome is an intermediate level of virulence. Which process dominates depends on the amount of variation among the viruses within rabbits, which is enhanced by a high viral mutation rate and by multiple infections, on the variation among rabbits in the severity of the disease, caused by infections by different viral strains, on the transmission efficiency of each viral strain, and on the evolution of resistance in the rabbits.

Genomic conflict may have been a driving force in many evolutionary transitions

● **KEY CONCEPT**

When there are conflicts between levels, both levels suffer. Resolution is not always possible, but at times conflict has had creative consequences.

What is the evolutionary significance of genomic conflict? Has it been a major force of evolutionary change, or has it played a modest role? It is too early for a definitive answer, for genomic conflicts have been recognized and studied only recently. Hurst et al. (1996) suggest that genomic conflicts had a role in the origin of chromosomes, mating types, sex, meiosis, sexual selection, diploidy, genome size, and speciation. Only future work can reveal how well these suggestions are supported by evidence. In the remainder of this chapter we discuss general principles of genomic conflicts and illustrate them with examples.

Selection in opposite directions at two levels creates genomic conflict

We first analyze the simplest type of multilevel selection, two-level selection, to see how it gives rise to genomic conflicts. Figure 9.3 shows a two-level hierarchy. The outer circle represents the higher level, for example a unicellular organism

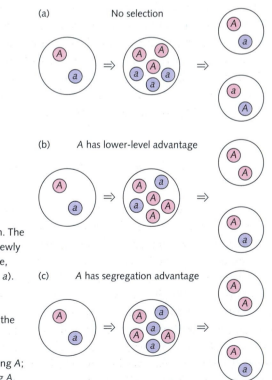

Figure 9.3 Two-level selection. The left-hand column represents newly formed cells; each, for example, with two mitochondria (A and a). The middle column represents mitochondrial replication. The right-hand column represents the result of cell duplication.
(a) No selection at lower level;
(b) lower-level selection favoring A;
(c) biased segregation favoring A.

or a cell in a growing multicellular organism. The inner circles represent the lower level, for example mitochondria within the cell. Two types of mitochondria, genetic variants marked *A* and *a*, occur within the same cell. Mitochondria replicate within the cell and are distributed at cell division to the daughter cells. Three outcomes are possible.

1. If both mitochondrial types replicate at the same rate and are fairly distributed to the daughter cells (Figure 9.3a), there is no natural selection at the lower, mitochondrial level.

2. If *A* replicates faster than *a* (Figure 9.3b), the proportions of *A* and *a* mitochondria change in the next cellular generation, and natural selection does operate on mitochondria. Whether it is in conflict with selection at the cellular level depends on the effects of *A* and *a* on the cell's phenotype. If *A* contributes more to the cell's fitness than *a*, there is no conflict, for selection works in the same direction at both levels. But if *a* benefits the cell more than *A*, there is genomic conflict because selection favors *a* at the level of the cell but *A* at the level of the mitochondria.

3. If *A* and *a* mitochondria replicate at the same rate but are distributed unequally—with biased segregation—to the daughter cells (Figure 9.3c), selection also operates on mitochondria. As in case 2, genomic conflict arises if the mitochondrial type favored at segregation is selected against at the cell level.

In cases 2 and 3 the ratio of *A:a* changes from one cellular generation to the next because mitochondrial transmission favors *A*, either because of faster replication or because of biased segregation. This illustrates the key role in the causation of genomic conflicts played by the transmission of the lower-level replicating units—the mitochondria—to the descendants of the higher-level replicator—the cell. Any systematic bias in this transmission creates genomic conflict. An example is provided by 'petite' mutations in yeast.

Mitochondrial 'petite' mutations in yeast replicate rapidly at the expense of their cells

Mitochondrial mutations in yeast and the yeast cells that contain them both experience two-level selection that produces genomic conflict. The cells of baker's yeast normally form large colonies on a solid culture medium, but occasionally a small, or 'petite,' colony is found. Such petite mutants often have defective mitochondria, suffer severe metabolic problems, and grow poorly, which explains their small colony size. Each yeast cell contains many mitochondria, and each mitochondrion contains many genomes. How can so many defective mitochondrial genomes accumulate in a yeast cell? Petite mutations, which are often deletions of large amounts of mitochondrial DNA, allow faster replication of their mitochondrial genome, which then out-competes the other mitochondria within the cell. All the cells derived in successive divisions from a cell with mutant mitochondria inherit the defective mitochondria and thus form a petite colony. This is a clear case of genomic conflict, for the mutation is favored in mitochondrial replication but is selected against among the yeast cells, whose growth it impairs.

What is the fate of petite mutants in a yeast population? Selection at the two levels—mitochondria and cells—occurs on different time scales. The process of mutation, increase in frequency, and fixation at the mitochondrial level takes place in only a few cell generations because there are many mitochondrial divisions between two successive cell divisions. Thus the petite mutation is successful in the short term. What are its chances in the long term?

First consider yeast cells reproducing by asexual division, as is normal under natural conditions. Because the mutation lowers the fitness of the yeast cells, petites are selected against in competition with normally growing yeast, and in time they must disappear if natural selection among the yeast cells is sufficiently effective. Taylor et al. (2002) tested this prediction (Figure 9.4) by varying population size to vary the strength of selection at the level of the yeast cells. They constructed yeast strains in which the cells carried both normal and petite mitochondria and followed the changes in petite frequency in cultures of different population sizes. They varied population size because they expected selection among the yeast cells (against the petite mutations) to be very effective in large populations but not in small populations. As population size decreases, sampling error increasingly obscures the fitness differences among the cell genotypes (Chapter 3, pp. 62 ff.), and cell genotype frequencies are increasingly determined by genetic drift instead of selection. The experimental results confirm the expectation: petites are victorious only in small cell populations; in large populations selection among the yeast cells is strong enough to eliminate petites.

Yeast cells can also become sexual. Then two cells of different mating types fuse, and the resulting diploid zygote undergoes meiosis to produce haploid asexual cells. Some petite mutations, called suppressive petites, show biased segregation in crosses between a petite strain and a normal strain: such crosses yield only petite offspring. The long-term success of suppressive petite mutations is more difficult to predict than the fate of petite mutations in asexual yeast cells. They are favored at the mitochondrial level, selected against at the cell

Figure 9.4 Frequencies of petites after 150 generations. Each bar represents the mean of five replicate populations of yeast cells founded by an ancestor who carried both normal and petite mitochondria. (From Taylor et al. 2002.)

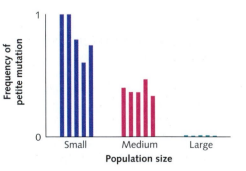

level, and favored in sex with non-petite cells. Sexual fusions enable these petite mutations to 'infect' normal lineages that under asexual reproduction can only become petites by mutation. For the petite mutation, sex resembles horizontal transmission from the point of view of a parasite. Whether suppressive petites can get established depends on the frequency of sex and on their negative effect on the cellular growth rate. The opportunities for suppressive petites clearly depend on the frequency of sexual reproduction of yeast cells.

> **Horizontal transmission:**
> *Transmission of parasites or pathogens to other organisms at times and places not necessarily associated with host reproduction or relationship.*

Genomic conflict in sexual and asexual systems

The yeast example shows that genomic conflicts arise less easily in asexual systems for two reasons. First, a genomic conflict presupposes selection acting at two or more levels. This implies that at the lower level there has to be genetic variation, in our example among the mitochondria, which in an asexual system can only arise by mutation. Because mutation rates are low, variation among the lower-level replicators is limited. Second, the long-term fate of a lower-level gene that causes a genomic conflict is linked in an asexual system to the fate of the higher-level replicator in which it finds itself. It cannot 'escape' by moving into another replicator, and because genomic conflict lowers fitness at the higher level, such a gene will be removed by selection.

> ● **KEY CONCEPT**
>
> From the point of view of a genomic parasite, sex allows both vertical and horizontal transmission, but asex only allows vertical transmission.

Sex opens the door to several types of genomic conflict

Neither reason why asexual systems are protected against genomic conflicts applies to sexual systems, for sex combines genomes from different lineages, and it can generate genetic diversity upon which natural selection can act at several replication levels (the input aspect of sex, see Figure 9.5). Then, when gamete fusion is followed by a meiotic division that produces haploid reproductive cells with genetic recombination and segregation (the output aspect of sex), genetic elements can become associated with replicators different from their original hosts. For example, in Figure 9.5 the lower-level variants *A* and *a*, originally associated with higher-level variants *B* and *b*, have changed their association after recombination.

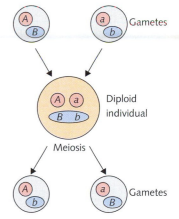

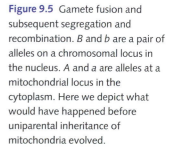

Figure 9.5 Gamete fusion and subsequent segregation and recombination. *B* and *b* are a pair of alleles on a chromosomal locus in the nucleus. *A* and *a* are alleles at a mitochondrial locus in the cytoplasm. Here we depict what would have happened before uniparental inheritance of mitochondria evolved.

Horizontal transmission creates genomic conflict between plasmids and their host bacteria

Many bacteria carry not only a chromosome but also smaller, circular DNA molecules called plasmids. These plasmids usually have some genes important for their own propagation, but they may also contain genes useful to the host cell, like those coding for antibiotic resistance. If plasmids were only transmitted vertically (i.e. at bacterial cell division to the daughter cells), their long-term fate would be strictly coupled to that of their bacterial host. If they lowered the cell's fitness, natural selection would remove them from bacterial populations by favoring cells that did not have such plasmids. However, many of these plasmids induce their host to conjugate with another, uninfected, bacterial cell. During conjugation the recipient acquires a plasmid while the donor cell retains a copy of the plasmid.

This horizontal transmission of plasmids is one form of bacterial sex, for bacterial chromosomal genes are sometimes also transferred and recombine with the recipient's chromosome. The consequences for genomic conflict are serious: the long-term fate of the plasmids is no longer coupled to that of their host, and if infection by horizontal transfer is frequent enough, plasmids can establish themselves in a bacterial population despite an average negative effect on bacterial fitness. Then there is genomic conflict because natural selection favors both maintenance of a plasmid via vertical and horizontal plasmid transmission and its loss via the fitness difference between plasmid-bearing and plasmid-free bacteria.

Several plasmids have evolved a remarkably clever mechanism to enhance their stable maintenance in the bacterial host population. They make the host cells addicted to their presence by producing both a toxin and its antidote. The antidote molecule is less stable than the toxin molecule. As long as the plasmid is present in a cell, there is no problem, for continuous production of both molecules guarantees that the toxin is ineffective. But if by chance a copy of the plasmid fails to get vertically transmitted to a daughter cell, that cell dies because the antidote has disappeared while toxin molecules are still present. Thus once a cell has acquired a plasmid, its descendants remain addicted to it. Although this addiction mechanism helps the plasmid to maintain itself, it would not guarantee maintenance in strictly asexual bacteria with only **vertical transmission**, where the host cell could lose in competition with uninfected cells. Sex with horizontal transmission overcomes, for the plasmid, the consequences of negative fitness effects on the bacterial cell.

Vertical transmission:
Transmission of parasites or pathogens from host parent to host offspring, often during host reproduction.

The *t* haplotype in mice exemplifies meiotic drive, widespread in many species

Segregation distortion:
Deviation from the Mendelian ratios that give equal chances to homologous alleles in meiosis; unfair ratios can be caused by nuclear genes that interfere with meiosis or with the products of meiosis to improve their own chances at the expense of their homologues.

Another feature of Figure 9.5 also creates an opportunity for genomic conflict: the segregation process itself offers a chance to gain a transmission advantage. A basic Mendelian principle of genetic transmission is that an individual heterozygous for a pair of alleles *A* and *a* produces 50% *A* gametes and 50% *a* gametes (the law of equal segregation). This is a direct consequence of the separation of each chromosome pair during meiosis. However, the law of equal segregation is broken (Lyttle 1991) when genes violate the Mendelian transmission rules by distorting segregation. **Segregation distortion** is also called **meiotic drive**.

Meiotic drive: *Genes distort meiosis to produce gametes containing themselves more than half the time.*

One of the best-studied cases of meiotic drive concerns the *t* haplotype in mice (haplotypes are groups of closely linked genes that tend to be inherited together). The name derives from the tailless phenotype that occasionally occurs in some species, including *Mus musculus* and *Mus domesticus*. Crossing experiments in the 1930s suggested that at the T locus three alleles were segregating, *T*, *t*, and +. The homozygotes *TT* and *tt* died before birth, while heterozygote *Tt* and *T*+ individuals were short-tailed or had no tail. Remarkably the *t* allele was transmitted much more frequently than the normal 1:1 ratio in crosses involving heterozygous *t*+ or *Tt* males.

Let us consider a simplified situation with only the *t* and the + alleles, which is enough to communicate the essence of meiotic drive. Heterozygous *t*+ males produce between 90 and 100% *t*-bearing sperm, depending on the particular *t* haplotype. Several genes within the *t* complex produce a 'toxic' effect on + chromosomes in heterozygous males, an effect to which they themselves are largely immune. This is again a poison/antidote system. *tt* homozygotes produce male sterility and are lethal or have severely reduced viability in both males and females.

As demonstrated by a population genetic model (question 9.2), a stable polymorphism can be expected for *t* and + chromosomes under broad conditions. The genomic conflict that is caused by opposing selection pressures at the level of the gametes (favoring *t*) and the individual (favoring +) results in both types of chromosome being maintained. This polymorphism may have been present in mouse populations for about 3 million years, predating the speciation events leading to the present species in the *Mus* subgenus (other trans-species polymorphisms are discussed in Chapter 14, pp. 334 ff.).

This type of genomic conflict does not always lead to stable coexistence of *t* and + chromosomes. For example, if the fitness reduction in *tt* homozygotes is small enough, then the segregation distortion in heterozygotes would be strong enough to drive *t* haplotypes to fixation. If this happened, all signs of abnormal segregation would disappear (question 9.3).

This genomic conflict compromises adaptation at the individual level: mean individual fitness is lower with the distorter than without it. The *t* haplotypes in mice show that segregation distortion can reduce individual fitness. Understanding the evolutionary stability of fair Mendelian segregation is therefore one of the basic challenges in evolutionary biology.

Until recently, segregation distortion could only be discovered when it was associated with a phenotype. Otherwise it may have been there, but we could not observe it. In mice, it was discovered through the associated tail abnormalities, which may have nothing to do with the cause of distortion. Another easy phenotype through which segregation distortion can be discovered is an abnormal male/female ratio in the offspring of a cross. This can indicate meiotic drive caused by a gene on a sex chromosome, although other explanations are possible. Several segregation distorters on sex chromosomes are known, such as the SR (sex ratio) trait in *Drosophila melanogaster*, which has been investigated in great detail.

Now that we have whole-genome sequences for an increasing number of species, genome-wide screening for transmission distortion of genetic markers has become possible. In human genomic data Zöllner et al. (2004) detected a modest but significant departure from Mendelian transmission proportions in the form of a small genome-wide shift towards excess genetic sharing among siblings, which means that siblings receive slightly more often the same alleles from their parents than expected under fair segregation. This result could be caused by meiotic drive, by differential success of gametes in achieving fertilization, or by viability selection among zygotes. In the latter case, there would be no genomic conflict, only individual selection against some alleles. The authors argue that probably viability selection against certain combinations of alleles is mainly responsible for the observed transmission distortion.

Meiotic drive is probably a rare exception. Most chromosomal genes have normal meiotic behavior. Because meiosis and mitosis are strictly regulated processes that prevent a biased distribution of chromosomes over the daughter cells, the evolution of a successful distorter is unlikely. In fact, even for *t* haplotypes and other well-investigated nuclear distorters, meiotic segregation seems to be normal, and the apparent segregation bias in heterozygous males is probably caused by post-meiotic poisoning of nuclei with the normal chromosome. Thus it appears to be very difficult to create effective distortion, possibly because evolution has long opposed it and put mechanisms in place to prevent it.

There is another reason why meiotic drive is rare. All chromosomal segregation distorters investigated so far consist of two genetic components: a gene producing a harmful effect in the homologous chromosome and a gene conferring resistance to the distorting chromosome. These two genes must be closely linked, for recombination would destroy their effect. Therefore, it is hard to see how segregation distorters evolve. Two simultaneous mutations producing closely linked genes with such functions are highly implausible, and if either mutation occurs separately, it has no selective advantage.

The cytoplasm is a battleground for genomic conflicts

A competitively superior cytoplasmic mutant has a segregation advantage

● **KEY CONCEPT**

Uniparental inheritance resolves conflicts among organelles but creates conflicts between the sexes.

Outside the nucleus there are more opportunities for biased segregation of genetic elements. Cytoplasmic genes, such as those in mitochondria and chloroplasts, are not distributed at cell division by a process as precise as meiosis. They occur in many copies per cell, and if a cell contains tens or hundreds of mitochondria, a precise segregation mechanism is not only difficult to evolve but also unnecessary, for if one daughter cell receives fewer mitochondria than the other, subsequent mitochondrial divisions easily compensate for the initial deficit. Other genomes, such as viruses or plasmids, may also occur in cytoplasm. They also occur in many copies per cell and lack a precise distribution mechanism at cell division. Evidence from many organisms suggests that an initial heteroplasmic condition, in which a single cell contains different alleles of a cytoplasmic gene, cannot be maintained through a series of consecutive mitotic cell divisions. After a few cell divisions the daughter cells become homoplasmic (containing copies of just one or the other allele). This suggests that cytoplasmic genes usually differ in their replication rates. Any cytoplasmic mutant that is competitively superior can, like the suppressive petite mitochondria in yeast, gain a segregation advantage in a heteroplasmic situation.

> **Heteroplasmy:** *A cell that contains genetic variants of a cytoplasmic genome.*

> **Homoplasmy:** *A cell that contains only one type of cytoplasmic genome, which may occur in multiple copies.*

Uniparental transmission prevents conflict between cytoplasmic and nuclear genomes

In sexual life cycles transmission of cytoplasmic genes is uniparental. Both the male and the female gamete may have cytoplasmic genomes, but the resulting offspring usually contain only the maternal cytoplasmic genes. Very few exceptions to this rule are known. In anisogamous species the maternal cytoplasmic contribution to the zygote outweighs the paternal contribution by so much that this alone could explain why only maternal cytoplasmic genes occur in the offspring. Active degradation of paternal cytoplasmic genomes has also been observed (Kaneda et al. 1995). Active elimination of the cytoplasmic genes from one parent also occurs in the zygotes of isogamous species in which both gametes contribute roughly equal amounts of cytoplasmic genomes but only those of one parental mating type are found in the offspring (Gillham 1994).

Why would the inheritance of cytoplasmic genes be uniparental, in contrast to the biparental inheritance of nuclear genes? The evolution of uniparental cytoplasmic inheritance is probably mediated by genomic conflict.

We can understand that conflict by asking, what would be the consequences of biparental cytoplasmic inheritance? Cytoplasmic genomes from both gametes then occur in the zygote, and no mechanism ensures that after cell division the daughter cells have exactly the same cytoplasmic composition. It is hard to see how such a mechanism could work because of the many copies of cytoplasmic genomes and their scattered location in the cell. Because the cytoplasmic genomes are not under the segregation control of the spindle apparatus, intracellular competition and sampling effects determine their distribution after cell division. In a multicellular organism originating from a heteroplasmic zygote, cell lineages will differ in the frequency of different organelle genomes. In particular, the genetic composition of the cytoplasm of the cells from which the gametes are derived is crucial, for only those cytoplasmic genes will be passed on to the offspring. Cytoplasmic genes will compete for this position. A cytoplasmic mutation with increased probability of winning the competition has a good chance of being selected, for it enjoys a transmission advantage—even if it lowers the fitness of the individual carrying it. Here again are the ingredients of genomic conflict: natural selection favors a gene at one level (the cytoplasmic genome) but selects against it at another level (the individual organism).

Now consider uniparental inheritance of cytoplasmic genes. If a cytoplasmic mutation occurs like the one postulated above, its fate will be fundamentally different. If it occurs in a male, it will not be transmitted to the offspring at all. If it occurs in a female, it will be transmitted to all offspring, male and female, but only to grandchildren through daughters. Direct competition with other alleles in the zygote and young embryo is prevented, for in half the cases the mutation comes from the mother and is exclusively transmitted irrespective of the paternal allele, and in half the cases it comes from the father and is never transmitted. The maternal cytoplasmic genome then has the same transmission pattern as the maternal nuclear genome. The genomic conflict that occurs with biparental inheritance is removed, and the fate of the cytoplasmic mutation only depends on its effect on the individual fitness of its carriers. If it enhances individual fitness it will increase in frequency; if it lowers individual fitness it will be eliminated.

Because biparental inheritance of cytoplasmic genes makes organisms vulnerable to genomic conflicts, whereas uniparental inheritance prevents them, it is attractive to suggest that this has been the main reason for the evolution of uniparental inheritance of cytoplasmic DNA from an ancestral pattern of biparental inheritance. Population genetic models suggest that this scenario for the evolutionary transition from biparental to uniparental inheritance can work under certain conditions (Hoekstra 1990, Law and Hutson 1992).

Uniparental inheritance of cytoplasm may have contributed to the origin of males and females

In discussing the evolution of uniparental inheritance of cytoplasmic genes we assumed the existence of males and females. Indeed, sexual differentiation seems to be essential for uniparental transmission. If both parental gametes were identical in type, it would not be clear beforehand which gamete would transmit its cytoplasmic DNA and which gamete would not. Then competition for transmission would arise and cytoplasmic mutants enforcing transmission would spread. Uniparental inheritance would not be stable, and the door would be open to genomic conflict. When the two gametes that form a zygote are always of different mating type, transmission or non-transmission can be coupled to mating type. So it seems likely that male/female dimorphism, or at least mating types, evolved before the evolution of uniparental cytoplasmic inheritance.

The interesting suggestion has been made that it may have been the other way round: that male/female or mating-type differentiation evolved to regulate uniparental cytoplasmic inheritance. In other words, selection to improve an imperfect mechanism of uniparental cytoplasmic inheritance triggered the evolution of two sexes or mating types. A supportive observation is that some organisms that do not have a system of two sexes, namely mushrooms and ciliates, accomplish sex without cell fusion, exchanging nuclei without exchanging cytoplasm. Because the cytoplasms of the two parents are not mixed, cytoplasmic inheritance is automatically uniparental. Therefore—so the proponents of this hypothesis argue—there is no need for the regulation of uniparental cytoplasmic inheritance and consequently no selection for sexual differentiation. In their view, this exception suggests the rule (Hurst and Hamilton 1992).

Uniparental transmission and anisogamy allow new genomic conflicts to arise

Uniparental cytoplasmic inheritance prevents the genomic conflicts that easily arise in a system of biparental transmission without precise segregation control. This solution is actually, however, a Trojan horse, for once it is established, uniparental inheritance creates a fundamental asymmetry in transmission that constantly favors any genetic element in the cytoplasm that enhances female fitness at the expense of male fitness or shifts the sex ratio towards more females. When this produces genomic conflict, nuclear genes will be selected to counteract the effects of the cytoplasmic gene.

In addition to the asymmetry in cytoplasmic transmission, sexual selection (see Chapter 11, pp. 251 ff.) has shaped additional asymmetries between males and females. Many cases of genomic conflict generated by the male/female asymmetry have been discovered. We discuss two examples: male sterility in plants and genomic imprinting in mammalian development.

Genomic conflict driven by mitochondrial genes explains male sterility in plants

About 5% of angiosperm species are gynodioecious, meaning that some individuals are male-sterile. Male sterility can be detected in phenotypes by the occurrence of normal hermaphroditic plants and plants that produce seeds but no viable pollen in the same population (see Figure 9.6). This trait has been investigated both in natural populations (e.g. *Plantago* and *Thymus*) and in several agricultural crops where the presence at appreciable frequencies of plants deficient in male function is puzzling. They can only reproduce as females and should have lower fitness than conspecifics that can function both as males and as females. Selection should eliminate the genotypes that cause male sterility.

> Gynodioecious: *Plant populations containing some individuals whose flowers only produce seeds and other individuals whose flowers produce both seeds and pollen.*

Figure 9.6 Normal hermaphrodite (right) and male-sterile (left) flowers in *Plantago coronopis*. (From Koelewijn 1993.)

The paradox disappears once we know that in most cases the male-sterile phenotype is caused by a mitochondrial mutation. For mitochondrial genes, which are inherited exclusively along the female line, a male is a 'dead end' from which no transmission to offspring is possible, and elimination of male function does not affect the transmission prospects of mitochondrial genes as long as there is enough pollen in the population. Indeed, a mitochondrial mutation that eliminates male function while enhancing the female fitness of its carriers will be favored, for the fate of mitochondrial genes is affected only by their carriers' female fitness.

In response (see Chapter 10, pp. 236 ff.), selection on nuclear sex-ratio genes will act to establish equal investment in offspring of both sexes. When the frequency of male steriles in a population increases because selection favors a mitochondrial gene causing male sterility, that increase in turn selects any nuclear mutation that suppresses cytoplasmic male sterility and restores a more equal sex ratio. Such nuclear 'restorer genes' are found in gynodioecious species, where selection for mitochondrial male sterility conflicts with selection for nuclear restorers of male fertility (e.g. Bentolila et al. 2002).

Evidence from inter-specific hybrids suggests that the nucleus can 'win' the conflict

The long-term outcome of conflict between mitochondrial male-sterility genes and nuclear-restorer genes is not easy to predict, for it depends on the precise relations between the fitnesses of the different genotypes. Indirect evidence suggests that the nuclear genome sometimes 'wins' the conflict, for male-sterile individuals can result from hybrid crosses between related species within which no male sterility occurs. This can be interpreted as evidence for complete within-species restoration of male fertility involving different mitochondrial mutations and nuclear restorers in the two species. In hybrids, a nuclear restorer from one species encounters a mitochondrial male-sterility gene from the other species, against which it is ineffective. The evolutionary dynamics of male sterility can be complex, for natural populations often contain several mitochondrial male-sterility mutations together with corresponding nuclear-restorer genes (Frank 2000).

● **KEY CONCEPT**

A conflict between the parents is expressed in the fetus during mammalian pregnancy.

Genetic imprinting: *The silencing of certain genes in the germ lines of parents prior to gamete production accomplished by methylating the DNA sequence.*

Genetic imprinting in mammals—a conflict over reproductive investment?

In mammals a remarkable phenomenon has been discovered called genetic imprinting, in which the sex of the parental genome determines the expression of genes in the offspring. For some genes the paternal copy is expressed and the maternal copy is inactive. For other genes the maternal copy is expressed and the paternal copy is inactive. Most genes are not subjected to imprinting; the few that are often affect growth. At loci where imprinting occurs, individuals are effectively haploid because only one allele is active. Imprinting is usually removed after the early developmental stages of the offspring. Note that imprinting prevents parthenogenesis in mammals: a parthenogenetic mutant would possess only maternal alleles, and development would be disrupted by the lack of expression of necessary genes coming from the male germ line (see Chapter 2, p. 49).

Mother and father may disagree over the growth rate of the fetus

Several hypotheses have been suggested for the evolution of imprinting, including Haig's idea that it may be the product of genomic conflict resulting from the different interests of the male and female parents (Moore and Haig 1991). In mammals, which have internal fertilization, a long pregnancy, and viviparity, paternal genes are expected to manipulate the mother to provide more nutrients to the fetus. Maternal genes are expected to resist. A gene enhancing the rate at which a mammalian embryo extracts resources from the mother may be selected in males but not in females. In the mouse, insulin-like growth factor 2 (*Igf2*) is expressed in the fetus and promotes the acquisition of resources from the mother across the placenta. The paternal copy of the *Igf2* gene is expressed, but the maternal copy is inactive. There is another gene, the insulin-like growth factor 2 receptor (*Igf2r*), which appears to inhibit the action of *Igf2*. This gene shows the reverse pattern: the maternal copy is expressed and the paternal copy is inactivated.

According to Haig, this makes evolutionary sense as follows (Haig and Graham 1991). If females can have offspring from several males, then it is not in the interest of a male to let the female hold resources in reserve for future offspring fathered by other males. Thus paternal genes are expected to manipulate the mother to supply more nutrients to the fetus, whereas maternal genes are expected to protect her from over-provision that would compromise future reproduction. Only with strict life-long monogamy would the interests of the father and mother coincide and would conflict not be expected. If Haig's explanation is correct, selection for these two genes acts in opposite directions in males and females, thereby creating a remarkable genomic conflict between paternal and maternal genes whose effects occur in the offspring.

Evidence from mice supports the conflict hypothesis

The long-term outcome of this conflict may well occur in mice. Mice with an inactivated paternal copy of *Igf2* are only 60% of normal size at birth, while mice with an inactivated maternal copy of *Igf2r* are born 20% larger than normal. This may be a stable reconciliation of paternal and maternal interests. It could also be interpreted as a tense standoff, a balance of interests in constant and enduring conflict. More examples, including a human genetic disorder related to this system of imprinting, and an elaboration of the theoretical background of the conflict hypothesis of genomic imprinting, are given in Haig (2002).

● SUMMARY

This chapter concentrates on multilevel selection within organisms that causes genomic conflicts because selection at different levels works in opposite directions.

- Organisms consist of a hierarchy of replication levels, at each of which natural selection may occur simultaneously.

- Replicators occur in groups, and under some conditions groups are also subject to natural selection, for they may form new groups that disappear at different rates depending on their composition.

- Replicating units that occur in few copies and whose replication and segregation are strictly controlled—such as cell nuclei and their chromosomal genes—do not easily cause genomic conflicts.

- Replicating units that occur in many copies and whose replication and segregation are not strictly controlled—such as cytoplasmic genetic elements—more easily cause genomic conflict.

- Sexual organisms are more prone to genomic conflicts than asexual ones.

- Conflicts in the interests of paternal and maternal genes can generate conflicts over fetal growth rate that affect birth weights and symptoms of pregnancy.

- Genomic conflicts can generate evolutionary change and may have been involved in several key evolutionary events, such as the evolution of the male/female distinction.

● RECOMMENDED READING

Haig, D. (2002) *Genomic imprinting and kinship*. Rutgers University Press, New Brunswick, NJ.

Hurst, L.D., Atlan, A., and Bengtsson, B.O. (1996) Genetic conflicts. *Quarterly Review of Biology* **71**, 317–64.

Keller, L. (ed.) (1999) *Levels of selection in evolution*. Princeton University Press, Princeton, NJ.

● QUESTIONS

9.1. Is two-level conflict as exemplified by petite mutants in yeast analogous to cancer? Just as mutant petite mitochondrial genomes spread within a yeast cell, out-competing normal mitochondria, mutant cancer cells spread within the body, out-competing normal cells. Cancer seems to be a case of genomic conflict because selection at the level of the somatic cells is in conflict with selection at the individual level. Is there some difference between petite yeast and cancer where the analogy breaks down?

9.2. The t haplotype polymorphism in mice involves segregation distortion favoring t-bearing chromosomes in heterozygous Tt males and lethality of both male and female tt homozygotes. Consider a theoretical case in which segregation distortion occurs both in males and females: suppose that of the gametes produced by A_1A_2 heterozygotes a fraction k has genotype A_2 and a fraction $1 - k$ has genotype A_1 ($k > 0.5$). Assume that A_2A_2 homozygotes are inviable or sterile. Show in a population genetic model (see Chapter 4, p. 79) that these conditions lead to a stable allele frequency equilibrium $\hat{q} = 2k - 1$.

9.3. If in question 9.2 the A_2A_2 homozygotes are normal (fitness 1), the distorting allele A_2 is expected to become fixed (A_1 will disappear from the population). When this has happened, will the segregation-distorting effect of A_2 still be observed? Why or why not?

9.4. Group and species selection are not likely to have shaped many patterns and certainly not many adaptations. Compare individual to group selection from the point of view of genomic conflict. What determines the power of selection at each level in a conflict? (see Price 1972).

CHAPTER 10

Life histories and sex allocation

● **KEY CONCEPT**

Natural selection shapes the
entire life cycle to maximize
reproductive success.

At what age and size should an organism start reproducing? How many times in its life should it breed? When it reproduces, how much energy and time should it allocate to reproduction, to growth, and to the maintenance of its body? Should it have a few, high-quality offspring or many, small offspring that are less likely to survive? Should it invest heavily in reproduction early in life and have a short life as a consequence, or should it put less into reproduction and live longer? How should it deal with risk? Should it hedge its bets and spread the risk of unsuccessful reproduction among many offspring or many reproductive events, or should it put all its eggs, literally or figuratively, into one basket? How many of its offspring should be male and how many female? Should that decision depend upon ecological or social circumstances or be fixed at birth? If it can be a sequential hermaphrodite, should it be born as a male and turn into a female later, or the other way around? These are some of the questions answered by biologists working on life histories and sex allocation.

Natural selection is made possible by variation in life-history traits

● **KEY CONCEPT**

Life-history traits are the
principal components of fitness;
all selection is mediated by
variation in life-history traits.

Life-history traits: *A trait
directly associated with
reproduction and survival,
including size at birth, growth
rate, age and size at maturity,
number of offspring, frequency of
reproduction, and lifespan.*

Life-history traits are directly related to natural selection, for survival, maturation, and reproduction determine reproductive success, and variation in reproductive success causes natural selection. It is variation in life-history traits that leads to natural selection. The evolution of life histories results in the basic characteristics of all species—how big they are, how long they live, how many offspring they have, how fast their populations can grow. And life-history traits participate directly in population dynamics: how many individuals of a species are present at any time. The population dynamics of interacting species—competitors, predators and prey, parasites and hosts, symbionts and mutualists—in turn contribute to the structure of biological communities. Thus life-history traits participate in key ecological interactions and their effects cascade from natural selection to the stability of ecological communities—all good reasons why evolution cannot be separated from ecology.

Tradeoffs constrain the set of possible solutions to life-history problems

The ideal life history is simply stated: a **Darwinian demon** matures at birth, immediately gives birth to an infinite number of offspring with the same characteristics, and lives forever. Such organisms would rapidly fill the universe, but they do not exist. Why not? The ideal organism does not exist because of tradeoffs, which determine the outcome of evolution for all life-history traits. Here are some of the tradeoffs involved in the evolution of the most important life-history traits.

> **Darwinian demon:** *An organism that matures at birth, gives birth to infinitely many offspring, and lives forever.*

At what age and size should an individual start to reproduce?

Being large is advantageous, for larger individuals can produce a greater total mass of offspring, but it takes time to grow large, and the longer one waits to mature while growing, the more likely it is that one will die without reproducing at all. Here the tradeoff is between the probability of surviving to a given age and the size—and therefore the **fecundity**—that can be attained at that age.

How should resources be divided among offspring?

> **Fecundity:** *A synonym for offspring production used for all organisms, not all of which give birth.*

How total offspring mass is divided into individual offspring determines the number of offspring and contributes to their 'quality,' their ability to survive to reproduce. Will lifetime reproductive success be greater if one only produces a few offspring that survive well, or if one produces so many, low-quality offspring that their sheer numbers ensure that several will survive? Here the tradeoff is between a few, high-quality offspring and many, low-quality offspring.

How should resources be divided between reproduction, growth, and maintenance?

Once reproduction begins, resources must be divided among three functions: immediate reproduction, growth or storage and thus future reproduction, and maintenance. If maintenance is neglected to increase reproduction, mortality rates will rise and lifespan will decrease, and if growth or storage are neglected to increase current reproduction, future reproduction will suffer. Here the tradeoffs are between survival and reproduction and between current reproduction and future reproduction.

Evolutionary optimization shapes traits involved in tradeoffs

Those are the problems of life-history evolution explained in terms of benefits, changes in traits that increase fitness (reproductive success), and costs, changes in traits that reduce fitness. Organisms are expected to evolve to the point where the net benefit, the positive difference between benefits and costs, is greatest. That point—not always attainable—is determined by the tradeoffs among life-history traits and by how the environment affects mortality and fecundity.

Frequency-dependent selection, not optimization, shapes sex allocation

The cost–benefit explanation assumes that when the optimal solution is found, it is optimal for the whole population. While that assumption has been successful up to a point, it does not hold in general, not even for core life-history traits like age and size at maturity. Nowhere is its violation clearer than in problems of sex ratio and sex allocation. If an organism can control how many female and male offspring to produce, then the best offspring sex ratio (sons/daughters) depends on the sex ratios produced by the other organisms in the population. If the others produce only males, it is best to produce only females. If the others produce only females, it is best to produce only males. The evolution of the sex ratio is the classical example of frequency dependence: the population sex ratio determines the mating opportunities for the offspring.

The distinction between optimality and frequency-dependence deeply affects our view of evolution. The optimality view is of a world at equilibrium in which individuals act independently to reach solutions. The frequency-dependent view is of a world dynamically changing in which the best strategy depends on what others do. This chapter offers an opportunity to see how well each approach performs in answering some of the main questions about life histories and sex allocation:

- at what age and size should an organism mature?
- how many offspring should it produce?
- how many times in its life should it attempt reproduction?
- how much of the reproductive investment should be allocated to male and how much to female offspring?

To explain life-history evolution, we combine insights from five sources

● **KEY CONCEPT**

All phenotypic evolution involves explanations with the same five elements used to explain life-history evolution.

In Chapters 3–5, we analyzed evolution as genetic changes in populations, but this is not the best way to explain the evolution of life-history traits. To do that, we have to combine information from five sources: (1) the demography of the population, (2) the risk of reproductive failure, (3) the heritability and plasticity of the traits, (4) tradeoffs among traits, and (5) the phylogenetic context of the species in question and how that affects the potential response to selection.

1. The demography of the population

In almost all organisms the probability of giving birth, and the probability of dying, vary with the age and size of the organism. It follows that the strength of selection on traits varies with the age and size of the organisms in which they occur. The field that explains this is *demography*. Demography connects age- and size-specific variation in survival and fecundity to variation in fitness and thereby tells us the strength of natural selection on life-history traits. For example, natural selection on reproductive performance is stronger in younger than in older adults (what is young and what is old depends on the species). Consider a population at evolutionary equilibrium in which the average values for age and size at maturity and survival and fecundity rates are close to the optimum for all age classes. Now let mortality rates increase in one age class, for example because of a new age-specific disease or size-specific predator. The change in mortality rates 'devalues' that age group and all subsequent ages because organisms are less likely to survive to those ages to reproduce. The evolutionary response will be decreased investment in older or larger age or size classes and increased investment in younger or smaller ones.

2. The risk of reproductive failure

If a bird puts all its eggs in one nest in 1 year, it loses them all if a predator finds the nest. If a fish in the North Atlantic concentrates all of its reproduction in a bad year for eggs and larvae, it loses them all. If all the seeds of a desert annual plant germinate in one very dry year, they all die. Two principles from probability theory tell us how the outcome of natural selection is shaped by risks like these:

- where fitness varies, measure it with the geometric, not with the arithmetic, mean;
- to reduce risk, spread it across a set of independent events.

The **geometric mean** is the nth root of the continued product of n terms. It measures the mean result of a series of events that are multiplied together. We use it to measure fitness because fitness is multiplicative, not additive: if each of two children has three grandchildren, then the original parent has 6 (2×3) not 5 ($2 + 3$) grandchildren.

As the variation in reproductive success between generations increases, fitness decreases, an effect measured accurately by the geometric mean. Consider two ways to achieve the same **arithmetic mean** fitness—2.5 per generation. In the first, reproductive success is relatively stable, alternating between two surviving offspring in one generation and three in the next: 2, 3, 2, 3, 2, 3 . . . . In the second, reproductive success is more variable, alternating between one surviving offspring in one generation and four in the next: 1, 4, 1, 4, 1, 4 . . . . After six generations, the first pattern has resulted in ($2 \times 3 \times 2 \times 3 \times 2 \times 3 =$) 216 descendants, the second in only ($1 \times 4 \times 1 \times 4 \times 1 \times 4 =$) 64—a striking contrast in success not

Arithmetic and geometric mean: *The arithmetic mean of $(2 + 3 + 2 + 3 + 2 + 3)/6 = 2.5$ and the geometric mean is $(2 \times 3 \times 2 \times 3 \times 2 \times 3)^{1/6} = 2.45$. The arithmetic mean of $(1 + 4 + 1 + 4 + 1 + 4)/6 = 2.5$ and the geometric mean is $(1 \times 4 \times 1 \times 4 \times 1 \times 4)^{1/6} = 2.0.$*

predicted by the arithmetic means, which are the same. The geometric mean does bring out the right contrast: 2.45 for the first, 2.0 for the second. The benefit of having more offspring in a good year does not balance the cost of having fewer offspring in a bad year, and the geometric mean accurately reflects the asymmetry in benefits and costs between good and bad years.

These two principles—spreading risk across independent events (not putting all your eggs in one basket) and measuring outcomes with the geometric mean—are also used to maximize returns from investment funds. The analogy between wealth and fitness is suggestive. A strategy that maximizes the geometric mean of financial returns has the highest probability of reaching or exceeding any given level of wealth in the shortest possible time. A strategy that maximizes geometric mean fitness has the highest probability of reaching or exceeding any given number of descendants in the shortest possible time. Microevolution is a bit like microeconomics.

Risk does not need to be variable in time or heterogeneous in space for selection to favor organisms that spread their risk. The environment simply needs to create some risk at all places and all times, and that is always the case. If two strategies have the same arithmetic mean fitness, but one of them spreads risk more effectively and has a higher geometric mean fitness as a result, then the risk-spreader wins and the risk-taker loses.

For example, annual plants face particularly striking risks. They live as adult plants for only part of a year; if there is a particularly bad year at their growing site, they might well leave no offspring. Plants in the genus *Heterotheca*, which germinate in late summer or autumn, overwinter as rosettes, grow flowering structures in early spring, start to flower in June, and continue to reproduce until the first killing frost, are certainly exposed to the type of risk just mentioned. They deal with it by producing two types of seed (Figure 10.1). One is

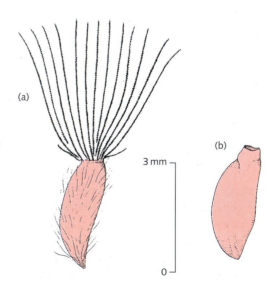

(a)

(b)

3 mm

0

Figure 10.1 The two seed types produced by *Heterotheca latifolia*. (a) The type adapted for dispersal in space; it germinates rapidly and completely. (b) The type specialized for dispersal in time; it germinates slowly and partially. Reproduced from Venable (1985) Ecology of Achene Dimorphism in Heterotheca-latifolia. Journal of Ecology, Vol. 73 with permission from Blackwell Publishing.

well adapted for wind dispersal, and it germinates rapidly—some seeds start germinating within 2 days, and nearly all seeds of this type have finished germinating within 39 days. This seed type spreads risk in space. The other seed type is poorly suited for dispersal in space but well adapted for sampling different seasons or years. Some seeds start to germinate after 8–10 days, but even after 99 days no more than 50% have germinated. This seed type spreads risk in time. The first type of seed allows rapid growth in a permissive habitat; the second type of seed provides insurance against extended draught. Thus risk is real, and adaptations to deal with it have evolved.

In contrast to a conservative bet-hedger like *Heterotheca*, the riskiest strategy of all would be an asexual organism that reproduces just once at old age and has only one offspring that does not disperse. Because it has waited a very long time to reproduce, it has run great risk that it will die without reproducing. Because it only reproduces once, it takes the risk that the year in which it reproduces will be a bad one, not a good one. Because it only has one offspring, it takes the risk that that offspring will not survive to reproduce for whatever reason—poor genotype, disease, predation, or accident. Because the organism is asexual, whatever offspring it has will be genetically identical to itself, and any pathogens that had specialized on that genotype will do well at infecting the offspring. Because the offspring does not disperse, it takes the risk that conditions will remain good where it is. This is the opposite of a Darwinian demon: it is a **Darwinian dolt**. Knowing what not to do sometimes makes clearer what should be done.

> **Darwinian dolt:** *An organism that delays maturation, reproduces asexually, has one offspring that does not disperse, and dies immediately after reproducing.*

3. The heritability and plasticity of the traits

What determines the inheritance and plasticity of life-history traits? Many genes influence life-history traits; they are quantitative traits (Chapter 4, pp. 86 ff.). The insights of quantitative genetics are important for life-history evolution. Recall that only a certain part of the genetic variation of a trait determines its reaction to selection; this part is the additive genetic variation. The proportion of the total phenotypic variation of a trait that is contributed by additive genetic variation is its heritability. When heritability = 1.0, the trait has exactly the same value in the offspring that it does in the average of the two parents; when heritability = 0.0, none of the phenotypic variation can be attributed to additive genetic variation, and the trait will not respond to selection. The heritabilities of life-history traits in many species are in the range 0.05–0.4. Thus most life-history traits that have been investigated could respond to selection (Chapter 4, p. 92).

In its original form, quantitative genetics was applied to laboratory organisms or domestic plants and animals, where one could assume constant environmental conditions. It has been extended to natural populations by including the effects of reaction norms, for the evolution of life-history traits is constrained both by the amount of genetic variation present in the population and by how that genetic variation is expressed in different environments. When the reaction norms of traits cross across environments, they alter the expression of genetic variation, changing both the strength of selection on traits and the capacity for a genetic response to selection across environments (Chapter 7, pp. 159 ff.).

4. The tradeoffs among traits

Life-history traits are connected by tradeoffs, which exist when a change in one trait that increases fitness is linked to a change in another trait that decreases fitness. The response of life-history traits to a new type of selection depends on the strength of the tradeoffs present. An improvement in one trait that is linked to high costs in connected traits cannot proceed very far. Important tradeoffs include those between the number of offspring and their survival to maturation and between reproductive investment and adult survival.

Tradeoffs have both a genetic and a physiological component. The genetic component can be expressed as a genetic correlation, which like heritability depends on the additive genetic variance of the traits. If there is a genetic correlation between two traits, then some genes affect both traits. If those effects are in the same direction for both traits, the genetic correlation will be positive. If the effects are positive on one and negative on the other, the genetic correlation will be negative.

The genetic correlation establishes a connection between two traits, but not all such connections imply tradeoffs. There are, in general, two possibilities, as follows.

- If both traits improve fitness when they are increased, and both traits reduce fitness when they are decreased, then a positive genetic correlation between the traits would not imply a tradeoff, but a negative genetic correlation would.

- If one trait improves fitness when it is increased, but the other trait improves fitness when it is decreased, then a positive genetic correlation between the traits would imply a tradeoff, but a negative genetic correlation would not.

The physiological component depends on how the organism is constructed and is a mixture of types of connections among traits. Some of those connections are the same for all individuals in a species, have been inherited from ancestors, reflect the phylogenetic history of the species, and differ among taxonomic groups. For example, C_3 plants and C_4 plants differ in metabolic pathways used to fix energy in photosynthesis. C_4 plants, such as maize, sorghum, and sugar cane, are able to utilize higher levels of incoming energy and can therefore grow faster than C_3 plants, but only if they have access to an abundant supply of water. They experience a very different tradeoff between growth and water stress than do C_3 plants. Other connections among traits vary among the individuals of a species for two reasons: developmental interactions with the environment, which are different for every individual, and variation in the genes that affect the traits involved in the tradeoff. For example, hole-nesting birds differ in their physical condition and in their ability to raise offspring for both genetic and environmental reasons.

Both causes of variation in traits—genetic and physiological—are constrained by features of development that are fixed within a lineage, for example how the nymphalid ground plan constrains genetic variation for wing-spot patterns in butterflies (Chapter 7, pp. 165 ff.).

5. The phylogenetic context of the species in question

Traits also need to be understood in phylogenetic context. Phylogenetic effects are the contribution to traits shared by all individuals because they belong to a species or larger taxonomic group. We normally think of them as 'the development' or 'the physiology' or 'the morphology' of a species or group. To understand how broadly those traits are shared and where in the history of the lineage they might have originated, we need to compare them with traits in close and distant relatives. The comparative method can help us to estimate how much of a pattern to attribute to history, and how much to attribute to microevolutionary processes that operated within the local population in the recent past (Chapter 13, p. 303).

Thus to explain the evolution of life-history traits, we need to understand how demographic selection operates on them; the role that risk minimization plays; the quantitative genetics that determines their response to selection; the tradeoffs, both genetic and physiological, that connect traits; and the phylogenetic context in which they sit. This is how we explain the evolution of any trait. The explanation has an intrinsic part—genetics, tradeoffs, phylogenetic effects—and an extrinsic part—selection expressed as effects on age- and size-specific mortality and fecundity rates. With that in mind, we discuss next the major life-history traits: age and size at maturity, clutch size and reproductive investment, and lifespan and aging.

The evolution of age and size at maturation

Age at maturity is a dividing line, for up to maturation natural selection for survival is strong, and after maturation aging begins. Fitness is often more sensitive to changes in age at maturity than to changes in other life-history traits, and at evolutionary equilibrium either earlier or later maturation would lower fitness. 'Small organisms are usually small not because smallness improves fecundity or lowers mortality. They are small because it takes time to grow large, and with heavy mortality the investment in growth would never be paid back as increased fecundity' (Kozlowski 1992).

> ● **KEY CONCEPT**
> At evolutionary equilibrium age and size at maturity have been adjusted to maximize reproductive success.

Early maturation reduces juvenile mortality and generation time but yields fewer offspring with higher mortality

Both early and delayed maturity have fitness costs and benefits. Early maturity has at least two benefits: it reduces the chance of dying before reproducing, and it

Figure 10.2 Infant mortality rate (deaths/thousand) as a function of mother's age in humans in the United States in 1960–61, based on 107 038 infant deaths documented by the National Center for Health Statistics. Each point is the mean of a 5-year age class. (After Stafford, unpublished data.)

reduces generation time. Shorter generations mean that offspring are born earlier and start reproducing sooner. One cost of early maturation is that the instantaneous mortality rate of the offspring rises when females mature earlier (Figure 10.2). A female that tried to mature very early would be so small and poorly developed herself that she could not produce any offspring at all. (Aphids are an exception.)

Two benefits of delayed maturation are often important. If delaying maturity permits further growth and fecundity increases with size, then delaying maturity leads to higher fecundity. If delaying maturity improves the quality of offspring or parental care, it improves offspring survival. Maturation will be delayed for these reasons until the fitness gained through increased fecundity and better offspring survival is balanced by the fitness lost through longer generation time and lower survival to maturity.

Thus earlier maturation brings shorter generation times and a shorter period of exposure to mortality before maturity; later maturation brings increased fecundity and lower juvenile mortality per time unit. At an intermediate age, an optimum should exist (Figure 10.3). Models making these assumptions accounted for 80–88% of the variation in natural populations of lizards, salamanders, and fish (Stearns and Koella 1986). Age at maturity appears to be adjusted to an intermediate optimum in many cases.

Because age and size at maturity vary among closely related species, among populations within species, and among individuals within populations, we know they can respond rapidly to natural selection. Age at maturity responds to artificial selection in flour beetles and fruit flies, and genetic variation for age and size at maturity has been documented in many species. Thus age and size at maturity can be adjusted by natural selection to local conditions within populations. There are also constraints on age and size at maturity imposed by history and design.

Risk-spreading adaptations can explain deviations from optimality predictions

The optimality view of the world assumes that one reproductive season is pretty much like the next. The risk-minimizing view of the world assumes that

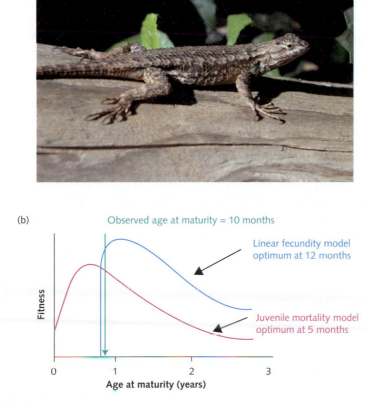

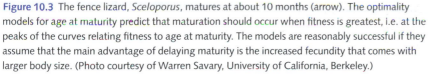

Figure 10.3 The fence lizard, *Sceloporus*, matures at about 10 months (arrow). The optimality models for age at maturity predict that maturation should occur when fitness is greatest, i.e. at the peaks of the curves relating fitness to age at maturity. The models are reasonably successful if they assume that the main advantage of delaying maturity is the increased fecundity that comes with larger body size. (Photo courtesy of Warren Savary, University of California, Berkeley.)

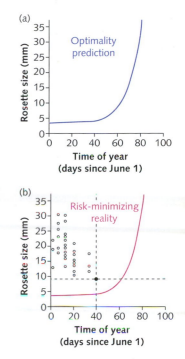

Figure 10.4 (a) An optimality model, based on the assumption that every year is the same, predicts that plants above and to the left of the line should mature in this growing season; those below and to the right of the line should delay maturation until the next year. (b) In fact, *L. inflata* is a conservative risk-minimizer; the only plants maturing in this season are considerably larger, and younger, than optimization would predict: the dots in the upper left indicate actual flowering times and rosette sizes. Earlier reproduction reduces the risk of encountering a shorter-than-average growing season. (From Simons and Johnston 2003.)

reproductive seasons vary greatly in quality, some being good, some being bad. It would be nice to know what difference those two views make. In the case of a semelparous flowering plant, *Lobelia inflata*, we have a fairly precise idea of the difference between optimality and risk-minimization (Simons and Johnston 2003). *L. inflata* flowers only once in its life, then dies, but the date of flowering within a season is quite variable, and flowering may be postponed a year: age at maturity is a plastic trait that depends on how rapidly the plant has grown and what size it has reached by a given point in the growing season. If we assume that one year is pretty much like another, then it is possible to calculate the optimal age at which to flower (Figure 10.4a) as a function of the size reached by a given date in the growing season. Plants that are above and to the left of the line have grown rapidly and should flower in this season; those that are below and to the right of the line have grown slowly and should postpone reproduction to the next season. This is not, however, what they do (Figure 10.4b).

If they are going to reproduce this year, rather than wait until next year, they finish doing so earlier, and at a larger size, than optimization would predict. This pattern is consistent with adaptations to minimize risk when the length of the season is unpredictable; at the site of the investigation in Nova Scotia, there is considerable variation in the length of the growing season.

The evolution of clutch size and reproductive investment

● **KEY CONCEPT**

Clutch size and reproductive investment evolve to an optimum determined by tradeoffs and constraints.

The second major feature of the life cycle is how reproductive investment varies with age. Reproductive investment is the product of the number of offspring and the investment per offspring. Thus we can ask two questions about its evolution: how large should each offspring be? and, how many offspring should be produced in each reproductive attempt? Investment in offspring includes not just the size of the offspring at birth but the parental care, if any, invested to bring the offspring to independence.

World records held by bats, kiwis, orchids, and dung beetles

Before looking at the theory, we look at some world records. Bats have the largest offspring for their body size in the mammals, the record being held by *Pipestrellus pipestrellus*. It bears twins whose combined weight at birth is 50% of the mother's weight after birth—and she must fly and feed while pregnant. The flightless New Zealand kiwi, a ratite bird the size of a large chicken, lays the largest egg for its size of any bird, more than five times larger than the largest eggs of domestic chickens. The caecilian *Dermophis mexicanus*, a tropical amphibian, gives birth to a clutch that weighs up to 65% of her post-birth weight. If we measure repro-

Altricial: *Born or hatched helpless, blind, and needing parental care for warmth and food.*

ductive effort by the weight of offspring when they become independent of their parents, rather than by their weight at birth, then altricial birds (birds that hatch naked, blind, featherless, and dependent) make amazing reproductive efforts. Their combined offspring can weigh at fledging up to eight times the weight of the parents, which must work very hard to feed that weight of young. The world record for smallest size and worst juvenile survival is probably held by an orchid seed weighing less than a microgram and with a chance of about one in a billion of surviving to reproduce. The insects with the best juvenile survival are probably species of dung or carrion beetles that lay only four or five eggs; their juveniles must have the same survival rate as juvenile elephants or whales.

The Lack clutch assumes that a few offspring flourish but many would starve

How many offspring of a given size should an organism produce in a given reproductive event? Attempts to answer this question began with David Lack,

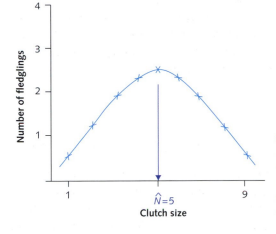

Figure 10.5 The Lack clutch: if the probability that a young bird will survive to leave the nest declines linearly with the size of the clutch, then it is best to lay an intermediate number of eggs.

who suggested that altricial birds should lay the number of eggs that fledge the most offspring. He assumed that only one tradeoff was important, the tradeoff between the number of eggs laid and the probability that offspring would survive until they left the nest. If the probability of surviving to fledge decreases linearly as the number of offspring increases, and if there are no tradeoffs with parental survival or with offspring survival to maturity, then the relation between clutch size and fitness—number of offspring fledged—is a parabola with an intermediate optimum (Figure 10.5). There have been many attempts to determine whether altricial birds produce clutches that are optimal in the sense just described. Clutches are often smaller than those that would be optimal if the only tradeoff were between number of eggs and survival to fledging. What are the reasons for deviations, positive or negative, from the Lack clutch? Answers for deviations in both directions may apply to all organisms, not just birds.

Effects that lead to optimal clutches smaller than the Lack clutch include these: (1) additional tradeoffs with parental and offspring life-history traits, (2) temporal variation in optimal clutch size, and (3) parent–offspring conflict won by the offspring:

1. Additional tradeoffs with parental and offspring life-history traits. Much is known about tradeoffs between clutch size and other life-history traits. By 1992 there had been 55 studies of the effects on various life-history traits of manipulating clutch size by adding or removing eggs or chicks (Stearns 1992). When clutches were enlarged, weight of fledglings was reduced in 68%, survival of fledglings to the next season was reduced in 53%, weight of parents was reduced in 41%, survival of the parents to the next season was reduced in 36%, and future reproduction of the parents was reduced in 57% of the studies in which they were measured. The reproduction of offspring from larger clutches was not measured often, but every time it was measured, it was reduced in larger clutches. Thus there are often good reasons not to lay more eggs.

Lack clutch: *The intermediate clutch size in altricial birds that produces the largest number of fledglings.*

2. *Temporal variation in optimal clutch size.* When offspring survival varies from year to year, producing a large clutch in a bad year will yield few survivors. According to the principles of risk minimization discussed above, when there is inter-annual variation in reproductive success, organisms that spread the risk of reproduction failure across a number of years will win and those that concentrate reproduction in one or a few years will lose. If spreading reproduction across several years means having fewer offspring in each year, then the risk minimizing effect will reduce clutch size. A tradeoff between reproduction this year and reproduction next year will add to that effect. If such a tradeoff exists, so that a large clutch in one year will be followed by a small clutch in the next, then optimal clutch size will be reduced, for the costs of the bad years will not be compensated by the benefits of the good years. Variation in year-to-year juvenile mortality rates appears to have selected for reproductive restraint and long lives in many marine fish, forest trees, desert plants, and marine invertebrates including corals, oysters, and many types of worm.

3. *Parent–offspring conflict won by the offspring.* Parent–offspring conflict can also reduce clutch size. In species that reproduce several times per lifetime and in which reproduction is costly, the parents will be selected not to invest all their resources in the current clutch but to save something for future reproduction. The offspring, on the other hand, will be selected to extract more investment from the parents than it is in the parents' interests to give. Such conflicts are often especially obvious at weaning or fledging, when parents may force offspring to become independent. If the conflicts are costly, clutch size will be reduced.

Effects that lead to optimal clutches larger than the Lack clutch include these: (1) inability to predict resource levels, and (2) ability to selectively abort less-fit offspring.

1. *Inability to predict resource levels.* If at the beginning of a season the parents cannot predict whether it will be a good or a bad year, and if newborn offspring are relatively cheap, then it may pay to have as many offspring as could survive in the best year possible, then reduce the clutch size to the actual resource levels encountered, either by neglecting some offspring, by letting the offspring fight with each other, or by directly killing or eating the weakest. Such brood reduction is common in birds of prey and fruit trees.

2. *Ability to selectively abort less-fit offspring.* If the fitness of offspring is variable, and if the variations in fitness among the offspring can be identified by the parents when the offspring are small, young, and therefore cheap, then it pays the parents to produce many zygotes, then selectively abort the less fit to concentrate investment on the more fit. This may explain the many zygotes fertilized in pronghorn antelope and nutrias, which greatly exceed, by 10–100 times, the number of offspring that could be born. It may

also explain the reproductive patterns of certain ascomycete fungi and the recurrent spontaneous abortions suffered by some Hutterite women (see Chapter 19, p. 492).

Clutch size is invariant within some lineages

Deviations from the optimal clutch may also result from constraints within lineages. For example, all birds in the order Procellariformes lay one egg, in some species not every year. For the largest species, the wandering albatross, which takes up to 33 days and travels up to 15 000 km on a single foraging flight, a clutch size of one may be optimal, for the single chick, which has a special starvation physiology, must wait up to a month for a meal, and two offspring probably could not be fed. However, the smaller, less widely foraging species in the order, including several species of petrel, also have clutches of one egg although they almost certainly could feed two or more chicks. In these smaller species, clutches appear to be phylogenetically constrained.

What happens when current reproduction reduces future reproduction and survival?

Lack assumed a single tradeoff between offspring number and offspring fitness and focused on a single clutch. In contrast, reproductive-effort models aim to predict the optimal reproductive effort over the whole lifespan taking all reproductive tradeoffs into account. They usually assume that if reproductive investment is increased at one age, then the probability that the parents will survive to reproduce again will be lowered, or, that if they do survive, their ability to reproduce in the next season will be reduced, or both. These models make several predictions, as follows.

- If mortality rates increase in one adult age class, then the optimal reproductive effort increases before that age and decreases after it (Michod 1979).

- If adult mortality rates increase, the optimal age for maturation decreases (Roff 1981).

- If mortality rates increase in all age classes, then optimal reproductive effort increases early in life and optimal age for maturation decreases (Charlesworth 1980).

Experiments confirm the predictions of reproductive-effort models

The assumptions and predictions of reproductive-effort models have been confirmed in experiments on kestrels in The Netherlands, on guppies in Trinidad, and on fruit flies in the laboratory.

The kestrel study (introduced in Chapter 2, pp. 35 ff.) focused on the idea that the fitness of parents consists of the contribution made to fitness by the offspring

Reproductive value: *The expected contribution of organisms in that stage of life to lifetime reproductive success.*

Residual reproductive value: *The remaining contribution to lifetime reproductive success after the current activity has made its contribution.*

they are currently producing—the reproductive value of their current clutch—plus all the later contributions to fitness that the parents could make in the rest of their life—their residual reproductive value. The current reproductive value of a female incubating its eggs is its clutch size times the reproductive value of one egg. The reproductive value of one egg is the sum of the probability that the egg becomes an adult that survives to reproduce once, times its first clutch size, plus the probability that it survives to reproduce twice, times its second clutch size, and so forth. The residual reproductive value of a parent is the probability that it will survive to reproduce once more, times the expected number of offspring that it will then have, plus the probability that it will survive to reproduce again a second time, times the expected number of offspring, and so forth.

Daan and his colleagues (1990) calculated the reproductive value of the clutch and the residual reproductive value of the parents for experimentally reduced and enlarged clutches. If kestrels correctly balance the costs and benefits of reproductive investment, then the total reproductive value of the clutches they actually laid (the control clutches) should be larger than that of either reduced or enlarged clutches. The main cost of producing a larger clutch was decreased parental survival, which decreased the residual reproductive value of parents of enlarged clutches. The offspring from control clutches also had better expectations of future reproduction than did those from enlarged clutches, but this effect was not as large as the effect on parental survival. When they accounted for these factors, they found that the clutch size actually laid was the one that yielded the highest reproductive value (Table 10.1).

Reproductive-effort models were also confirmed in the field manipulation experiments done by Reznick and his colleagues (1990) on guppies living in shallow streams in Trinidad (this example was introduced in Chapter 1, pp. 8 f.). At some sites their main predator is a cichlid fish that can eat large, sexually mature guppies and causes high mortality rates in all size classes. At other sites, their main predator is a killifish that eats mostly small, juvenile guppies and causes lower mortality rates than the cichlid in all size classes.

Table 10.1 The results of clutch-size manipulation in kestrels.

	Reduced	Control	Enlarged
Number of broods	28	54	20
Mean clutch size	5.25	5.19	5.40
Mean number fledged	2.60	3.95	5.84
Reproductive value of clutch	2.52	4.20	5.59
Local parental survival	0.65	0.59	0.43
Residual reproductive value	9.88	8.89	6.49
Total reproductive value	12.40	13.09	12.08

Table 10.2 Divergence of life-history traits in guppies after field manipulations.

Life-history trait	Control (cichlid)	Introduction (killifish)
Male age at maturity (days)	48.5	58.2
Male weight at maturity (mg of wet weight)	67.5	76.1
Female age at first birth (days)	85.7	92.3
Female weight at first birth (mg of wet weight)	161.5	185.6
Size of first litter	4.5	3.3
Offspring weight (mg of dry weight)		
Litter 1	0.87	0.95
Litter 2	0.90	1.02

In 1976, before the experiments began, guppies from cichlid sites matured earlier, made a larger reproductive effort, and had more and smaller offspring than did guppies from killifish sites. The differences were heritable, and they fit the predictions of reproductive-effort models. To demonstrate that predation caused the pattern, predation was manipulated to decrease the mortality rates on all age classes. After 11 years, or about 20 generations, significant evolution was observed as predicted (Table 10.2). Age and size at maturity increased when guppies were placed in killifish sites, size of the first brood decreased, and size of offspring in the first two broods had already increased and corresponded qualitatively and quantitatively to the differences found in the unmanipulated populations. Changes in the intensity of mortality rates caused by a manipulation of predation led to rapid evolutionary change in life-history traits in the direction predicted. The speed of evolution was striking. If the unmanipulated populations were at evolutionary equilibrium, then it took just 7–18 generations after the manipulation for most of the traits to reach equilibrium again. The traits that changed most occurred early in life. This was one of several recent studies demonstrating significant, rapid evolutionary change in ecologically important traits and suggesting that separating evolution from ecology would be artificial.

Using fruit flies in a laboratory experiment, Stearns and colleagues (2000) tested the reproductive-effort model with two treatments. The treatments differed in the adult mortality rates administered by killing a percentage of flies after counting them twice per week: high in the first treatment (the probability of dying within 1 week as an adult was 99%), low in the second treatment (the probability of dying within 1 week as an adult was 36%). Both juvenile and adult densities were the same in all treatments.

Fruit flies are subject to a tradeoff that determines much of their response to such differences in selection. They can decrease development time by pupating and eclosing earlier, but to do so they have to pay the price of being smaller, and

Eclosion: *Emergence from a pupa; insects eclose from pupae.*

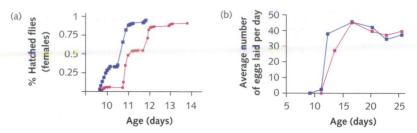

Figure 10.6 Fruit flies, *Drosophila melanogaster*, that have evolved in response to high adult mortality rates (a) develop more rapidly and eclose earlier and (b) lay more eggs early in life, than do flies that have experienced lower adult mortality rates. In both panels, the blue curve corresponds to the flies that encountered higher adult mortality rates.

smaller flies have lower fecundities. Under these conditions wild flies reach peak fecundity when 16–17 days old. In the experiment, they were placed into the population cage when they were 14 days old, after which most of those encountering the high-adult-mortality treatment only had 1 day in which to lay eggs before they were killed. Thus the flies evolving under high adult mortality should increase the number of eggs laid on that one day, the 14th or 15th day of life. They had two options. They could develop faster, eclose earlier, start to reproduce earlier, and thus achieve their peak reproduction by the 15th day of life, but pay the price of being smaller and having lower fecundity. Or they could develop more slowly, spend more time eating as larvae, eclose later, have higher fecundity, but pay the price of not yet being at their peak fecundity by the 15th day of life.

Analysis of the development time–body size–fecundity relationship suggested that they should choose the first option, and they did. Within 2 years of selection they had evolved precisely the differences predicted by a reproductive-effort model, and, like the guppies, traits expressed early in life, development time and early fecundity, were the first to change (Figure 10.6).

The guppy and fruit fly examples confirm a basic tenet of the evolutionary theory of aging, our next topic: selection on reproductive performance earlier in life is stronger than selection on reproductive performance later in life.

Lifespans evolve, and so does aging

● **KEY CONCEPT**

Selection optimizes lifespan at an intermediate value, but mutational effects shorten it below that optimum.

It is not immediately obvious why organisms age and die, and why different species have different maximum lifespans (Figure 10.7). The longest-lived invertebrates are sea anemones, lobsters, and bivalves: some clams can live 220 years. The shortest-lived are rotifers, insects, and small crustaceans: some live less than a week. The record for mammals in zoos is held by the African elephant at 57 years, with the domestic horse and the spiny echidna tied for second at 50 years. Plants vary in lifespan from days to many centuries. The variation in lifespan among species makes it clear that lifespan has evolved. Cells that all

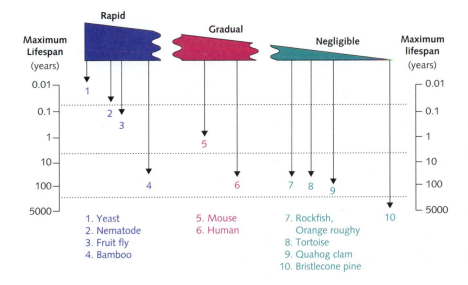

Figure 10.7 Sexually reproducing species differ greatly both in lifespan and in rate of aging. Some asexual clones have even greater longevity. (Reprinted from Finch and Austad (2001) History and prospects: symposium on organism with slow aging. *Experimental Gerontology* **36**, 593–7. Copyright (2001), with permission from Elsevier.)

share the same toolkit construct organisms that age at very different rates, some of them—such as quahog clams and bristlecone pines—aging so slowly that we have difficulty detecting any increase in mortality rates or decrease in reproductive rates even in very old individuals.

Why is the soma mortal and the germ line immortal?

However, the most striking puzzle is not why clams live longer than lobsters, or why humans live longer than chimpanzees. The most striking puzzle is why our germ line is so well maintained that it is potentially immortal, connecting us through an unbroken sequence over 3.5 billion years long to the origin of life, while we are so poorly maintained that we age and die even if we are protected from accidents and given optimal conditions. Our germ line is part of our own body, determined by the same genes, built with the same biochemistry. Why can it survive indefinitely while we must die? The answer is a triumph of evolutionary thought.

Death at old age makes little difference to fitness

To see why death late in life makes little difference to fitness, we compare two ways that a fly might live its life. In both cases, we follow the fly from the moment it emerges from its pupa. We assume that mortality is 80% per day and that the fly lays 10 eggs per day. In the first case (Figure 10.8a) some flies could survive a very long time. We can calculate its lifetime egg production from the infinite series

$$1 + a + a^2 + a^3 \ldots = (1 - a)^{-1}$$

when $|a| < 1$. In this case $a = 0.8$, the daily survival rate.

(a)

Age (days)	1	2	3	...	19	20	21
Survival probability	1	0.80	0.64	...	0.018	0.014	0.012

Figure 10.8 Why death at old age makes little difference to fitness. (a) This fly can potentially live a very long time; its expected reproductive success is 50 eggs per lifetime. (b) This fly dies after 19 days of reproduction. Its expected reproductive success is 49.3 eggs per lifetime. The difference between (a) and (b) is small. (Courtesy of Martin Ackermann.)

(b)

Age (days)	1	2	3	...	19	20	21
Survival probability	1	0.80	0.64	...	0.018	0.014	

What we want to know is the sum over all ages of the probability of surviving to that age times the number of offspring produced at that age; that is, the number of offspring produced, on average, per lifetime. In this case, with $a = 0.8$, $10(1 + 0.8 + 0.64 \ldots) = 10 (1/0.2) = 50$. Thus this strategy leaves 50 progeny.

In the second case (Figure 10.8b), all flies die after having laid their eggs on their 19th day of life. This strategy leaves 49.3 progeny. Because very few flies survive after day 19, the loss of all age classes older than 19 has resulted in a reduction in total lifetime reproduction of only 0.7 offspring, a very small impact. That is why events that happen only in older age classes have very little effect on fitness.

Any population can be divided into young and old organisms such that the contribution to reproductive success of younger organisms will be greater than that of older organisms. Even in species with indeterminate growth that gain fecundity with size, at some age so many will have died that an older age class can no longer contribute as many offspring as a younger age class that contains more survivors. For this reason, selection on survival rates and fecundities must always decline with age after reproduction begins. They decline at different rates in organisms with different life histories. In humans in industrialized countries, the selection to improve survival has dropped almost to zero by the time one is 50 years old, a fact that does not lighten the hearts of the middle-aged (Figure 10.9).

The evolution of lifespan is a problem defined by life-history tradeoffs

Extrinsic mortality: *Mortality caused by environmental factors, such as predators, disease, and weather.*

In considering the evolution of lifespan, it is helpful to distinguish between **extrinsic mortality** and **intrinsic mortality**. Extrinsic mortality is imposed by the environment—for example, by predators, diseases, or bad weather. Intrinsic mortality is mortality caused by the degradation of physiology and biochemistry as organisms age.

Intrinsic mortality: *Mortality caused by neglect of maintenance leading to degradation of essential functions.*

Two ways of thinking about lifespan both give important insights. On the one hand, we can think of the evolution of lifespan as resulting from selection

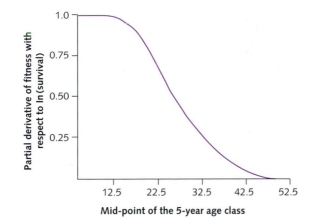

Figure 10.9 The sensitivity of fitness to changes in survival rates declines with age. The case depicted corresponds to the population of the United States in about 1940. Sensitivity has been standardized so that the greatest sensitivity is 1.0. (From Charlesworth and Williamson 1975.)

acting directly on reproduction and indirectly on adult mortality rates through tradeoffs with reproduction. Longer lifespans will evolve if extrinsic mortality rates decrease in older organisms, increasing the value of older organisms because of their increased contribution to reproductive success. Longer lifespans will also evolve if extrinsic mortality rates increase in younger age classes, decreasing the value of younger organisms because of their decreased contribution to reproductive success. Thus one can think of average lifespan as the result of interactions between age-specific mortality rates and a set of reproductive tradeoffs. Such tradeoffs are in part fixed effects—effects present in all individuals in the population—caused by the common development and physiology that characterize the species and represent part of its phylogenetic heritage.

> **Fixed effects:** *Biological features that do not vary and that are shared by all organisms within a lineage.*

Thus life-history theory views the evolution of the reproductive lifespan as a balance between selection to increase the number of reproductive events per lifetime and tradeoffs that increase the intrinsic sources of mortality with age. The first lengthen life, the second shorten it. They combine to adjust the length of life to an intermediate optimum. The factors causing selection to lengthen life decrease the reproductive value of juveniles and increase the reproductive value of adults. These include lower adult mortality rates, higher juvenile mortality rates, increased variation in juvenile mortality rates, and decreased variation in adult mortality rates. Thus age-specific selection will adjust the length of life to an intermediate optimum determined by the interaction of selection with tradeoffs intrinsic to the organism and viewed as physiological effects between age classes.

The evolution of aging is a problem defined by gene expression and mutations

We can also think of the evolution of lifespan as the byproduct of genetic effects. Consider a gene that improves the reproductive success of the younger organisms at the expense of the survival of the older organisms. It has positive effects

Age (days)	1	2	3	...	19	20	21
Survival probability	1	0.80	0.64	...	0.018	0.014	

Figure 10.10 An example of how small the increase in fecundity early in life must be to more than compensate for certain death at old age. This genotype produces one more egg in the first day of life than those depicted in Figure 10.8. Like the fly in Figure 10.8(b) it dies after its 19th day of life. This fly can expect to have 50.3 eggs per lifetime. (Courtesy of Martin Ackermann.)

Pleiotropy: *One gene has effects on two or more traits.*

Antagonistic pleiotropy: *One gene has positive effects on fitness through its impact on one trait but negative effects on fitness through its impact on another trait.*

early in life and negative effects late in life. A gene that affects two or more traits is called a pleiotropic gene. Effects that increase fitness through one trait at the expense of decreased fitness through another trait are antagonistic. Therefore such genes have antagonistic pleiotropy. A mutation that has such effects, positive enough early in life, not too negative later in life, will be favored by selection, increase in frequency, and in most cases go to fixation.

In the fly example above (Figure 10.8), the gene might increase reproduction by just one egg to 11 in only the first day of life, while killing all flies after the 19th day (Figure 10.10). Such a fly could expect to lay 50.3 eggs per lifetime, more than the 50 laid by a fly that could potentially live forever but laid only 10 eggs per day. This pleiotropic gene would invade and spread to fixation in a population consisting of the other type, even though the difference between Figure 10.10 and Figure 10.8b is so small that you might have trouble seeing it at first.

Antagonistically pleiotropic genes are thus the first sort of gene important for the evolution of aging and lifespan. There is indirect evidence that they exist; the search to locate specific examples is intensifying. Two traits in humans, one in males and one in females, suggest how such genes might act. Prostate cancer occurs at high frequency in males over 70, but it can be prevented by treatment with female hormones or castration. It appears to be a consequence of a long period of exposure to testosterone, a hormone absolutely necessary for male reproductive performance. Osteoporosis, or a loss of bone density, is mediated by estrogens in older women, but estrogens are essential for reproduction in younger women. In both cases the old-age pathologies are associated with age-related changes in the responses of tissues to hormones which are essential to reproduction.

The second kind of genetic effect that may be involved in aging is the accumulation of mutations with age-specific expression. Consider a gene that is only expressed in a certain age group. Selection against a deleterious mutation in such a gene is stronger if it is expressed in younger organisms that contribute more to reproductive success. In a population at evolutionary equilibrium, the number of mutations present for a given trait depends on the per-trait mutation rate and the strength of selection operating on the trait. If we assume that the per-trait mutation rates are the same for traits expressed in young and old organisms, then only the strength of selection differs. The strength of selection is stronger on traits only expressed in younger organisms (see Figure 10.9), where it reduces genetic variation. Thus we expect to find more mutations with age-specific expression present for traits only expressed in older organisms, where selection (Chapter 5, pp. 111 ff.) is relaxed. This is the mutation-accumulation effect. It can only work if the expression of mutations is restricted to specific age classes. The evidence for mutations with age-specific effects is not yet convincing.

Fitness is maximized at a level of repair less than that needed for indefinite survival

Thus the evolutionary answer to the question, why age?, has two parts. The force of selection declines with age; after a certain age organisms are irrelevant to evolution. Given this decline, two sorts of genetic effect become possible, the accumulation of genes that benefit younger age classes at the expense of older ones, and the accumulation of mutations with stronger effects on older age classes than on younger ones. Maturation is the point where these effects should start to occur; before then they should not be seen. Aging should follow the onset of reproduction with diffuse erosion of physiological and biochemical functions caused by many genes that produce aging as a byproduct, not as an adaptation. Aging caused by a few genes with large effects is not ruled out but is not expected to be the usual case.

It follows that fitness is maximized at a level of investment in repair that is less than would be required for indefinite survival. That is why we grow old and die. Aging results from the accumulation of unrepaired somatic damage, and species with different longevities should exhibit corresponding differences in their levels of somatic repair (Kirkwood 1987). Repair is costly. More than 2% of the energy budget of cells is spent on DNA proofreading and repair, on processes that determine accuracy in protein synthesis, on protein turnover, and on the scavenging of oxygen radicals that damage biological structures. Decreases in external sources of adult mortality will increase the value of investment in repair; coupled to that should be decreased investment in growth and reproduction, leading to longer lifespans and lower reproductive efforts. Increases in extrinsic adult mortality will devalue investment in repair, and then we should see less repair and higher reproductive investment.

Lifespan responds rapidly to experimental evolution in a manner consistent with theory

These ideas have been tested on bacteria, fruit flies, nematodes, and mice with consistent results: when the contribution of older age classes to reproductive success is increased by only allowing older organisms to reproduce, aging is postponed. When the contribution of younger age classes to reproductive success is increased by increasing the mortality of older organisms, aging becomes more rapid. With respect to tradeoffs, in some cases, increased lifespan is accompanied by reduced fecundity early in life, in others, by changes in larval growth rates, survival rates, and competitive ability. Lifespan responds rapidly to selection in the laboratory in a manner consistent with evolutionary theory, and longer life must be paid for by reductions in performance early in life, either in lower fecundity, smaller body size, or lower juvenile survival and competitive ability.

How should parents invest in male and female offspring or function?

● **KEY CONCEPT**

Because fitness is gained both through male and female function, the allocation of reproductive effort to male versus female function has major impact on fitness.

Dioecy: *Having separate sexes; individuals are either males or females; used for plants.*

Gonochorism: *Having separate sexes; individuals are either males or females; used for animals.*

Simultaneous hermaphrodite: *An organism with fully functional male and female reproductive organs producing both eggs or ovules and sperm or pollen. Called* **monoecy** *in plants.*

Sessile: *Permanently attached, therefore immobile.*

Many organisms can control the sex of their offspring. Others can mature as one sex, reproduce, then change sex and reproduce as the opposite sex, functioning for part of their lives as males and for part as females. Sex-allocation theory predicts the allocation of lifetime reproductive effort among male and female offspring or between male and female function. It makes strikingly successful predictions and unites previously unrelated patterns as aspects of a single explanation.

Dioecy (gonochorism) may be advantageous because it is expensive for an organism to maintain both male and female organs and less costly to specialize in one sexual function. Simultaneous hermaphroditism (monoecy) may, however, be advantageous when mates are hard to find. It is common among sessile or slow-moving animals and plants. Hermaphrodites can fertilize themselves, but this generally happens only if mates are not available, for self-fertilized hermaphrodites often suffer from inbreeding.

Male and female function are equivalent paths to fitness

Patterns of sex allocation raise two central questions that connect to the evolution of mating systems and social behavior:

* What is the equilibrium sex ratio for organisms with separate sexes? How many sons and how many daughters should be produced?

* For sequential hermaphrodites, what sex should the organism be born as and how old and large should it be when it changes sex?

One fact is the key to answering both questions: every diploid, sexually produced zygote gets half its autosomal genes from its father and half from its mother. Because of this fact, there are two paths through which fitness can be gained: through male function or male offspring and through female function or female offspring. This holds for sequential and simultaneous hermaphrodites as well as for organisms with separate sexes. Thus we can only judge the success of a particular sex-allocation strategy by taking into account how well it succeeds in gaining fitness through both male and female paths. It will not be a better strategy than an existing one unless the amount of fitness gained through improved function in one path more than compensates for the amount of fitness lost through decreased function through the other path.

The evolutionarily stable sex ratio is 50:50 under simple assumptions

What should be the sex ratio favored by selection under the simplest assumptions? Fisher (1930) answered this question as follows: consider a large population with two sexes, well mixed, with random mating (no social structure), external fertilization, and no parental care. In such circumstances, each male has the same chance to mate as any other male, and each female has the same chance to mate as any other female. In a population that consists of more females than males, on average each male gets more mates than each female, thus favoring a gene for male bias; in a population that consists of more males than females, some males will not find mates, and on average fewer males will contribute offspring to the next generation, thus favoring a gene for female bias. In this manner deviations from equal frequencies of the two sexes produce frequency-dependent selection that leads to a stable 50:50 sex ratio. It can be maintained either by having half the females in the population produce only male offspring and half only female offspring, or by having each female produce half male and half female offspring.

> **● KEY CONCEPT**
>
> We expect 50:50 sex ratios under random mating with external fertilization and no parental care. If sex ratios are not 50:50, then one of those assumptions is probably violated.

The genetic mechanisms that produce that sex ratio are diverse

Selection for a 50:50 sex ratio probably led to the evolution of sex chromosomes, which in the simplest XX/XY system constrain the offspring sex ratios of individual females to average 50:50. In mammals and fruit flies, females are XX (**homogametic**) and males are XY (**heterogametic**). In birds and butterflies, females are heterogametic and males are homogametic. Genetic sex-determining mechanisms can even vary among populations within a species. In the house fly, for example, females are heterogametic south of the Alps and males are heterogametic north of the Alps (Franco et al. 1982). These examples show that selection acting on phenotypes may favor a 50:50 sex ratio in many species, and that the genetic mechanisms that produce that sex ratio can be quite diverse.

> **Homogametic:** *The sex having two similar sex chromosomes; for organisms with chromosomal sex determination; females are XX in humans.*

> **Heterogametic:** *The sex having two different sex chromosomes; for organisms with chromosomal sex determination; males are XY in humans.*

Sex can be determined by genes, environment, or parasites

Whereas chromosomes determine sex in birds, mammals, and many insects, sex in many other species is determined by quite different mechanisms (Bull 1983). Sex can be determined by genes or the environment, and the sex of some hosts is determined by parasites. Sex chromosomes are just one mechanism of genetic sex determination. Two others are also important. In all of the Hymenoptera—the ants, bees, wasps, and sawflies—and in some beetles and mites, females are diploid and males are haploid, having hatched from unfertilized eggs. Other species have sex-determining genes, not chromosomes. In some species these are single loci; in others sex is determined by many loci that may be spread over several chromosomes.

Sex can also be determined by the environment. In some reptiles—turtles, crocodiles, and alligators—sex is determined by the temperature at which the eggs are incubated. In turtles, males develop at cool temperatures ($<27°C$) and females at warm temperatures ($>30°C$). Alligators do the opposite: females develop when it is cool and males when it is warm. In crocodiles, females develop at both cool and warm temperatures and males at intermediate ones.

In many reef fish that change from one sex to the other as they age—primarily wrasses and parrotfish—sex is determined, at least in part, by social interactions. Large, old, high-ranking individuals are male, and younger, smaller, low-ranking individuals are female. The marine worm *Bonellia* and the rhizocephalans, relatives of barnacles that parasitize crabs, have extremely flexible sex determination. Both have planktonic larvae, and if a larva settles on substrate rather than another member of its own species, it metamorphoses into a female. If a larva settles on a female, in the case of *Bonellia*, or on a hermit crab already parasitized by a female, in the case of the rhizocephalans, it metamorphoses into a dwarf, parasitic male consisting of little more than a testis and a duct connecting it to the female reproductive system.

Sex can also be determined by cytoplasmic parasites that are vertically transmitted from parents to offspring through the gametes. A bacterium living in cytoplasm will leave no descendants if it occurs in a male host, for sperm transmit no cytoplasm to the next generation (see Chapter 8, pp. 207 f.). It is therefore in the interests of cytoplasmic parasites to occur in a female, and some of them have evolved the ability to feminize their hosts, turning males into functional females that reproduce as females. Bacteria in the genus *Wolbachia* are the best understood. They occur widely in arthropods and have been identified as feminizing factors in wood lice, wasps, and ladybird beetles. In experiments with dramatic results, wasps that only produced female offspring were 'cured of their disease' by treatment with antibiotics. Treated females produced offspring with normal 50:50 male/female sex ratios; untreated females produced only female offspring (Stouthamer et al. 1990). If *Wolbachia* spread unchecked through a host population, their hosts will eventually produce only female offspring and go extinct. Some populations of one host, a wood louse, appear to have solved this problem by incorporating the sex-determining portion of the bacterial DNA into the nuclear genome, resulting in a new sex-determining wood-louse gene (Juchault et al. 1993) and 50:50 sex ratios.

● **KEY CONCEPT**

When a sex ratio is not 50:50, then enough fitness must be gained through the rarer sex to balance the fitness gained through the commoner sex.

The Shaw–Mohler theorem explains a huge diversity of sex-allocation patterns

The natural history of sex determination reveals a wonderful diversity of mechanisms with many consequences. Now we move from the mechanisms of sex determination to the sex ratios that we expect them to produce. Unusual sex ratios can be understood as implications of a single, unifying idea. Recall that in analyzing the standard 50:50 sex ratio, Fisher assumed that males and females are equally good at producing male and female offspring at all ages and sizes and that mating is at random in a thoroughly mixed large population. When those assumptions do not hold, we need a more general answer to the question, what sex ratio is expected?

The answer was given by Shaw and Mohler (1953). They reasoned as follows. Both mothers and fathers should gain the same fitness through sons as through daughters. Consider a population with some sex ratio. Will a new mutant that produces a different sex ratio increase in frequency and invade the population?

The Shaw–Mohler Theorem states: for a mutant with a different sex allocation than the resident population to invade, its total fitness must be greater than the residents'. It must increase the fitness gained through one sex function more than enough to compensate for fitness lost through the other sex function.

If a population has long been tested by such mutants, most of the mutants that increase total fitness have already occurred and have been fixed. The population should therefore be close to evolutionary equilibrium.

We now discuss how this key idea explains age and size at sex change in sequential hermaphrodites, offspring sex ratios that depend on social rank, highly skewed sex ratios occurring under local competition among siblings for mates, and flexible sex allocation in parasitoid wasps encountering different prey sizes.

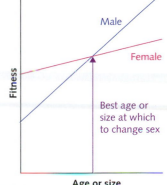

Figure 10.11 The size-advantage model for sequential hermaphroditism: it is best to be born as the sex that loses less by being young and small, and to change to the sex that gains more by being old and large, at the age and size where one would gain more fitness by being the other sex. (After Ghiselin 1969.)

Start in the sex that loses less by being small, change to the sex that gains more by being large

Sequential hermaphroditism should evolve, given sufficient flexibility in the mechanisms of sex determination, where the advantage of being a particular sex changes with age or with size (Figure 10.11). Organisms should be born into the sex that loses less by being small or young, and they should change into the sex that gains more by being old or large. Sex change should occur when the fitness advantage of being one sex when old or large becomes greater than the fitness advantage of remaining the other sex that was small and young.

Protandric organisms begin their reproductive career as males, then switch sex and function as females. **Protandry** has evolved where a large female enjoys an advantage in caring for offspring and where females gain more reproductive success as size increases than do males. A small male is able to fertilize a female with a modest amount of sperm and is relatively mobile. A female that cares for developing eggs and embryos may be able to accommodate more offspring as she grows larger.

For example, the gastropod *Crepidula*, found in rocky inter-tidal habitats, is a sequential hermaphrodite that copulates in 'daisy chains' in which the bottom

> **Protandry:** *Individuals are born as males, reproduce as males, then change sex and reproduce as females. In plants, individuals express male function prior to female function, producing pollen before being pollinated.*

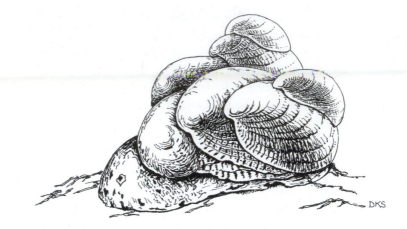

Figure 10.12 Copulating *Crepidulas*: the bottom organism is female, the top organism is male, and some of the intermediates are changing from male to female function. A case of protandrous sequential hermaphroditism where females gain more fitness by being large than do males. (By Dafila K. Scott.)

Protogyny: *Individuals are born as female, reproduce as females, then change sex and reproduce as males. In plants, individuals express female function prior to male function, being pollinated before producing pollen.*

organism is old, large and fully female, the top organism is young, small and fully male, and the intermediate organisms are in various stages of changing from male to female function, still producing a mixture of both kinds of gamete (Figure 10.12).

Protogynous organisms begin reproducing as females, then switch sex and function as males. For **protogyny** the size-advantage model suggests that small females should have more offspring than small males, large males more offspring than large females. These conditions are realized when big males can win fights to monopolize opportunities to mate with many females. The winning strategy is then to reproduce as a female until large enough to win the fights, then change sex. Such males often control harems of females, and when the male dies, the largest female in the group becomes a male. Most species in the marine fish family of wrasses, which includes the cleaner fish and the anemone fish, are protogynous. The male and female forms that one individual displays in the course of its development are so strikingly different that taxonomists have often classified them as separate species (Figure 10.13).

High-ranking females should have male-biased litters

Sex allocation can also vary as a function of social rank (Trivers and Willard 1973). In polygynous species, one male controls access to and mates with several females. Low-ranking females or females in poor condition should have female-biased litters or clutches. High-ranking females or females in good physiological condition should have male-biased litters or clutches. The reasons are these: if the social rank of the mother affects the condition of the offspring, offspring will tend to inherit their mother's social rank. Whereas female offspring in poor condition will always bear offspring, male offspring in poor condition will probably not father any offspring because they will not be able to compete successfully for access to females. If female offspring in good condition cannot have many more offspring than female offspring in poor condition, whereas male offspring in good

Figure 10.13 Male and female blueheaded wrasses: the brightly marked fish is a harem-holding male, the smaller fish are female, and if the male were removed, the largest female would change sex and start to function as a male within days, attaining normal male function within weeks. This is a case of protogynous sequential hermaphroditism where males gain more fitness by being large than do females. (Photo copyright Alexis Rosenfeld / Science Photo Library.)

condition will father many offspring by copulating with many females, then the relative reproductive success of male versus female offspring should increase with the social rank of the mother. Therefore low-ranking females should invest more in daughters, and high-ranking females should invest more in sons.

Clutton-Brock and Iason (1986) summarized the evidence for adaptive variation in the sex ratio of offspring in more than 30 mammal species. Many claims of sex-ratio variation were based on inadequate evidence: only a few mammal populations had sex-ratio variation strong enough to rule out random causes, and trends that appeared in one population often proved to be inconsistent when checked in other populations. However, at least one example of adaptive variation in the sex ratio of offspring is found in red deer, where high-ranked females produce more sons and low-ranked females produce more daughters (Figure 10.14). Part of the reason is that sons of subordinate females suffer higher juvenile mortality than sons of high-ranked females, and part is that most stags with high reproductive success are the sons of dominant mothers (Clutton-Brock and Iason 1986). In other words, sons inherit some of their mother's social rank. The mechanism that adjusts the sex ratio is not known, but only two are plausible: either sperm are selected before fertilization, or offspring of one sex are selectively aborted after conception.

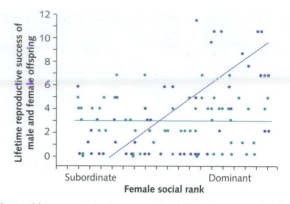

Figure 10.14 Offspring lifetime reproductive success, daughters versus sons, for female red deer of different social ranks. Daughters, green circles and line. Sons, blue circles and line. Mother's social rank has no effect on the reproductive success of daughters, but the sons of high-ranking mothers have significantly higher reproductive success than do the sons of low-ranking mothers. (From Clutton-Brock and Iason 1986.)

Under local mate competition produce few sons and many daughters

Unusual sex ratios result when all grandchildren stem from matings between brothers and sisters, an extreme case of local mate competition among siblings—brothers compete with each other for all possible matings. Because one son can inseminate all the daughters, a second son would be wasted, and the mother can get more grandchildren by producing another daughter than a second son. The optimal sex ratio is then one son and as many daughters as possible. An extreme example is *Acarophenax*, a haplo-diploid parasitic mite, in which the one son fertilizes all the daughters inside the mother, who is then eaten by her offspring. Fitness gained through male function is the same whether there is one son or many, but fitness gained through female function increases in proportion to the number of daughters, so the number of sons should be minimized and the number of daughters maximized.

Local mate competition: *The offspring of one parent compete with each other for mates.*

A parasitoid wasp flexibly adjusts the sex of its offspring to the size of its host

Some parasitoid wasps display striking short-term flexibility in sex allocation, which depends on prey size in a relative way. If a female encounters a series of prey, such as insect larvae, that are of different sizes, she will lay female eggs in the larger ones and male eggs in the smaller ones. The definition of what is large and what is small depends on the sizes encountered recently, for example in the last hour. If she encounters, say, 3 and 4 mm larvae, she will lay male eggs into the 3 mm larvae and female eggs into the 4 mm larvae. If she encounters 4 and 5 mm larvae, she will lay male eggs into the 4 mm larvae and female eggs into the 5 mm larvae. This resembles the size-advantage model. Male offspring lose less by being small than female offspring, which gain more by being large.

In these examples the connections between sex allocation, mating systems, sexual selection, and life-history evolution are strong. There is no distinction here between ecology, evolution, and behavior. All three fields are associated in how the examples described are explained.

● SUMMARY

This chapter analyzes how natural selection interacts with tradeoffs to design life histories and sex allocation for reproductive success.

- At what age and size should an organism mature? The answer depends on how likely it is that it will die as a juvenile, whether that risk differs from the risks it will encounter once it has matured and is adult, and the rate at which it gains potential fecundity as it grows and ages.

- Once it matures, how many offspring should it have? The answer depends on tradeoffs between offspring number and the expected lifetime reproductive success of offspring, and offspring number and its own subsequent reproductive success—in short, on fitness gained through current offspring and fitness gained through the rest of its reproductive activities.

- How long can it expect to live if it is not eaten by predators, killed by disease, or subject to accident? The answer depends on the kinds of tradeoffs it has inherited. They determine how much it can neglect maintenance to invest more in offspring. It also depends on how many genes with positive effects early in life and negative effects late in life have accumulated in its population.

- How much should it invest in male as opposed to female offspring, and if it is a sequential hermaphrodite, as what sex should it start life and at what age and size should it change sex? The answer depends on the sex ratio that it encounters in its local environment and on its capacity to acquire resources and social rank.

The answer to each of these questions could be 'whatever maximizes lifetime reproductive success.' When these questions have been answered, some important ones remain. How much should the organism invest in structures and behaviors that help it to win the competition for mates? How much should it invest in structures that make it attractive to the opposite sex? What is the opposite sex looking for in a mate? These questions are addressed in the next chapter, on sexual selection.

● RECOMMENDED READING

Charnov, E.L. (1982) *The theory of sex allocation*. Princeton University Press, Princeton, NJ.

Stearns, S.C. (1992) *The evolution of life histories*. Oxford University Press, Oxford.

● QUESTIONS

10.1. When Fisher imagined the simplest possible mating system and from that predicted a 50:50 sex ratio, he imagined organisms that are not at all like us. List all the ways in which Fisher's imaginary population differs from a real human one. Nevertheless, almost all mammals have a sex ratio that is very close to 50:50 at birth, including humans. Why is that so?

10.2. If we compare red deer to protogynous wrasses, it appears that the red deer have made the best of a bad situation. If they were sequential hermaphrodites, all would be born as females, and only those that grew rapidly and acquired good physiological condition and large body size would change sex and reproduce as males. The population would contain fewer males and more females, and the risk that any given individual would not reproduce at all would be lower. Why are red deer not sequential hermaphrodites?

10.3. In the kestrel study, the control birds had an average selective advantage of about 0.07 over the birds with manipulated clutches (13.09/12.08 = 1.08, 13.09/12.40 = 1.06). Assume that the population consists only of individuals producing one egg less than the optimum, that there is a single gene that can affect clutch size, and that a dominant mutant allele arises at this locus that increases clutch size by one egg. Using methods from Chapter 4 (p. 79), and assuming that selection is directional and that the frequency of the mutant at the start is 0.01, calculate the increase of the mutant. How long does it take the population to reach the optimum? What difference would it make if clutch size were polygenic with a heritability of 0.3 and selection were directional? How would the response change if selection were stabilizing?

10.4. Imagine a large population that consists of 25% adult males and 75% adult females. Mating is at random; mating success depends directly on the frequency of appropriate partners; both males and females mate only once. Now contrast two sex-allocation strategies. One type of female gives birth to three male and one female offspring. The other type of female gives birth to two male and two female offspring. Survival of offspring is independent of the sex-allocation strategy of the parent. All females in the population produce four offspring per lifetime. Which type will have more grandchildren?

CHAPTER 11

Sexual selection

Every morning before sunrise, from January to May, male sage grouse gather on their traditional display grounds, called leks, in eastern Oregon. As dawn breaks the males fan their tails, puff their chests, and arch their wings, strutting and emitting penetrating popping sounds, cooing between the pops (Figure 11.1).

They fight one another for the central display sites, and they remain on the lek despite strong winds and sub-zero temperatures. The females lurk in the bushes around the lek, occasionally entering to mate with a central male. The males have large, elaborately ornamented tails that they display in a fan and large chest pouches connected to their lungs from which they release air explosively, producing popping noises that can be heard far away (Figure 11.1). Their extravagant displays are expensive and dangerous. It takes a lot of energy to display vigorously in cold weather, coyotes and golden eagles are common in

● **KEY CONCEPT**

Sexual selection is the component of natural selection associated with mating success. It improves mating success at the cost of other components of reproductive success, such as adult survival.

Lek: *A traditional display site, often used for generations, where males gather to defend mating territories and females come to choose mates.*

Figure 11.1 Male sage grouse displaying on a lek. (Photo courtesy of Beverly Stearns.)

that habitat, and displaying males are conspicuous and relatively immobile. They get up before dawn to make love in the snow at the risk of being eaten. Why would such behavior evolve?

Key questions about sexual selection

This chapter addresses the following questions.

- How did sexual selection originate? It was made possible by anisogamy.
- How does sexual selection work? The main ideas are competition for mates and choice of mates.
- Is there good evidence for sexual selection? We discuss the spermatophores of katydids, the badges of status in male red-winged blackbirds, the long tails of male widow birds, and the colorful spots and large tail fins of male guppies.
- What determines the strength of sexual selection? Parental care and mating systems influence which sex has the greater potential reproductive rate. That influences the operational sex ratio, which contributes to the strength of sexual selection. The other chief determinant of the strength of sexual selection is variation among males in the number of mates that they have, which in turn is influenced by how the distribution of receptive females is clumped in space and time and by the life histories of females.
- How does sexual selection work in plants? Pollen scramble for fertilization and flowers compete for pollinators. Competition among pollen to fertilize ovules is analogous to competition among sperm to fertilize eggs, but pollinator choice is not analogous to mate choice, for the genes for flower morphology (in plant genomes) do not become associated in the offspring with the genes for flower choice (in pollinator genomes).

Operational sex ratio: *The local ratio of sexually active males to receptive females.*

Sexual selection explains the existence of costly mating traits

Mating success can be improved by competing for or being chosen by mates

● **KEY CONCEPT**

Choice of and competition for mates are the processes that drive sexual selection.

Traits like the displays of male sage grouse puzzled Darwin, for they obviously increase mortality rates. Darwin's solution was sexual selection, the component of natural selection represented by success in mating. The male sage grouse are taking a risk by displaying in the lek, and they are investing a lot of energy in feathers and vocalizations. The risk is worth it, for those that can produce a convincing display in the center of the lek will mate with many females. Those that cannot will not have any offspring at all.

There are two key processes that often determine mating success, both at leks and elsewhere: competition for and choice of mates. Usually males compete with other males for opportunities to mate, and females choose the males that they prefer. However those sex roles are sometimes reversed, with females competing and males choosing. Which sex plays which role depends on which sex limits reproduction. Usually males can mate and have offspring many more times per lifetime than can females. Under those circumstances, males compete for females, are eager to mate, and are not particularly discriminate in their partner choice; females choose mates carefully and are coy. However, in some species females can mate and have offspring more times per lifetime than can males, and in those cases females compete for mates and are eager to mate, whereas males choose mates and are coy.

Note that the word choice in the context of sexual selection does not necessarily refer to a conscious mental event. It refers to anything intrinsic to an individual that makes that individual more likely to mate with some partners than others. Think of it as a signal–receiver system: one partner makes a signal, and the other partner receives it and then reacts in a way that depends on the information contained in the signal. In the case of the female sage grouse, we are not sure what they are responding to. Is it the vocalizations, the location of the male within the lek, his ability to display vigorously for a long time, or some combination of all of these?

Sexual selection is not the only reason for sexual dimorphism

Some differences between males and females are not the product of sexual selection for improved mating success; they result from selection that has nothing to do with mating success but that still differs in the two sexes. We consider three cases, one resulting from differences in gamete size, one from differences in feeding ecology, and one from primary sex differences.

Differences in gamete size

For example, it is often energetically more costly for a female to produce eggs than for a male to produce sperm. And as we saw in Chapter 10 (p. 240) when considering sequential hermaphrodites, like *Crepidula*, that were evolving without a social system, females gain fecundity more rapidly by increasing size than do males. In such cases small males have higher fecundity than small females, and large females have higher fecundity than large males. We should therefore expect females often to be larger than males in species that are not sequential hermaphrodites as well as in those that are. We can think of that as the default condition, the condition that arises in the simplest situation.

Differences in feeding ecology

When males and females have different ecologies, their different morphologies may be caused by natural selection. Female mosquitoes suck blood from vertebrates, males feed on flower nectar, and the sexes have different mouth shapes for reasons that have nothing to do with mating success.

Primary sex differences

Sexual dimorphism may also reflect primary sex differences, the differences in morphology that are directly associated with reproduction rather than mating success. However, even primary sex differences may be subject to sexual selection. For example, human breasts, whose primary role is the nourishment of offspring, have acquired a secondary role in mate attraction. In such cases the distinction between primary and secondary sex characters breaks down.

The limiting sex can be choosy; the other sex competes for mates

If sexual selection has been responsible for producing sexual dimorphism, then two types of interaction were probably involved. Mate choice is one. The other is competition for mates. Competition for mates can occur both between males competing for females and between females competing for males. Mate competition will be stronger in the sex with the greater reproductive potential, which competes for the sex with the lesser reproductive potential. Usually male reproductive success is determined by how many females they can mate with, whereas female reproductive success is determined by how many offspring they can produce. That is why males usually compete and females usually choose.

While males are competing with each other for females, females are competing among themselves for the most attractive males. Similarly, while it is usually females that choose males for some reason, males may also choose females. Competition for mates is usually more important in males, and mate choice is usually more important in females, but both processes can occur at the same time in both sexes.

Mates are chosen for direct benefits, good genes, and attractive sons

The consequences of mate choice are subtler and more surprising than those of competition for mates. One reason is that choice involves a signal sent by one sex and received by the other sex. The signal communicates characteristics of the potential mate, and choosing a mate is one of the most important decisions a sexually reproducing organism makes. On what criteria should the choice be based? Some are straightforward. Does the potential partner control a superior feeding territory? When both parents care for the offspring, or when one partner feeds the other during part of the reproductive period, is the potential partner a superior forager?

Other criteria are not so easily detected. Is the potential partner healthy? Does it carry parasites that might infect oneself or one's offspring? Does it carry genes that would make offspring more susceptible or more resistant to infection? All populations contain some individuals infected with diseases or parasites, but not all infections are obvious. (Parasites may conceal their presence to enhance their probability of transmission.) If a partner uses a reliable signal to advertise that it is healthy and free of parasites, that should create a mating advantage, for mates would be selected to recognize and prefer such signals.

Reliable, honest signals should be costly. If they were not costly, cheaters could imitate the signal, deceive the partner, and produce mistakes in mate choice that, by resulting in less-fit offspring, would remove the reason for choosing partners on this basis. The organism receiving the deceptive signal would then evolve to ignore it. Thus potential mates should evolve signals that honestly indicate their state and cost something to produce, and potential partners should choose their mates on the basis of a costly trait.

The reason that sexual selection by mate choice can lead to surprising results is that the genes that determine mate preference and the genes that determine the traits preferred come together in the offspring. The genes for the preferred trait and for the preference then increase in frequency together, in self-reinforcing fashion. This sets in motion an evolutionary process with special features, including the possibility of arbitrary preferences for costly traits, traits that reduce survival while improving mating success.

Sexual selection is a component of natural selection

Sexual selection is thus selection for traits associated with mating success and partner choice. Mating success and partner choice trade off with other components of fitness, primarily adult survival, as do other reproductive traits, such as fecundity. If a change in a trait increases lifetime reproductive success sufficiently by improving the ability of an individual to attract mates and fertilize them, it will be favored by selection even though it lowers survival probability. Sexual selection will change traits influencing mating success until the improvement in mating success is balanced by costs in other fitness components; then the response will stop. That explains why males take risks to mate, why juvenile males develop their secondary sexual characters only on maturation, and why sexually dimorphic characters expressed in males are not expressed in females. If juveniles developed them before maturation, or if females developed them at all, they would suffer the costs without enjoying the benefits, which are restricted to mating males. Many sexually dimorphic traits that reduce the survival of the individuals that carry them can be explained with a tradeoff between mating success and survival. Thus sexual selection involves a tradeoff, as does selection on life-history traits. However, because the tradeoff is influenced by interactions between two or more individuals, it produces special features not found in standard life-history tradeoffs.

Selection for mating success can outweigh selection for survival

When we contrast the macroevolutionary pattern of sexually selected traits with the microevolutionary process that produces them, we notice an apparent

Figure 11.2 Birds of paradise. (a) and (b) Photos courtesy of David Rimlinger; (c) Photo copyright Douglas Janson; (d) Photo copyright Hans & Judy Beste/Ardea.

paradox (Shuster and Wade 2003). The macroevolutionary pattern indicates that closely related species can differ dramatically in sexually selected traits; there is no better example than the strikingly different plumage of closely related male birds of paradise (Figure 11.2). This suggests that sexual selection is one of the fastest and strongest types of selection.

Now we turn to the microevolutionary process. Sexual selection operates via mating success. Why is this single positive component of male reproductive success sufficient to outweigh the sum of the several negative components in the evolution of these exaggerated male traits? How can sexual selection be one of the strongest evolutionary forces when it affects primarily one sex, produces a trait in that sex whose expression is detrimental in the other sex, affects only one fitness component, and is exposed to selection for only part of the lifetime in the favored sex? The answers are to be found in the way that the sexes differ in variation in lifetime reproductive success (Chapter 2, pp. 28 ff.).

There are often large differences between the sexes in variation in lifetime reproductive success. Here is how those differences arise. If the breeding sex ratio is one male to one female, then every mating success for one male is a mating missed by another male. The least-successful males have less reproductive success than the least-successful females, and the most-successful males have greater reproductive success than the most-successful females. The fundamental cause of the sex difference in variation in reproductive success is variation in the number of mates per male. When this difference is large enough, selection for mating success in males will be stronger than the opposing viability selection in females and in males at earlier life stages.

How did sexual selection originate?

The road from the origin of sexual selection to the sage grouse lek has been a long one. When life originated about 3.8 billion years ago, reproduction was asexual, and it remained so for about 2 billion years. Sexual reproduction with meiosis and recombination originated with the eukaryotes about 1.5–2.0 billion years ago. The first sexual organisms were single-celled and produced gametes of equal size: they were isogamous.

The first step in the differentiation of sexes was the origin of mating types, which occur today in fungi, algae, and ciliates. In a species with mating types, gametes are of the same size, and there is no sexual dimorphism in the organisms producing the gametes. The only restriction on mating partners is determined by mating type: an individual can only mate with partners that are not of its own type. Selection for mating types is driven in part by inbreeding avoidance. It is also driven by frequency-dependent selection, for rare types can easily find many partners, but common types have a hard time finding an appropriate partner because they frequently encounter their own type.

Anisogamy was fundamental for the origin of sexual selection

The evolution of mating types can proceed in two directions. In the first, the number of mating types present in a population increases without limit,

> ● **KEY CONCEPT**
>
> The evolutionary sequence was first sex, then mating types, then anisogamy, and then sexual selection, which produced gender.

creating a situation in which almost every individual encountered is a potential mate. This may have happened in ciliates, in which 40 or more mating types have been found. When evolution takes the second direction, the number of mating types is reduced to two. The reduction to two mating types was necessary for the evolution of anisogamy, the critical step on the path to sexual selection.

Anisogamy: *Having gametes of different sizes; large eggs and small sperm.*

Differences in gamete size mean that females and males invest differently in reproduction

Once some isogamous population evolved two mating types, selection could change those mating types into organisms that differed in the size of their gametes. Larger gametes are thought to have been selected in one mating type because they resulted in zygotes with greater energy stores and better survival probabilities and because they produced more of the pheromones that attract gametes. Smaller gametes are thought to have been selected because they could be produced in greater numbers, were more mobile, and could actively find the large, pheromone-producing gametes. This led to specialization either on large or on small gametes, for intermediate forms paid the costs of being mediocre at both functions without realizing the full benefits of either. Thus anisogamous populations evolved with two types, one that produced large gametes, eggs, and another that produced small gametes, sperm. The individuals producing eggs were females and those producing sperm were males, by definition.

A female cannot have as many offspring per lifetime as a male can

Now, a key principle of sexual selection came into play. The lifetime reproductive success of females is limited by the number of eggs they can produce, that of males by the number of eggs they can fertilize. Females became a limiting resource for males, setting off competition among males for mates and allowing females to choose their partners. Both primary and secondary sexual characters then evolved. (Primary sexual characters have functions necessary for reproduction but not directly involved in mating success, for example, differences in the biochemistry of the ovaries and testes. Secondary sexual characters have been shaped by sexual selection for success in mating, for example the antlers of male deer, used in male–male competition for mates, and the extravagant display of the peacock's tail, which peahens use in choosing mates.) Males and females with striking secondary sexual characters are the product of a long history of sexual selection in anisogamous organisms.

Organisms compete for mates in contests, scrambles, and endurance rivalries

The main forms of direct competition for mates are contests, scrambles, and endurance rivalries.

- In a contest, the rivals display or fight directly with one another over mates or over the resources needed to attract mates. Bull elephant seals fighting for a stretch of beach, red deer stags fighting with their antlers for control of a harem, and male great tits fighting to defend their territories are all engaged in contest competition for mates.

- In a scramble, finding a mate quickly is crucial for success. Often the successful male is the first one to arrive. Pollen germinating on a flower style scramble to reach the ovum first.

- In endurance rivalry, persistence brings rewards. In frogs and toads with an extended mating season, the ability to keep calling night after night for many weeks can strongly affect reproductive success. Persistence also affects mating success in species with direct male–male contest competition. It can determine the amount of time that a male can display without leaving for food or water, and it can determine the length of a reproductive season during which a male can maintain his top rank in repeated fights with other males.

Mate competition co-occurs with mate choice whenever females use competitive ability as the criterion for choice.

The limiting sex is the one with the lower maximum reproductive rate

In katydids, or bush crickets (family Tettigoniidae), reproduction has an unusual feature that was used to demonstrate that the limiting sex is the one with the lower maximum reproductive rate. In katydids the sex in which mate competition is stronger alternates between male and female depending on the food available. Males transfer their sperm to the female in a large, nutritious spermatophore. The female eats the nutritious part of the spermatophore (but not the sperm, which fertilize her), and the nutrients in it increase her fecundity. When food is scarce, females are reproductively limited by the availability of male spermatophores. When food is abundant, males are reproductively limited by the availability of females. Changing food supply should therefore change the sex that must compete more strongly for mates.

The sex that competes for mates invests less in each reproductive attempt

As predicted, where food was limited, male bush crickets engaged less in courtship, female bush crickets fought over males, males were more discriminating in their

● KEY CONCEPT

Members of the sex with the higher potential reproductive rate, usually males, compete for matings with the sex with the lower potential reproductive rate, usually females.

choice of mates, preferring large, fecund females, and males invested more per reproductive attempt than did females. In contrast, when extra food was supplied, many males courted, females were more discriminating in their mate choice, and females invested more than males. When food is scarce, the male reproductive rate declines because they cannot produce spermatophores rapidly. When food is abundant, spermatophore production is rapid, and females have less reason to compete for males. Thus the sex that experiences greater competition for mates is the one that invests less in each reproductive attempt (Gwynne and Simmons 1990), and the limiting sex is the one with the lower maximum reproductive rate (Clutton-Brock and Vincent 1991).

Mate competition explains large, well-armed males

Mate competition is the main explanation for the evolution of sexual size dimorphism and the weapons used in fights over mates. The most striking examples of sexual size dimorphism in mammals occur in pinnipeds—seals and their relatives (Figure 11.3). Males can be up to six times as heavy as females, and there is a strong relationship between the degree of sexual size dimorphism and the number of females controlled by a breeding male (Figure 11.4).

Figure 11.3 Southern elephant seal bull and cows. (Photo copyright John Beatty/Science Photo Library.)

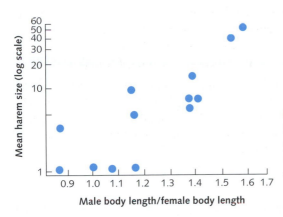

Figure 11.4 Sexual size dimorphism in pinnipeds increases with the number of females controlled by a breeding male. Each point represents one species. Elephant seals are at the upper right; harbor seals are at the lower left. (From Alexander et al. 1979.)

Elephant seals exemplify contests and endurance rivalry

In elephant seals the reasons for the striking size dimorphism appear to be male contests and endurance rivalry. Unlike whales, pinnipeds cannot give birth in the water. During the 3-month-long reproductive season, females haul out on beaches where they give birth and mate before returning to sea. The large concentrations of females are a resource that a male can defend, and males fight with each other for stretches of beach with females on them. Large males win most fights and access to females. Less than one-third of all males copulate at all during a breeding season, and just a few males account for most of the matings. During the mating season, males do not feed. Only those in good condition have the endurance to defend a harem for an entire season.

When a large bull controls a harem of many females, a small bull can sometimes sneak a copulation without being noticed, making possible some female choice. Female northern elephant seals protest against copulation by emitting loud calls and moving their hind quarters, attracting dominant males who chase off smaller males. Females protest less when approached by dominant males. While female choice has some role, male contests and endurance rivalry appear to explain most of the size dimorphism in pinnipeds.

Weapons have evolved in many species where males fight for mates, including deer and antelope; scarabid, lucanid, and cerambycid beetles; certain fish, which use their lips in wrestling; and narwhals, whose males carry a tusk up to 2.5 m long that can inflict serious injury on rivals. Experimental removal of antlers reduces reproductive success in male red deer and reindeer, for they use their antlers as both offensive and defensive weapons in contests over mates. Male beetles use their horns to pry up rivals and push them off mating sites or to grasp, lift, and throw their rivals to the ground.

Choosing mates can increase fitness, but choice has costs

● KEY CONCEPT

Organisms should choose their mates whenever they can using criteria that make evolutionary sense, but discovering the criteria they are actually using has often been a challenge.

Organisms should be careful in choosing a mate, but not too careful. Carelessness will result in hybridization with other species and offspring of low fitness; excessive discrimination will take so much time that the opportunity to mate will disappear before the choice can be made.

Many organisms choose mates for immediate benefits

Mates should either be chosen because they provide immediate phenotypic benefits or because they provide genes that increase offspring fitness. Some immediate benefits are these: the mate has higher fecundity, is a better provider of food to the partner or care to the offspring, defends a breeding place that is safer, richer in food, or both, or offers better protection against predators or other potential mates that might harass, than do other mates. For example, in scorpion flies, females choose males based on the quality of the food that the males bring to them. In mottled sculpins, which have male parental care, females prefer large males to small ones, for large males are better able to defend nests. In field crickets, females receive direct benefits from both repeated matings (with the same male) and multiple matings (with different males), apparently being able to absorb seminal fluid and use it as nutrition to produce eggs (Wagner et al. 2001).

Choice based on immediate phenotypic reward can explain size dimorphism and some exaggeration of traits that produce particular benefits, but it cannot explain the extravagant morphologies and displays of peacocks, sage grouse, and birds of paradise (Figure 11.2) that led Darwin to suggest sexual selection in the first place. In lekking species, the basis of the choice must be primarily the partner's genes, for there is no male parental care. Mating occurs swiftly on the lek, after which the female leaves to raise her offspring on her own. Except for temporary protection on the lek, the only things that a lekking male gives to the female are his genes. In lekking species females will be selected to choose indicators of male genetic quality that predict higher offspring fitness, including the ability of male offspring to attract mates.

Three other reasons for choice are good genes, attractive offspring, and sensory bias

There are three main ideas on how female preferences for male indicators of genetic quality should evolve. The first is that females should prefer males displaying honest, costly signals that suggest they contain genes for superior survival ability. This idea has been labeled either as the 'good genes' or as the 'handicap' hypothesis, the first stressing the fitness advantages, the second emphasizing the costliness of the honest signal. Andersson (1994) uses the neutral term 'indicator mechanisms'. Choice based on indicator mechanisms is expected to rapidly evolve into the next possibility, choice based on arbitrary preferences (Shuster and Wade 2003).

The second idea is that when females prefer males with higher fitness, their preference genes will be united in their offspring with the male's genes for higher fitness. The female preference genes will then spread in the population because they hitch-hike on the success of the male's fitness genes. Once female preferences are established, they select for male traits that would otherwise be neutral or disadvantageous—except that females prefer them. Females increase their long-term reproductive success by selecting mates with heritable traits that make their sons attractive to females in the next generation. This idea is called the Fisherian runaway hypothesis after Fisher, who had the idea.

The third idea is that females inherit sensory capacities from ancestors that bias the traits that they select in males. For example, color-blind females might select striking black and white patterns but not striking colors, and females incapable of hearing low-frequency sounds will not choose males that emit such sounds. This idea is called the sensory bias hypothesis.

We now examine each of these ideas in more detail.

It pays to advertise the ability to resist parasites and pathogens

If females select males because males signal that they can sire offspring with superior fitness, then a male trait that should interest females is resistance to infection by parasites and disease (Hamilton and Zuk 1982). Structures that might have this function are the red belly of the male stickleback, throat wattles in turkeys, and eye color in pheasants. If males produce ornaments to advertise their resistance to disease, then:

- male fitness should decrease with increased parasite infection;
- ornament condition should decrease with increased parasite burden: maintaining a parasite-sensitive ornament in good condition must be costly;
- there must be heritable variation in resistance;
- females should choose the most-ornamented and the least-parasitized males.

These conditions for the evolution of female preferences for males with superior resistance to diseases and parasites appear to be met in guppies, sticklebacks, and pheasants, but they have not yet been demonstrated in careful studies of several other species. Females do sometimes select males for parasite resistance (see Chapter 19, p. 492; Hutterite women appear to choose mates with genes that will help offspring to resist disease), but they also select males for other reasons.

Sexually selected traits balance the costs and benefits of sexual and natural selection

Guppies illustrate the balance of natural and sexual selection on the same trait. The natural enemies of guppies in their native streams are two species of predatory fish whose density increases as one goes from the headwaters of the streams down into larger rivers. As the density of predators increases, the number and size of the colored spots on male guppies decreases. Females prefer males with large orange spots. The orange color in the spots comes from a pigment acquired by eating crustacea; thus larger spots indicate better foraging ability. Females also prefer males with color patterns that make them stand out from the complex, pebbly background of the stream bottom, with its shifting patterns of light and shade.

Males from populations with low predation pressure engage in more-complex courtship maneuvers than do males from populations with high predation pressure. When predators are introduced to previously predator-free populations in the field, male ornaments rapidly become less striking and courtship behavior rapidly becomes less complex. Thus natural selection and sexual selection through female preference exert opposing pressures on male coloration and courtship behavior in guppies.

Female sticklebacks choose males by smell to get offspring of intermediate MHC diversity

Vertebrates react to infection with both innate and acquired immunity. The innate response is immediate and energetically quite expensive; the acquired response is more efficient but takes some time to elicit. The genetic basis of the acquired immune response resides in the major histocompatibility complex (MHC), a set of genes in vertebrates that produces the proteins from which antigens are constructed. The complex consists of several genes that form two classes. The antigens are formed by combining a molecule produced by a class I gene with a molecule produced by a class II gene. Because the antigens are produced by combining the products of genes, the range of antigens that can be produced depends upon having a variety of MHC alleles to work with, and some MHC genes are the most polymorphic known, having up to 200 alleles. The number of MHC molecules expressed in one individual is much smaller than the number existing in the population; humans express on average six different MHC class I molecules and eight different MHC class II molecules. It is advantageous for an individual to have a diversity of MHC alleles from which it can assemble a broad range of antigens with which to fight many types of infections.

Fish have a vertebrate immune system and an MHC complex, like humans. In a natural population of three-spined sticklebacks, individuals had from two to eight MHC class II alleles. Female sticklebacks choose males using olfactory cues on the basis of the male's MHC diversity (Figure 11.5a). Most females choose males that have high MHC diversity, but females who themselves have high MHC diversity tend to choose males with low diversity. By taking their own MHC diversity into account when choosing a mate, females appear to be aiming to produce offspring with five or six MHC alleles. Sticklebacks with five or six MHC alleles have the lowest parasite loads, both in natural populations and after experimental exposure to three common parasites (Figure 11.5b). Sticklebacks with lower MHC diversity not only suffered more from parasite

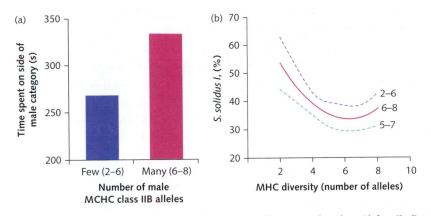

Figure 11.5 (a) When presented with a choice between spending time with males with few (2–6) or many (6–8) MHC alleles, female sticklebacks choose to spend time with males with more MHC alleles. (Reproduced with kind permission from Reusch et al. (2001) Female sticklebacks count alleles in a strategy of sexual selection explaining MHC polymorphism. *Nature* 414, 300–2.) (b) When experimentally exposed to the infectious stages of tapeworms, sticklebacks with 5–7 MHC alleles suffered less from infection than did those with either fewer or more MHC alleles. The y axis is a percentage index of tapeworm infection; the tapeworm is called *Schistocephalus solidus*. Numbers on the curve are sample sizes. (Reproduced from Kurtz et al., Major histocompatibility complex diversity influences parasite resistance and innate immunity in sticklebacks. The Proceedings of the Royal Society of London, Vol. 204, 2004, 197–204.)

infection after experimental exposure, they also had symptoms of a more intense—and expensive—innate immune response (Kurtz et al. 2004). Thus female sticklebacks appear to be choosing males by using very refined criteria, to yield not only offspring that are resistant to parasites, but offspring that are resistant in the most efficient manner possible.

Extra-pair matings in blue tits are an opportunity to study the reasons for mate choice

Blue tit females definitely choose mates; they may do it both for immediate benefit and to avoid inbreeding. Kempenaers et al. (1992) identified the fathers of blue tits in a Belgian forest by DNA fingerprinting. About one-third of the nests contained offspring sired by more than one male. Some females visited neighboring territories and solicited copulations; some males were often solicited; others were never solicited. Females paired with attractive males (males that were often visited by other females) rarely visited neighboring territories; those paired with unattractive males (males rarely visited by other females) often visited neighboring territories for extra-pair copulations. Unattractive males were smaller and died younger than attractive males, and the offspring of attractive males lived longer than the offspring of unattractive males. This could be explained by direct phenotypic benefit if attractive males had better territories and were better parents. If the extra-pair offspring of attractive males also had better survival, which is not known, it would be hard to explain the results only in terms of direct phenotypic reward—there would then have to be a genetic component to male attractiveness itself.

Female blue tits do increase the average heterozygosity of their offspring through extra-pair matings. Extra-pair offspring have better survival, and the males have more elaborate crown color, a secondary sexual trait that improves their mating success (Foerster et al. 2003). Thus female blue tits are making choices somewhat similar to the female sticklebacks discussed above: they are choosing mates with different and complementary genes.

When preferences and preferred traits coevolve, extravagant ornaments can result

Fisher suggested another idea for how female preferences evolve. The Fisherian process can start with mate choice for any reason—direct benefits, good genes, or sensory bias—and once it gets started, it can rapidly take on a life of its own that produces mate choices for new reasons. Consider the beginning of a Fisherian process started by the choice of good genes in guppies. The males have brightly colored spots. Suppose that guppy females start to select males with a trait, such as orange spots, that varies genetically and is an indicator of the ability to acquire food and resist parasites. Females choosing males with larger spots will have sons with better survival. The genes for large orange spots in males will spread in the population because they are associated with better survival, and the genes that make females prefer large orange spots will also spread. So far, the argument is strictly in terms of the evolution of preferences for good genes.

Once this process starts, it makes a new effect possible, for a female that chooses a male with an extreme ornament (the orange spot in this case) will tend to produce both daughters with extreme preferences and sons with extreme ornaments. Because the male offspring are preferred by females and therefore have superior fecundity, genes for orange spots and for the preference for orange spots spread through the population. The process is self-reinforcing. Now every time a mutation arises that increases female preference for orange spots, it will spread, and the stronger the preference in the females for orange spots, the stronger the selection for big orange spots in males. The ever-stronger preference in the females causes the ornament in the males to evolve with ever-increasing speed. For this reason the process is called Fisherian runaway selection.

Fisher argued that this process would progressively exaggerate male ornaments until the sexual preference was balanced by a reduction in male survival. In fact, if the female preferences are strong enough, then reductions in male survival are not a large enough cost to stop the process before the population is driven to extinction. Some other factor must be involved to explain how populations with strong female preferences survive. That factor appears to be the cost of having a preference, the cost of taking the time—and the risks— necessary to choose a male (Bulmer 1989). When the female preference is costly, a Fisherian process can exaggerate the male trait, but not without limit, for it will be stopped by the costs of choice. It could still produce impressive ornamentation before it was stopped. The spectacular plumage of birds of paradise is largely the product of sexual selection by female choice, and some of it is so exaggerated (Figure 11.2) that a Fisherian runaway process must have been involved.

In sticklebacks, male color is genetically correlated with female preference for that color

The theory of the Fisherian process is more complete than are the experimental tests of its assumptions and predictions. The problem is to find genes for female preferences and male ornaments and to document the dynamics of their coevolution under experimental conditions. Good evidence for one assumption was provided by Bakker's (1993) study of sticklebacks. Stickleback males guard nests into which females lay eggs; the male develops red coloration on his belly in the breeding season. Both the red coloration of the males and the preference of females for males of different degrees of coloration were found to vary genetically, and the sons' intensity of red coloration was genetically correlated with the daughters' preference for red (Figure 11.6). Such a genetic correlation of male trait and female preference, a necessary condition for the Fisherian process, could also result from selection for indicators of parasite resistance. In fact, by choosing males with brighter-red bellies, females avoid parasitized males (Milinski and Bakker 1990). Red bellies may be an indicator mechanism, but the Fisherian process may also be involved. Andersson (1994) summed up as follows: 'No critical test has been performed that supports Fisherian sexual selection and excludes the alternatives, or estimates their relative importance.'

Figure 11.6 (a) Male sticklebacks develop red coloration during the breeding season. (Photo copyright Laurie Campbell/NHPA.) (b) The correlation among stickleback fathers between the intensity of red coloration in sons and the preference of daughters for red males. (From Bakker 1993. Reproduced by kind permission of the author and *Nature*.)

Mate choice may depend on pre-existing features of the female sensory system

The third hypothesis for the evolution of female preferences focuses on their origin. Which male trait is exaggerated by sexual selection may depend on its fit to pre-existing features of the female sensory system (Ryan 1985). The cases that best support this notion are the mating calls of male tungara frogs and the preference of female platyfish for males with sword-like tails.

Male tungara frogs call from pools visited by females. Males emit two calls, 'whines' and 'chucks.' The whine appears to be a species-identification call and is given by isolated males. When a male begins to hear competition from other males, he starts to emit chucks, which are attractive to females. Isolated males do not emit chucks because they attract the fringe-eared bats that eat displaying males, making chucks costly. However, the tradeoff is that females select males that have deep chucks, for the offspring of such fathers have better survival (Ryan 1985). So far, it appears that female tungara frogs prefer deep chucks because they indicate good genes, and that may well be the mechanism that now maintains the trait. The neurobiology of the female ear in the tungara frog and its close relatives suggests, however, that another explanation for the origin of the preference is plausible.

The female ear is biased toward the low-frequency components of the chuck, and females of a closely related species have a similar sensitivity to the chuck, even though males of that species do not produce it. There are two ways to interpret this evidence. We can postulate that the two species had a common ancestor whose ears were biased to hear the chuck but whose males did not yet produce it. In that case, the male chuck first evolved in the tungara frog in response to a pre-existing female bias. Or, we can postulate that the common ancestor of both species had both the male chuck and the female sensory bias, but that the related species lost the male chuck. Because closely related species do not have the chuck but do have the preference, the first scenario is more plausible.

Genetic variation among males is required for female preferences to evolve

There is a potential logical problem with the evolution of female preferences for indicators of male genetic quality. It requires that males vary heritably in fitness. If they did not, there would be no reason to choose among them. However, fitness is certainly under directional selection to increase, and under continued strong directional selection, the genetic variation for fitness should be reduced to near zero (recall the discussion of selection–mutation balance in Chapter 5, pp. 111 ff.). It would appear that female preferences for indicators of male genetic quality could not evolve because there would not be enough genetic variation among males to give females any benefit from their choice.

In fact, three factors could maintain enough genetic variation for fitness to allow sexual selection to work: spatial variation, temporal variation, and mutations affecting fitness. If conditions vary from place to place, or from time to time, so that the most-fit genotype at one time or place has lower fitness at another time or place, and if gene flow and environmental variation cause changes in fitness rank frequently enough, then considerable genetic variation for fitness can be maintained in a population. Also, if there is a steady flow of favorable mutations into a population, which will happen if the population is large enough, then natural selection will be continually pulling some favorable mutations through to fixation. At any time we would always find some genetic variation for fitness, for some favorable mutations would be increasing in frequency and would not yet have gone to fixation.

Thus there is an objection to the theory of sexual selection: genetic variation for fitness in males is expected to vanish. However, this objection is not fatal, for it is based on a simplified view of how selection works in natural populations. How much genetic variation for male fitness can be maintained in natural populations by these mechanisms is a question not yet answered; it needs to be enough to allow female choice to work as a mechanism of sexual selection.

There is a great deal of evidence for sexual selection

Andersson (1994) cites studies of 186 species, mostly insects, fish, anurans, and birds, in which sexual selection has been demonstrated 232 times. The most common mechanism, female choice, occurred in 167 cases. In 76 of those 167 cases any influence of male contests was excluded. In 30 cases male choice, usually of large females, was documented, and in 58 cases males contested access to females. The trait most commonly selected was male song or display, followed by male body size, male visual ornaments, female body size, male territory, and other material resources. Because of the theoretical interest in mate choice, results on mate choice are probably over-represented and results on male contests are probably under-represented in the literature. Little is known about the genetics of most traits studied.

> ● **KEY CONCEPT**
>
> That sexually selected traits are among the most striking and costly products of evolution emphasizes the central role of reproductive success.

An extravagant ornament can be maintained by male–male interactions

Redwing blackbirds are strikingly dimorphic, the females being brown and streaked, the males mostly black with bright red shoulder patches with lower yellow borders. These patches, called epaulets, are displayed in territorial contests with other males. In an experiment, the epaulets were painted black, with clear paint used on controls, or presented in dummies with epaulets of different sizes. Males with epaulets painted black lose their territories to males with normal patches, and dummies with larger epaulets provoke stronger aggression in territory-holding males than do dummies with smaller epaulets. Males that are just passing through a territory conceal their epaulet, but those prepared to fight for a territory expose it. Thus it appears to function as a threat display. Thus an extravagant ornament can be maintained in a population by male–male interactions.

While female choice is not necessary to explain the existence of traits selected through male–male competition, once evolved they may be good traits for females to use in choosing, for it is difficult to fake the quality of weapons and badges of status that are constantly on trial in male–male competition. The evolutionary sequence could also go in the other direction for badges of status, which may be derived from signals that originally evolved in the context of female choice, then became appropriated for male–male interactions (Berglund et al. 1996).

Experiments on African widow birds show female preference for males with long tails

Male widow birds establish breeding territories on the grasslands of East Africa. They have tails that are often more than half a meter long. The way that males use their tails suggests that they function in female choice rather than male contests, for males do not expand their tails during territorial contests, but they do expand them into a deep keel during the advertising flights they perform when females visit their territories. Andersson (1982) tested the function of these long tails by experimentally shortening and lengthening them in the field. He set up nine groups of four males each, matched for territory quality and tail length. He took one male at random from within each group and cut its tail to 14 cm. The piece removed was then glued to the tail of another of the four males in that group. The other two males served as controls. One was not manipulated; the other had its tail cut at the midpoint and re-glued.

To measure the mating success of the males, Andersson counted the number of nests on each territory. Before the manipulations (Figure 11.7), all males had on average 1.5 nests on their territories. After the manipulations, plus sufficient time for females to build new nests, the males with shortened tails averaged only half a nest per territory and the males with lengthened tails averaged nearly two nests per territory. Evidently some of the females from the territories of the males in the first three treatments had moved into the territories of the males with artificially lengthened tails. Thus female widow birds prefer to build their nests on the territories of males with longer tails. Natural selection may be preventing further increases in male tail length, for females preferred much longer tails than are found in natural populations. This case demonstrates female choice for an exaggerated male ornament in a field manipulation experiment.

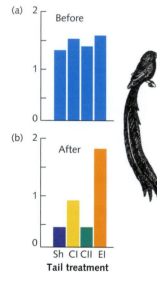

Figure 11.7 Experimental changes in the length of the tails of male long-tailed widow birds cause changes in the numbers of nests that females build on their territories. Panel (a) shows the number of nests per territory before tails were artificially shortened or lengthened; panel (b) shows the number of nests after the manipulation. Sh, tails shortened; El, tails elongated; CI, unmanipulated control; CII, re-glued control. (From Andersson 1982. Reproduced by kind permission of the author and *Nature*.)

What determines the strength of sexual selection?

Darwin saw a relationship between patterns of parental care, the mating system, and the strength of sexual selection. For analysis in depth we refer you to books by Clutton-Brock (1991), Andersson (1994), and Shuster and Wade (2003). Here we mention only the key points.

● KEY CONCEPT

The sex with the higher potential reproductive rate competes more strongly for mates and is the sex subject to stronger sexual selection.

The strength of sexual selection is determined by the ratio of potential reproductive rates

Bateman (1948) demonstrated that male fruit flies (*Drosophila*) can increase their lifetime reproductive success by increasing the number of females that they mate with, whereas female *Drosophila* achieve most of their reproductive success with a single mating. Bateman concluded that this was why males competed for mates and females were the scarce resource. Bateman's principle, that *male reproductive success depends on number of matings per lifetime, female reproductive success depends on number of offspring produced per lifetime*, holds for species that have no parental care and where the male does not feed or protect the female. For more complex mating systems, we need a more general principle (Clutton-Brock 1991): *the sex with the higher potential reproductive rate competes more strongly for mates and is the sex subject to stronger sexual selection.*

The principal factors determining the potential reproductive rates of the two sexes are the mating system and the pattern of parental care. How these factors interact to influence the strength of sexual selection can be seen in the determinants of the operational sex ratio, the ratio, at any given place and time, of receptive females to sexually active males. This ratio depends:

- on how individuals of the limiting sex form groups;
- on differences between the sexes in survival rates; and
- on differences between the sexes in the time that it takes to find a mate, breed, and care for the offspring.

The operational sex ratio is an important determinant of the opportunity for sexual selection.

However, the strength of sexual selection does not depend just on the local operational sex ratio. That is only the sole determinant of the difference between the sexes in selection when each male has the same number of mates. When the variation in number of mates per male is large, then that variation contributes strongly to the sex difference in the opportunity for selection. The chief factors contributing to variation in number of mates per male are the spatial and temporal aggregation of receptive females and female life histories (Shuster and Wade 2003).

- *Spatial aggregation.* The variation in number of mates per male will be large when females are spatially clumped, so that the male that manages to get into an aggregation of females will get many matings, and males elsewhere will get fewer or none.

- *Temporal aggregation.* Mating opportunities vary in time just as much as they do in space. When females are synchronously receptive, the ability of a few males to mate with many females decreases, the variation among males in reproductive success declines, and sexual selection becomes less strong. When female receptivity is spread out over a long mating season, then some males may succeed, and others fail, in serial polygamy, and the variation in number of mates per male can be large. (Variation in the mating success of a single male from one part of the mating season to the next—for example, some good early but bad late, others bad early but good late—will decrease the variation in mating success among all males over the entire season.)

- *Female life histories.* Female life history also influences the strength of sexual selection and the criteria that females use in choosing males. When females are polygamous—mating several times with different mates—and iteroparous—they have several sets of offspring per lifetime—the differences between the sexes in the opportunity for selection (the variation in reproductive success) are reduced. Then, because of the potentially large differences in the quality of clutches resulting from different mates, the genetic quality of males may become an important criterion to females (good genes).

Thus the availability of receptive females in space, in time, and over the course of the female's life determines, in a broad sense, the mating opportunities for males. Within that broad structure, specific mating systems determine the details, which we now examine.

Mating systems describe which sex has more offspring or more mating partners

The key impact of space, time, and female life history on sexual selection and on the difference in the opportunity for selection in the two sexes is realized through the number of mates per lifetime of males and females. Therefore the classification of mating systems is based on those numbers (Shuster and Wade 2003). The opportunity for sexual selection is weakest under monogamy and strongest under polygyny and polyandry.

Monogamy: *Each sex has a single mate for life.*

In **monogamy**, equal numbers of males and females have offspring, and each sex has one mate for life, e.g. moorhens, albatrosses, and swans (Figure 11.8); dung beetles; and some cichlid fish. Usually both parents care for the offspring. Experience with the partner in a monogamous relationship can improve performance. In the kittiwake and the Manx shearwater, the reproductive success of pairs that had been together for several years was greater

Figure 11.8 Monogamous trumpeter swans. They live 15–25 years on average, lay 4–6 eggs per year, and mate for life. (Photo courtesy of Jim Zipp.)

than that of pairs that had formed more recently. In apparently monogamous song birds the reason for mate fidelity is probably not only the need for two parents to rear the offspring but in some cases the prevention of extra-pair copulations by aggressive interactions. That apparent monogamy may conceal serious conflict is suggested by genetic determination of paternity, which has revealed that some species that had been thought to be monogamous are in fact promiscuous.

In polygyny, the commonest mating system in mammals, females mate with a single male for life, males may mate with more than one female, and more females than males have offspring. Usually the females care for the offspring. Polygynous species include those where males hold harems, such as gorillas, elephant seals, horses, and red deer, and those where males display on leks, including grouse, cotingas, birds of paradise, eight species of mammals, including Uganda kob and fallow deer, and at least one insect, a Hawaiian *Drosophila*. Thus the term polygyny covers a great diversity of situations. When the females occur in social groups with a limited home range defensible by the male, as in Hanuman langurs (a ground-dwelling monkey native to India), or wander in a group within a limited area to forage and drink, as in Grevy's zebra (Figure 11.9), or come together in groups during a mating season, as in red deer and elephant seals, they can be controlled as a harem by an aggressive male whose task is exhausting. Elephant seal males playing the role of harem master often only manage to hold it for a year or two before dying.

Polygyny: *Females mate with a single male but males may mate with more than one female.*

Figure 11.9 A harem of zebras. (Photo courtesy of Beverly Stearns.)

Figure 11.10 A polyandrous female red-necked phalarope in Iceland. (Photo copyright Jim Zipp)

Polyandry: *Males mate with a single female, but females may mate with more than one male.*

In **polyandry**, males mate with a single female for life, but females may mate with more than one male, and more males than females have offspring, e.g. pipe fish and sea horses, some plovers and sandpipers, and jacanas (birds found in South and Central America). Polyandry in birds probably evolved from monogamy with biparental care, followed by a transitional state in which the female occasionally deserted, leaving the male to sit on the nest, followed by male-only parental care. In the phalarope (a relative of plover; Figure 11.10), sex dimorphism has changed with sex role. The females are larger and more brightly colored than males, they compete for males to incubate their clutches, and when a female phalarope performs her mating display, her ovaries secrete testosterone. Polyandry can be either sequential or simultaneous. The female phalarope lays a series of clutches in sequence, moving from one male to the next and leaving each with eggs to incubate. Female spotted sandpipers and jacanas defend large territories in which several males incubate clutches simultaneously.

Polygynandry is exemplified by the dunnock, a small brown bird that early naturalists thought was monogamous and that was, ironically, occasionally cited as an example of marital fidelity in church sermons in England. Genetic fingerprinting has revealed an extremely flexible mating system, including monogamous pairs, males with two females, and females with two unrelated males. Females have least success in polygyny, more in monogamy, and most success in polyandry. Males have most success in polygyny and least in polyandry. These conflicts of interest are mirrored in the types of aggressive behavior displayed in each mating situation. In polygyny, the male tries to keep both females and the dominant female tries to drive the subordinate female away. Monogamy is better for a male than polyandry because the dominant male suffers from interference costs incurred by the subordinate male in polyandry. Therefore in polyandry dominant males try to drive off subordinate males, and whenever a subordinate male has the opportunity, he leaves the polyandrous territory to join a single female on a neighboring territory and become monogamous (Davies 1992).

These labels describe rough patterns. Precise comparisons require quantitative measurements of the effect of the mating system on the number of mates per male and per female and on the operational sex ratio. When the full variation of receptive partners in space and time is taken into account, we find that there are many categories of mating system, among which the strength of sexual selection and the difference in the opportunity for selection in the two sexes vary enormously. That variation makes clear:

- that mating systems have evolved in many different directions, becoming quite diverse and difficult to fit into simple categories;
- that there is great variation in the mating systems of insects, crustaceans, and other invertebrates, which offer many opportunities for research; and
- that the traditional categories of monogamy, polygyny, polyandry, polygynandry, polyandrogyny, and polygamy fail to capture much of the important variation in the spatial and temporal structure of mating systems.

> **Polygynandry:** *Both sexes are variable in their mate numbers, but males are more variable than females.*

> **Polyandrogyny:** *Both sexes are variable in their mate numbers, but females are more variable than males.*

> **Polygamy:** *Both sexes have several partners and approximately equal variation in mate numbers.*

Sexual selection is strongest in species with harems or leks, with exceptions

Sexual selection has apparently been strongest in species with harems or leks where the operational sex ratio is strongly female-biased and where receptive females are concentrated in both space and time. In harem-holding pinnipeds, the more females a male can control, the more intense the fights among males over females, and the greater the difference in the sizes of the sexes. Striking examples of sexual dimorphism in plumage ornaments are found in lekking birds: grouse, peacocks, and birds of paradise. Not all species that hold harems are strikingly size dimorphic: horses and zebras are examples. Not all species with leks have males with exaggerated ornamentation: one-quarter of all lekking birds are sexually monomorphic.

For several reasons, the match between mating system and sexual dimorphism is not precise.

- The classification of the mating system could be wrong. This can be checked with genetic fingerprinting of offspring to determine the real parents.

- Opportunities for mate choice and the reasons for choosing mates may not match the classification of mating systems precisely. This can happen when the mating system is incorrectly classified, as was the case with polygynandrous dunnocks before they were carefully investigated.

- The strength of sexual selection can vary among the species within a mating-system category. Rather than dealing with qualitative categories and expecting them to match the diversity of nature, it is better to measure the strength of sexual selection and the difference in the opportunity for selection in the two sexes. The natural variation is quantitative, not qualitative, and the natural diversity is not easily fitted into categories, no matter how numerous.

Sex-role reversal is an exception that confirms the rule

In some species—giant water bugs, some other hemipteran insects, some cichlid fish, pipefish, and seahorses, and some shore birds and jacanas—only males care for offspring. They provide a test case for sexual-selection theory: males are the limiting resource, so females should compete for males and males should choose mates. In spotted sandpipers (Oring et al. 1991), each male rears one clutch of four eggs per season, but a female can lay up to five clutches per season. This reverses Bateman's principle: here female reproductive success increases with the number of mates, but male reproductive success does not. Females arrive earlier than males on the breeding grounds and fight among themselves for breeding territories, for the number of males that a female can attract increases with the size of her territory and the length of foraging beach that she controls. Thus males make greater parental investment, have the lower potential reproductive rate, and are the limiting resource for the females, which compete more strongly for mates (Andersson 1994).

Sexual selection in plants involves pollen and pollination

Sexual selection in plants occurs through competition among pollen grains to reach the ovary and through competition among flowers for pollinators. Sexual selection differs in plants because many plants are hermaphrodites, with male and female functions combined in a single individual, and because many plants use insects as pollinators. Sensory bias operates in pollinator–flower coevolution, influencing flower morphology by restricting the types of signals that result in an improvement in pollination success. For example, bees can see in the ultraviolet, and many bee-pollinated flowers are decorated with designs leading to the nectary and only visible at ultraviolet wavelengths. Showy floral displays are exaggerated ornaments used to compete for scarce pollen—they are cases of sex-role reversal where female functions compete for scarce male services. Sometimes the exaggeration of female traits is driven by the avoidance of nectar-stealers that do not pollinate rather than by sexual selection, for example, the Madagascar star orchid with its 30 cm floral tube and pollinator with a 30-cm-long tongue (Figure 2.8).

● **KEY CONCEPT**

Many plants have sex-role reversal: female functions compete for scarce male services.

Sexual selection also occurs in gametes

Up to now we have discussed sexual selection as a process operating on the large, diploid stage of the life cycle. It also operates on gametes (Bernasconi et al. 2004). In organisms where multiple insemination occurs, there will be competition among the gametes donated by different fathers. The females may be passive, but it is more likely that females will actively choose sperm or pollen if they carry information on male quality. Female choice can be exercised either by the inseminated females, through morphological structures and physiological and molecular mechanisms, or by the eggs themselves.

● **KEY CONCEPT**

Sperm compete and females (eggs or adults) choose.

Male–male competition at the gamete stage

In angiosperm plants, pollen compete for eggs whenever the number of pollen deposited on a pistil is greater than the number of egg cells that can be fertilized. In most plants that are pollinated by wind or by insects, paternity analysis has shown that the seeds within a fruit are often sired by more than one pollen donor. Competition among sperm occurs whenever a female mates with more than one male. Adaptations to mediate sperm competition are widespread in animals and include sperm-storage organs, copulatory plugs, modifications of genitalia to aid in the removal of rival sperm, and mate guarding.

One of the more intriguing adaptations mediating sperm competition is sperm heteromorphism—sperm from a single donor that differ in morphology and genetic content, one type being fertile and the other sterile (Swallow and Wilkinson 2002). Such sperm are widespread, if not common, in arthropods, occurring in butterflies, moths, flies, bugs, and spiders. The sterile sperm are thought to aid in transporting fertile sperm and in displacing rival sperm; if that is the case, then they resemble an altruistic caste of worker insects. They also occur in some mammals, where they may sensitize the female's immune system to the presence of a particular paternal genotype in her body and thus help to keep a zygote expressing that genotype from being expelled as a foreign body.

Some plants produce heteromorphic pollen. They do so in two ways. In a few cases fertile pollen is produced on some anthers and sterile pollen on others; the sterile pollen is used to feed pollinators and the fertile pollen to produce offspring. More commonly all the pollen is fertile and the heteromorphism is a risk-spreading strategy like the seed heteromorphism discussed in Chapter 10 (p. 218). For example, in the violet *Viola diversifolia* pollen may have few or many apertures. Pollen with many apertures germinates more quickly and is more competitive on freshly opened stigmas; pollen with few apertures survives longer and can sample more pollination opportunities in time.

Female choice at the gamete stage

Females can promote competition among male gametes by having an elongated, morphologically complex, and physiologically challenging reproductive tract. Selecting among gametes will cost females less than selecting among zygotes or early embryos. Recipient plants influence the growth rates of pollen tubes; many also have mechanisms to prevent self-fertilization. In animals with sperm storage, some females display a last-in, first-out pattern, others mix sperm and use them at random, and still others expel sperm from earlier mates and use sperm from later mates, thus using sperm manipulation as part of the mechanism of mate choice. Whether eggs or ovules themselves can choose among several sperm or pollen has not yet been clearly established, but it seems likely.

● SUMMARY

This chapter considers the evolutionary consequences of competition for and choice of mates: sexual selection.

- Sexual selection is a component of natural selection in which mating success trades off with survival.

- Sexual selection accounts for many of the attractive ornaments of plants and animals.

- Contest competition for the mate that is the scarcer reproductive resource explains much of sexual size dimorphism, particularly in polygynous species. Active choice of mates of the non-limiting sex has also been well documented.

- The strength of sexual selection is determined by the operational sex ratio, by the spatial and temporal aggregation of receptive females, and by the number of times females reproduce per lifetime.

- The huge natural diversity of mating systems is not easily fitted into rigid categories, but such categories can be constructed and are somewhat informative. To understand the strength of sexual selection and the opportunity for natural selection in both sexes, one cannot rely on predictions made from somewhat questionable categories—one must observe the spatial and temporal distribution of matings in both sexes for individuals followed until they die, so that lifetime mating success can be measured.

- Alternatives for sexual dimorphism should be considered in the following order: (1) primary sex differences, (2) ecological sex differences, (3) choice for direct phenotypic benefit, (4) the Fisherian process and choice for good genes.

Sexual selection often involves mate choice, and mate choice is also involved in speciation, which is discussed next.

RECOMMENDED READING

Andersson, M. (1994) *Sexual selection*. Princeton University Press, Princeton, NJ.

Darwin, C. (1871) *The descent of man, and selection in relation to sex*. John Murray, London.

Mead, L.S. and Arnold, S.J. (2004) Quantitative genetic models of sexual selection. *Trends in Ecology and Evolution* **19**, 264–71.

Shuster, S.M. and Wade, M.J. (2003) *Mating systems and strategies*. Princeton University Press, Princeton, NJ.

QUESTIONS

11.1. Wild turkeys are dramatically sexually dimorphic; domestic turkeys are less so. Suppose that wild female turkeys chose males on the basis of expensive traits that indicate disease resistance and that artificial selection for rapid weight gain in domestic turkeys destroyed female choice. What would you predict about the evolution of disease resistance in domestic turkeys? If we observe that domestic turkeys are less resistant, does that necessarily mean that wild females had been choosing more-resistant males, or are other hypotheses equally plausible?

11.2. Butterfly fish are sexually monomorphic, apparently monogamous, and strikingly colored. Can their striking coloration be explained by sexual selection involving mate choice? An alternative explanation is that the striking colors are species-recognition mechanisms, that they help butterfly fish to avoid choosing the wrong mate. If that is the case, then what is the difference between sexual selection by mate choice and species recognition?

PART 3

Principles of macroevolution

Parts 1 and 2 discussed microevolution; Parts 3 and 4 discuss macroevolution. Microevolution occurs within populations and species, involves changes in gene frequencies, genetic drift, and phenotypic design for reproductive success, and occurs rapidly. Macroevolution occurs among species and clades, involves speciation, extinction, and biogeographic patterns, and occurs slowly. Whereas microevolution concentrates on process, macroevolution focuses on pattern. We can watch microevolution occurring, but we must usually infer macroevolution indirectly from fossils and the comparative method.

Microevolution influences macroevolution by designing traits that affect the probability of extinction and rates of speciation. Macroevolution influences microevolution by providing a context of older traits that constrain microevolution's course. These include modes of growth and development; body plans; ability to live in water or on land; the ability to produce wood, to branch, to fix carbon dioxide with alternative biochemical pathways, and so forth. Many of these constraints are not absolute but relative—the slowly evolving process constraining the rapidly evolving one.

Development and drift link micro- to macroevolution. The macroevolutionary aspect of development resides in the deep conservation of developmental control genes; its microevolutionary role arises most clearly in phenotypic plasticity and reaction norms. While genetic drift is a key process in population genetics, it is also a necessary assumption for the methods of molecular systematics that reconstruct the Tree of Life.

Speciation also connects micro- and macroevolution. A by product of selection and drift, it produces the species whose differences generate macroevolutionary patterns. We therefore begin our discussion of macroevolution with speciation (Chapter 12), then examine how to infer the relationships of species—phylogeny and systematics (Chapter 13). Given relationships, we can use comparative methods to investigate history (Chapter 14). With macroevolutionary principles in hand—speciation, phylogenetics, and comparative methods—we go on to consider, in Part 4, The history of life.

CHAPTER 12
Speciation

Speciation connects micro- to macroevolution

Chapters 2–11 described the microevolutionary processes that occur within populations. In this chapter we start to make the transition to macroevolution, the patterns in fossils and phylogenies above the species level. The bridge between micro- and macroevolution is speciation, which is responsible for the diversity of life.

And life forms are incredibly diverse. They range in size from viruses that can be seen only with electron microscopes through 100 ton whales up to the clones of some trees that may cover a square kilometer. Some live under the Antarctic ice, others in hot thermal vents on the ocean floor at temperatures well above the boiling point of water. Present evidence indicates (see Chapter 13, p. 303) that all organisms now alive—with the possible exception of viruses—shared ancestors that lived more than 3500 million years ago (earlier forms may have left no descendants). Thus there has been enormous diversification during evolution. The basic unit in which most life forms are classified is the species, of which about 10 million currently exist. Only about 1.4 million species have been described and named. The fossil record shows that many species have existed that are now extinct.

> ● **KEY CONCEPT**
>
> Life presents itself as a set of discrete groups. A species is one such basic natural unit.

Why does life present itself as a set of discrete groups?

This enormous diversity is largely a diversity of discrete groups, not a continuum. Living things tend to occur in groups such that those within a group resemble each other more than those from different groups. Species are a particularly important type of natural group. They are usually separated from other species by differences in appearance, behavior, ecology, genetics, and many other biological characteristics. Why is the living world structured into discrete groups? How did this diversity come about? How are species defined, and how do they originate?

What is a species?

In practice taxonomists use all sorts of differences to identify species

● KEY CONCEPT

While different criteria can be used to define species, the diversity of nature challenges all simple definitions of species.

Anyone can discriminate between a leopard and a tiger, although the individuals of these related species have many traits in common. This means that some traits—including coat pattern—vary less within each species than between them (see Figure 12.1a). Thus the species unit seems to be a natural one, not just an arbitrary invention of biologists who need a classification system to communicate about the organisms they study. But species differences are not always so clear cut. Taxonomists use all sorts of differences—morphological, behavioral, and genetic—to identify species. Occasionally they have serious problems deciding how much a group must differ to be classified as a separate species. Sometimes the discriminating traits between two species partially overlap (Figure 12.1b).

Sibling species and hybrids pose some problems

Sibling species: *Species that resemble each other so closely that they cannot be distinguished by external morphology.*

In extreme cases, species that are reproductively isolated can be so similar in their morphology that they can hardly be distinguished simply by looking at them carefully. Such groups of species are called **sibling species** or cryptic species. For example, *Drosophila melanogaster* and *Drosophila simulans* are sibling species whose females cannot be distinguished and whose males can be distinguished only by experts. Populations of the same species may also differ, particularly when they live far apart. Across their geographic range species are often subdivided into a mosaic of subspecies or races, and where two subspecies meet and hybridize a so-called hybrid zone occurs. For example, the crow *Corvus corone* has two subspecies in Europe. The all-black carrion crow *Corvus corone corone* occurs in much of western Europe, and the hooded crow *Corvus corone cornix*, which is gray with a black head, wings and

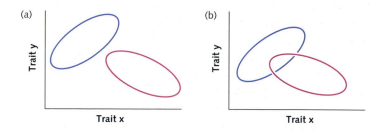

Figure 12.1 Hypothetical distributions of the values of two traits in two different species. In both diagrams the within-species variation is smaller than the between-species variation. In (a) the trait value distributions are completely separated, while in (b) there is some overlap.

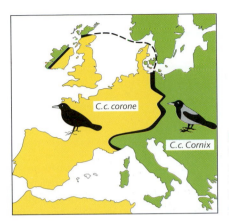

Figure 12.2 Map of Europe with the distribution of the carrion crow *Corvus corone corone* and the hooded crow *C. c. cornix* and the hybrid zone between them. (From Skelton 1993.)

tail, inhabits eastern and northern Europe, northern Scotland and western Ireland (Figure 12.2).

Where the subspecies meet there is a hybrid zone varying in width from 20 to 200 km in which they freely interbreed. All sorts of intermediate pheno-types can be found within (but rarely outside) the hybrid zone, which is thought to be old and stable, having existed since the last Ice Age. We do not know why it is stable. Hybrid zones are much studied because they may pro-vide insights into processes that contribute to speciation, as we will see later in this chapter.

Evolutionary biologists agree on how to recognize species but disagree on what they are

How should one define species? What characteristics do species have in com-mon? Evolutionary biologists still disagree on this issue. Remarkably, judging from the extensive scientific literature on this topic, abstract species concepts are more problematic than the practical identification of species. However, bio-logists do use some general species criteria to judge the validity of the various species concepts. We introduce these criteria, discuss the major species concepts and their application, then return to the species criteria to see how they help us reconcile the differences among the species concepts.

Separation. For populations of organisms to qualify as species, they must be somehow separated from each other regardless of the level at which they are separated (morphological, behavioral, genetic) or the mechanisms that pro-duced the separation (geographic or behavioral).

Cohesion. The populations must be internally cohesive, both genetically and ecologically. By genetic cohesion we mean that they must actually or potentially interbreed; by ecological cohesion we mean that the individuals of each popu-lation must actually or potentially occupy the same habitat at the same time. Ecological cohesion is necessary but not sufficient for genetic cohesion.

Species concept: *How we define what a species is.*

Species criteria: *Criteria used to recognize species when we see them.*

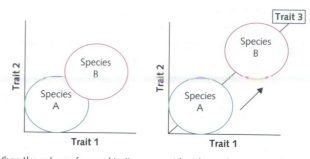

Figure 12.3 When the values of several traits are considered together, species that overlap for one or two traits can often be distinguished when a third or fourth trait is added to the analysis. Similarly, species that largely overlap in DNA sequences for one gene can be distinguished when enough other genes are added to the analysis. In practice, advanced statistical methods are used to see whether the clusters exist, and the axes are transformations that combine the original traits.

Monophyly. The organisms and populations within a species must share a single most recent common ancestor.

Distinguishability. There are three levels of distinguishability ranging from easy to hard to distinguish. If the species is *diagnosable*, then trait states have been fixed within it that are not present in any other species—every organism in the species has those trait states, and no organism not in the species has them. That is the simplest situation, the one exploited in taxonomic keys.

The second level of distinguishability is that of *phenetic clusters*, by which we mean that although many traits vary within the species, few of them being reliably diagnostic, and although many of the traits overlap with those of other species, when the values of several traits are considered together, the species form clusters of trait values with little or no overlap. For example, two species might overlap to some degree for two traits but when a third trait is added, they separate into two clusters of variation (Figure 12.3).

The third level of distinguishability is that of *genetic clusters*, by which we mean that although the species may be morphologically indistinguishable, when their DNA is sequenced and the sequences of several different parts of the genome are considered together, they form clusters with little or no overlap among species (like the phenetic clusters in Figure 12.3).

Thus to qualify as a species, a population of organisms must be separate, cohesive, monophyletic, and distinguishable. Bear these criteria in mind as we work through the concepts.

Monophyletic: *All species in a monophyletic group are descended from a common ancestor that is not the ancestor of any other group and no species descended from that ancestor are not in it.*

According to the biological species concept the decisive criterion is the ability to interbreed

The most influential species concept has been the biological species concept (BSC), proposed by Dobzhansky (1937) and propagated by Mayr: 'species are groups of actually or potentially interbreeding natural populations that are

Biological species concept (BSC): *Species are actually or potentially interbreeding natural populations that are reproductively isolated from other such groups.*

reproductively isolated from other such groups' (Mayr 1963). According to the BSC the decisive criterion is successful sexual reproduction: the ability to produce fertile offspring. The popularity of the BSC stems from two biological insights. First, sexual reproduction promotes uniformity within a species by genetic recombination. All individuals that can interbreed share a common gene pool; copies of genes in one individual may end up in future descendants of any other conspecific individual, but not in descendants of an individual of a different species. Genetic recombination within the common gene pool thus prevents strong divergence of any sub-group of individuals. Second, if two groups do not interbreed, there is no gene flow between the gene pools, allowing further genetic divergence between the groups by natural selection and genetic drift. The inability to interbreed prevents species from merging together at a later time when conditions have changed. Such species remain distinct in sympatry, when they occur in the same habitat at the same time, and this criterion is often used in practice to define 'good species.' Thus—according to the BSC—the number of good species cannot decline through hybridization because species that can fuse would, by definition, not be good species.

Asexual organisms and hybrids pose problems for the biological species concept

Not all organisms occur in groups within which there is sexual interbreeding and between which gene flow is prevented. We consider two types of exception: asexual organisms and inter-species hybrids.

Asexual organisms

Absence of sexual recombination is rare but does occur in some groups of organisms. For example, the imperfect fungi appear to have lost the sexual life cycle and reproduce exclusively through spores formed by mitosis. They are classified into species because within-group variability in so-called diagnostic traits is small compared to between-group variability. Thus the species *Aspergillus niger* is recognized by the traits it shares with other *Aspergillus* species and by a few special characters, of which the black color of its spores is the most important. This species consists of a very large collection of asexual clones, and each clone is reproductively isolated from all other clones. Strict application of the BSC would force us to consider each clone a separate species, which is neither practical nor meaningful. Because the BSC refers by definition to sexual reproduction, it is best not to apply this species concept to the imperfect fungi. Another example is *Taraxacum officinale*, the common dandelion. The species consists of sexual, diploid plants and asexual, triploid plants. Experts can distinguish some asexual clones or groups of clones from each other and from diploids on the basis of morphological details. Thus one could argue that in accordance with the BSC the asexual triploids and the sexual diploids should not be grouped together and each clone (or group of similar clones) should be given a separate name. Indeed, some systematists have constructed long lists of so-called *Taraxacum* microspecies. A similar situation occurs within the genus of water fleas, *Daphnia*, where molecular systematics has revealed a complex network of hybridization events and where some clones are obligately asexual while other groups of clones engage in intermittent sexual reproduction. In none of these cases is the BSC particularly helpful.

Inter-species hybrids

The BSC is also problematic when species are sexual, but barriers to inter-species breeding are not strong. Inter-species matings producing fertile hybrids are not common among related animal species (although they do occur), but they are frequent in plants and fungi. The difficult question then arises, with how much gene flow between gene pools can the BSC retain its meaning? For the carrion crow and the hooded crow, the current opinion is that both belong to the same species and the BSC would apply, but wolves and coyotes are considered separate species, despite the fact that they can and do hybridize and that some wolf populations are threatened with extinction through hybridization. Darwin considered a species definition like the BSC based on sterility barriers but rejected it because so many recognized species can hybridize in nature (Mallet 1995).

Species concepts have been developed for bacteria, asexual eukaryotes, and cryptic species

A special situation occurs in bacteria and viruses. Bacteria have traditionally been classified into species on the basis of their pathology, morphology, and antigenic properties. DNA sequencing of bacterial genes shows that individual genes may be composed of bits and pieces of different—sometimes of very different—origin (Maynard Smith et al. 1991). Such mosaic gene structure is probably a consequence of 'localized sex,' the exchange by homologous recombination of small stretches of DNA transferred into recipient cells by conjugative plasmids, transformation, or transduction. There is similar evidence for mosaic genes in viruses. The implication is that pieces of DNA can travel by lateral gene transfer between related (and sometimes between unrelated) species. Thus bacterial and viral species are not genetically isolated from each other, and gene pools are generally wider than named species. Here the BSC is not much help, at least not in its original form.

One way around this problem was suggested by the observation that the bacterial genome can be conceptually divided into core genes and auxiliary genes (Lan and Reeves 2001). The core genes include housekeeping genes, genes involved in basic cell metabolism. There is little selective advantage in exchanging such genes with distantly related taxa, for the functions they perform are general. Because they rarely transfer, they diverge among taxa, and eventually that divergence will prevent recombination among the separating taxa—the genes become so different that they are no longer recognized as homologous by the machinery of recombination. Auxiliary genes, in contrast, are involved in adapting to local conditions; they include genes for pathogenicity, antibiotic resistance, toxins, and novel metabolic functions. Unlike the core genes, there are advantages to exchanging auxiliary genes, which can contain useful solutions to novel problems. The core genome hypothesis predicts that the barriers to interspecies recombination for core genes are not shared by auxiliary genes: we should be able to recognize bacterial species if we look only at the core and are not distracted by the periphery.

Wertz et al. (2003) applied this insight to several samples of each of seven taxa of bacteria that live in the intestine. They defined the core as six housekeeping genes from whose sequences they constructed a phylogenetic tree in which the seven taxa were clearly separated (Figure 12.4). The levels of diversity within named species were less, often much less, than the levels of diversity between species, and the distances between samples from different species were greater, usually much greater, than the distances between samples within

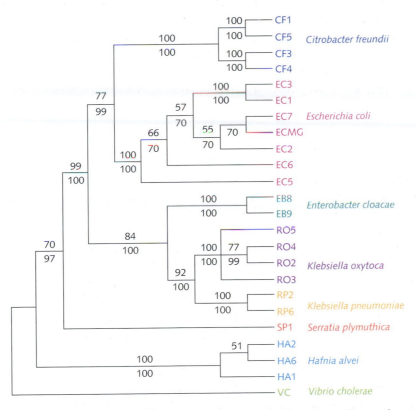

Figure 12.4 A phylogeny of seven bacteria using *Vibrio cholerae* as the outgroup. The use of an outgroup (a species which clearly does not belong to the group being analyzed) is necessary to interpret the phylogenetic tree as an evolutionary sequence (see Chapter 13, p. 303). The phylogeny is based on the sequences of six core genes; molecular systematics using only that part of the genome yields a phylogeny that groups species of bacteria just as well as it groups species of sexual eukaryotes. The numbers on the phylogeny represent two ways of evaluating the reliability of each branch, with 100 being most reliable. The horizontal axis represents time, and the further to the left that a join occurs, the further back in time that the split in lineages occurred. (Reproduced from Wertz et al. (2003) A molecular phylogeny of enteric bacteria and implications for a bacterial species concept. Journal of Evolutionary Biology, Vol. 16 with permission from Blackwell Publishing.)

species. It may appear that some named species contain more genetic diversity than others, but this probably simply the result of having more samples for some of them. Here the named species form genetic clusters whose relationships to one another have been determined by comparing all possible relationships and choosing the one that maximizes the probability that the observed DNA sequences would be observed, starting from a common ancestor and assuming that subsequent mutations are fixed at random (see Chapter 13, p. 303).

Thus it appears valid to speak of bacterial species as clusters defined by the genealogies of a sample of core housekeeping genes. This approach can be used not only for asexual eukaryotes, such as the bdelloid rotifers whose ancient asexuality became famous in the context of the evolution of sex, but also for cryptic species, species that cannot be distinguished morphologically. In fact, it was in cryptic species that this approach was first used.

Perhaps the most striking group of cryptic species is formed by members of the *Tetryhymena pyriformis* complex, a group of ciliate protists related to *Paramecium* (Figure 12.5). This complex consists of at least 15 and possibly as

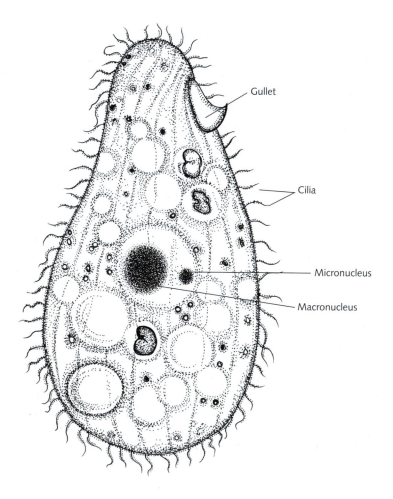

Figure 12.5 *Tetrahymena*, a ciliate lineage that has formed a large, genetically diverse group of phenotypically indistinguishable species.

many as 100 species that are totally isolated from each other genetically but that cannot be distinguished phenotypically despite the fact that they have many easily recognizable morphological traits both on and within the cell. In contrast to their phenotypic indistinguishability, they differ impressively in DNA sequences and in the amino acid composition of their proteins. Their DNA sequences indicate that the members of the group diverged a long time ago, perhaps as much as 100 million years ago. Analysis of the proteins out of which their cellular structures are built has shown that the building blocks have changed while the structures have remained visually indistinguishable. The simplest interpretation is that the common structure was established long ago, and that its molecular components have changed independently in each of the lineages (Nanney 1982). The species concept most easily applied to *Tetrahymena* is that of discrete genetic clusters.

Cryptic species remain a major source of undescribed biodiversity; they are diagnosed as genetic clusters

Another impressive array of cryptic species is found in the slender salamanders of the American West. By morphological criteria, only three or four species can be recognized. Molecular methods have uncovered a huge array of hidden diversity; to date 18 species are recognized, and several remain to be described (Jockusch and Wake 2002). Cryptic species are also common in many major marine groups, including corals, barnacles, and copepods (Knowlton 1993). All cryptic species are diagnosed as discrete genetic clusters using molecular systematics (Chapter 13, p. 303).

The phylogenetic species concept emphasizes monophyly and diagnosability

The BSC lacks a historical dimension, for it can only be applied to contemporary organisms. Indeed, since every living sexual organism is linked via an uninterrupted chain of successive ancestors to totally different life forms hundreds of millions of years ago, there must have been a continuity of sexual fertility between ancient ancestors and the present organism. The BSC provides no criterion for where to draw the lines between successive species along such a line of descent. Such a phylogenetic perspective is provided by the phylogenetic species concept (PSC), defined by Cracraft (1983) as a monophyletic group composed of 'the smallest diagnosable cluster of individual organisms within which there is a parental pattern of ancestry and descent.' Within such a group organisms share certain derived characters that distinguish them from other such groups. This species concept avoids the problems associated with the strict reproductive isolation required by the BSC. However, it has some problems of its own.

Phylogenetic species concept (PSC): *A monophyletic group composed of 'the smallest diagnosable cluster of individual organisms within which there is a parental pattern of ancestry and descent.'*

High-resolution molecular characterization poses problems for the phylogenetic species concept

It is not clear how many shared derived characters a monophyletic group of organisms should have to be classified as a separate species. If one searched hard enough with high-resolution molecular methods, an established species could be split up into many very small groups of individuals that each shared a common derived character. Clearly, giving species status to all such small groups is not meaningful, for in this way any newly derived trait would produce a new species and the number of species would explode. Recently the PSC has started to be modified to avoid extreme division of species.

Species concepts can be reconciled with species criteria

● **KEY CONCEPT**

Species originate in stages and acquire their distinguishing criteria in steps.

Although there has been disagreement over species concepts, most biologists have the same thing in mind when they think of a species: a species is a discrete evolutionary lineage, or more precisely, an entire population-level lineage segment—a segment of a phylogenetic tree—from origination to extinction (Figure 12.6).

Species have extension in time; they come into being in a series of stages. In passing through those stages they acquire the species criteria discussed above—separation, cohesion, monophyly, and distinguishability. The criteria mark stages in the origin of a species; they do not determine species status (de Queiroz 1998). From their initial separation each lineage, taken as a whole, is subsequently monophyletic. After some time, differences are fixed in some traits, not necessarily at first those concerned with reproductive isolation. From that point on, the species can be distinguished using fixed trait differences. After some time, differences accumulate and enough are fixed to cause the species to become reproductively isolated. After a long-enough period, most genes sampled will have alleles whose most recent common ancestor was within the new lineage, and from that point onward, the history of the genes corresponds with the history of the species.

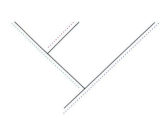

Figure 12.6 Each dotted line represents a species from origination to the present; in this segment of a phylogenetic tree, there are three species.

Sequential, cumulative genetic changes are associated with speciation

As lineages diverge, they steadily acquire genetic differences (Wu 2001). Before they separate completely, while they are still existing as populations or races among which interbreeding occurs and gene flow is substantial, a few genes differentiate because of differences in selection among the populations, but the majority of genes for which there are no selective differences among populations mix freely (Figure 12.7; stage I). After separation, fixed differences start to appear in more genes, causing some degree of reproductive isolation; if the populations come back into contact and mix freely, they could still fuse (stage II). Later, the populations will have diverged so much, will have accumulated so

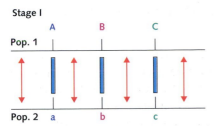

Stage I

Population/races with differential adaptation; reproductive isolation not apparent; three divergent loci shown here.

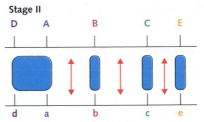

Stage II

Transition between race and species with some degree of reproductive isolation; populations may fuse or diverge.

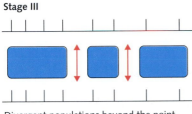

Stage III

Divergent populations beyond the point of fusion but still share a portion of their genetics via gene flow: good species.

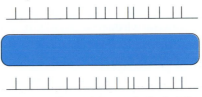

Stage IV

Species with complete reproductive isolation.

Figure 12.7 When the process of speciation is viewed genetically, it can be conceptualized as four stages in which an increasing portion of the genome is progressively cut off from gene flow. Arrows represent gene flow; boxes represent barriers to gene flow; the whole genome is represented as a single line on which a few representative genes have been picked out and labeled with letters; genes labeled with lower-case letters are homologous to but different from those labeled with upper-case letters—they have diverged through mutation and selection. (Reproduced from Journal of Evolutionary Biology, Vol. 14, Wu 2001, The genic view of the process of speciation, with permission from Blackwell Publishing.)

many fixed genetic differences, that they can no longer fuse if they come into contact, but limited hybridization maintains limited genetic exchange in a small part of the genome (stage III). Finally, with complete reproductive isolation, the entire genomes are separated. It is at this point (stage IV) that almost any gene sampled will have alleles whose most recent common ancestor occurred within the new lineage.

How long does it take to proceed through these four stages? We do not know in general, but for a few well-studied populations and species we have some estimates (Wu 2001). *D. melanogaster* has two races, Z and M, which differ in about 15 genes for behavior but not in any genes for male or female sterility or genital morphology. They separated about 100 000 years ago and are at stage I. *D. melanogaster* has two sibling species, *D. simulans and D. mauritiana*. They differ in about 120 genes for male sterility, fewer than 10 for inviability or female sterility, and more than 19 for genital morphology. Hybrid females do not suffer much fitness reduction and are inseminated by the males of both pure species with little discrimination; first-generation hybrid males are sterile. They separated 0.3–1.0 million years ago and are at stage III. Stage IV is represented by *D. melanogaster* and *D. simulans*, which produce only sterile or inviable hybrids. They differ in more than 200 genes for male sterility and more than 10 genes for inviability and female sterility. They separated 1–2 million years ago and are at stage IV.

We do not have as many details for the recently uncovered differences between savannah and forest elephants in Africa (Roca et al. 2001); their limited hybridization and large genetic distance suggest that they are at stage III or between stage III and stage IV. They are thought to have separated about 2.6 million years ago. Now, there are 300–400 fruit fly generations (3 weeks) per elephant generation (20–30 years). When we estimate the rate of speciation in generations rather than years, African elephants appear to speciate faster than fruit flies. The rapid speciation of elephants in Africa is matched by their performance in the fossil record, where one of the fastest rates of morphological change is that of the dwarfing of elephants on Mediterranean islands in the Pleistocene era.

Species originate as byproducts of intra-specific evolution

● **KEY CONCEPT**

Speciation, the elemental macroevolutionary event, is a byproduct of standard microevolution—selection and drift.

Speciation is a central process in evolution. A theory of evolution that could not explain speciation would be seriously flawed. Do species originate as an inherent consequence of microevolution, or must different, additional processes be involved? Is speciation caused by natural selection acting on variations produced by mutation and recombination and by neutral genetic drift, the same processes that drive microevolutionary change within populations? At present there are no reasons to think that speciation requires mechanisms beyond those that generate change within species: speciation appears to be a byproduct of intra-specific evolution.

Populations diverge geographically, by mate choice, or by habitat choice

● **KEY CONCEPT**

There are three scenarios for speciation: *allopatric* speciation where populations diverge in geographic isolation; *sympatric speciation*, where species diverge in the same location; and *parapatric speciation*, where species diverge in geographically adjoining populations.

Populations can be genetically separated because they are either geographically or reproductively isolated. With geographical isolation interpopulational matings do not occur because of the physical separation, but matings might occur and be fertile if the populations did mix. Upon secondary contact interpopulational matings could also be sterile because the formerly isolated populations had diverged genetically or would be avoided because of behavioral divergence. Then reproductive isolation would be complete and the gene pools would be separated. Reproductive isolation may also originate within a population when differences in mating behavior between subpopulations in the same area lead to isolation. The role of geographical isolation and the origin of reproductive isolation have generated much discussion.

Is complete geographic isolation of populations, with no gene flow, necessary to start speciation? Some think that most speciation events occurred in allopatry,

Allopatry: *Occurring in two or more geographically separate places.*

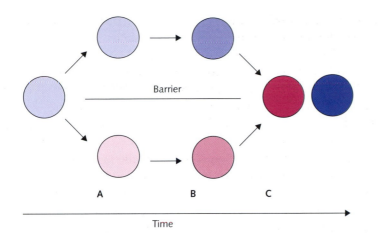

Figure 12.8 Schematic representation of allopatric speciation. The circles represent populations and the shading represents divergent selection in one of the geographical areas. (A) Splitting up into geographically isolated populations. (B) Divergent selection. (C) Reproductive isolation at secondary contact.

with different populations completely isolated in space. Others hold that speciation in **sympatry**, where the subpopulations diverge while continuing to live in the same place, has also played a significant role.

Sympatry: *Occurring in the same geographic area.*

In allopatry populations accumulate independent genetic differences

It is easy to imagine that speciation starts when populations become geographically isolated, are exposed to divergent selection, and evolve independently. After enough time they will have accumulated so many genetic differences that they will be reproductively isolated if they come into contact again. If isolation is complete, speciation has occurred. This is the allopatric model of speciation (Mayr 1963), represented in its simplest form by Figure 12.8.

According to this model, the first step in speciation is for one population to split into two or more completely isolated subpopulations. Such splitting may be caused by migration, by local extinctions of intervening populations, or by geological events. The barriers separating the populations may be geographical or ecological. Examples are populations on islands, in lakes, on mountain tops, in patches of forest surrounded by savanna, or in fields surrounded by forest.

Darwin's finches exemplify allopatric speciation

The finches on the Galápagos Islands suggested to Darwin the idea of descent with modification from a single ancestor species. They are thought to have speciated allopatrically (Lack 1947, Grant 1986), for the Galápagos Islands formed within the last 5 million years as volcanoes emerged from the sea. The islands have never been connected to a continent or to each other. Thirteen different species are recognized, with up to 10 on a single island. The initial stages of the process were presumably as follows. About 3 million years ago a small group of birds from South or Central America colonized at least one of the islands. After the population had established itself, dispersers colonized other

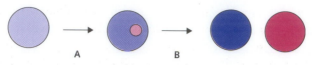

Figure 12.9 Schematic representation of sympatric speciation. (A) Small divergence leading to some degree of genetic separation within a single population. (B) Further differentiation and genetic separation produce complete reproductive isolation.

Secondary reinforcement: *The reinforcement of prezygotic barriers to hybridization after secondary contact between isolated populations.*

islands. Because ecological conditions varied among the islands, the genetically isolated populations encountered different selective forces and differentiated. The next stage was the establishment of secondary contact, through dispersal, between the differentiated populations. If birds from two populations did not interbreed, or if their offspring were inviable or sterile, speciation had been completed in allopatry. If some successful interbreeding was still possible, the outcome would not have been clear. This aspect—**secondary reinforcement**—is further considered below (p. 296), where we discuss the origin of reproductive isolation during speciation.

Sympatric speciation occurs despite continued gene flow between the diverging groups

Often ecological conditions cause partial isolation of subpopulations, reducing gene flow between subpopulations but not stopping it completely. Does incomplete isolation between subpopulations under divergent selection allow speciation? This model is called sympatric speciation (speciation occurring within a single population; see Figure 12.9) in contrast to allopatric speciation, which assumes complete geographic isolation at the start of the speciation process.

While allopatric speciation is undisputed, whether sympatric speciation is likely and common, or can only occur under restrictive conditions, was long controversial (Bush 1994). The main problem was whether subpopulations can become reproductively isolated and differentiate despite the presence of some gene flow. Recently both theory and data have accumulated that make sympatric speciation appear to be both plausible and, in some cases, the speciation scenario most likely to explain the facts (Via 2001).

Sympatric speciation works when disruptive selection coevolves with assortative mating

The key new feature of the theoretical models is the combination of disruptive selection with assortative mating (Dieckmann and Doebeli 1999, Kondrashov and Kondrashov 1999). Consider a single population of a single species living in a single habitat. Suppose that habitat contains two resources whose efficient exploitation requires different adaptations. Part of the population will primarily use one of the resources, another part of the population will primarily use the other. Selection will start to adapt each part of the population to its resource, but because interbreeding is frequent, adaptation will be quite imperfect. In particular, the offspring of crosses between parents with adaptations to different resources will not be able to exploit either resource efficiently because they will have a mixture of traits. That creates a situation where assortative mating becomes advantageous. If one part of the population not only specializes on one of the resources but also has a tendency to mate with others specializing on that resource, then the two processes—adaptation to efficient exploitation of the resource and mate selection to avoid producing offspring that are inefficient at exploiting the resource—will reinforce each other. As the traits involved in resource exploitation diverge, their divergence increases the selective advantage of mating assortatively. And as the precision of assortative mating increases, the divergence of the ecologically important traits accelerates. It is now well accepted that in theory at least this process will cause sympatric speciation. It is a process in which intraspecific competition drives the separation of one population into two coexisting populations that then through improved assortative mating become reproductively isolated. Schluter (2000) therefore calls it *competitive speciation*.

Competitive speciation in lake fish exemplifies sympatric speciation

One of the most plausible examples of competitive speciation is the differentiation of fish living in lakes into forms specialized either on the shore and pelagic habitats or on the deep and shallow habitats. This may have happened in sticklebacks and salmonids; a good example is that of cichlids of the genus *Tilapia* living in a small lake in Cameroon (Schliewen et al. 2001). Five forms of *Tilapia* live in Lake Ejagham. They originated from a nearby river and shared a common ancestor about 10 000 years ago. Two of the forms appear to be nearing the end of the process of speciating. They share breeding coloration and morphologies that suggest that they are each other's closest relatives. One of them is large, is found inshore, and has relatively small eyes; the other is smaller, pelagic, and has relatively large eyes. They feed on different prey in deep and shallow water, but their breeding sites overlap in time and space. Analysis of their nuclear genomes reveals that there is some gene flow between them, but it is quite restricted. The mechanism that restricts the gene flow appears to be size-assortative mating: large with large and small with small. Mixed large/small mating pairs are only rarely observed (Figure 12.10).

Flies and wasps that shift hosts may speciate sympatrically

Another convincing set of cases of sympatric speciation involves host shifts in parasites or phytophagous insects. Some individuals start to lay their eggs in a new species of host, perhaps because it has recently arrived or increased in numbers. Such a situation has been studied by Bush and Smith (1998) in *Rhagoletis polmonella* flies. These flies used to lay their eggs in hawthorn, whose fruits are

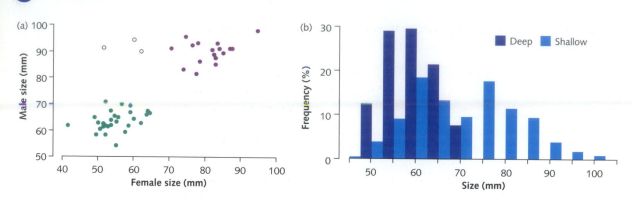

Figure 12.10 Size-assortative mating and habitat distribution by size in two *Tilapia* forms living in a small lake in Cameroon. (a) With a few exceptions (open circles), assortative mating by size is quite precise: small mates with small (green dots) and large with large (purple dots). (b) Small, nonbreeding fish (<70 mm) are found in both deep (>4 m) and shallow (<4 m) water. Large, nonbreeding fish are found only in shallow water. (Reproduced from Tautz (2001) Genetic and ecological divergence of a monophyletic cichlid species pair under fully sympatric conditions in Lake Ejagham, Cameroon. *Molecular Ecology* **10**, with permission from Blackwell Publishing.)

eaten by the larvae. In 1864 *Rhagoletis* was found in apples, later it parasitized other fruits, and now one can discriminate between an apple race and a hawthorn race that differ genetically, being characterized by different frequencies of enzyme variants.

Rhagoletis share with many parasitic insects a behavior that strongly favors differentiation in sympatry: females prefer to lay eggs in the type of fruit from which they themselves hatched. Females that hatched from apple prefer to lay eggs on apples, and females that hatched from hawthorn prefer hawthorn. Males are also true to the type of fruit they came from in their mating behavior: males that hatched from apple tend to mate on apple, and the same holds for males from hawthorn. Thus matings are mostly between males and females from apple and between males and females from hawthorn.

The breeding times of the two races have also diverged, contributing to their sympatric reproductive isolation. In the laboratory, however, flies from both races still interbreed freely. Therefore, although in nature considerable reproductive isolation has evolved, the potential for full interbreeding remains. Speciation is not yet complete, but the genetic differentiation and partial reproductive isolation observed in *Rhagoletis* did occur in sympatry. Whether speciation will be completed under these conditions, given sufficient time, remains an open question; the recent theoretical work mentioned above suggests that it is likely.

The phylogenetic distribution of some plant-eating insect taxa suggests sympatric speciation by host shifts. For example, there are hundreds of species of fig wasp, each breeding on its own species of fig. Allopatric speciation seems implausible here, for it would require a history of many geographic isolation events, whereas many fig species and their wasps are now sympatric.

Sympatric speciation can also occur through divergence in flowering times of plants

In plants, partial reproductive isolation may occur following the evolution of different times of flowering. The grass species *Agrostis* and *Anthoxanthum* are able to grow near spoil heaps from mines despite the high concentrations of copper, lead, and zinc in the soil (see Chapter 2, pp. 45 f.). Particular types within these species are more tolerant of metals than the normal types. On the spoil heaps only the tolerant types can survive, while the normal types dominate the vegetation just outside the spoil heaps. The tolerant and normal types show signs of incipient reproductive isolation because they flower at slightly different times and the tolerant types have a higher degree of self-fertilization. The differences in flowering time are genetic and may represent adaptations to the local conditions. This example parallels the *Rhagoletis* case discussed above: adaptation to different conditions may produce sympatric reproductive isolation as a byproduct.

Sudden sympatric speciation by polyploidization is undisputed

A special and undisputed form of sympatric speciation is caused by a change in the genetic system that produces sudden reproductive isolation. Several mechanisms are known, most involving major chromosomal alterations. The commonest is **polyploidization**, a doubling of the complete set of chromosomes, which arises due to irregularities in cell division. If the chromosomes double, but the cell fails to divide, then the cell contains twice the normal number of chromosomes. This is called **autopolyploidy**. Further normal divisions transmit the polyploid condition to all descendant cells. In somatic tissue this leads to a polyploid region. In reproductive tissue the outcome depends on details; one possibility is the production of gametes containing twice the normal haploid number of chromosomes.

Commoner than autopolyploidy is **allopolyploidy**, resulting from hybridization between related species followed by doubling of the chromosomes (Figure 12.11).

Polyploidization: *A doubling of the complete chromosome set.*

Autopolyploidy: *A doubling in the number of chromosomes when a gametocyte fails to divide in meiosis; all the resulting chromosomes come from a single parent.*

Allopolyploidy: *An increase in the number of chromosomes involving hybridization followed by chromosome doubling. The offspring then contain chromosomes from both parental species.*

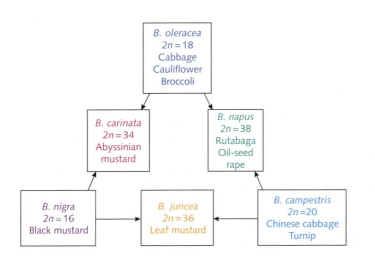

Figure 12.11 Allopolyploidy in the genus *Brassica*. Three different parent species have hybridized in all possible pair combinations. Subsequent chromosome doubling in the hybrids has produced three new polyploid species.

Hybrids often suffer because quasi-homologous chromosomes do not pair properly during meiosis. When the complete hybrid chromosome set doubles, every chromosome has a homologous partner, and meiosis can proceed normally.

Once a polyploid individual has arisen by whatever mechanism, it is reproductively isolated from the parental types because a mating with a parental diploid individual would yield triploid offspring. Triploids are almost always sterile because of severe problems in meiosis. They may, however, be able to propagate asexually, and in fact many asexual clones in plants (and some animals) are triploid.

70–80% of plant species probably originated as polyploids

Speciation by polyploidization has been particularly important in flowering plants, where 70–80% of the species are thought to have originated as polyploids. Many crop plants recently became polyploid, perhaps as part of the process of domestication. These include wheat, oats, potato, tobacco, cotton, and alfalfa. Further in the past apples, olives, willows, poplars, and many genera of ferns may have originated as polyploids.

Stable, viable polyploidy is most likely to arise in organisms capable of self-fertilization and asexual reproduction. Self-fertilization enhances the probability that two unreduced gametes unite to form a polyploid individual; two diploid gametes would fuse to form a tetraploid, for example. Asexual reproduction avoids problems in chromosome pairing and segregation during meiosis. Speciation by polyploidy is less significant among animals, because animal mating systems often do not allow self-fertilization and vegetative reproduction. Exceptions are found in some fish, earthworms, crustaceans and a few others, where polyploidization has been common, and in groups with parthenogenetic females, including beetles, moths, and pill bugs.

Parapatric speciation occurs along a border shared by two populations

Parapatry: Occurring in adjacent areas sharing a border.

Parapatry refers to the situation where two populations live next to each other, sharing a common border. There may be some gene flow between them. One scenario for parapatric speciation is basically allopatric; it starts with the founding of a daughter population near to the parental population but geographically isolated from it. In isolation the daughter population accumulates genetic changes. It then expands until it comes into contact with the parental population. At that point reciprocal competition prevents further expansion, and some gene flow may occur along the border if hybridization is still possible. Often the genetic changes involve chromosome rearrangements, which lead to defective segregation in hybrid zygotes. Well-studied cases include deer mice in Mexico, house mice in Europe, and grasshoppers in Australia.

Another scenario for speciation starts with range expansion. For example, a population of frogs may move steadily northward through the eastern United States following the retreat of the glaciers; a population of seagulls may expand

from a single island to a ring of populations around the Arctic Ocean; a population of salamanders may expand from one valley to a ring around the Central
Valley of California. Such broad geographic ranges are too great for free gene
flow across all the populations. In the case of the frogs, the populations in the
south become effectively isolated from the populations in the north; similarly
with the seagulls and salamanders. Even though contiguous populations can
interbreed across the entire range, populations widely isolated from one other
suffer from hybrid infertility or reduced viability when crossed.

Reproduction isolation is a criterion of speciation

Reproductive isolation may be prezygotic or postzygotic

The physical separation of allopatric populations guarantees zero gene flow
between them and thus their reproductive isolation. If they come into contact
again they may or may not be reproductively isolated. According to the
allopatric speciation model, the divergence built up during allopatry will have
caused intrinsic reproductive isolation as a byproduct that only becomes
apparent in secondary sympatry. This reproductive isolation may be prezygotic
or postzygotic. Prezygotic reproductive isolation occurs when individuals from
different populations do not mate because of behavioral differences, most
importantly mate choice, mating at different times or seasons, or mating in
different habitats. Postzygotic reproductive isolation occurs when the different
populations do interbreed but without success: the matings are sterile or the
offspring inviable. Allopatric speciation leading to complete prezygotic or
postzygotic reproductive isolation is unproblematic. How frequently speciation
has followed this course remains to be seen.

In bacteria and yeast sequence divergence is the basis for genetic isolation

Genetic recombination causes the exchange of genetic information between
lineages, and the amount of genetic recombination determines the extent of
genetic isolation. Genetic recombination among enterobacteria depends on the
difference in their DNA sequences (Vulic et al. 1997). The more genomes
diverge in sequence, the less likely it is that they will recombine. More precisely,
genetic recombination appears to require enough blocks of sequences that are
identical in the two mating partners, and these blocks must be large enough to
allow recombination to begin. The extent of genetic isolation increases
exponentially with increasing sequence divergence. Thus in bacteria sequence
divergence offers a structural basis for genetic isolation. Similar results have
been found in yeast, which has a typical eukaryotic sexual cycle with meiosis,
and some data on recombination in mice also point in this direction.

● KEY CONCEPT

Reproductive isolation separates
evolving lineages by cutting off
gene flow between them.

Prezygotic: *Before fertilization,
e.g. mate choice.*

Postzygotic: *After fertilization,
e.g. early development.*

Thus sequence divergence could be the general barrier to recombination that characterizes fully completed speciation.

What happens to reproductive isolation in secondary contact?

But what happens when sexual allopatric populations show only partial reproductive isolation when they come back into contact in secondary sympatry? Or when sympatric populations have developed partial reproductive isolation, like the apple and hawthorn races of *Rhagoletis*? How much gene flow between the different populations or races will stop the speciation process? Under what conditions will reproductive isolation become more complete and when will it break down?

Secondary reinforcement of reproductive isolation in hybrid zones is problematic

Because postzygotic isolation has high costs—matings are sterile or wasted on hybrid offspring of poor viability—Dobzhansky thought that mechanisms would be selected to convert postzygotic into prezygotic isolation. This process is known as secondary reinforcement. Reinforcement is particularly relevant in hybrid zones, where hybrid progeny often have low fitness. Here reinforcement mechanisms would help to keep species separate, whereas hybridization destroys the distinction between them. Although reinforcement seems plausible at first sight, many experts doubt that it has an important role. One problem is that hybrid progeny may backcross to either of the parental types, so that within a few generations a whole range of hybrid types will be formed, some of them very similar to one parental type, others resembling the other parental type. Then hybridization between the two parental types will be rare because they will be separated by the hybrid zone, and reinforcement would be very weak. We return to the role of reinforcement below when we discuss experimental evidence on speciation.

Sexual selection may have driven speciation in Hawaiian flies and African fish

Two well-investigated and spectacular cases of rapid and profuse speciation suggest an important role for sexual selection in causing prezygotic reproductive isolation.

Speciation of drosophilids (fruit flies in the genus *Drosophila*) in the Hawaiian archipelago has been studied since the 1960s by Carson and Kaneshiro and their coworkers. About 800 *Drosophila* species are endemic to the Hawaiian islands (compared to about 2000 species in the rest of the world). Speciation among the Hawaiian drosophilids has probably followed a course similar to Darwin's finches on the Galápagos islands. A few fruit flies must have arrived on one of the islands and founded populations that dispersed from there to other valleys, mountains, and islands, with allopatric diversifying selection followed by secondary sympatry. The males of different species often differ

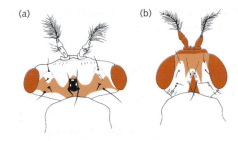

Figure 12.12 *Drosophila heteroneura* and *D. silvestris* are closely related sympatric species with similar ecology but dissimilar courtship behavior. (a) The hammer-shaped head of male *D. heteroneura*. (b) The head of male *D. silvestris*, which closely resembles the heads of females of both species. (From Skelton 1993.)

greatly in body and wing patterns and have unusual modifications of mouth-parts and legs, but females of different species are often similar (Figure 12.12); the remarkable morphological features of the males are the byproducts of genes controlling courtship behavior (Kaneshiro and Boake 1987). Here prezygotic reproductive isolation appears to originate through allopatric changes in courtship behavior, possibly as a chance event in a small founder population, and has often initiated speciation in Hawaiian *Drosophila*, preceding adaptation to different food sources and divergence in other characters.

The explosive speciation of cichlid fishes belonging to the genus *Haplochromis* in Lake Victoria in Africa is another example of sexual selection driving speciation. The lake contains an estimated 500–1000 haplochromine cichlid species that evolved since the last Ice Age when the lake was completely dry, perhaps less than 13 000 years ago. Males of related sympatric species always differ in color (red, blue, or yellow); intraspecific male color variation also sometimes occurs. Females do not have bright colors. Seehausen et al. (1997) have shown that female choice of male color causes reproductive isolation between sympatric species and between intraspecific color morphs. In laboratory experiments females of a sympatric red/blue species pair preferred males of their own species over those of the other species under broad-spectrum illumination. Under monochromatic light, where color differences are masked, mating preferences disappeared. Water pollution caused by humans may actually threaten species diversity. The increased turbidity of the lake impairs visibility to such an extent that in some areas females can no longer distinguish males of related species from their own. Since the hybrids are fully fertile, species diversity may decline as a consequence—although this effect on cichlid diversity in Lake Victoria is small compared to the impact of introduced Nile perch and pollution-induced anoxia on the breeding grounds.

As in the Hawaiian drosophilids, sexual selection for particular types of males (here males with striking colors) may be the first stage in speciation. This could occur in sympatry given some spatial heterogeneity. For example, depth affects color perception, and new mate preferences may stabilize in a subpopulation that stays long enough at a different depth in the lake. In the first stage of

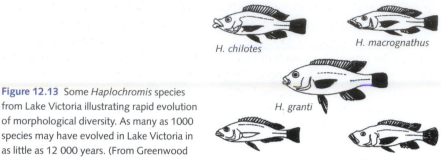

Figure 12.13 Some *Haplochromis* species from Lake Victoria illustrating rapid evolution of morphological diversity. As many as 1000 species may have evolved in Lake Victoria in as little as 12 000 years. (From Greenwood 1974.)

speciation many incipient species could coexist, separated by mate choice. The second stage in the speciation of haplochromines in Lake Victoria is thought to be diversifying selection on feeding habits and other specializations. The resulting differentiation can be rapid because the jaw apparatus of the cichlids consists of a large number of bony elements that allow independent modifications (see Figure 12.13). Indeed, before many of them became extinct, the species in the lake had extremely diverse feeding habits, including specialists on eating fish scales, on eating eggs and larvae sucked out of the mouths of mouth-brooding cichlids, and on eating phytoplankton, snails, fish, or insects.

Gametes can also choose mates

It has recently been recognized that much of the action in mate recognition occurs in the gametes; genes directly related to fertilization evolve particularly rapidly (Table 12.1). Theory suggests that competition for fertilizations will increase the

Table 12.1 Some of the genes involved in gamete recognition that evolve particularly rapidly (from Swanson and Vacquier 2002).

Gene	Function	Organism
Phb.3.2 and others	Mating-type pheromone	Basidiomycete fungi
SCR	Sporophytic self-incompatibility	Brassicaceae
Lysin	Dissolves egg envelope	*Tegula* and abalone (gastropod mollusks)
Bindin	Adheres sperm to egg	Sea urchins
Ph-20, β-fertilin	Sperm-surface recognition	Mammals
ZP2	Egg envelope, sperm binding	Mammals
ZP3	Egg inducer of sperm acrosome reaction	Mammals
Zonadhesin	Sperm surface	Mammals

variation in male gamete-recognition proteins; this initiates sympatric speciation. Avoidance of competition between the incipient species would then reinforce the rapid evolution of gamete-recognition proteins. This idea is consistent with the fact that the rate of divergence of genes related to fertilization is not constant but highest between the pairs of species that are most closely related (Van Doorn et al. 2001). Thus recent progress, both theoretical and empirical, supports the idea that sexual selection has an important role in speciation and suggests that a significant element of mate choice occurs at the gamete stage.

Experiments on speciation yield two important results

It is often impossible to reconstruct the speciation processes that have lead to presently existing species. Even though much circumstantial evidence might point to sympatric speciation, how could we be sure that there never was an allopatric stage? For this reason, experiments designed to duplicate part of the speciation process under controlled laboratory conditions help to obtain information about the feasibility and relative importance of the various aspects of speciation. When Rice and Hostert (1993) reviewed the results of laboratory experiments on speciation, they came to two important conclusions.

> ● **KEY CONCEPT**
> The entire speciation process may be difficult to reconstruct in the laboratory, but key elements of it have been tested experimentally.

Divergent selection can produce reproductive isolation despite some gene flow

Many experiments (mostly with *Drosophila*) confirm that divergent selection applied to allopatric populations can produce pre- and postzygotic reproductive isolation as a byproduct. Reproductive isolation is probably caused by pleiotropic effects of genes that were selected during adaptation to the different environmental conditions used in the experiment. When postzygotic isolation was observed, it appeared to be environment-dependent. Only rarely did unconditional postzygotic isolation evolve.

There is also experimental support for the evolution of reproductive isolation between sympatric populations connected by gene flow, provided that divergent selection is strong relative to the gene flow. Evolution of reproductive isolation in sympatry was particularly successful when divergent selection was applied to several characters simultaneously and when hybrid viability was zero.

Reinforcement is probably unimportant in generating prezygotic isolation

Prezygotic isolation between a pair of related species is often stronger when the individuals tested come from an area where the species are sympatric than when they are collected in allopatry. This has been thoroughly confirmed for

Drosophila species by Coyne and Orr (1989); the pattern has also been observed in other species, including fish and frogs. These observations are consistent with the idea that prezygotic isolation has evolved to prevent the production of hybrids with low fitness (reinforcement), but they do not prove it. An alternative explanation is 'reproductive character displacement,' which assumes that speciation was already complete when the species came into secondary contact, but that prezygotic isolation mechanisms diverged in secondary sympatry simply to reduce the amount of time spent mistakenly courting partners that were already reproductively isolated. Reinforcement, in contrast, presupposes ongoing gene flow while prezygotic isolation evolves. Of many experiments designed to demonstrate reinforcement, only one or two were successful. The great majority failed. Thus the importance of reinforcement in producing prezygotic isolation is doubtful.

Speciation is the birth, extinction the death of a lineage

As stated at the start of this chapter, speciation connects micro- to macroevolution. Much of this chapter has been concerned with describing the present state of our understanding of the microevolutionary processes that lead to reproductive isolation and the separation of evolving lineages. We now consider the consequences of speciation in preparation for the next two chapters. When we look at an entire evolving clade, we can see that some of its lineages are speciating and branching, increasing the number of species in the clade, and others are losing diversity because extinctions are more frequent than speciations. The balance of these two processes determines both the number of species extant in the clade at any point in time and the topology of the clade—its geometric shape, for example whether each of its major branches has the same number of minor branches and twigs, or whether some branches consist only of one species and others of many. There is thus a kind of demography at the species level, a balance between speciation (birth) and extinction (death), which underpins patterns of radiation and determines how much biodiversity the clade generates. In the next chapter we turn our attention to how we can infer the shape of phylogenetic trees and, in subsequent chapters, how we can use that knowledge to infer life's history.

● SUMMARY

This chapter discusses the definition of species and how species originate.

- Several species concepts have been developed. The biological species concept (BSC), which groups organisms on the basis of interfertility, has been the most influential. The core genome concept adapts the BSC to bacteria. The lineage-segment concept adapts it to phylogenies.

- Species concepts are problematic, for none fits all groups of organisms, but application of species criteria to the process of speciation shows that many of the apparent differences among species concepts disappear once we see that many of them simply mark different stages through which most species pass.

- Speciation is characterized by genetic separation and morphological differentiation. These processes can interact to enhance each other's effects.

- Several scenarios have been proposed for the start of speciation. The genic view of speciation suggests that at first only a small portion of the genomes of the diverging species is isolated. Speciation progressively increases the proportion of the genome that is no longer experiencing gene flow, until eventually, with completion of speciation, the entire genome becomes isolated.

- Some believe that allopatric speciation between populations that are physically separated is the rule, and sympatric speciation between populations connected by gene flow is an exception.

- Those who think that sympatric speciation need not be rare find support in the results of experiments that mimic speciation processes in the laboratory, in recent theoretical progress, and in the increasing number of case studies best explained by sympatric speciation.

- Also controversial is the significance of reinforcement, or selection for the prevention of crossbreeding between two populations that produce hybrids of low viability. Reinforcement has encountered theoretical objections and is not supported by the results of experiments that were designed to discover it.

- Evidence on two spectacular and profuse speciations—Hawaiian fruitflies and cichlid fishes in Lake Victoria—suggests that sexual selection has an important role in bringing about reproductive isolation.

- A significant element of sexual selection occurs in the gametes, where mate-recognition proteins are known to evolve particularly rapidly.

- The balance between speciation and extinction determines the biodiversity of clades.

Speciation produces branches in phylogenetic trees. How we infer the branching patterns in those trees, the patterns that document the relationships of species, is discussed in the next chapter.

● RECOMMENDED READING

Howard, D.J. and Berlocher, S.H. (eds) (1998) *Endless forms: species and speciation*. Oxford University Press, Oxford.

Schluter, D. (2000). *The ecology of adaptive radiation*. Oxford University Press, Oxford.

Trends in Ecology and Evolution (2001) Special Issue on Speciation **16**: 325–413.

● QUESTIONS

12.1. There are well-studied examples of 'half-way' speciation in the form of partially reproductively isolated populations, and cases of completed speciation demonstrated by closely related sibling species, but very few documented examples of the complete process of speciation. What is required to demonstrate speciation, and why is this apparently so difficult?

12.2. The main processes involved in speciation are genetic separation and phenotypic differentiation. List plausible causes for both processes.

12.3. The following type of experiment has been used to support the reinforcement model of speciation. Equal numbers of male and female virgins were collected from two strains of *Drosphila* and mixed to allow mating. Genetic markers allowed offspring to be classified as coming from homotypic or heterotypic matings. Only males and females from the homotypic matings were used for the next generation, in which the procedure was repeated. All hybrid progeny died. After 20 generations increased prezygotic isolation was observed. Do you think this is a proper way of testing the possibility of speciation by reinforcement? If not, how would you modify the experiment?

CHAPTER 13
Phylogeny and systematics

Phylogeny refers to the history of a species, to its relationships to other species (in Greek *phyl-* refers to tribe; *gen-* refers to origin or descent). Systematics refers to the methods used to discover that history (in Greek *systematos* refers to a complex whole put together). At the largest scale, the group studied is all of life on Earth, and the goal is to discover the entire pattern of relationships of all things of whose existence we have some evidence. As Darwin put it (1859):

It is a truly wonderful fact—the wonder of which we are apt to overlook from familiarity—that all animals and all plants throughout all time and space should be related to each other in groups . . . The several subordinate groups in any class cannot be ranked in a single file, but seem to be clustered around points, and these around other points. If species had been independently created, no explanation would have been possible of this kind of classification . . . The affinities of all the beings of the same class have sometimes been represented by a great tree. I believe this simile largely speaks the truth . . . The green and budding twigs may represent existing species; and those produced during former years may represent the long succession of extinct species . . . As buds give rise by growth to fresh buds, and these, if vigorous, branch out and overtop on all sides many a feebler branch, so by generation I believe it has been with the great Tree of Life, which fills with its dead and broken branches the crust of the earth, and covers the surface with ever-branching and beautiful ramifications.

So important was this concept to Darwin that he expressed it in the only illustration in *The Origin of Species* (Figure 13.1).

The Tree of Life has three main branches: Bacteria, Archaea, and Eukaryota

Today the methods of molecular systematics, introduced below, have allowed us to reconstruct the major features of the Tree of Life. Any complete branch of the Tree of Life, large or small, recent or ancient, is called a clade—a natural group of related organisms that all share a most recent common ancestor. At the largest scale, the Tree of Life is primarily a tree of bacteria with three clades—the Bacteria, the Archaea, and the Eukaryota—rooted at about 3.7 billion years

● **KEY CONCEPT**

All living things are related. By inferring their relationships, we can reconstruct the history of life on this planet.

Systematics: *The branch of biology that investigates relationships among species to understand the history of life.*

Clade: *A natural group of related species containing all descendants of the most recent common ancestor of the group.*

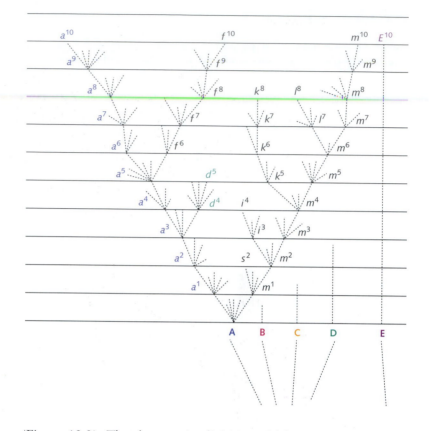

Figure 13.1 Darwin's picture of a phylogenetic tree. Time runs from bottom to top, a convention preserved to this day. Of the five species present at the start—A, B, C, D, and E—only A and E have left living descendants. The others went extinct in the order B, C, D. Of the many descendants of A, only a^{10}, f^{10}, and m^{10} are living in the 10th time interval. (From Darwin 1859.)

ago (Figure 13.2). The three main divisions of life are well supported; the position of the root of the tree, marking the origin of life, is difficult to infer. There is some support from ancient gene duplications for placing the root on the branch leading to the Bacteria, which would make the Archaea and the Eukaryotes sister groups. At this scale the plants, animals, and fungi are small branches that you can locate to the left of the Eukaryote stem.

The Bacteria are common prokaryotes living in virtually all environments; they include the human gut commensal *Escherichia coli*, soil bacteria like *Bacillus subtilis*, as well as pathogens like *Salmonella*, *Staphylococcus*, and *Helicobacter*. The Archaea were discovered relatively recently; like Bacteria, they are small, single-celled prokaryotes. Some inhabit extreme environments with high temperatures or high salt concentrations; others inhabit normal seawater. The Eukaryotes are derived from a symbiotic event in which the proto-eukaryote ancestor, which probably had already evolved a nucleus and a cytoskeleton, acquired a proteobacterium—the group of bacteria that contains *Escherichia* and *Agrobacterium*—that evolved into the mitochondrion. (Look for *mitochondrion* in the lower right of Figure 13.2.) Later the chloroplast, which originated as a cyanobacterium, was independently acquired at least three times: in the lineages leading to the green algae and higher plants, to the red algae, and to an obscure group called the glaucocystophytes. (Look for

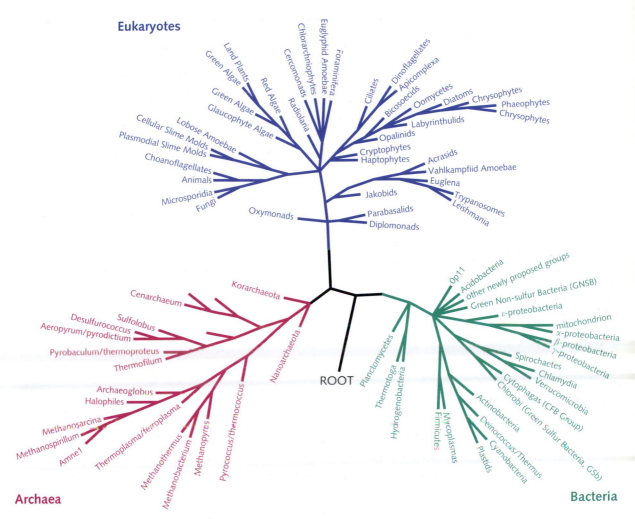

Figure 13.2 The Tree of Life on planet Earth begins about 3.7 billion years ago. There are three major branches—the Bacteria (e.g. *Bacillus*), the Archaea (e.g. *Methanobacterium*), and the Eukaryota (e.g. fungi). At this scale the plants, animals, and fungi represent minor, late radiations. (From Baldauf 2003.)

plastids in the bottom right of Figure 13.2.) Thus the ability of eukaryote cells to respire and to photosynthesize was acquired from bacterial symbionts.

Multicellular organisms form three main groups: fungi, plants, and animals.

In the **radiation** of multicellular organisms, three of the major modern groups— plants, animals and fungi—share a common ancestor at about 800–1000 million years ago (Figure 13.3). The times at which various groups originated are given by the concentric circles, one for each 100 million years from 800 million

> **Radiation:** *The diversification and divergence of species within a group with a single common ancestor (a clade).*

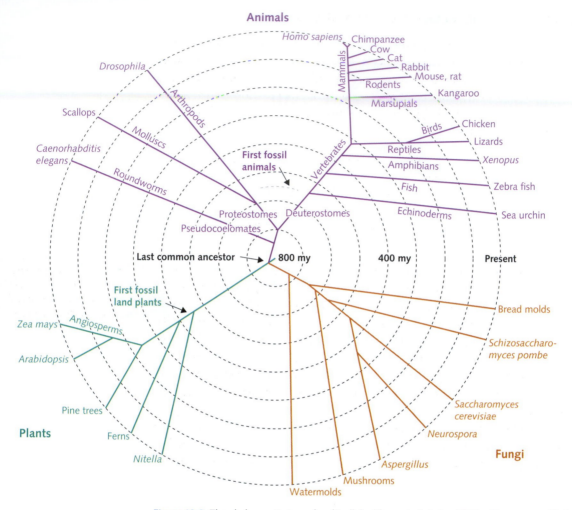

Figure 13.3 The phylogenetic tree of multicellular life, rooted at about 800 million years, with three main branches: plants, fungi, and animals. (Courtesy of T.D. Pollard.)

years ago to the present. Fossils of animals with skeletons are first available in the Cambrian era, about 550 million years ago. They include arthopods, mollusks, echinoderms, and chordates. The first fossil land plants are known from about 440 million years ago, and the angiosperms diverged from the conifers about 220 million years ago. Several major groups—the roundworms, mollusks, and arthropods—appear to have originated before the Cambrian era, as did the line leading to the echinoderms and vertebrates. The branch leading to the mammals split off from the other 'reptiles' about 300 million years ago, while the branch leading to the birds split off from the lizards about 200 million years ago. Some of these dates, estimated from molecular divergence, are older than those estimated from fossil evidence.

The Tree of Life is not obvious; it must be discovered

Without phylogenetic trees—pictures like Figures 13.2 and 13.3—we cannot make much sense of the pattern of life. Related organisms are similar in many important ways, and members of one group differ from members of other groups in just as many important ways. Thus the Tree of Life is essential to understanding much of biology, but it is not given to us: it must be discovered. As Darwin (1859) put it:

Our classifications will come to be, as far as they can be so made, genealogies; and will then truly give what may be called the plan of creation. The rules for classifying will no doubt become simpler when we have a definite object in view. We possess no pedigree or armorial bearings; and we have to discover and trace the many diverging lines of descent in our natural genealogies, by characters of any kind which have been long inherited.

Early evolutionary biologists could only use morphological characters to work out relationships. Today those characters that 'have been long inherited' include DNA sequences, which systematic biologists supplement with data of many other sorts, mainly morphological and developmental. The use of molecular techniques has proven to be extremely powerful in systematics, for it has provided methods to rapidly measure and properly compare a vast number of characters—long DNA sequences from many species—that yield estimates of relationship of unprecedented accuracy reaching in some cases far back into geological time.

> **Character:** *A trait that varies among taxa and that in any given taxon takes one out of a set of two or more different states.*

Before considering the methods of systematics, we look at some striking recent discoveries powered by these new methods that modified or overturned previous ideas of relationship or otherwise provided unexpected insights.

Molecular systematics has yielded surprising insights

Relatives of jellyfish evolved into intracellular parasites

Myxozoa or myxosporidians are single-celled parasites of invertebrates and fishes that cause serious damage. After penetrating the skin they have a complex fungus-like life cycle in the cells of their hosts, where they eat cytoplasm, reproduce sexually, and form spores. The spores resemble the stinging cells of coelenterates, such as jellyfish, not fungal spores (Figure 13.4). Myxozoa were traditionally classified with another group of single-celled parasites, the microsporidia, which are close relatives of fungi, and in 1989 they were placed in a major group of their own. Now molecular systematics identifies them as highly modified cnidarians, placing them within the group that contains the jellyfish, corals, and sea anemones.

> ● **KEY CONCEPT**
>
> Much of the history of life that has been obscured by the evolution of morphology can be recovered from DNA sequences.

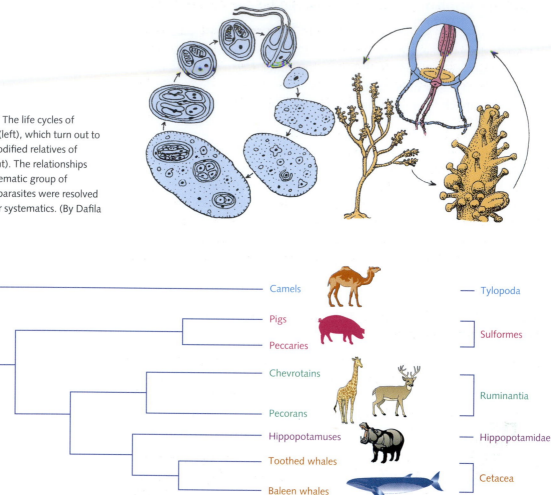

Figure 13.4 The life cycles of myxozoans (left), which turn out to be highly modified relatives of jellyfish (right). The relationships of this problematic group of fungus-like parasites were resolved by molecular systematics. (By Dafila K. Scott.)

Figure 13.5 The closest relatives of whales are hippopotamuses, and whales are clearly ungulates—highly modified for life in the sea. (Reproduced from Nikaido et al. (1999) Phylogenetic relationships among cetartiodactyls based on insertions of short and long interpersed elements. *Proceedings of the National Academy of Sciences of the United States of America* **96**, 10261–6. National Academy of Sciences, USA.)

Whales are ungulates; their closest relatives are hippopotamuses

Whales are mammals that re-entered the sea after a long period on land. Because their highly modified morphology makes their relationship to other mammals obscure, they are prime candidates for molecular detective work. Recent systematic analysis of DNA sequences has revealed that whales are ungulates—not carnivores related to seals and otters—and relatively recently derived ungulates at that. Their closest relatives are hippopotamuses, and their next-closest relatives are deer and antelope (Figure 13.5).

Giant pandas and lesser pandas are not each other's closest relatives

The giant panda, a bamboo-eating specialist now found only in western Szechwan, looks like a bear, but its genital anatomy and vocalizations are unlike any bear, and it has an extra digit, derived from a wrist bone, found in no other animal, that it uses to grasp bamboo shoots. Its closest relative was long thought to be the lesser panda, which also lives in Asia, eats bamboo, and whose molar teeth resemble those of the giant panda. The lesser panda has a long, ringed tail and resembles a raccoon or coatimundi. Molecular systematics have now shown that the giant panda is most closely related to the bears, and the lesser panda is now believed to be more closely related to the mustelids (the weasels and their relatives) and to the procyonids (the raccoons and their relatives) than it is to the bears. The similarity of their molar teeth, adapted to bamboo-feeding in both species, misled us about their relationships, for their teeth resembled each other because natural selection had shaped them to similar tasks, not because that morphology had been inherited from a common ancestor (Figure 13.6).

Who transmitted HIV to a rape victim? Systematics can help to identify criminals

The methods of systematics are not limited to basic questions about the Tree of Life. They can be applied to very practical problems. Suppose you are a detective asked by the police to identify which of two suspects transmitted the HIV virus to a rape victim who developed AIDS after the rape was committed. You

(a)

(b)

Figure 13.6 The giant panda and the lesser panda, two mammals that both eat bamboo and live in Asia. Molecular systematics have shown that the giant panda is related to the bears, whereas the lesser panda is the sister group to all canoid carnivores. Thus the pandas is not a natural group.
(Photo (a) copyright Pat & Tom Leeson/Science Photo Library. Photo (b) copyright Martin Wendler/NHPA.)

take blood samples from the victim, the two suspects, and an AIDS patient whom you are certain had nothing to do with the rape case. From those blood samples you isolate the RNA genome of the HIV virus and confirm that all four persons are infected. From the RNA you prepare a DNA copy that you sequence. The critical part of the four sequences is 30 bases long and is shown in Table 13.1.

Using a computer program, you prepare a phylogenetic tree based on these four sequences. The program examines all the possible trees and delivers the one that implies fewer mutations in nucleotides than any other (Figure 13.7).

This tree suggests that the second suspect infected the rape victim. The conclusion is supported by four changes in sequence that are shared by the second suspect and the victim: A → G at position 1, A → T at position 2, G → A at position 12, and A → G at position 20. HIV is an RNA virus that evolves rapidly, and since the rape occurred, the virus has continued to evolve in both suspect and victim. In the suspect, there has been a change at position 25, C → T, and in the victim there have been two changes, at position 21, T → C, and at position 28, A → G.

This example is artificial, but similar methods were used to identify the sailor that introduced HIV to Scandinavia, the dentist with HIV who infected several

Table 13.1 DNA copies of hypothetical HIV RNA sequences taken from the rape victim, two suspects, and one unrelated person.

		1	5	10	15	20	25	30
Unrelated person		A	AGCT	TCATA	GGAGC	AACCA	TTCTA	ATAAT
Suspect 1		A	AGCT	TCACC	GGCGC	AGTTA	TCCTC	ATAAT
Suspect 2		G	TGCT	TCACC	GACGC	AGTTG	TCCTT	ATAAT
Rape victim		G	TGCT	TCACC	GACGC	AGTTG	CCCTC	ATGAT

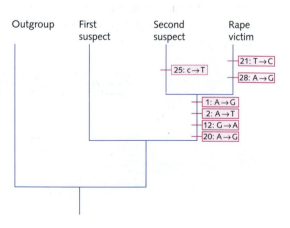

Figure 13.7 The phylogenetic tree based on the nucleotide sequences in the text that implies the fewest changes overall. The tree is rooted using the unrelated person, and it identifies the second suspect as the probable rapist. The character changes that support each branch unambiguously are listed. (Prepared by C. Baroni-Urbani.)

of his patients in Florida (both the sailor and the dentist died before this was discovered), and in criminal cases to show that a gastroenterologist injected his girlfriend with blood taken from a patient with HIV (Metzker et al. 2002) and that cases of encephalitis in New York and New England were caused by the first strains of West Nile virus to appear in the western hemisphere (Lanciotti et al. 1999).

Thus systematics has discovered many surprising and important relationships ranging from the strains of HIV found in a small group of humans to the largest features of the Tree of Life. How could it do this? To answer that, we now consider the concepts, terminology, and methods of systematics.

How phylogenetic concepts are defined

Relationship is defined by how recently common ancestors were shared

For a long time after people began trying to classify organisms in a systematic fashion, they used a variety of definitions of relationship. One definition was, 'things that look like each other are more closely related to each other than they are to things that look different.' This is perhaps logical, but it is wrong, for some things that resemble each other superficially—such as African euphorbias and American cacti—are not closely related at all, and other things that appear to be quite different—such as the myxozoans and jellyfish discussed above—turn out to be related. Zimmermann (1931) took the critical step of defining relationship as the sharing of a recent common ancestor. For example, apples are more closely related to magnolias than they are to ginkgos because apples and magnolias share a more recent common ancestor with each other than either does with ginkgos (Figure 13.8).

By defining relationship as sharing a most recent common ancestor, we can then identify groups that are natural—groups that accurately reflect genealogy—and groups that are unnatural or incorrectly identified—groups that distort genealogy. Systematics has developed special terminology to describe natural and unnatural groups.

> ● **KEY CONCEPT**
>
> Phylogenetic concepts are defined to be consistent with and useful in the logical discovery of natural groups, such as species and clades.

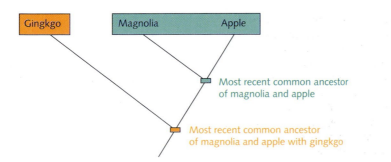

Figure 13.8 Phylogenetic relationship is defined as recency of common ancestry; the more recent the common ancestor, the more closely related the groups.

Essential terminology includes monophyletic, paraphyletic, and polyphyletic

Monophyletic: *All species in a monophyletic group are descended from a common ancestor that is not the ancestor of any other group and no species descended from that ancestor are not in it.*

Paraphyletic: *A group that does not contain all species descended from the most recent common ancestor of its members; some of those species are outside it.*

Polyphyletic: *A group that contains species descended from several ancestors from which members of other groups also descended.*

All species in a **monophyletic** group are descended from a common ancestor that is not the ancestor of any other group: no species descended from that ancestor are not in the group. Thus monophyly technically defines what we mean by a natural group. The dog clade, Canidae, is a monophyletic group. All of its members are more closely related to each other than they are to any species outside the Canidae, and there are no species descended from the common ancestor of the canids, to our knowledge, that are not included in the group.

A **paraphyletic** group is an unnatural group that does not contain all species descended from the most recent common ancestor of its members. One now-classical example of a paraphyletic group is the reptiles. The word reptiles refers to an unnatural group because it does not include within it the birds and the mammals, both of which are descended from branches of the phylogenetic tree located well within the reptiles (Figure 13.9). Another paraphyletic group is the fish, which does not include the tetrapods, which are descendants of fish.

Another unnatural group is called **polyphyletic** if its species are descended from several ancestors that are also the ancestors of species classified into other groups. Figure 13.10 depicts a polyphyletic group—the homeothermia or warm-blooded tetrapods—that is polyphyletic because it does not include the several groups of reptiles that share an ancestor with the birds and mammals. Other labels for unnatural polyphyletic groups are worms, algae, and protozoa.

Examples of convergence are found in plants, animals, and molecules

In addition to the terminology that describes natural—monophyletic—and unnatural—paraphyletic and polyphyletic—groups, we also need terms to describe the two major reasons that traits and DNA sequences can look the same: either because they are descended from common ancestors, or because natural selection or drift shaped them in similar ways so that they now look the same although they are descended from ancestors who were unrelated and looked different. We have abundant evidence that things descended from

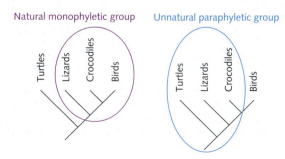

Figure 13.9 There is no common term for the natural monophyletic group on the left; the term reptiles refers to the unnatural paraphyletic group on the right, a group that does not include the birds.

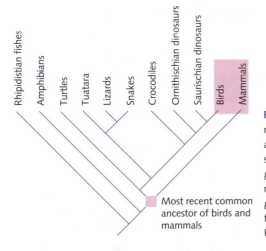

Figure 13.10 Polyphyletic groups contain members descended from several ancestors that are also the ancestors of species classified into other groups. The group homeotherms, uniting birds and mammals, is an unnatural polyphyletic group. (Reproduced with kind permission from Jonathan Armbruster, Auburn University.)

common ancestors can now look different—through evolutionary divergence—and that things descended from different ancestors can now look similar—through evolutionary convergence. Thus both similarities and differences can mislead us with respect to the real genealogy. Consider a few examples.

The Old and New World succulent plants, which have adapted to arid conditions with similar morphology, are a spectacular example of convergence. Some New World cacti (family Cactaceae) are so similar in shape to some Old World euphorbs (family Euphorbiaceae) that only an expert can tell them apart (Figure 13.11). In other groups flowers and fruits have converged because plants with different ancestors are pollinated by similar insects or dispersed by similar birds.

> Convergence: *Two species resemble each other not because they shared common ancestors but because evolution has adapted them to similar ecological conditions.*

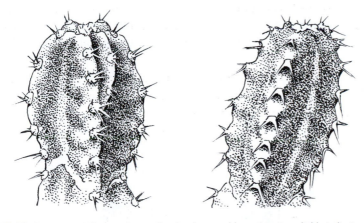

Figure 13.11 Convergence is one reason why simple resemblance is not a reliable indicator of systematic relationships. The cacti of the New World and the euphorbs of Africa have evolved similar morphologies to deal with similar ecological problems. Left: the cactus *Cereus validus* from Argentina, half size. Right: the euphorb *Euphorbia resinifera* from Morocco, double size. (By Dafila K. Scott.)

When they encounter phenotypic convergence, biologists must look at DNA sequence data or at additional morphological data to discover the underlying relationships. In animals, striking examples of convergence include the bills of Old World sunbirds, New World hummingbirds, and Hawaiian honey creepers, adapted to extract nectar from deep flowers; the fusiform shape of porpoises, tunas, sharks, and ichthyosaurs, adapted to fast swimming (see Figure 13.14 below); and the wings of birds and bats, which have independently evolved similar adaptations to rapid long-distance flight or to hovering in front of flowers.

Adaptive convergence also occurs in molecules, such as the hemoglobin of birds that fly at high altitudes. The bar-headed goose lives at altitudes of 4000 m and flies up to 9200 m in the Himalayas; the Andean 'goose,' a relative of the mallard duck, lives at 6000 m in the Andes. Both species have hemoglobin molecules with higher oxygen affinity than those of their lowland relatives, and one mechanism of increased oxygen affinity is precisely the same in both species— disruption of a contact between the α and β chains through the same amino acid substitution in the β chain (Gillespie 1991). Thus exactly the same modification of the hemoglobin molecule occurred at least twice in distantly related species whose nearest relatives have hemoglobin molecules with different structures.

> **Divergence:** *Related species no longer resemble each other because evolution has adapted them to different ecological conditions.*

Divergence, which occurs when originally similar species become dissimilar, is well illustrated by the Hawaiian lobelias (Givnish et al. 1995). One of the six native genera of Lobeliaceae, *Cyanea*, contains 55 species that constitute 6% of the endemic flora of Hawaii. They are restricted to particular islands or parts of islands. All *Cyanea* descend from a single ancestor; many have undergone striking changes in growth form, leaf size and shape, and flower morphology (Figure 13.12).

They vary from 1 to 14 m in height and have leaves that can be simple, compound, or doubly compound, ranging from 0.3 to 25 cm in width and up to 1 m in length. The flowers, which coevolved with endemic birds, have corolla tubes that range in length from 15 to 85 mm. The genus includes shrubs, trees, and a vine. Any classification based solely on growth form or leaf morphology would

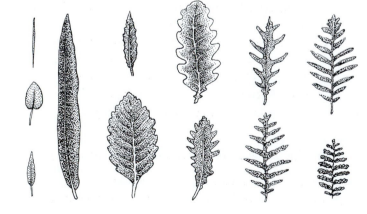

Figure 13.12 Divergence is another important reason why simple resemblance does not indicate relationship. The Hawaiian lobelias have undergone a dramatic radiation in which their leaves have evolved many different forms. Their flowers and DNA sequences continue to indicate that they are closely related. (By Dafila K. Scott.)

not reflect phylogeny, placing members of this genus into several unrelated families. The traits that group them are flower structure, fruit color, and DNA sequence.

Among animals, the radiations of cichlid fish in the great lakes of Africa, of finches in the Galápagos and drepaniid birds in Hawaii, of land snails in Polynesia, of rodents in South America, and of amphipods in Lake Baikal all show striking recent divergence. Among plants, the huge radiation of flowering plants, and within that the radiations of the grasses, the composites, and the orchids, do the same. On a larger scale, divergence is the reason for the diversity of life.

The many cases of convergence and divergence make clear that it is not easy to see when similar traits in two species are similar because those species shared an ancestor in which that trait occurred. However, that is precisely what we need to know if we want to build reliable phylogenetic trees: we need to establish that morphological traits and DNA sequences in two or more species are similar because of shared ancestry. If that were not a problem, systematics would be simple.

Homology and orthology describe structural and molecular similarity due to ancestry

Homology is the term used by biologists to indicate that a trait in two or more species is descended from a common ancestor. Two morphological structures are called homologous by morphologists if they are built by the same developmental pathway and share the same relative position to other structures, such as nerves and blood vessels (see Figure 13.13). The hypothesis for their similarity is derivation from a common ancestor from which similar developmental mechanisms were inherited. Two genes are called orthologous if they have DNA sequences so similar that it is very likely that they derive from a common ancestor. Such molecules are similar by inspection, orthologous by hypothesis. The determination of orthology is more reliable for long DNA sequences, where it is very improbable that random mutation would yield similar states in two organisms. It is less reliable for short sequences.

> **Homology:** *An hypothesis that similarity of a trait in two or more species indicates descent from a common ancestor.*

> **Orthology:** *An hypothesis that the similarity of DNA sequences is explained by ancestry.*

There can be a connection between molecular orthology and morphological homology, but the connection is neither necessary nor reliable. During evolution genes can acquire new roles in new structures, and cases are known where structural homology has been preserved over long periods of time while both DNA and protein sequence homology have been destroyed (recall the *Tetrahymena* example; Chapter 12, pp. 284 ff.). When DNA sequence similarity and morphological homology have different phylogenetic patterns, the difference tells us to look for something interesting in the evolution of development (Wagner 1989).

Convergent structures are analogous, not homologous

Just as homology describes underlying similarity despite divergence, analogy describes superficial similarity despite lack of common ancestry (Figure 13.14).

> **Analogy:** *Convergent traits whose similarity is caused by shared selection.*

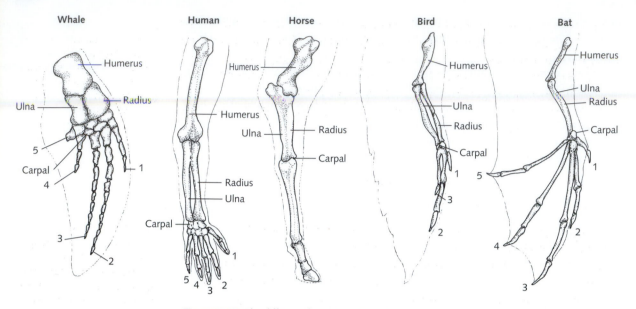

Figure 13.13 The different elements of the vertebrate limb are a classical example of morphological homology. The underlying similarities in construction can be recognized despite divergence into arms, paddles, wings, forelegs, and fins.

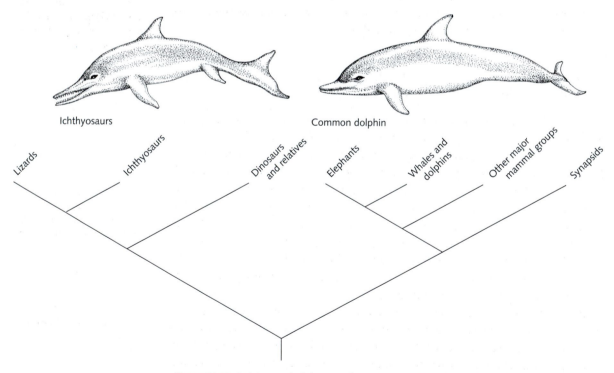

Figure 13.14 Dolphins and ichthyosaurs have similar fusiform shapes, fins and flippers, and jaws filled with teeth adapted for rapid swimming and catching fish and squid. The members of the many lineages between them do not have these adaptations. Although they look similar, they are not closely related.

It is a synonym for morphological convergence, which is one reason for homoplasy, a term broader than analogy that indicates similarity for *any* reason other than common ancestry, including drift. The cacti and euphorbs discussed above exemplify analogous shapes; so do the dolphins and the ichthyosaurs, which share similar shapes, live birth, and precocious offspring.

With that introduction to some important phylogenetic concepts, we now turn to tree building.

> **Homoplasy:** *Similarity for any reason other than common ancestry. The commonest cause of homoplasy in morphological traits is convergence, in DNA sequences mutation.*

How to build a phylogenetic tree

It is clear from the examples above that phylogenetic trees can give us many insights. But how are they properly constructed?

To build a phylogenetic tree, you first need a collection of specimens of all the species in the tree. For each of these specimens, you must determine the states in which many traits are found. Those trait states are called characters.

Note that we assume that you are able to recognize the same traits in different species. This is not a problem when you are working with a collection of butterflies, but it is a problem if you are working with a collection so broad that it includes, for example, elephants and trees. At that scale, DNA sequences are much easier to interpret than morphological traits.

> ● **KEY CONCEPT**
>
> To infer relationships properly, one must measure many characters accurately for all the species concerned, establish the homology of those characters, use them to construct all the phylogenetic trees that are consistent with them, then choose the tree (or trees) that satisfies a logical and agreed-upon criterion, of which there are several.

First construct a matrix with species in rows and characters in columns

From such information you build a character matrix with, for example, species in the rows and characters in the columns. Now you need to decide which characters tell us something useful about relationships—which characters are, in that sense, informative. One of the most important insights of systematics is that it is only shared derived characters that are informative: characters that are shared by all the species in the group you are focusing on—the ingroup—but not by the species in their relatives—the outgroup—because they originated on the branch of the phylogenetic tree leading to the ingroup. Characters that are shared between the ingroup and the outgroup because they originated in the more distant past are not informative about the relation of this group to the outgroup: there will be more about this below.

You would therefore like to know which characters are derived, or relatively recent, and which characters are primitive, or relatively ancient. But you cannot make that determination of what is primitive and what is derived until you have a phylogenetic tree on which to map the characters, and the tree is precisely what you are trying to get. Thus a dilemma: without a tree, you cannot determine what came earlier and what came later; and without knowing what came earlier and what came later, you cannot describe the tree of relationships.

> **Ingroup:** *An assumed monophyletic group, normally made up of the taxa of primary interest.*

> **Outgroup:** *One or more taxa assumed to be phylogenetically outside the ingroup.*

Then choose either the simplest or the most likely tree corresponding to this matrix

There are several ways out of this dilemma (Holder and Lewis 2003), as follows.

Parsimony: *The principle that things should be kept as simple as possible.*

- Try all possible trees and choose those that are simplest, those that imply the fewest changes in characters. This is the principle of **parsimony**, a principle of logic called Ockham's razor and stated by William of Ockham (d. 1347): 'Entities should not be multiplied unnecessarily.' In systematics this translates into the criterion that the best tree is the one with the fewest changes in **character states** and the least convergence.

Character state: *A specific value taken by a character in a specific taxon.*

- Choose the tree that would make it most likely that you would observe the characters that you actually did observe. This is the principle of **maximum likelihood**. To use it, one must assume a model of how evolutionary change occurs; the basic model assumes the same type of random change down all branches of the tree.

Maximum likelihood: *A method of inferring the process that would make the data observed the most likely of all possible data sets.*

- Choose the tree that would make it most likely that you would observe these branches and branch lengths, and character distributions, given a prior expectation of what the tree should look like. This approaches the problem of phylogenetics using **Bayesian inference**, a statistical method that first establishes a basic expectation (the prior probability), and then estimates the likelihood of observing the data given that expectation (the posterior probability).

Bayesian inference: *A method that focuses on the likelihood of observing one thing given that another thing has already occurred.*

- Use various combinations of these three basic ideas.

We now explore these issues in greater detail.

Synapomorphies contain information about relationships; symplesiomorphies do not

Synapomorphy and symplesiomorphy are essential systematic terms derived from Greek roots. We introduce them through the classical example of morphological homology, the vertebrate limb. Having a forelimb with humerus, radius, ulna, carpals, and metacarpals does not help to distinguish bats from turtles, because within the tetrapods that complex trait is shared among all groups, telling us nothing about their ancestor–descendent relations. However, having a limb with that complete structure does help to distinguish tetrapods from lobe-finned fishes, for in that context it is a shared, derived trait, a **synapomorphy** (shared = *syn*-, derived = *-apo*-, trait = *-morph*: synapomorphy), a trait that originated once in their common ancestor, which is shared by all of them and is not found in their closest relatives. If some member of the group does not have it, then it is because it has been lost since the group originated. Figure 13.15 distinguishes between informative, shared,

Synapomorphy: *A shared, derived character state indicating that two species belong to the same group.*

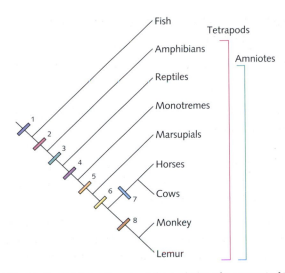

Figure 13.15 Eight key traits are informative about the evolution of a segment of the vertebrate clade: (1) the vertebral column, (2) lungs, (3) an amniotic egg, (4) lactation, (5) eggshell absent, (6) complex placenta, (7) hooves on finger and toe tips, (8) opposable thumb and fingernails. The vertebral column is a synapomorphy of the vertebrate clade but a symplesiomorphy of every vertebrate group—fish, amphibians, reptiles, and so forth. Lungs are a synapomorphy of the tetrapod clade but a symplesiomorphy of the amniotes.

derived traits—synapomorphies—and uninformative, shared, ancestral traits—symplesiomorphies (shared = *syn-*, recent = *-ples-*, trait = *-morph*).

There are three important points here:

(1) simply looking similar is not necessarily informative;

(2) informative traits are shared, derived traits;

(3) what is shared and what is derived, and therefore what is informative, depends on the context, on what part of the tree you are looking at.

> **Symplesiomorphy:** *A character state shared by all members of this group as well as with members of related groups.*

To infer the tree from the character matrix follow changes in character states between species

We start to build a phylogenetic tree by considering the elemental step—the change of a trait from one state to another (Figure 13.16). We denote the things whose relationships are being analyzed—the genes, species, or larger groups—by capital letters, A, B, C . . ., and the traits being used to determine relationship by numbers, 1, 2, 3 . . ..

The problem is how to infer the correct tree from the character matrix. Figure 13.17 shows how a tree is associated with a character matrix. Each change in character state on a tree is associated with a coded change in a character matrix.

We begin not with the tree, which we are trying to infer, but with the character matrix, which we can observe. Given the character matrix depicted in

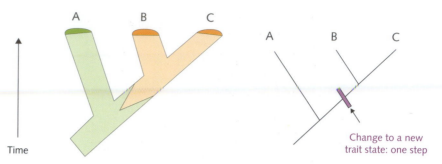

Figure 13.16 The elemental event in systematics is the evolutionary change of a single trait from one state to another. This is normally marked on the tree with a bar at the point where it happened (see Figure 13.15) and is coded 0 for the ancestral state and 1 for the derived state. Note the simplification of a complex microevolutionary dynamic (left) into a single step (right). A, B, and C are things related by ancestry—genes, species, or larger groups.

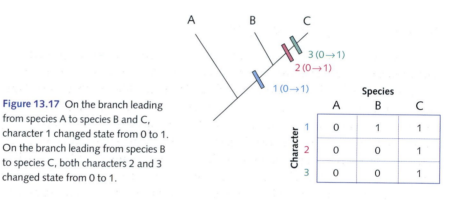

Figure 13.17 On the branch leading from species A to species B and C, character 1 changed state from 0 to 1. On the branch leading from species B to species C, both characters 2 and 3 changed state from 0 to 1.

	Species		
Character	A	B	C
1	0	1	1
2	0	0	1
3	0	0	1

Figure 13.17, there are two possible ways to draw the tree (Figure 13.18). The first (Figure 13.18a) emphasizes overall similarity, including the shared ancestral states (symplesiomorphies) indicated as 0s, and the second (Figure 13.18b) emphasizes derived similarity (synapomorphies) indicated as 1s.

In Figures 13.17 and 13.18 the characters changed consistently and it was possible, using a clear set of rules, to build a single best tree from the character matrices. But what should we do when the characters conflict, when they appear to be telling different stories, as is the case in the matrix in Figure 13.19, where character 3 conflicts with characters 1 and 2? To deal with character conflict we can apply the principle of parsimony, which means looking at all ways of placing the characters on all the possible trees, then choosing the tree or trees that imply the fewest changes. In this case, there is one tree that implies four changes (two different ways), and one tree that implies five changes (Figure 13.20). We choose the tree that implies fewer changes, the tree that groups B with C.

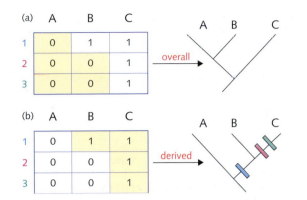

Figure 13.18 (a) If we use overall similarity as a criterion for relationship, then the ancestral states of all three traits, coded 0, predominate, and we judge A to be more similar to B than either is to C. (b) If we use derived similarity, coded 1, as a criterion, then we correctly group B with C. Note that we have indicated the emergence of the derived traits by placing markers on the tree in (b).

Trees are rooted by the choice of outgroup

After we have decided either which tree is the simplest, using the principle of parsimony, or which tree would have been most likely to produce the pattern actually observed, using the principle of maximum likelihood, there is still one more critical step to be taken. We have to choose an outgroup, a group closely related to the entire clade we are analyzing. By making an outgroup comparison, we are then able to root the tree (Figure 13.21). For example, an appropriate outgroup for a phylogenetic analysis of the flowering plants would be a conifer. Rooting the tree establishes the direction of character change within the ingroup; the character states in the outgroup are assumed to be ancestral, and changes in character states within the ingroup are taken as derived by comparison to the ancestral state. Note that phylogenetic trees are like mobiles—think of them as objects that you can pick up by the root and spin freely in the air as they hang from your hand. For example, in the tree on the right in Figure 13.21, it makes no difference whether A is above B or B is above A. Both express the same relationship.

	A	B	C
1	0	1	1
2	0	1	1
3	1	1	0

Figure 13.19 A case of character conflict, which occurs frequently. Character 3 conflicts with characters 1 and 2. Characters 1 and 2 group species B and C together; character 3 groups species A and B together. We assume that the common ancestor was in state 0 for all three traits.

Outgroup comparison: *A method used to root phylogenetic trees and thus to establish the direction of character change.*

Maximum likelihood methods find the tree most likely to produce the data observed

In the maximum likelihood approach a tree is judged by how well it predicts the observed data; the best tree is the one with the highest probability—the greatest likelihood—of producing the observed pattern. To use this method, you need a way to calculate the probability of a data set given a phylogenetic tree. For sequence data, this is usually done using a model based on the probabilities of point mutations occurring at random. The most likely tree is found by considering many candidate trees—all possible trees or a representative sample of

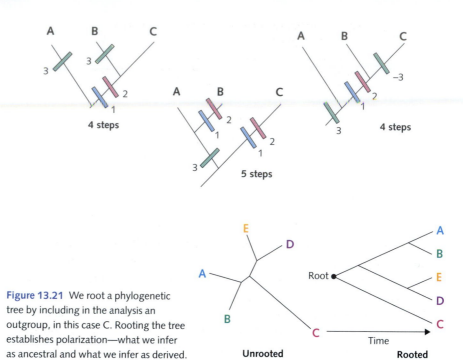

Figure 13.20 Using parsimony to decide which tree is the best when characters are in conflict. We assume here that the ancestor was in state 0 for all three traits. We choose the tree that implies the least change, in this case the tree that groups B with C, not the tree that groups A with B.

Figure 13.21 We root a phylogenetic tree by including in the analysis an outgroup, in this case C. Rooting the tree establishes polarization—what we infer as ancestral and what we infer as derived.

them. For each candidate tree the probability is calculated of finding any two sequences at opposite ends of a branch (for example a T at one end and a C at the other). This is done for all branches, the probabilities are multiplied together to get the likelihood for the whole tree, and the best tree is then the one with the maximum likelihood (Felsenstein 1988). The method is logically appealing and computationally expensive. Its range of application is increasing as computers improve.

Constructing a tree for even a modest number of species requires enormous computations

Thus far it appears that constructing a phylogenetic tree is a simple task. You get a matrix with species in columns and characters in rows, place all the characters on all the possible trees, and either pick the simplest or the most likely. It is true that the task is logically straightforward, but it rapidly becomes a computational nightmare as the numbers of species and traits increase (see Table 13.2).

To put the numbers in Table 13.2 in perspective, there are thought to be about 7–10 million species alive on the planet today, the number of protons, electrons, and neutrons in the universe is on the order of 10^{130}, and the number of trees that can be evaluated with a reasonable number of characters running a computer for 9 months in the year 2001 was about 24 billion, enough to

Table 13.2 As the number of taxa increases, the number of possible trees explodes.

Number of taxa	Number of possible binary trees
3	1
4	15
10	34 459 425
20	8 200 794 532 637 891 559 375
500	$1.0084917894 \times 10^{1280}$

analyze a tree with 10 species but not enough to handle even 20 species. The computational burden can be somewhat reduced by using a parsimony tree as the starting point for a maximum likelihood analysis, but computation remains a serious problem. If quantum computing becomes a reality, it may become possible to take the brute-force approach and evaluate all alternatives even for trees with thousands of species. Until then, we shall have to be satisfied with various approximations for trees with more than 15–20 species.

The names of groups should reflect relationships

How to construct a phylogenetic tree is one issue, and it is separate from the issue of how best to name the groups that are implied by the structure of the tree. Suppose now that we have constructed a tree and want to give names to the various groups that we can recognize on it. It is important that the names correspond to natural groups, that they reflect genealogy. Names should not be given to paraphyletic or polyphyletic groups, for they would then imply that things existed which in fact did not—they would be unnatural names. Unfortunately, the naming of names occurred before modern systematics had clarified the underlying logic of relationships, leaving us with terms like reptiles for a paraphyletic group and homeotherms for a polyphyletic group. Figure 13.22 illustrates, on the left, the traditional names for various vertebrate clades and, on the right, technical terms that properly recognize their genealogical relationships. For example, fish is not a proper phylogenetic term because it does not describe the entire clade, including tetrapods. Similarly, reptiles is not a proper phylogenetic term because it does not include all the other amniotes, including mammals.

● **KEY CONCEPT**

The names given to groups have not always kept up with the changes that modern systematics has made in the Tree of Life. Names should correspond to natural, monophyletic groups.

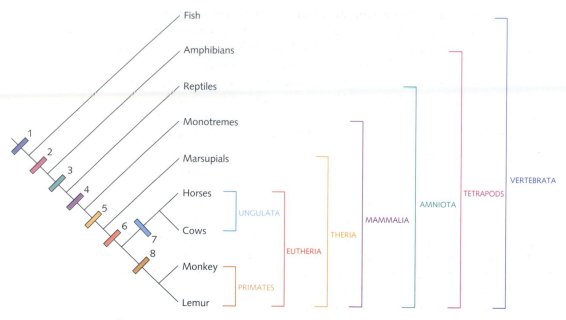

Figure 13.22 A rough outline of the vertebrate clade showing the informal and sometimes logically incorrect names of the groups on the left and the technically correct terms for the larger groupings on the right. (Numbered characters as in Figure 13.15.)

Important issues in molecular systematics

Alignment must be used to establish the homology of sequences

● **KEY CONCEPT**

DNA sequences are a powerful tool for identifying related species, but molecular systematics must be approached critically to be used reliably.

Because descendants inherit traits from their ancestors through genes, the history of descent is recorded in changes in the DNA sequences. Molecular data on sequences in genes are a simple form of character data: the characters are positions in the sequence, and the character states are the nucleotides at those positions. This sounds simple but assumes that the positions compared are homologous, that they derive from the same positions in a common ancestor. There are two problems with this assumption.

First, with four nucleotides, the probability that two nucleotides are the same simply because of mutation is high.

Second, in all but the most highly conserved sequences, insertions and deletions have occurred since the species being analyzed diverged from their shared ancestor. These insertions and deletions cause the overall lengths of orthologous DNA sequences in related species to differ. They also remove and add nucleotides at various places within the sequences. To make the sequence homology consistent along the entire length of the sequences being compared, gaps have to be inserted into the sequences where deletions have occurred. Gaps also have to be inserted in the sequence of one species opposite the positions where insertions have been added to the sequence of the other species. This is done so that the rest of the positions—the majority of them—that are thought to be homologous can be aligned into the same column. Only after that alignment has been done can we compare nucleotides at the same position to see if any of them have changed.

This alignment process will be done differently depending on the assumption one makes about the ancestral sequence from which the observed sequences are thought to be derived. Thus alignment is a critical step that involves assumptions about homology and phylogeny. There are algorithms that align sequences automatically, introducing some objectivity, but the selection of an algorithm can be subjective, and the algorithms are not always reliable. In practice many alignments are performed manually.

Even after sequences have been aligned, homoplasy—sequences that are similar not because of shared ancestry but because of some process that has occurred since they last shared ancestors—remains common. Homoplasy can be reduced but not eliminated by selecting and weighting characters. Because of the problem of homoplasy in sequence data, methods based strictly on parsimony cannot extract all the information available in sequences. The desire to gain access to that additional information is one reason that molecular systematists often use maximum likelihood methods.

Thus aligning DNA sequences is equivalent, at the molecular level, to establishing the homology of morphological characters at the phenotypic level.

The neutral model and the molecular clock

The dating of lineages starts with a fossil whose age marks, at least approximately, the divergence of the lineages. Many methods in molecular systematics assume that mutations are then fixed at the same overall rates in each lineage. This assumption connects evolutionary genetics (Chapters 3, 4, and 5) to systematics. The important part of the assumption is the regular rate of substitution of nucleotides. Molecular evolution does not have to be neutral for the methods to work, but it does need to have a strong form of statistical regularity that is most plausibly supplied by neutrality. Mutations occur independently, different nucleotides are fixed in each lineage, and as time goes by, differences in sequences accumulate. That is not controversial.

What is controversial is the assumption of a molecular clock: the claim that each lineage accumulates changes in sequences—substitutions, or mutations that have been fixed—*at the same rate*. The number of changes that have accumulated then estimates the time elapsed since the lineages shared common ancestors.

Molecular clock: The approximately constant rate of nucleotide substitution for particular genes and classes of genes within particular lineages.

There are several problems with this idea. First, the genomes of all organisms within a group share a similar structure that determines both the mutation rate and which parts of the genome are exposed to selection. Thus the rate of nucleotide change should be similar within large groups—eukaryotes, prokaryotes, RNA viruses—but not between them. The clock should tick at different rates in groups with different genomic structures.

Second, lineages differ in generation time. Small organisms usually have shorter generations than large ones, and body size varies dramatically among lineages. Since the mutation rate is a rate per generation, not a rate per year, one would expect changes to accumulate more rapidly in lineages with short than with long generations. This effect has been demonstrated in several cospeciating groups, such as gophers, with longer generation times, and their lice, which have shorter generation times.

Third, if the group being analyzed met all the assumptions required for a molecular clock, then the phylogenetic tree produced by the analysis would be clock-like, meaning that the distance from the common ancestor to the tips of each of the branches would be the same: each path would have the same number of codon or amino acid substitutions. However, this is rarely the case. The lengths of the branches connecting one ancestor to several descendants often differ. One might want to accept a rough correlation between divergence times and number of differences in substitutions between lineages, rather than a precise fit to a clock-like tree. However, even there problems arise, for the confidence limits that one can place on such relations are so broad that in practice the resulting clocks are imprecise.

For all these reasons, many studies try to avoid the assumption of a molecular clock. However, they are used when no other method is available to estimate the timing of an event, and when compared with the times delivered by fossils, the times delivered by molecular clocks often suggest interesting disparities. Recall that orthologous sequences are sequences in two or more related species derived from what was a single sequence in the most recent shared ancestor. A good predictor of the amount by which orthologous sequences in two species have diverged remains the time that has elapsed since they existed in a common ancestor. A longer branch in a molecular phylogeny suggests that more time has elapsed along that branch than along a shorter one—except, of course, when they are sister-branches.

Use the right molecule for the problem at hand

Each type of molecule and method of analysis is best suited to a certain range of problems. Hillis et al. (1996) and Avise (1994) discuss in depth the advantages and disadvantages of various molecules and methods. Here we simply make the point that one should choose the right combination for the problem at hand.

Molecules are like radioisotopes: they change at different rates. Uranium[238] has a half-life of 4.5×10^9 years, which makes it useful for dating objects of about the age of the Earth or the moon, whereas carbon[14] has a half-life of 5730 years, which makes it useful for dating archaeological objects from a few hundred to about 20 000 years old. The genomes of RNA viruses like HIV change so quickly that every person infected soon carries an identifiably different strain. Mitochondrial DNA, which is haploid, has a relatively fast substitution rate. It evolves rapidly enough to be useful for comparisons of lineages that diverged recently, but it can also be used to establish relationships among groups that are several million years old. Beyond that point it becomes so altered by repeated mutations that useful information is obscured by noise.

To get good molecular information on events that occurred in deep time, we need highly conserved genes, genes that change very slowly, such as the DNA that codes for the small subunits of ribosomal RNA. Such genes contain useful information about events that occurred 500–1500 million years ago. They can be used, for example, to test the idea (Margulis 1970, 1981) that mitochondria and chloroplasts are intracellular symbionts derived from prokaryote ancestors, an idea now strongly supported by sequence data:

- the nucleotide sequences of the 16 S RNA gene from chloroplasts indicate that chloroplasts are more closely related to photosynthetic cyanobacteria than to the nuclear genome of maize;

- similar analysis suggests that mitochondria are derived from the α subdivision of the purple bacteria.

Comparing the phylogenies of organelles and nuclei reveals striking differences. The nuclear sequences support the traditional view that the plants, fungi, and animals form a group distinct from the protists. The mitochondrial sequences suggest that the plant mitochondria were independently derived from the purple bacteria much more recently than the mitochondria found in fungi, ciliates, green algae, and animals. Chloroplasts appear to have been more recently acquired than mitochondria, for their sequences have diverged less from their bacterial ancestors, less of the chloroplast than of the mitochondrial genome has been transferred to the nuclear genome, and they entered hosts that did not yet exist when mitochondria entered the eukaryote lineage.

Thus highly conserved DNA sequences record the ancient history of organisms that have left no fossils, from a time in Earth history when we have very little other information, and allow us to test a major evolutionary hypothesis, the symbiotic theory for the origin of eukaryotic cells.

The genealogy of genes can differ from the phylogeny of species

Genes in the same organism can have different evolutionary histories

Molecular systematics is not only used to build trees relating species; it is also used to construct the history of single genes. The trees constructed from different genes in the same organisms often have different structures because each gene has had a different evolutionary history. A gene genealogy can differ from a species phylogeny because mutations do not occur simultaneously and are not constrained to occur during speciation (Figure 13.23). One gene may have diverged prior to a speciation event, another gene may have diverged after that speciation event. Thus genes have different genealogies, and only some genealogies have the same structure as the phylogeny of the species in which the genes occur. For events occurring within a species, the recovery of a reliable gene genealogy must be done in sequences with little or no recombination, such as the mitochondrial genome, because recombination produces nets, not branches.

● KEY CONCEPT

Because mutation is not synchronized with speciation, genes and species have histories that are similar in general but can differ in important details.

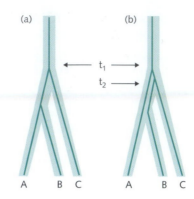

Figure 13.23 Phylogenies of species and genealogies of genes: The species tree is described by the large outer figure and is the same in both cases. The gene genealogy is described by the lines contained within the tree. Whenever more than one line is present, the gene is polymorphic. In both lineages there is a first mutation prior to the first speciation event. (a) The gene tree and the species tree have the same branching pattern, i.e. the second mutation occurs between the first and the second speciation events. (b) The genes and species have different branching patterns. The second mutation event occurs shortly after the first mutation event and before the first speciation event; thus here the genealogical split pre-dates species divergences. (From Avise 1994.)

A time to the most recent common ancestor is estimated from the difference in DNA sequences

Consider the DNA sequences of two copies of the same gene. They could be two alleles from a single population or from two related species. Assume that there has been no recombination, and that mutations have been neutral. That would be the case if the DNA were sampled from mitochondria or other haploid asexuals and if mutations produced no change in protein function. Such changes should be neutral or nearly so.

The two copies of the gene differ at several neutral sites. At some time in the past, when both copies of the gene derived from a common ancestor, there were no differences between them. How long did it take for this many differences to accumulate? To make that calculation, we must assume a constant mutation rate, but we do not have to make any assumptions about population size or selection at nearby loci, for neutral mutations accumulate within genes at rates that do not depend on these factors. Other important genetic properties of the population do depend on population size and selection; these include the number of mutations that will be fixed in the entire population and the amount of polymorphism that exists in the population at any time. But the number of mutations that have been fixed along an individual lineage since the last common ancestor depends only on the mutation rate and the time elapsed (Hudson 1990).

Mutation rates for single nucleotides are about 10^{-8} to 10^{-9} per organism per generation. With that information, we can use the neutral theory of evolution (Chapter 3, p. 54) to estimate how long ago the common ancestor existed. The error in the estimate depends on the lengths of the DNA sequences and the

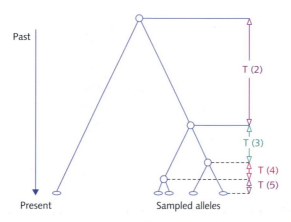

Figure 13.24 In coalescence analysis one uses a mathematical model of evolution that assumes a similar rate of fixation of mutations in all branches of the tree, and calculates backwards from an array of existing species until all branches of the tree coalesce in a single root. T(5) is the estimated age of the youngest allelic relatives, T(4) the next youngest, and so forth. (From Hudson 1990.)

number of mutations detected in them. We build the tree starting from the tips of the branches in the present, then work back in time, calculating the points of coalescence of the branches into a common ancestor. The trees that result are not phylogenies of species but genealogies of genes, and the process is called a coalescent process because the calculations yield the age at which the differences coalesce into the same ancestral sequence (Figure 13.24).

Once a reliable genealogy is obtained, the number of mutations separating the molecular ancestor from its tips, and the assumption of a molecular clock, can be used to date when the most recent common ancestor lived. Reliable genealogies allow us to compare old and young polymorphisms. The age of polymorphisms can be either greater or less than the age of the common ancestor of two species.

We have here described only the simplest possibility—no recombination, no geographical structure, and no selection—but the methods have been extended and can, to some extent, deal with more-complex scenarios.

> **Coalescence:** *A process of inference from different existing DNA sequences in related species back to the single ancestral sequence of the shared ancestor.*

● SUMMARY

This chapter discussed what phylogenetic trees are, how to construct them, and some issues that arise in their construction.

- The Tree of Life is not obvious, it must be discovered. Because much of the information that we need to discover it has been modified by the evolutionary process, we have to use methods that compensate for those modifications.

- Modern phylogenetic methods are making many changes in traditional views of the Tree of Life, some at the largest scale, such as the discovery of Archaea as the third major group of life, and some at the smallest scale, such as changes in which species is considered to be the closest relative of some other species.

- Reliable phylogenetic methods are based on using characters that are informative about relationships. The informative characters are shared, derived characters, not characters that are shared because they were present in distant ancestors and have not changed since.
- The starting point for a phylogenetic analysis is a character matrix—a matrix or table that lists the series of species to be analyzed and the characters being used. Many trees can be built from the same character matrix. We prefer either the simplest, the one implying the least change, or the tree that maximizes the probability of observing the distribution of character states actually seen, or some combination of these and other criteria.
- Trees are built from data using methods that can produce different trees. If all methods yield the same branch of a tree in a large and reliable data set, then that branch can be regarded with confidence. If the data are equally consistent with several branching patterns, judgement about relationships should be suspended.
- Genetic drift in microevolution produces a rough molecular clock in macroevolution. If a tree is clock-like, then the length of a branch is proportional to the time that elapsed along that branch. This is often roughly the case but rarely precisely the case.
- The genealogy of genes also produces trees that trace the history of genes, but those trees may differ from the phylogenetic trees of species within which they are embedded.

Given a reliable phylogenetic tree, we can then conduct a comparative analysis of trait evolution and historical biogeography, the topics of the next chapter.

● RECOMMENDED READING

Hillis, D.M., Moritz C., and Mable, B.K. (eds) (1996) *Molecular systematics*, 2nd edn. Sinauer Associates, Sunderland, MA.

Kitching, I.L., Humphries, C.J., Williams, D.M., and Forey, P.L. (1998) *Cladistics: the theory and practice of parsimony analysis*. Oxford University Press, Oxford.

● QUESTIONS

13.1. Why is it more reliable to interpret fossils as the tips of dead branches than as the direct ancestors of living species?

13.2. Can a phylogenetic tree ever be anything more than just a working hypothesis?

13.3. If every branch in the Tree of Life defines a new natural group, and if the Tree of Life includes everything from bacteria to whales, then how useful is the Linnean system of taxonomy, which tries to place all organisms into nested categories, species within genera within families within orders within classes within phyla? If that system does not work well, what would you recommend to replace it? A taxonomy should accurately reflect all the information in a phylogenetic tree. Is that goal possible?

CHAPTER 14

Comparative methods: trees, maps, and traits

In Chapter 13 we discussed how to build phylogenetic trees. We saw that methods that combine the logic of parsimony and maximum likelihood with DNA sequences and morphological data are continuing to improve our understanding of the relationships of organisms and, therefore, of their history. In this chapter we continue that train of thought by asking, what can we do with a phylogenetic tree? There are many answers. Four have led to particularly interesting insights:

(1) place the phylogenetic tree on a map to see how history correlates with geography;

(2) place traits on the phylogenetic tree to understand the history of a particular trait;

(3) do both to see how the history of a trait correlates with geography;

(4) place two traits on the tree then use the structure of the tree to remove history from their correlation.

We consider each of these methods in turn, concentrating on the insights provided by the examples and simplifying the technical details, which can be studied in depth in the works cited.

Putting trees on to maps reveals history

We begin with the insights gained from placing phylogenetic trees on to maps, and we proceed from the relatively recent into the relatively ancient past.

● **KEY CONCEPT**

Combining phylogenetics with geology and geography yields powerful insights into history.

The European flora and fauna were assembled from glacial refuges in Spain, Italy, and the Balkans

During the Pleistocene era, continental glaciers covered large areas of Europe and North America. Plants and animals persisted in refuges to the south,

Colonization from the Balkans

Grasshopper
(*Chorthippus parallelus*)

Colonization from Spain, Italy & the Balkans

Hedgehogs
(*Erinaceus spp*)

Colonization from Spain and the Balkans

Bear
(*Ursus arctos*)

Alder
(*Alnus glutinosa*)

Oaks
(*Quercus spp*)

Shrew
(*Sorex araneus*)

Figure 14.1 Three pairs of European post-glacial colonization pattern. (Reproduced from Hewitt (2001) Speciation, hybrid zones and phylogeography—or seeing genes in space and time. *Molecular Ecology* **10** with permission from Blackwell Publishing.)

and when the glaciers melted about 12 000 years ago, they moved north, repopulating the areas freed of ice. By studying the phylogeny of the resulting populations, primarily with mitochondrial DNA sequences, it has been possible to reconstruct the post-glacial history of much of Europe (Hewitt 2001). There were three main glacial refuges: in Spain, in Italy, and in the Balkans. And there are three major patterns in the recolonization of Europe (Figure 14.1).

- In the first, exemplified by grasshoppers and alders, the colonization came mostly from the Balkans, sweeping north to England and Scandinavia and meeting the Spanish and Italian populations in the Pyrenees and Alps, where hybrid zones were formed.

- In the second, exemplified by hedgehogs and oaks, the colonization came from all three refugia, with colonists from Spain invading France and England, colonists from Italy invading Germany and Scandinavia, and colonists from the Balkans invading eastern Europe and the Baltic states.

- In the third, exemplified by bears and shrews, most of the colonists came from Spain and the Balkans with little input from Italy, and hybrid zones formed where the Spanish and Balkan populations came into contact in Germany and Sweden.

The ancestors of human mitochondria and Y chromosomes lived about 200 000 years ago

Recently many studies have reported on the molecular diversity of human mitochondrial DNA (mtDNA). It represents only 1/200 000th of our genome, is haploid, is transmitted through females, and experiences little recombination. Because it also mutates at least 10 times faster than nuclear DNA, it can be used to track the divergence of populations that separated recently. An analysis of 144 mtDNAs sampled in different human groups (Cann et al. 1987) suggested, on the assumption that humans separated from chimpanzees 4–5 million years ago, that the molecular ancestor of all human mtDNA molecules lived about 200 000 years ago. Vigilant et al. (1991) confirmed this result and estimated the coalescence time at 166 000–249 000 years (Figure 14.2).

Similar studies done on the human Y chromosome, which is inherited through males as a haploid that does not recombine, yielded a similar estimate of coalescence time: 270 000 years ago (Dorit et al. 1995).

There are two sources of error in such estimates. The first error results from the sampling process itself, which yielded a certain number of sequences of a certain length. The second error arises in the calibration of the molecular clock, for the date of the human–chimpanzee split is not known accurately. When both errors are taken into account, then the coalescence times for the Y chromosome

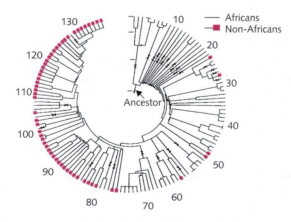

Figure 14.2 Coalescent analysis of human mitochondria suggests that the common ancestor of all extant mitochondria existed in a woman who lived about 200 000 years ago. The diagram shows a mitochondrial phylogenetic tree wrapped into a circle with the root closest to the middle. The numbers are humans sampled, numbered from closest (1) to most distant (130) from the common ancestor. (From Vigilant et al. 1991.)

and the mitochondrial data cannot be estimated precisely (Hillis et al. 1996), and the estimate of 200 000 years must be treated with caution. The branching pattern in the genealogy is estimated more reliably than the date of the last common ancestor, and results like Figure 14.2 do contain information about relationships.

Our ancestors came from Africa, but the molecules do not yield a precise date for their emergence

The initial claim that the ancestral mtDNA molecule was present in an African woman (the African Eve theory) was controversial, for equally plausible mtDNA phylogenies suggested a European or Asian origin (Hillis et al. 1996). More recent studies support the idea of an African origin with deep branches of the mtDNA phylogeny within Africa. They estimate the time of emergence of modern humans from Africa somewhere between 50 000 and 150 000 years ago. They also report a burst of divergence in the populations that left Africa 35 000–70 000 years ago (Horai et al. 1995, Ingman et al. 2000). An African origin is the best way to reconcile paleontological, archaeological, and genetic data, with good support for an African origin coming from fossils (Barbujani and Excoffier 1998). The mitochondrial data do not indicate that we all share a single common ancestor, for these data do not tell us where our nuclear DNA came from or where its many ancestors were located. The human population at that time was not necessarily small; the data are consistent with large populations.

Some human MHC polymorphisms are so old they are shared with chimpanzees

What do other human genes tell us about the same period of human history? The major histocompatibility complex (MHC) consists of 30–5 genes that code for polypeptide chains. These chains combine to form a huge variety of antigens. The genes fall into two classes (I and II) that each has two subclasses (A and B), a classification based on the structure of the antigens produced by the combination of the polypeptide chains. The products of the A gene subclass are labeled α polypeptides and those of the B gene subclass are labeled β polypeptides. The human leukocyte antigen (HLA) proteins consist of a combination of any α with any β polypeptide from any of the genes within an MHC class. The genes within one class all produce polypeptides that combine with those from the other class to form antigen molecules.

Some of these genes have many alleles; up to 200 have been identified so far for the highly polymorphic loci and it is likely that many more exist. These are the most highly polymorphic genes known. Other MHC genes are represented by only one or a few alleles (Figure 14.3). The reason for this diversity is thought to be the role these molecules play in immune defense and in the

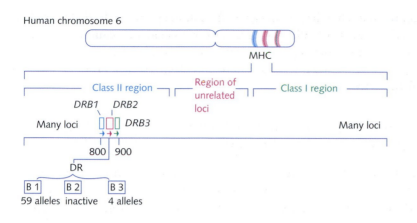

Figure 14.3 The human MHC complex is located in two regions of chromosome 6. The Class II region is shown in some detail. Within that region 3 MHC genes—the *DRB* genes—have been selected to illustrate the polymorphism of MHC loci. For example the *DRB1* gene is present in human populations in at least 59 different allelic forms. (From Klein et al. 1993.)

discrimination of self from nonself. Every individual can be effectively discriminated on the basis of combinations of MHC polypeptides, and some of the MHC genes are associated with particular infectious diseases, including tuberculosis, leprosy, dengue fever, HIV, hepatitis, malaria, leishmaniasis, and schistosomiasis (Singh et al. 1997).

Even more remarkable, some of the alleles at the same locus differ by up to 90 nucleotide substitutions and their products by up to 30 amino acid substitutions. It takes a long time for that many differences to accumulate. They must be old. Comparing the DNA sequences for the alleles at one locus between humans and chimpanzees yielded a striking result: some human alleles are more closely related to chimpanzee alleles than they are to other human alleles (Figure 14.4). Remember, this is a comparison only of alleles within a single locus. At least some of the MHC **polymorphism** must be older than the speciation event that separated humans from chimpanzees.

The molecular clock for primates has been estimated at $(1.2–1.56) \times 10^{-9}$ substitutions per site per year. If so, the youngest alleles in Figure 14.4 separated 2.7–3.5 million years ago, most of the branches separated 5–15 million years

Polymorphism: *The existence in a population of two or more alternative forms of a trait or gene.*

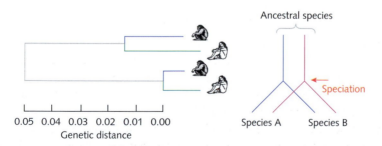

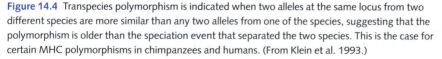

Figure 14.4 Transpecies polymorphism is indicated when two alleles at the same locus from two different species are more similar than any two alleles from one of the species, suggesting that the polymorphism is older than the speciation event that separated the two species. This is the case for certain MHC polymorphisms in chimpanzees and humans. (From Klein et al. 1993.)

ago, probably before the human–chimpanzee separation, and the main branches separated 15–35 million years ago, perhaps before the separation of Old World and New World monkeys.

This result constrains minimum population sizes over those periods. To pass 100 alleles from one generation to the next, one needs a minimum of 50 individuals, each heterozygous for that MHC gene and each heterozygote different from the other 49. The minimum number, however, is a serious underestimate for several reasons. First, some individuals share alleles. Second, the minimum estimate assumes just one locus, but at least four MHC loci are highly polymorphic, and all of them contain polymorphisms older than the species. Third, even if the minimum number of individuals did, at one point in time, actually contain the maximum possible number of alleles, random drift would have led to the loss of most of the polymorphisms in a small population. To avoid such loss, the polymorphisms must have been maintained by selection pressure.

To avoid losing polymorphisms, ancestral populations must have been fairly large

Computer simulations show that even with a population of 10 000 individuals, most neutral alleles initially present would be lost within 10 000 generations, which is roughly the time elapsed since *Homo sapiens* branched off from *Homo erectus*. With selection for heterozygotes with an advantage of 0.3 in a population of 1000 individuals, most alleles are lost within 1000 generations (Klein et al. 1990). But MHC loci are highly polymorphic for many alleles that are millions of years old. This suggests that since we shared common ancestors with chimpanzees, the effective human population size has not dropped much below 10 000 individuals.

Polymorphisms shared among species because of descent from common ancestors are called trans-species polymorphisms. Another well-studied example concerns polymorphisms in the self-incompatibility loci of flowering plants. These evolved to prevent self-fertilization and promote outcrossing and, like the MHC loci, some have large numbers of alleles, more than 30 in tobacco and more than 100 in cabbage and its relatives. Some alleles in tobacco are more closely related to alleles in petunias than they are to other alleles in tobacco, confirming the trans-species nature of the polymorphism (Klein et al. 1998).

We now all share a tiny sample of the mitochondria from a large ancestral population

The MHC results suggest that human populations for the last several million years were moderately large, and certainly not very small. There were at least 10 000 people alive 200 000 years ago, probably many more, and each of us probably has a nuclear gene from many of them. In contrast the mitochondrial results suggest that all human mitochondria descend from one female who lived less than 1 million years ago; the other mitochondria then in the population have since disappeared, as would be expected with random drift of mitochondria. Since the ancestral populations were moderately large, the drift was probably caused by variation in female reproductive success rather than by founder effects and genetic bottlenecks. The surviving mitochondria may have spread in part because they contributed to reproductive success, but we do not have to posit an adaptive advantage to explain the result. We might have the best of the mitochondria present long ago, but we might also have just some random sample of the mitochondria then present.

Because the common ancestor of the alleles of a polymorphic gene can be much older than the population in which it is found (see gene geneologies in Chapter 13, pp. 327 ff.), gene trees do not reflect the branching history patterns of populations until those populations have been separated for very long times, about six times their effective population size in terms of the number of generations. For human populations with an effective size of 10 000 individuals and a generation time of 20 years, this means separation times of about 1.2 million years, which is longer than the estimated age of our species. It is not surprising that we find related alleles in very divergent populations, and their presence does not necessarily indicate recent episodes of gene flow.

The Hawaiian arthropods have colonized and speciated on new islands as they formed

Now we turn to an example that goes a bit deeper into time (Figure 14.5). The Hawaiian Islands formed over a hot spot beneath the Pacific Plate. The hot spot pushes lava through the crust, forming huge volcanoes that emerge from the 5-km-deep ocean as islands. These islands are then carried northwest on the Pacific Plate away from the hot spot. As they move, the crust bends beneath their weight, they sink into the sea, and rain erodes them. These processes eventually reduce high islands first to low islands, then to atolls, then to submerged seamounts. By the time they near the intersection of the Kuril and Aleutian Trenches 8300 km away and 200–400 million years later, islands that had once reached 3–5 km above sea level may be 1.5 km below sea level. Importantly, their lavas date the islands, giving us a geological timeline determined independently of any inferences with molecular clocks.

Given that mechanism of island formation, in which islands form on a conveyor belt and move off the hotspot from south-east to north-west, we would expect that the organisms living on the islands would have moved down the island chain in the opposite direction, from older islands colonized earlier to younger islands that emerged from the ocean more recently. In the terrestrial arthropods—primarily insects and spiders—that is precisely the pattern indicated by placing phylogenetic trees on to a map of the archipelago (Figure 14.6).

Three patterns were formed. In the first, there is a single speciation event when the dispersers arrive on a new island. In the second, there are multiple

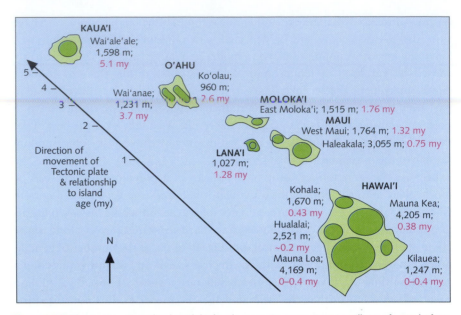

Figure 14.5 The six Hawaiian islands with high volcanoes. Ages are given in millions of years before present (my); heights of volcanoes are given in meters. Volcanoes are marked with dark green circles. The oldest volcanoes are furthest away from the current hot spot under Hawai'i. (Reproduced from Roderick and Gillespie (1998) Speciation and phylogeography of Hawaiian terrestrial arthropods. *Molecular Ecology* **7**, with permission from Blackwell Publishing.)

speciation events upon arrival associated with local geographic isolation, sexual selection, sympatric speciation, or all three. In the third, the species divide up on arrival into separate populations, one on each volcano, each of which then eventually becomes a species. Note that the sequence of phylogenetic events inferred from molecular systematics matches the sequence of geological events inferred from the mechanism of island formation and the ages of the islands.

Madagascar carnivores are all mongoose relatives that speciated in Madagascar

Only four of the roughly 20 orders of mammals live on Madagascar, which has been isolated from Africa by a deep channel for at least 88 million years. Those four orders—carnivores, primates, rodents, and insectivores—all originated after Madagascar had become an island. Thus the mammals living on Madagascar had to get there by crossing the Mozambique Channel, which is now 400–800 km wide. They may have made the crossing on floating mats of vegetation or on trees ripped from the banks of the Zambezi or Limpopo Rivers by monsoon floods.

How many times did that happen? The phylogenetic evidence indicates that there was only a single colonization event for the carnivores (Figure 14.7), and that the arrival of the carnivores on Madagascar about 18–24 million years ago

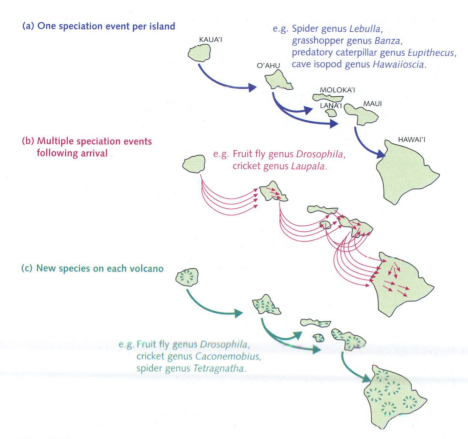

(a) One speciation event per island

KAUA'I

e.g. Spider genus *Lebulla*,
grasshopper genus *Banza*,
predatory caterpillar genus *Eupithecus*,
cave isopod genus *Hawaiioscia*.

O'AHU

MOLOKA'I

LANA'I   MAUI

HAWAI'I

(b) Multiple speciation events
following arrival

e.g. Fruit fly genus *Drosophila*,
cricket genus *Laupala*.

(c) New species on each volcano

e.g. Fruit fly genus *Drosophila*,
cricket genus *Caconemobius*,
spider genus *Tetragnatha*.

Figure 14.6 The progression of arthropod lineages down the Hawaiian chain resulting in distinct species on each island or volcano. Arrows indicate the direction of speciation events. (a) One speciation event per island. (b) Multiple speciation events following arrival but not necessarily associated with particular volcanoes. (c) New species form on each volcano. (Reproduced from Roderick and Gillespie (2001) Speciation and phylogeography of Hawaiian terrestrial arthropods. *Molecular Ecology* **10**, with permission from Blackwell Publishing.)

came much later than that of the lemurs, who got there about 62–66 million years ago. This suggests both that oceanic dispersal over such distances is rare, and that it is not plausible that Madagascar was ever connected by a land bridge to Africa after breaking away, for if she had been, then the dispersers should have walked across the bridge together and arrived on the island at about the same time, rather than roughly 40 million years apart (Yoder et al. 2003).

Plants moved between east Asia and eastern North America via two routes

Botanists have long known that in several groups the closest relatives of plants living in east Asia are in eastern North America—far away, not nearby on the same continent. Such **disjunct distributions** are a biogeographic puzzle. Clues to solving the puzzle come from the histories of the continents and oceans.

Disjunct distribution:
Geographical distribution in which the closest relatives of a group are found far away with more distantly related groups occupying the intervening territory.

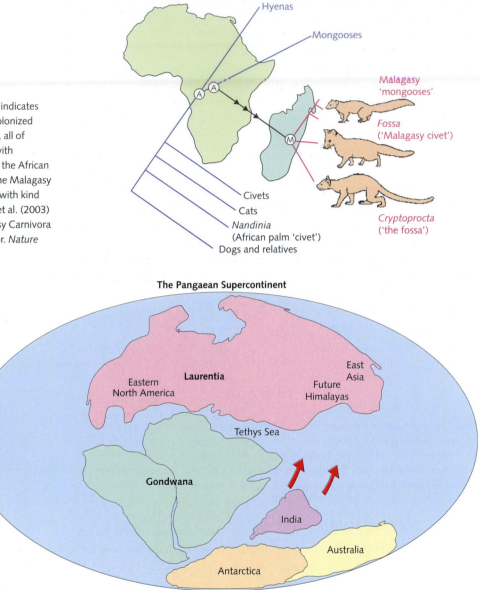

Figure 14.7 Phylogeny indicates that Madagascar was colonized only once by carnivores, all of which share ancestors with mongooses. A indicates the African radiation. M indicates the Malagasy radiation. (Reproduced with kind permission from Yoder et al. (2003) Single origin of Malagasy Carnivora from an African ancestor. *Nature* **421**, 734–7.)

Figure 14.8 The Tethys Sea formed a tropical pathway along whose coast plants could disperse between eastern North America (ENA) and east Asia (EA). It existed from 200 to 23 Ma. When North America broke off from Eurasia and rotated north-westwards, it came into contact with Siberia, and there have been several times when plants and animals could move across a land bridge between east Asia and North America. (Marshak 2001.)

Since plate tectonics was discovered, we have known that after the breakup of Pangaea about 200 million years ago, the Tethys Sea stretched from eastern North America through the tropics, first north of Gondwana, then, after the breakup of Gondwana, north of South America, Africa, India, and Australia (Figure 14.8).

The Tethys Sea endured for roughly 180 million years, closing when Africa moved north to contact Asia 23 million years ago in what is now the Arabian Peninsula and the Middle East. After the breakup of Pangaea and the separation of North America from Africa and Eurasia, there were several periods when a land bridge existed across what is now the Bering Straits, connecting Siberia to Alaska. Thus the question arises: did plants move between east Asia and eastern North America by dispersing along the margin of the Tethys Sea, or did they make that move by crossing the Bering land bridge (Donoghue et al. 2001)?

If plants dispersed along the margin of the Tethys Sea, they would have left relatives in Europe and western Asia, near the remnants of the Tethys Sea—the Atlantic Ocean, Mediterranean Sea, Black Sea, Caspian Sea, and Aral Sea. That would constitute an Atlantic Track. If they moved across the Bering land bridge, they would have left relatives in east Asia, Japan, and, possibly, western North America. That would constitute a Pacific Track (Figure 14.9).

All the following groups have disjunct distributions: *Liquidamber* and *Cercis*, small ornamental trees and shrubs, are clades whose species both appear to have taken the Atlantic Track; *Hamamelis* (witchhazel), *Triosteum* (horse gentian; used in homeopathic medicine), *Buckleya* (a rare semi-parasite on hemlock roots), and *Torreya* (stinking cedar) all appear to have taken the Pacific Track. Estimates of the divergence times based on molecular and fossil evidence differ strikingly for the two tracks; the Atlantic Track being consistently old,

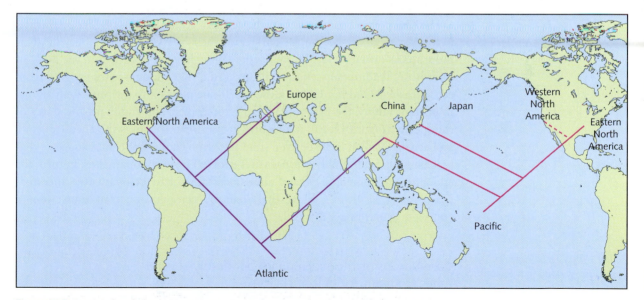

Figure 14.9 Present-day phylogenetic patterns expected in plants that dispersed from east Asia to eastern North America via the margins of the Tethys Sea (Atlantic Track) or via the Bering land bridge (Pacific Track). (Reproduced with permission from International Journal of Plant Sciences, Vol. **162**, Donoghue et al., Phylogenetic patterns in Northern Hemisphere plant geography, 2001, University of Chicago Press.)

and the Pacific Track containing a mixture of divergence times. This is consistent with the closure of the Tethys Sea 23 million years ago and with the repeated appearance and disappearance of a Bering land bridge.

Thus we can use phylogenetic trees to illuminate historical biogeography by placing trees on to maps. They can also be used to understand trait evolution.

Plotting traits on to phylogenetic trees reveals their history

● KEY CONCEPT

Reconstructing the history of trait evolution can overturn received opinion.

By plotting traits on to trees we can infer the sequence of evolution, the probable state of the traits in ancestors, and the number of times that a particular condition evolved. Here we examine two cases that changed ideas: the parasitic wasps, whose phylogeny suggests that parasitic life styles evolved in a direction opposite to that previously assumed; and the carnivorous plants, whose phylogeny shows that certain styles of capturing insects evolved independently several times.

In one family of Hymenoptera, endoparasitism is ancestral and ectoparasitism is derived

There are about 115 000 described species of Hymenoptera (ants, bees, and wasps). They include most of the social insects, except the termites, which are in a group related to roaches. The parasitic Hymenoptera are among the important sources of mortality for many insect species, including crop pests.

Hymenopteran parasites can be either ectoparasitic or endoparasitic. Ectoparasite larvae live on the surface of their host and feed by burying their mouthparts into its body. Endoparasite larvae live and feed within the body of their host. Traditional scenarios all suggested that endoparasites evolved from ectoparasites (Godfray 1994). However, if one plots on to the phylogeny of one important family, the Braconidae, the ectoparasitic and endoparasitic states of the species (Figure 14.10), it appears that endoparasitism is the ancestral state and ectoparasitism is the derived state (Dowton et al. 1998). This suggests that at least within this family ectoparasites evolved from endoparasites. In turn, the endoparasitic braconid wasps almost certainly had ectoparasitic ancestors (Dowton et al. 1997), suggesting that the ectoparasitic trait underwent a complete evolutionary reversal.

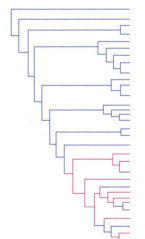

Figure 14.10 The phylogeny of parasitic wasps with endoparasitism plotted as blue lines and ectoparasitism plotted as pink lines. It appears that ectoparasitism is a derived state. (From Dowton et al. 1998.)

Carnivorous plants evolved independently several times from autotrophic ancestors

Carnivorous plants have radically modified leaf and stem morphology, including pitcher and flypaper traps for catching insects and other small animals. Are they all descended from a common ancestor, or did they arise more than once, independently? Are pitcher and flypaper traps evolutionarily independent, or is one type more likely to arise in a lineage in which the other is already present? Darwin thought that all flypaper traps were descended from a single common ancestor. Was he correct?

Albert et al. (1992) answered these questions with a molecular phylogenetic analysis of taxa from 72 plant families using a chloroplast gene that codes for a photosynthetic enzyme. The rates of nucleotide substitution in this gene make it appropriate for analysis of phylogenetic events at this depth in time. The results were surprising. Flypaper traps appear to have evolved independently at least five times, and pitcher plants at least three times. Two of the pitcher plant clades appear to be sister groups of two of the flypaper trap clades. For example, the Old World pitcher plants, *Nepenthes*, seem to be the sister group to the sundews and the venus flytraps, the Droseraceae. Pitcher-plant anatomy has thus probably evolved from the simpler, flytrap anatomy at least twice. Thus the structural similarity of pitcher plants misleads us about their relationships, for some pitchers evolved by convergence from dissimilar ancestors.

Anole lizards repeatedly evolved similar ecomorphs on different islands

This next example combines phylogeography with comparative analysis of trait evolution. The lizards in the *Anolis* clade radiated on Caribbean islands and adapted independently on each island to local conditions. In the Greater Antilles (Cuba, Hispaniola, Jamaica, and Puerto Rico) they have specialized in similar ways on each island. Some have become large and live in the crowns of trees; some inhabit grass and bushes; some forage anywhere on tree trunks; others spend most of their time on trunks near the ground, others on trunks near the canopy. The morphology of lizards living in similar habitats but on different islands has evolved to similar states, called ecomorphs, a term indicating that morphology is associated with habitat (Figure 14.11).

There are two basic ways this could have happened. In the first, the lizards speciated and radiated into different habitats on one island, forming ecomorphs there. Those species then dispersed to other islands, retaining their morphology and habitat preference. In this first scenario, each ecomorph evolves once. In the second, the lizards disperse among the islands, and then the speciation and differentiation into ecomorphs happens independently on each island. This second case would provide much stronger evidence for an association between morphology and habitat, for in the first, that association arises only once; in the second, it arises repeatedly. To discriminate between these two possibilities, Losos et al. (1998) first established the reality of ecomorphs by demonstrating that their specimens fell into morphological clusters associated with habitats (Figure 14.11a). They then plotted the ecomorphs on to the molecular

● **KEY CONCEPT**

Historical biogeography informed by phylogenetics sheds light on community assembly.

Ecomorph: *A morphology associated with a particular ecological habitat.*

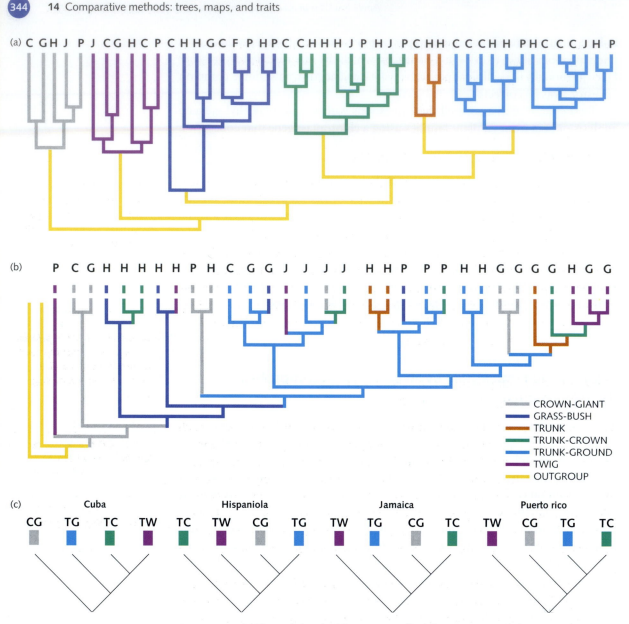

Figure 14.11 (a) The Anole lizards of the Greater Antilles fall into six ecomorph classes associated with the habitat in which they live and forage. (b) These ecomorphs originate independently in the phylogeny several times. (c) On four islands the same ecomorphs have evolved independently from three different ancestral states. (Reproduced with permission from Science, Vol. **279**, Losos et al., Contingency and determinism in replicated adaptive radiations of island lizards, pp. 2115–18, 1998, AAAS.)

phylogeny of the lizards (Figure 14.11b). There was no evidence for the first and much support for the second scenario. They then isolated the portions of the phylogeny that represent the radiations on each of the islands (Figure 14.11c). There you can see that on each island, each of the ecomorphs evolved. Moreover, the colonizing ecomorph and the sequence in which the series of ecomorphs evolved was different on each island. This example nicely combines the contingency of history with the determinism of convergence: different historical paths with contingent origins all led to similar current states.

Species are not independent samples

In the *Anolis* example just discussed, the two scenarios differed in the number of independent evolutionary events. In the first, each ecomorph evolved only once. In the second, the one that actually appears to be the case, each ecomorph evolved independently several times. The reliability of the inference of an association of morphology with habitat depends upon the number of times it happened. If it happened only once, the association could be fortuitous. If it happened repeatedly, with the same result each time, the association probably exists for good causal reasons. This is not just a problem in the *Anolis* clade; it is a general problem with any phylogenetic tree. We solve this problem by using independent contrasts, explained next.

● **KEY CONCEPT**

Shared histories at variable depths in time introduces special problems for statistical analysis of comparative patterns.

The method of independent contrasts controls for shared ancestry in isolating independent changes

The number of independent evolutionary origins of a trait is normally much smaller than the number of species in a sample. For example, there are more than 68 species of lizards in the genus *Sceloporus*, and at least 28 of them are viviparous. However, when Shine (1985) plotted the trait viviparity on to a phylogeny of the genus, he saw that the trait had arisen at most four to six times within the genus. Any analysis that counts the number of species rather than the number of independent evolutionary events on the phylogenetic tree may thus greatly inflate the sample size and give a misleading impression of how strong a pattern really is.

The solution, suggested by Felsenstein (1985), is depicted in Figure 14.12. The key insight in this figure is that the *change* that occurs after a speciation event in one daughter species is independent of the *change* that occurs after that event in the other daughter species. The figure depicts eight related species, A, B, C, D, E, F, G, and H. A and B, C and D, E and F, and G and H form pairs of closest relatives. The changes that occurred between A and B after they speciated are independent of the changes that occurred between C and D after they speciated, although all four species share a common ancestor. Thus if we measure a trait x on each species, then the difference between A and B in trait x is independent of the difference between C and D in x. Differences so calculated are called independent contrasts ($x_1 \leftrightarrow x_2$ is independent of $x_3 \leftrightarrow x_4$), and this method of analysis is called the *method of independent contrasts*. It assumes that all branch lengths are equal; if they are not, appropriate corrections must be included.

Independent contrasts:
Contrasts that do not contain any confounding influence of shared history because they evolved after that history was no longer shared.

Contrasts: *Differences in trait states between two taxa.*

Viviparity: *Giving birth to fully formed offspring, as opposed to eggs.*

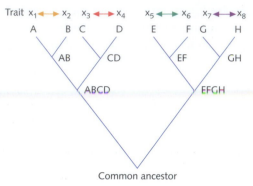

Figure 14.12 The method of independent contrasts. The key idea is that changes that occur after one speciation event, such as $(x_1 \leftrightarrow x_2)$, are independent of changes that occur after another speciation event, such as $(x_3 \leftrightarrow x_4)$. (From Felsenstein 1985.)

The method can be applied to contrasts formed all the way down the phylogenetic tree. One can infer the value of a trait in a presumed ancestor by calculating the mean value of the trait among all the surviving descendants to associate a value of the trait with every node on the tree. The contrasts—or differences—are then calculated just as they are for the species at the tips of the tree. The method is most reliable, and the contrasts are easiest to interpret, when one only calculates the contrasts as high on the tree, as close to the tips of the branches, as possible, where the species being compared share much of their ecology, and where ancestral states can be inferred more reliably.

Plant genera living in the shade make larger seeds than plant genera living in the sun

The seeds of trees span many orders of magnitude in weight, from milligrams to kilograms, and botanists have long sought to understand the causes of this enormous variation in offspring size, a trait directly related to fitness. One hypothesis is that the seeds of shade-tolerant trees, which must germinate and establish themselves on leaf litter beneath the forest canopy, should be larger than the seeds of light-demanding trees, which germinate in the open and have access to more light energy as soon as they produce their first leaves. Grubb and Metcalfe (1996) tested this idea with an insightful application of the method of independent contrasts. They first contrasted the seed size of shade-tolerant and light-demanding species within a genus, then they contrasted the mean seed sizes of genera all of whose species were either shade-tolerant or light-demanding. They compared species with their closest relatives within a genus and genera with their closest relatives within a family. In other words, they worked downwards from the reliable, directly observed tips of the phylogenetic tree to the less-reliable, inferred ancestral states.

Their results (Figure 14.13) were intriguing. The comparison of species within genera revealed no tendency for the shade-tolerant trees to have larger seeds than the light-demanding trees. Paired relatives living in different habitats

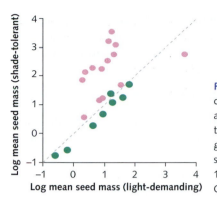

Figure 14.13 Phylogenetically controlled comparisons of the dry seed mass of shade-tolerant and light-demanding trees. The pink circles compare the mean values of related genera within families; the green circles compare the mean values of related species within genera. Points above the 1:1 line indicate larger seeds in shady habitats. (From Grubb and Metcalfe 1996.)

had seeds of nearly the same size. However, the comparison of genera within families supported the hypothesis. Most of the shade-tolerant genera had larger seeds than their closest light-demanding relatives; only one did not.

The authors interpreted this as follows. The interspecific, intrageneric differences evolved relatively recently. That the comparisons of genera within families reveal an apparent response to the shade–light contrast may simply reflect the longer time that the difference has had to evolve. Those genera might also have evolved their seed sizes in other habitats before they invaded the environments in which they are currently found, and their seeds may have those sizes because their ancestors lived in other habitats and evolved those seed sizes for reasons that have nothing to do with the shade–light contrast. This case thus exemplifies both the power and the limits of the method of independent contrasts.

Trees living in the savannah allocate more to roots; those in the forest, more to leaves

How must a tropical tree change to respond to new problems encountered when it moves from savannah to forest or forest to savannah? To answer this question, Hoffmann and Franco (2003) chose nine congeneric pairs, each containing one savannah and one forest species, and grew all 18 species under two light regimes and two nutrient regimes (in all four combinations). They then measured how they had allocated energy into roots and leaves. This experimental design gave them nine independent opportunities to see how the trees had managed the transition.

They found that savannah species allocated more of their growth to roots and maintained lower leaf area per unit plant mass. The savannah species were also more phenotypically plastic in response to the experimental differences in light intensity: when shaded they increased leaf area per unit plant mass enough to compensate for the reduction in net assimilation rate per unit leaf area and to maintain their relative growth rates virtually unchanged. Overall, they found that more of the variation in the traits was associated with phylogeny and thus history than it was with the difference between savannah and forest. Thus each of the nine clades sampled did respond to the habitat difference, but that response came on top of even larger, conservative differences associated with the overall phylogeny.

Longer-lived albatrosses are more faithful to their mates

Mate fidelity is thought to be advantageous in birds because pairs that have interacted with each other have learned to adjust to individual particularities in mating, nest building, foraging, and provisioning offspring. They have also been able to test each other's reliability in providing those services. The longer a species lives, the more advantageous mate fidelity should be, for if life is short, there is little opportunity for the hypothesized advantages to accumulate, but if life is long, they not only accumulate but provide benefits for a long time. The hypothesis of an association between lifespan and mate fidelity has, however, been difficult to test for reasons that apply in many comparative studies. First, mate fidelity tends to vary much more between species than within species. It is therefore not a trait that can be readily subjected to experimental manipulation. Second, several other factors influence both mate fidelity and longevity and must be properly controlled in order to isolate the proposed association of fidelity and longevity. Large organisms live longer than small ones, and they breed less frequently. Thus at least those two confounding factors must be controlled for.

Using the method of independent contrasts on data that had been adjusted for body size and frequency of breeding attempts, Bried et al. (2003) showed that mate fidelity and adult life expectancy were strongly correlated in the Procellariiformes, the bird clade that contains albatrosses, petrels, and their relatives (Figure 14.14). Species in this clade are also known for their site fidelity, returning year after year to the same areas on the same islands to nest. Divorce

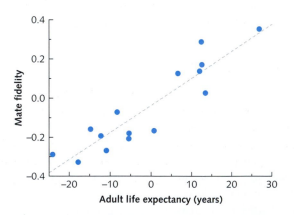

Figure 14.14 The association between mate fidelity (expressed as deviations from the mean number of mates per lifetime) and life expectancy in albatrosses and their relatives is strong. Here both traits have been statistically corrected to adjust for body size and mating frequency. (Reprinted from Bried et al. (2003) Mate fidelity in monogamous birds. *Animal Behaviour* **65**, 235–46. With permission from Elsevier.)

is known to be costly—it takes a new couple several years to regain the level of performance enjoyed by faithful pairs. It is therefore thought that site fidelity is a method of achieving mate fidelity—they return to the same place to find the mate they had last year, or 2 years ago, after spending the intervening months or years foraging over broad reaches of the world's oceans.

Thus comparative methods yield interesting insights across a broad range of biological problems, including historical biogeography, life histories, behaviour, ecological plant physiology, and ecology.

General comments on comparative methods

All comparisons should start with a reliable phylogeny

There are many more comparative methods than we have described here. Many fall into two broad categories: those designed to control or remove the effects of phylogeny, and those designed to infer correlated change or the origin of functional trait complexes (Miles and Dunham 1993). Most rely on methods of phylogenetic reconstruction or statistical analysis that can be traced through the recommended reading at the end of the chapter. Several general points about all comparative methods are important.

First, life has a history that is ignored at the peril of anyone comparing two or more species. A proper comparison is based on a reliable phylogeny constructed with modern systematic methods (Chapter 13, p. 303). A good start is to plot the traits compared on to such a phylogeny. That immediately reveals how many times the traits evolved in the clade and suggests hypotheses about primitive and derived states. This simple procedure is ignored by anyone who compares just a nematode, a fly, and a mouse—three popular model systems for molecular genetics and developmental biology sampled from widely separated twigs on the Tree of Life—and dares to talk about ancestral and derived states.

> ● **KEY CONCEPT**
> History cannot be ignored.

Species are not independent samples

Second, species are not independent samples, for all the descendants of a single ancestral species share some of that species' traits. Only by plotting traits on to reliable phylogenies can we infer how many times a trait originated in a given clade, and thus the number of independent events. If that means that something very interesting only happened a few times, then we have to accept that we may not be able to decide whether trends associated with such events are significant.

Even the best comparisons can only suggest, not reveal, causes

Third, even the best controlled and most reliable comparative inferences cannot reveal causes. At their best they reveal strong correlations. For example, all species of macaque forage in groups with a similar social structure, but they occupy a very wide range of habitats. Does that mean that their social structure was inherited from their common ancestor and has not yet adapted to each habitat inhabited by macaques, or does it mean that the social structure is a superbly flexible adaptation capable of dealing with all the problems posed by a wide range of habitats? To answer such questions we need more than comparative information.

The historical component of a pattern is a mixture of adaptation and constraint

Fourth, comparative methods cannot, by themselves, tell us how much of a pattern to ascribe to adaptation and how much to constraint, for the historical component of a pattern is usually a mixture of the two. However, they can increase the enlightenment gained from mechanistic approaches by placing the experimental results for single species into a broad phylogenetic context. Comparisons always test and can expand the generality of the questions asked and the conclusions reached in experimental research programs.

● SUMMARY

This chapter describes some important comparative methods with examples.

- Comparative methods partition the components of variation among species into the influence of phylogeny—resemblance to ancestors—and the effects of selection and drift—divergence from ancestors.

- There are two basic methods plus their combination: plot trees on maps, plot traits on trees, or do both. All have yielded important insights.

- A technical problem in comparative trend analysis is to get data that are independent, where measurements are not influenced by a shared ancestor. The method of independent contrasts solves that problem.

- Comparative methods deliver answers to big questions that experiments cannot address, but they cannot determine causation. Their strengths are reliable description of large-scale patterns and the power to pose questions.

This chapter completes our discussion of macroevolutionary principles. We next examine the history of life.

● RECOMMENDED READING

Avise, J.C. (2000) *Phylogeography: the history and formation of species*. Harvard University Press, Cambridge, MA.

Crisci, J.V., Katinas, L., and Posadas, P. (2003) *Historical biogeography: an introduction*. Harvard University Press, Cambridge, MA.

Harvey, P.H. and Pagel, M.D. (1991) *The comparative method in evolutionary biology*. Oxford University Press, Oxford.

QUESTIONS

14.1. You have the opportunity to do two research projects, but you only have time and money for one of them. The first would be a comparative study of sexual selection in lizards in which you would apply the method of independent contrasts to look at evolutionary changes in the skin color of males associated with changes in mating systems. For this project you would use data gathered by other people and already available in the literature. The second project would be an experimental study of one lizard species in which you would manipulate parasite loads in males and study their changes in skin color and mating success. Which project would you choose and why?

14.2. Marsupials are found primarily in Australia and South America (the opossum is a relatively recent arrival in North America). You could use the geological estimate of the timing of the breakup of Gondwana to date the split between marsupial (metatherian) and placental (eutherian) mammals, or you could use a molecular estimate. Look up both estimates in the recent scientific literature and discuss why they differ and what that might mean.

PART 4

The history of life

Part 3 of this volume discussed what species are and how they evolve (Chapter 12), how we infer the relationships among species and clades (Chapter 13), and how we can use this kind of information to investigate history and biogeography (Chapter 14).

In Part 4 we discuss the history of life from three perspectives.

First we ask, in Chapter 15, what were the key events in evolution that had the greatest consequences? Many of these had something to do with changes in the way that information was inherited or with the origin of new units of selection. Often a genetic conflict within or among units of selection had to be resolved before the key event could be completed.

Second we ask, in Chapter 16, what were the major events in the geological theater that shaped life's evolution? Here we deal with both global and local catastrophes, and we point out the extent to which life has also shaped the planet, just as the planet has shaped life.

Third we ask, in Chapter 17, what are the major patterns of macroevolution, as revealed first in fossils, then in the radiations of the major clades? Here we highlight the messages of macroevolution and discuss some of its major controversies—stasis, punctuation, large-scale trends, and the issue of progress.

Together Parts 3 and 4 constitute an overview of evolution at and above the level of species where the processes of speciation and extinction over long periods of time have produced major patterns in the history of life.

When we contrast the perspectives of microevolution (Parts 1 and 2) with those of macroevolution (Parts 3 and 4), it is natural to wonder how they fit together. They are not independent. They influence each other, and the arrows of causation point in both directions, from micro to macro and from macro to micro. How they do so is explored in Part 5.

CHAPTER 15
Key events in evolution

Evolutionary biologists study both the history of life and the mechanisms that drive evolutionary change. Like geologists and astronomers they want to know what happened in the past, and they want to explain history with general principles. In the next two chapters we will discuss some highlights in the history of the planet and of life. Just as the history of the Earth was shaped by key events, key events in evolution shaped the life we observe today. In this chapter we consider the origins of fundamentally new things that had major impact on subsequent evolution, drawing material from Maynard Smith and Szathmáry (1995). First we make some cautionary remarks.

Key event: *An event involving the origin of something fundamentally new with a large impact on subsequent evolution.*

The importance of a key event can only be judged in retrospect

According to Jacob (1977), evolution proceeds by tinkering: finding an *ad hoc* solution to an immediate problem. Evolutionary change occurs either because of the short-term advantage of genetic variants in a specific environment, or by chance. No mechanism can cause evolutionary change because of beneficial effects in the distant future. But we judge the importance of key events retrospectively by their long-term consequences. For example, the origin of sex had many long-term consequences, but it did not originate for those reasons, and many consequences could not have been predicted at the time sex originated.

Our perception of important events is biased by anthropocentrism

When thinking about evolution as history, it is tempting to start with our species and work backwards through a series of imagined ancestors to simple early forms. That script suggests that evolution has produced ever-increasing complexity and that our species is the culminating point in the evolution of life. When we view evolution from the bottom rather than the top of the Tree of Life (Figures 13.2 and 17.7), the progressive interpretation becomes questionable. All other existing species have an evolutionary ancestry as old as our own, and some originated later than *Homo sapiens*. We are further biased by having much more information on the few lineages that produced plants and animals than on the many lineages that produced the enormous variety of microorganisms. In lesser-known parts of the tree key evolutionary events may have occurred that are unknown to us. This bias explains why most of the evolutionary events discussed in this chapter occurred in the lineages containing multicellular organisms: plants, fungi, and animals.

Our interpretation of key events is no better than constrained speculation

Many key events in evolution occurred long ago. We can infer them from differences between existing organisms and from the fossil record, but what happened and why it occurred must remain speculations. The chance of reconstructing the evolution of novelties is small, for present-day structures have been extensively modified since their origin. Thus what we observe today is not directly informative about origins. Repeating key evolutionary events in experiments is difficult, for we do not know, or cannot create, the appropriate starting conditions. Speculations about the origins of key evolutionary novelties are, however, constrained by physics, chemistry, and biology.

Key events have been few in number and large in impact

Some key events in evolution were these (Maynard Smith and Szathmáry 1995):

- the origin of replicating molecules;
- the sequestration of replicating molecules in compartments;
- the condensation of independent replicators into chromosomes;
- the transition from RNA as gene and enzyme to DNA with its genetic code for proteins;
- the symbiotic origin of eukaryotes;
- the origin of sexual populations from asexual clones;
- the origin of multicellularity in plants, animals, and fungi;
- the origin of societies with reproductive castes;
- the origin of language.

Rather than discussing all these events, we describe cases that illustrate principles involved in the evolution of novelty and then discuss the principles.

The origin of life

● **KEY CONCEPT**

Life emerged from abiotic matter through chemical evolution.

The most fundamental and perhaps the most intriguing key event was the origin of life. To be clear about its origin, we first have to define life. As for many other concepts in biology, a precise definition is difficult. A rough definition characterizes life by two of its essential features: a living thing should have metabolism—a coordinated system of chemical reactions contributing to its maintenance, a system that imports energy to maintain order—and hereditary replication—a system of copying in which the new structure resembles the old.

> **Life:** *A process involving metabolism and hereditary replication.*

Prebiotic chemical evolution produced a diverse array of organic compounds

In the 1920s Oparin and Haldane independently suggested that in aqueous solutions under a low-oxygen atmosphere with energy supplied by lightning or ultraviolet radiation (the 'primitive soup'), a variety of organic molecules would be synthesized. In a famous experiment Miller and Urey (Miller 1953) mimicked these conditions. Using a gas mixture of CH_4 (methane), NH_3 (ammonia), and H_2 (hydrogen), simulating lightning by electric discharges between two electrodes, in the presence of a water solution containing some simple inorganic molecules, Miller found that several important biological compounds formed, notably amino acids. Such experiments have been repeated with similar results, including the production of nucleotide precursors. They demonstrate the possibility of chemical evolution of a diverse prebiotic chemical environment. However, many biologically significant compounds have never been obtained in such prebiotic synthesis experiments, including nucleic acids, the chemical basis for heredity in all existing organisms.

More recently it has been suggested that prebiotic chemically diverse environments may have evolved on the surfaces of pyrite crystals. Another possibility is chemical evolution on the surface of droplets in clouds. Both can be defended on theoretical grounds. Neither rules out the other, or the primitive soup.

Irrespective of the precise conditions of prebiotic chemical evolution, a collection of organic compounds does not imply life. In particular, we need to understand the origin of hereditary replicators in that chemical environment. Most experts think that RNA is a good candidate for the primitive hereditary replicator for two important reasons. RNA can act as a template for replication, and it also can function to some extent as an enzyme, assisting the replication process. However, we do not yet know how RNA could be formed in the prebiotic environment.

Eigen's paradox: large genomes require enzymes, but enzymes require large genomes

Even if we could find a plausible scenario for the evolution of simple replicators like RNA, the next problem is Eigen's paradox. Eigen (1971) noted a problem with the accuracy of replication. Suppose that the replicator is a polymeric molecule like RNA, a chain of nucleotides, and that a particular molecule is optimal. During replication, occasional errors will occur, generating a family of molecules. This family will contain some molecules identical to the original (replicated without errors) and a collection of molecules similar but not identical to the optimal molecule (those replicated with errors). Selection favoring the optimal type (perhaps because it permits faster replication) should counteract the accumulation of non-optimal types. If errors in replication occur with a constant probability per subunit, then long molecules are replicated with more mistakes than short molecules. It can be shown that a critical size of a replicator molecule exists (the so-called error threshold), above which larger molecules cannot be maintained because they are replicated with too many mistakes. Eigen's paradox then follows. Non-enzymatic replication has low accuracy, so that only small molecules can be maintained. For accurate replication enzymes are needed, but the primitive genomes are too small to code for them. This is the catch 22 of prebiotic evolution: large genomes are only possible with replication enzymes, but replication enzymes require large genomes.

Theoretical solutions of Eigen's paradox have been proposed (see Maynard Smith and Szathmáry 1995). None as yet are fully satisfactory, but the fact that RNA molecules can function as enzymes is almost certain to be an element of the solution.

The origin of the genetic code

RNA came first; it can function both as a genetic material and as an enzyme

● **KEY CONCEPT**

The first genetic code was not DNA but RNA. It originated when ribozymes acquired amino acid cofactors while acting as enzymes, as well as stores of genetic information.

RNA was probably the hereditary material before DNA, but with the exception of the RNA viruses all organisms now use DNA as the genetic material. The most important difference between RNA and DNA is that DNA normally exists in cells as two paired strands, the double helix; RNA is usually found as a single strand. This has two consequences. First, when RNA is replicated, there can be no proofreading, for there is no complementary strand against which to compare it. The per-base error rate is therefore in the range 10^{-3}–10^{-4}, and this high error rate limits RNA to being the genetic material only in small genomes. Second, RNA can form many stable three-dimensional structures, whereas DNA almost always occurs as a double helix. This is what allows RNA to function as an enzyme as well as a genetic material. Thus RNA was not only able to serve as the genetic material in very small, simple organisms; in those organisms it could also function as an enzyme. From that point onwards variation and selection could start to create large, DNA-based genomes and the machinery needed to translate nucleic acid information into protein sequences.

The evolution of a DNA-based genome involved the process of translation

Translation is sketched in Figure 15.1. The process begins outside the figure with transcription, the copying of DNA into a messenger RNA (mRNA) molecule,

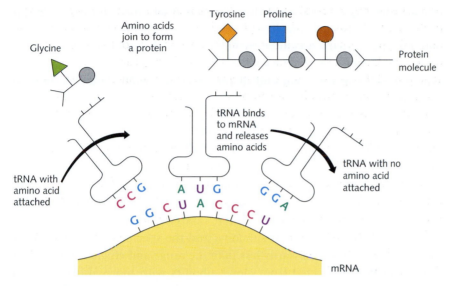

Figure 15.1 Translation as it occurs today. The codons in messenger RNA (mRNA) pair with specific codons in transfer RNA (tRNA), bringing into close proximity the amino acids bonded to the other ends of the tRNAs in the special microenvironment created by the RNA-based ribosome. (From Maynard Smith and Szathmáry 1999 by permission of Oxford University Press.)

which then moves into the position shown. The reason that we can conceive of the mRNA molecule as a sequence of triplet codons is evident from the figure: the transfer RNAs (tRNAs) that implement the genetic code have two key parts, one that binds to the mRNA and one that binds to a specific amino acid. The part that binds to the mRNA consists of three nucleotides, forming a codon. The mRNA is fed through the ribosome, and as it moves it presents its codons sequentially for tRNA binding. The tRNAs carrying their amino acids line up along the mRNA in sequence, bringing their amino acids into proximity with each other. The joining of the amino acids into a protein occurs within a special microenvironment created by the ribosome, which is a large RNA–protein complex. One significant argument for the idea that RNA came first is that all the major elements of the translation machinery—mRNA, tRNA, and ribosomes—are either pure RNA or RNA–protein complexes.

The keys to the genetic code are enzymes that bind specific amino acids to specific tRNAs

The genetic code associates nucleotide sequences—triplet codons—with amino acids. As can be seen in Figure 15.1, the triplet codon UAC specifies tyrosine, CCU specifies proline, and GGC specifies glycine. Three of the 64 codons are used to stop translation and release the amino acid chain from the ribosome as a completed protein. The remaining 61 codons specify 20 amino acids, most of which are therefore coded by more than one codon, making the genetic code redundant. A critical step in the assignment of the code is carried out by the enzymes that bind the amino acids to the tRNA, which occurs at a site distant from the codon-binding site. There is therefore no chemical reason associated with the codon itself why any particular amino acid should be assigned to any particular codon—why tyrosine should be coded for by UAC, for the amino acid is binding at the other end of the molecule, which could be modified without changing the codon.

The code originated when ribozymes acquired amino acids as cofactors

In the early RNA world all reactions were catalyzed by RNA molecules functioning as enzymes—ribozymes. The number of different kinds of chemical reaction that could have been catalyzed by a macromolecule that consists of only four different subunits, and the specificity and efficiency of those reactions, was necessarily less than the range and efficiency of enzymes that could be constructed from more subunits with more, different, three-dimensional distributions of shape and charge. Maynard Smith and Szathmáry (1999) therefore suggest that the transition from the RNA to the DNA world began when ribozymes acquired amino acids as cofactors, thereby increasing their range, specificity, and efficiency as enzymes. And amino acids would have been abundant, forming easily under the conditions of primitive Earth.

The easiest way to link an amino acid to a specific site on a ribozyme is to have it bind first to a short nucleic acid, an oligonucleotide, which forms a sticky handle that binds specifically to a site on the ribozyme. That ribozyme catalyzed some chemical reaction. As the process continued, the ribozyme was gradually transformed into a protein enzyme through a series of hybrid intermediates. We do not yet, however, have a genetic code. To make that step, we can imagine that the amino acid is attached to the oligonucleotide by another ribozyme. That ribozyme could catalyze reactions, producing a pool of many cofactors, each consisting of an amino acid attached to an oligonucleotide. Today a critical feature of the genetic code is implemented in assignment enzymes, the enzymes that attach amino acids to particular tRNAs. The precursors of those assign-ment enzymes were the ribozymes that attached amino acids to oligonucleotides to convert them into cofactors for other ribozymes.

The amino acid cofactors grew into polypeptide cofactors that made mRNA advantageous

Thus far we have precursors of tRNAs carrying amino acids that were bound to ribozymes as cofactors, still quite a distance from the complicated present-day machinery of protein synthesis. We are, however, not very far from it in logic. All we need to imagine is that it was at some point, in some ribozymes, advantageous to link the amino acid cofactors to each other; first two, then three, then more, until instead of a single amino acid cofactor a polypeptide cofactor was formed. Correspondingly specific enzymes would have evolved to catalyze that reaction efficiently. In the process, it would become advantageous to link together the oligonucleotides that coded for the emerging polypeptide. That then became the precursor of the first mRNA molecule. Once mRNA existed, coding for enzymes, there would have been selection for templates with lower mutation rates and a gradual transition to DNA.

mRNA made DNA advantageous, and stable DNA genomes could become large

While speculative, this scenario is plausible, for it shows that the bridge that had to be crossed to connect the RNA world to the modern world can be crossed in a series of small steps, each of which is consistent with chemistry and biology if not demonstrable without a time machine. The basic idea deserves repetition: the genetic code is thought to have originated in a gradual switchover from ribozymes with amino acid cofactors to protein enzymes with nucleotide cofactors, a process in which tRNAs emerged naturally from the oligonucleotides that endowed the amino acid cofactors with specificity. Proto-mRNAs followed, coding for polypeptides of increasing length, eventually by DNA as a template coding for mRNA, selected because of its much lower mutation rate. That key event enabled larger genomes coding for more complicated organisms to emerge, for the lower mutation rate of DNA allowed longer messages to be transmitted reliably.

The code suggests that all living organisms had one ancestor, not that all life had one origin

The genetic code is nearly universal; the deviations from the standard code in mitochondria and elsewhere are fairly minor. This is strong evidence that at one point all life on Earth traces back to a single common ancestor that had that code. Such evidence does not, however, mean that life had a single origin. Life may have had many origins, but the type of life that survives today is the one that had this genetic code, this type of information transmission, and this way of ensuring stability of the genetic material combined with the production of an ensemble of enzymes that could catalyze a broad range of reactions, each with specificity and efficiency. Those may have been the reason that it won out over competitors operating on other design principles.

The evolution of chromosomes

Chromosomal linkage of genes promotes their synchronous replication and fair segregation

● KEY CONCEPT

Chromosomes evolved to facilitate synchronous, fair segregation that suppressed conflict.

In all known organisms genes are linked on chromosomes, but early in evolution there must have been a transition from unlinked to linked genes. Chromosomal linkage has two important consequences. Genes no longer replicate autonomously but in synchrony, and a type of cell division becomes possible that produces daughter cells with copies of all genes.

Synchronous replication of genes reduces the scope for competition between genes within cells. If genes were not linked, synchronous replication would be much harder to achieve and control, particularly with thousands of genes. Although a similar argument applies to the chromosomes, which must also replicate in synchrony, the smaller number of chromosomes makes control of replication feasible.

Fair segregation inhibits conflicts, and linkage helps ensure fair segregation

The second consequence of linkage is that it contributes to orderly segregation of genes at cell division. Each daughter cell must get the full complement of genes through a mechanism ensuring chromosome duplication at cell division and proper segregation of the chromosomal copies into the daughter cells. Even a primitive chromosome would be better in this respect than a collection of unlinked genes. Fair segregation of genes at cell division is crucial in preventing the detrimental effects of genomic conflicts (Chapter 9, p. 197). Linkage of genes on chromosomes helps to suppress such conflict but does not completely prevent it, for some genes can distort segregation.

B chromosomes and transposons escape chromosomal replication and segregation control

Interestingly, some features of present-day genetic systems—although not of immediate relevance for understanding the evolutionary origin of chromosomes—show that escape from replication and segregation control can nevertheless occur. Many species contain chromosomes called B chromosomes that accumulate through successive divisions due to abnormal segregation. B chromosomes are not transcribed and do not contain information vital to the organism; they are genomic parasites. Excessive accumulation of B chromosomes reduces individual fitness and is countered by individual natural selection. The example of B chromosomes shows that unequal chromosomal segregation within a cell can be a real danger.

That the evolution of chromosomes does not protect completely against mutant nuclear genes evading strict replication control is illustrated by the transposons ('jumping genes') which inhabit the genomes of both prokaryotes and eukaryotes. Transposons are either a gene, a small group of linked genes, or a stretch of DNA that is not transcribed. They can move to new positions on the same or on a different chromosome. Transposition often involves replication of the transposon, leaving a copy behind at the original site and increasing the number of transposons within the genome. Because transposition can reduce individual fitness—it induces mutations—several mechanisms have evolved to suppress it. Transposable elements seem to be present in all species, making up at least 10–20% of the genome. In humans they may account for 45% of the genome (Lander et al. 2001). Transposons illustrate genomic conflict (Chapter 9, p. 197) between selection favoring mutants that increase the replication rate of transposons and selection favoring the suppression of transposons through stronger replication control. Such conflicts arise as soon as a new level of replication originates, in this case linked genes on chromosomes.

Eukaryotes differ from prokaryotes in key organizational features

There is as yet no clear picture of the early evolution that eventually gave rise to the three current cellular domains Archaea, Eubacteria (together forming the prokaryotes), and Eukaryota. A common assumption is that the prokaryotes, characterized by the absence of a membrane-bound nucleus, are the oldest existing life forms on Earth, but other views are possible (Poole et al. 1999). Prokaryotes and eukaryotes differ consistently in several features that led to the hypothesis that the eukaryote cell has evolved from prokaryotic ancestors.

Among those striking and consistent differences are these (see Figure 15.2).

● **KEY CONCEPT**

Eukaryotes originated from a symbiotic fusion of prokaryote ancestors. They differ from prokaryotes both morphologically and in the organization and transmission of genetic information.

- Prokaryotes have a rigid outer cell wall; eukaryotes have an internal cytoskeleton.

- Prokaryotes have a single cellular compartment; eukaryotes have a complex set of internal membranes that subdivide the cell into microenvironments specialized on particular chemical functions.

- Prokaryotes have a single circular chromosome attached to the cell wall; eukaryotes have a set of linear chromosomes located within a nuclear membrane.

- In prokaryotes the transcribed mRNA is translated directly into proteins; in eukaryotes transcription occurs in the nucleus and translation occurs in the cytoplasm.

- While both prokaryotes and eukaryotes are inhabited by genomes that are independent of their primary genomes, those of prokaryotes are simple,

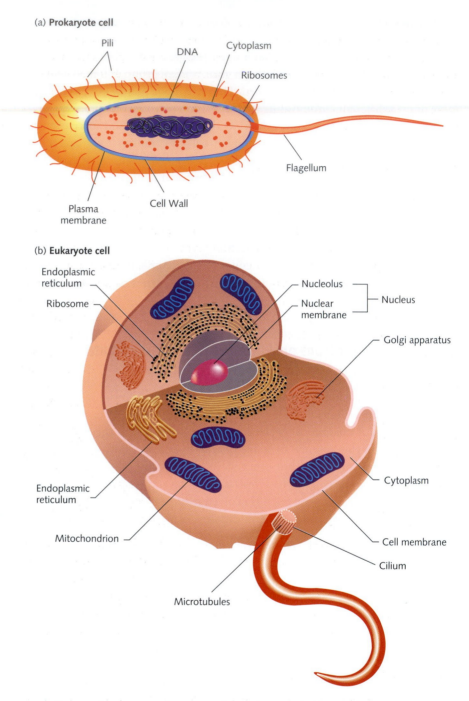

Figure 15.2 The major differences between (a) prokaryotes and (b) eukaryotes. (From Maynard Smith and Szathmáry 1995 by permission of Oxford University Press.)

small, circular pieces of DNA—plasmids—while those of eukaryotes are in complex organelles—mitochondria and chloroplasts—whose ancestors were themselves free-living prokaryotes.

Prokaryotes move with a flagellum anchored in the cell wall; eukaryotes move with a cilium composed of 11 filaments arranged in a ring of nine outer filaments and two central filaments. However, the existence of some 350 proteins in eukaryotic cells that have no homology to proteins in Archaea and Bacteria has suggested an alternative hypothesis. This hypothesis states that the eukaryotic nucleus is an endosymbiont with inputs from Bacteria and Archaea into a host cell that was called a chronocyte; this chronocyte was not prokaryotic but had a cytoskeleton and an extensive membrane system (Hartman and Fedorov 2002). In this view eukaryotes have not evolved from prokaryotes, but from proto-eukaryotes.

The possible transition from cell wall to cytoskeleton

The traditional view that eukaryotes derive from prokaryotes suggests the following scenario for the evolution of the cytoskeleton from the cell wall. However, alternative views are quite possible. The rigid cell wall of prokaryotes solves a critical problem for tiny organisms immersed in water: it keeps them from swelling up and bursting like a balloon as water moves across the concentration gradient from the external medium into the cytoplasm. That is one major benefit of a cell wall. The major cost is that if you have a rigid cell wall, you can only feed off molecules that can move through it; you cannot eat larger objects. The first organism that found a way to keep from bursting without a cell wall was also the first organism with the opportunity to become a predator, to ingest other organisms either whole or in large pieces by folding its cell membrane around them—by phagocytosis. The solution to osmotic challenge without a cell wall was a cytoskeleton, a network of protein filaments anchored to the cell membrane that stabilizes the cell from within like the steel girders of a modern building. The cytoskeleton probably did not originate in order to keep the cell from bursting; it probably originated for other reasons. But once it was in place, the cell wall, which costs a lot to produce, could be lost with little penalty, and once the cell wall was gone, predation became an option.

The loss of the cell wall was associated with another important opportunity. In prokaryotes the circular chromosome is anchored to the cell wall at two points. When the cell wall was lost, it became easier to form multiple chromosomes with multiple starting points for DNA replication. Without that limit on places to start replication, there was less of a time constraint on the amount of DNA that can be copied, and thus less of a limit on the size of a genome that can be copied within a reasonable fraction of a generation. This opened the way to larger, more-complex organisms that could still have a fairly short generation time.

The acquisition of organelles

There is no controversy on the evolutionary origin of the cytoplasmic organelles in eukaryotic cells. Mitochondria derive from an ancestor belonging to the α subdivision of the protobacteria, with closest extant relatives being obligate intracellular symbionts of the order Rickettsiales (Emelyanov 2003). Chloroplasts descend from a cyanobacterium (Cavalier-Smith 2000). There are at least two scenarios for these symbiotic origins.

The first envisages a proto-eukaryotic cell that is a predator. It feeds regularly on such bacteria, but there is variation among these predators in their digestive efficiency, and some of the bacteria survive, encased in an invagination of the predator's cell membrane, for a fair amount of time. Some of those bacteria provide the predator with useful molecules as byproducts of their own metabolism, and some of those predators are not only inefficient at digestion, they also happen to provide the encased bacterium as byproducts of their own metabolism some of the metabolic precursors—the food—that it needs to make the products that the predator finds useful. From selection on such variants evolved a lineage of former predators that found it more useful to domesticate and farm the bacteria they had eaten than to digest them and use their component parts. In favor of this scenario is the fact that the tapping proteins that are inserted into the organelle membranes to facilitate the transfer of nutrients are encoded in the host nucleus and plausibly evolved from host proteins. The farmers thus appear to have engineered some plumbing that allows them to feed and harvest their crop more efficiently.

The second scenario envisages two organisms, one a bacterium, the other a proto-eukaryote, sitting side-by-side on some surface. Each produces as a byproduct of its metabolism some metabolic precursors—some food—that the other needs. They are thus complementary specialists on different kinds of metabolism. For example, the excretory products of a eubacterium—H_2 and CO_2—could have been the food of a proto-eukaryote. This works well until one or both reproduce. One of the progeny cells will then be located at a distance from the source of food. The solution to this problem is to coordinate reproduction in such a way that each daughter cell maintains a relationship with a daughter cell of the partner. That relationship is best preserved if one partner engulfs the other, sequestering it within its cell membrane.

Thus the two scenarios differ both in how they envisage the original relationship—predation versus mutualism—and also in how they see the timing of the origin of metabolic cooperation. In the predation scenario metabolic cooperation evolves long after the predator has first started to feed on the prey. In the mutualistic scenario metabolic cooperation evolves while both partners are still independent organisms, and it is the existence of that cooperation that then provides a reason for one to engulf the other.

The transfer of organelle genes to the nucleus

During the course of their evolution mitochondria and chloroplasts have had most of their genes transferred to the nucleus by a process called endosymbiontic gene transfer. Evidence for this stems from evolutionary sequence comparisons, and obvious remnants of mitochondrial and chloroplast genomes have been found in nuclear chromosomes. Recently gene transfer from the tobacco chloroplast genome to nuclear chromosomes has been observed to happen under laboratory conditions at a very high rate (Stegemann et al. 2003), suggesting that endosymbiontic gene transfer is still occurring.

What might be the advantages of transferring organelle genes to the nucleus? The organelles exist in the cytoplasm in many copies; in diploids the nuclear genome exists in only two copies. Thus the first advantage is that when a cell divides, if most of the organelle genes have been moved to the nucleus, a great deal of time and material can be saved, for each gene only needs to be copied twice, rather than hundreds or thousands of times. A second advantage is that if many of the genes whose products are used in organelles no longer exist in the organelles but only in the nucleus, then the nucleus can control the multiplication rates of the organelles. The nucleus has thus gained control over organelles with which it may come into conflict, as is the case with the nuclear–mitochondrial conflict over sex allocation in gynodioecious plants like *Plantago* (see Chapter 9, p. 197).

The reasons for keeping some genes in the mitochondria and chloroplasts probably have to do with the fact they are energy-transducing organelles. Their function causes the production of oxygen radicals, which are highly reactive and mutagenic molecules. Minimization of the production of these toxic molecules requires optimal redox control, and for this the organelles need to be in control of the expression of genes encoding components of their electron-transport chain so that feedback can be rapid and precise (Allen 2003).

Mitochondria lost photosynthesis and chloroplasts lost respiration

The ancestors of the mitochondria— α protobacteria—and the ancestors of the chloroplasts—cyanobacteria—were both independent photosynthetic organisms that could also respire. They were both able to perform the functions carried out in eukaryotic cells by their descendant organelles. Why was one of them then not used for both purposes? The mitochondrial ancestors appear to have originated and evolved their photosynthetic capacity before the concentration of oxygen in the planet's atmosphere and oceans had reached significant levels. They cannot photosynthesize in the presence of oxygen. Therefore as eukaryotes invaded oxygenated environments there was no reason for their mitochondria to maintain their now-useless and still expensive photosynthetic machinery. So mitochondria lost the ability to photosynthesize. Cyanobacteria, on the other hand, use the same machinery for both functions: they burn sugars with oxygen to produce ATP with the same biochemical pathways that they use to capture light energy in combining CO_2 with H_2O to make sugars and oxygen. Thus cyanobacteria do not have a means of controlling photosynthesis and respiration separately. When their chloroplast descendants found themselves living in eukaryotic cytoplasm inhabited by mitochondria that were specialized on respiration, the chloroplasts could lose respiration without significant penalty and concentrate on photosynthesis.

The origin of multicellularity

Cell clusters may have originated to exclude cheaters from groups of cooperators

● **KEY CONCEPT**

Multicellularity originated to exclude cheaters, suppress genetic conflict, and reduce predation risk.

The first step in the transition from single, independent cells to multicellularity was arguably the origin of cell clusters. But why should cells cluster, and would clustering be stable? One possibility is that clustering would be beneficial to any cells that were beginning to cooperate, for it would allow them physically to exclude non-cooperators and thus not have to subsidize the costs imposed by cheating (Pfeiffer and Bonhoeffer 2003). Once they had evolved the ability to form stable clusters, it would then have been much easier to evolve exchanges of specific resources among cells.

Boraas et al. (1998) studied the evolution of multicellularity in a model system. They inoculated chemostat cultures of the unicellular green alga *Chlorella vulgaris* with a predator. Within a few generations globular clusters of tens to hundreds of *Chlorella* cells appeared. After about 20 generations most clusters were eight-celled, and within 100 generations eight-celled colonies dominated the cultures. These colonies stably replicated their characteristic eight-celled form both in continuous culture and when plated on solid medium. They were too large to be eaten by the predator but small enough for each cell to be in direct contact with the nutrient medium. In the absence of the predator *Chlorella* did not evolve eight-celled colonies and remained unicellular.

This experiment shows that a simple form of multicellularity easily evolves in response to predation. It also suggests that large multicellular forms are only viable if structures exist to transport the substances needed by every cell; simple diffusion from the medium into every cell is no longer sufficient. Because the eight-celled *Chlorella* form evolved so quickly, the transition from single-celled to eight-celled probably required only a few mutations. There was no sign of cellular differentiation: all eight cells were equivalent.

Starting development from a single cell suppresses genetic conflict

Even in these simple multicellular forms the cells have to cooperate as a functional unit. A cell that continued to divide would increase its representation in the colony but would lower colony performance: genomic conflict within such a colony is possible. However, conflict is unlikely because all eight cells descend from a single cell and are genetically identical except for mutations occurring during divisions from the single- to the eight-celled stage. Genomic conflict between genetically identical cells is by definition impossible: any advantage or disadvantage to one of the cells is automatically an advantage or disadvantage to the others, because fitness gains or losses are translated into frequency changes of genes. If the colonies were formed by sticking together eight genetically different cells, there would be some scope for genomic conflict within colonies. Then selfish behavior of one cell could increase the frequency of genes causing such behavior because they are present in that cell line but not others. This may explain why development in multicellular organisms usually starts from a single cell, guaranteeing maximal relatedness between the cells of an individual.

Differentiation requires regulation of gene expression and some form of cellular memory

A transition to multicellularity with different specialized cell types requires two things. First, for cells with identical genes to develop into types with different functions, genes expressed in some cell types must be shut off in others. This requires the evolution of mechanisms for gene regulation. In a simple form of gene regulation, a gene product directly influences the activity of its gene; other mechanisms also exist. Genes were already regulated in single-celled organisms, which must express different genes at different times. Therefore, the first requirement for differentiated multicellularity need not have been difficult.

The second requirement is some form of cellular memory or epigenetic inheritance that maintains the differentiated state through cell divisions (see Jablonka and Szathmáry 1995). In simple epigenetic inheritance, the state is transmitted by distributing a regulatory substance to daughter cells at cell division. In eukaryotes the functional state is often inherited through DNA methylation or proteins bound to DNA. If a stretch of DNA containing a gene is methylated or bound to a protein, the gene cannot be expressed. This epigenetic system may have evolved from defense systems found in bacteria. Specific DNA sequences on bacterial chromosomes are methylated to protect them from destruction by the hosts' restriction enzymes, which degrade foreign parasitic DNA, such as a viral genome, when it enters the host cell.

> **Epigenetic inheritance:**
> *Somatic inheritance of the differentiated state of the cell through cycles of cell division.*

The evolution of germ line and soma

The germ line–soma distinction is clear in many animals, but not in plants and fungi

Many multicellular organisms are organized into a reproductive and a non-reproductive part. In most animals the separation between reproductive and non-reproductive cells is nearly absolute and almost always occurs early in development. In plants and fungi differentiation into reproductive and somatic cells also occurs, but it happens later during development and is less absolute, for somatic cells may dedifferentiate into stem cells that can form reproductive cells; the reverse is also possible.

> ● **KEY CONCEPT**
> There is no genomic conflict over reproductive specialization in an organism where all cells share the same genotype.

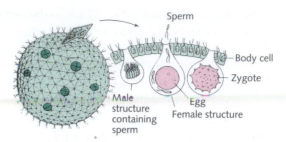

Figure 15.3 Reproduction in *Volvox*. Left: a few gonidia have produced daughter colonies by asexual division. These daughters are still attached to the mother colony. Right: a cross section through part of the maternal colony shows sexual reproductive structures formed by gonidia after induction by sexual hormone. Here a monoecious (homothallic) species is shown, but other *Volvox* species are dioecious (heterothallic), each colony producing only one kind of gamete. (From Gould and Keeton 1996.)

Volvox is a simple organism with reproductive and non-reproductive cells

Simple multicellular organisms with reproductive and non-reproductive cells occur in the order Volvocales. Among these green algae are several colony-forming genera including *Volvox*. *Volvox* forms spherical colonies of 500–50 000 cells of two types. Most cells are somatic. A few cells, the gonidia, are much larger than the others and are specialized for reproduction. In the absence of a sexual hormone the gonidia reproduce asexually, forming small daughter colonies that are released from the mother colony. In the presence of a sexual hormone the gonidia differentiate into sexual structures (see Figure 15.3).

Selection at the colony level more than compensates for selection at the cell level

The differentiation into germ line and soma is widespread and must have evolved independently in many lineages. What selected this differentiation? We assume that division of labor enhances individual fitness because it allows more-efficient conversion of resources into reproductive capacity. Bell (1985) did find that *Volvox* colonies produce more offspring than the equivalent number of unicells. But is their germ line–soma differentiation stable? After all, only gonidia are allowed to form copies of themselves, while somatic cells do not divide and have a cellular fitness of zero. If altruism is behavior that benefits others at a cost to oneself, somatic cells are altruistic. They contribute resources from which the reproductive cells profit by producing offspring. However, this is not altruistic behavior in the *genetic* sense because the somatic cells have the same genes as the gonidia and therefore the same genetic fitness. A mutation in a somatic cell that converted it into a reproductive cell would be selected against, for descendant colonies carrying this mutation would have more than the optimal number of gonidium cells. A mutation in a somatic cell causing it to continue cell division (a sort of cancer) would be selected against because it would produce colonies of suboptimal shape and size.

Social insect colonies resemble *Volvox* but differ because their units are not genetically identical

In lineages such as the eusocial insects the differentiation into reproductive and non-reproductive units has occurred at a level higher than the cell. Honeybee workers normally do not reproduce and perform non-reproductive tasks in the colony, contributing to the reproductive success of the queen. The potential for genomic conflict is greater here than in *Volvox*, where reproductive and non-reproductive units have the same genotype, for worker bees are not genetically identical to the queen. How can we explain this apparent reproductive altruism? We address this important question at the end of the next section.

Principles involved in key evolutionary events

Several general principles appear to be involved in many key evolutionary innovations.

Previously independent units join a larger whole, losing independence in a new unit of replication

Units capable of independent replication become part of a larger whole, thereby giving up their independence (see Figure 15.4). In Figure 15.4 units A and B may be as similar as unlinked genes joining to form linked genes on a chromosome, or they may be as different as the independent organisms that merged to form the eukaryotic cell (see above, pp. 362 ff.). Key evolutionary events that resulted in a higher-level regulation of replication include the transition from unicellular to multicellular organisms and the origin of social groups.

Hierarchical replication structures emerge with multilevel selection and genomic conflicts

The transition to higher replication levels requires higher-level replication structures and coordination and control of replication between levels. For example, genes replicate as parts of chromosomes, chromosomes replicate as part of a cell, cells replicate as part of a multicellular organism, and sexual organisms replicate as part of a sexual population or social group. Such a hierarchical structure allows multilevel selection (see Chapter 9, p. 197), which can generate genomic conflicts. Thus the problem of controlling genomic conflict often arises in evolutionary transitions to higher replication levels.

● **KEY CONCEPT**

Many key evolutionary events involve the origin of a more-inclusive unit of selection made up of previously independent units in which there is a new division of labor and often a new organization of information transfer.

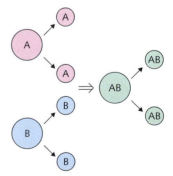

Figure 15.4 Two independent replicators A and B merge into a unit AB in which their replication is no longer independent.

Can developments that increase the potential for conflict be defined as progress?

The addition of replication levels creates a hierarchical structure. Evolutionary 'progress' might be defined as a tendency to expand replication hierarchies, at least in the lineage that produced multicellular organisms. We should, however, be careful about proposing general evolutionary progress if the trend for which progress is claimed implies increased conflict. In lineages that produced unicellular microorganisms a trend to add replication levels is less clear, although different levels of replication do occur. For example, in bacteria replication occurs at the levels of transposable elements, plasmids, the bacterial chromosome, and the cell. Whether developments that increase the potential for conflict can be defined as progress is open to discussion.

Units specialize on different functions, achieving a division of labor

Tasks that were carried out by one unit become distributed over different units. In some sense this is the reverse of the above process. In Figure 15.5 units A and B have arisen that are independent replicators undergoing separate adaptive evolution. They are however not fully independent because they continue to share a common gene pool and interact, having to function within the same individual or population of individuals. Some key origins that involved this process include the differentiation of cell types in multicellular organisms, the origin of anisogamy and male–female differentiation, the evolution of the germ line–soma distinction, and the evolution of castes in social insects. Functional differentiation also creates scope for conflict, because genes good for one differentiated type may be bad for another and because uncontrolled replication of one type will jeopardize cooperation between types.

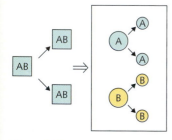

Figure 15.5 A unit AB that carries out two functions differentiates into two units, A and B, each performing one of the functions. The box surrounding A and B indicates that they remain contained within an individual organism or population of organisms.

Some key events are marked by a change to a new system of information transmission

Hereditary information has to be transmitted in some form. Information storage and transmission varies from simple systems with few possibilities, like chemical autocatalysis in which a compound helps to form copies of itself, to complex systems such as genetic coding in nucleic acids and human language, which can transmit an almost infinite number of different messages. Transitions to new systems of information transmission have greatly affected subsequent evolution. Examples include the origin of the genetic code, the origin of epigenetic inheritance, and the origin of human language.

The evolution of cooperation

The evolutionary origins of differentiated multicellular organisms and of groups of social individuals show the importance of cooperation between parts of organisms and types of individual. In such systems selfish mutants can 'cheat' by no longer providing resources for the reproductive parts or individuals but by reproducing themselves instead. Thus the stabilization of cooperation is a necessary feature of several key events in evolution. The basic idea explaining the evolution of cooperation is that being cooperative results in greater reproductive success than being non-cooperative. It can, however, be difficult to see how cooperation yields more fitness than non-cooperation. Does a sterile worker bee have higher fitness than she would if she stopped working and became reproductive herself? The key to such questions is the concept of kin selection, introduced in Chapter 4 (pp. 85 f.).

If you help a relative to increase its reproductive success, you also enhance your own fitness because some of the genes in your relative are identical to your own genes. The closer the relatedness, the more likely it is that this will be the case. Recall (Chapter 4, p. 86) that the condition under which cooperative behavior is expected to evolve is a simple inequality, Hamilton's rule:

$$b \cdot r > c$$

where b is the fitness benefit to the recipient, r is the relatedness of the recipient to the helper, and c is the fitness cost of the behavior to the helper.

Let us apply Hamilton's rule to germ line–soma differentiation in *Volvox*. The somatic cells are helpers and the gonidia are recipients. The genetic relatedness between the two types of cells is 1.0 (they are identical), and cooperation will evolve if $b > c$, if the fitness benefit is greater than the cost. This is the case if a colony consisting of N cells produces more offspring than a collection of N unicells, as was found.

Thus the theory of kin selection predicts that cooperation between different levels of replicators in a system will be stabilized by high relatedness between replicators. This is probably why natural selection has produced reproductive systems in which offspring start developing from one or a very few cells: all the descendant cells are identical except to the degree that they differ by somatic mutation. When somatic mutation does become important, then conflict emerges, as it does with cancer.

Other processes lead to the evolution of cooperation among unrelated individuals (see, for example, Frank 1998).

● **KEY CONCEPT**

The simplest explanation for the evolution of cooperation is kin selection. Other processes explain cooperation among unrelated individuals.

● SUMMARY

This chapter discusses a small number of key changes that have occurred in evolutionary history with long-lasting effects on subsequent evolution.

- Key events included the origin of life, cells, a genetic system, the genetic code, eukaryotic cells, sexual reproduction, and differentiated multicellular organisms.
- The principles often involved in these events include the addition of a new level of replication, functional specialization, and the stabilization of cooperation between different parts.
- Three important events within the lineage that produced multicellular organisms are the evolution of chromosomes, the evolution of multicellularity, and differentiation into reproductive and non-reproductive parts.
- These events show a tendency to add new replication levels, which creates the need for stable cooperation and the danger of genomic conflict between levels.
- Genetic relatedness and kin selection are key concepts in the understanding of the evolutionary stabilization of cooperation.

Thus key events can be deduced from the existing organization of life. The next chapter discusses insights into major historical events that can be induced from molecular data.

● RECOMMENDED READING

Maynard Smith, J. and Szathmáry, E. (1995) *The major transitions in evolution.* Oxford University Press, Oxford.

Maynard Smith, J. and Szathmáry, E. (1999) *The origins of life: from the birth of life to the origin of language.* Oxford University Press, Oxford.

● QUESTIONS

15.1. Explain why genomic conflict is relevant for understanding the origin of many evolutionary novelties.

15.2. Chemical communication often occurs between the cells in bacterial colonies, resulting in coordinated behaviors and cellular division of labor and differentiation into distinct cell types. Would you therefore conclude that bacteria can be viewed as multicellular organisms? Is the concept of kin selection relevant here?

15.3. Evolution depends on storage and transmission of information. It has been argued that human language, a relatively recent novel information-transmission system, has greatly influenced our evolution and will continue to do so. Can you think of examples that show the role of language in evolution?

CHAPTER 16

Major events in the geological theater

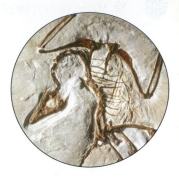

This chapter describes major events in the history of life and the planet, the events that structured the changing theater in which evolution plays. We begin with a look at an organism, a community, and a landscape as mosaics of parts with quite different ages. Organisms, communities, and landscapes have been assembled over time, and some parts of them have been around much longer than others. We then ask, how has the planet shaped life? We live on a particular planet, with a certain surface temperature, a certain distribution of abundances of atoms and molecules of various sorts, with continents that move steadily across its surface, with large, deep oceans of liquid water, with volcanoes that occasionally erupt catastrophically, and with ice ages in which continental ice caps form. All have had major consequences for life. The chapter closes with a consideration of how life has shaped the planet. It has done so by modifying the composition of the atmosphere and by creating huge deposits of iron ore, limestone, oil, and coal, shaping a planetary environment that is so far from the equilibrium it would achieve without life that extraterrestrial astronomers would infer that something special is occurring on planet Earth.

Organisms and landscapes are historical mosaics

On a wooded shoreline in south-east Alaska (Figure 16.1) we see maple trees, bracken fern, moss-covered rocks, orchids, dragonflies, a frog, a mule deer, and a Great Blue Heron. In the intertidal zone, barnacles, mussels, and a starfish are visible. A loon swims on the water surface, beneath which we can see anemones, an octopus, a crab, sea urchins, and algae. In deep water, a family of humpback whales attacks a school of herring.

Life is built from recycled stars

The organisms and landscape consist of elements, parts, and forms that appeared in the history of the universe and planet at vastly different times. The

> ● **KEY CONCEPT**
> The organisms currently on Earth represent groups and forms that originated over billions of years. The geological structures currently forming the surface of the planet represent events that also span billions of years.

Figure 16.1 A shoreline in south-east Alaska. The elements, the organisms, and the landscape itself are all mosaics of parts and forms that originated at very different times in the history of the planet. (By Dafila K. Scott.)

cells of the maple tree are constructed from molecules consisting of atoms with nuclei. Hydrogen—the most common element—formed in the first milliseconds of the Big Bang about 13 500 million years ago (Ma). Helium and lithium appeared after 3 min, but all the heavier elements, including the light elements that form the building blocks of most biological molecules—carbon, nitrogen, oxygen, sulfur, phosphorus, sodium, calcium, and the other elements up to and including iron—were created much later through thermonuclear fusion in stars, then expelled into interstellar space in novas or supernovas. Almost all elements heavier than iron were created in supernova explosions (Mason 1992). Most elements found on Earth today originated 5000–13 500 Ma, before the solar system formed from recycled starstuff.

Some of the elements formed in supernovas—including copper, zinc, selenium, molybdenum, and iodine—have important biochemical roles (Stryer 1988). Copper occurs in plastocyanin, part of the photosynthetic mechanism that transforms light energy into chemical energy. Copper thus helped to oxygenate the planet. Zinc is an essential part of carboxypeptidase A, an enzyme that breaks down proteins. Zinc aids digestion. Selenium is part of an enzyme that protects cells from poisons. Molybdenum forms part of the enzyme in nitrogen-fixing bacteria that converts molecular nitrogen into usable ammonium ions. Molybdenum helps to fertilize the biosphere. Iodine, part of the vertebrate hormone thyroxin, helps to control the expression of genes involved in growth and energy metabolism. Both the maple leaf and your body are built from recycled stars.

The names and dates given by geologists to the major divisions of Earth history are listed in Table 16.1. Please read that table now and refer to it while reading the rest of this chapter.

Life forms have a mosaic history

If we could let evolutionary history unfold before us, how far back in time would we have to go to see the first maple tree? (The fossil record gives a conservative answer, for the dates of first appearances in the fossil record can only be pushed

Table 16.1 Overview history of the Earth and life.

Period	Duration (Ma)
Big Bang, stellar recycling, nucleosynthesis	13 500–5000
Solar System forms, planetoid bombardment	5000–4600
Archaean: origin of life	4600–2500
Proterozoic	2500–540
Paleoproterozoic: origin of eukaryotes	2500–1600
Mesoproterozoic: protists radiate	1600–1000
Neoproterozoic: multicellularity	1000–540
Phanerozoic	540–0
Paleozoic	540–250
Cambrian: most animal phyla appear	540–500
Ordovician: jawless fish, plants and arthropods on land	500–435
Silurian: fungi	435–410
Devonian: first forests, fish radiate	410–355
Carboniferous: seed plants, insects radiate	355–295
Permian: cycads, reptiles, mass extinction	295–250
Mesozoic	250–65
Triassic: gymnosperms, teleosts, crocodiles	250–203
Jurassic: dinosaurs, frogs, turtles, birds	203–135
Cretaceous: angiosperms, lizards, mass extinction	135–65
Cenozoic	65–0
Paleogene	65–23
Paleocene: mosquitos, whales, grasses	65–53
Eocene: bats, beeches, primates, ungulates	53–34
Oligocene: grasslands	34–23
Neogene	23–1.6
Miocene: grazers radiate	23–5.3
Pliocene: hominids, orchids	5.3–1.6
Quaternary	1.75–0
Pleistocene: language, art, burial, ice ages	1.75–0.01
Holocene: agriculture, writing, domestic animals	0.01–0

back by new discoveries.) According to the fossil record, we could have seen a maple leaf 60 Ma. A maple is an angiosperm, a flowering plant, and traces of angiosperm pollen have been found in the early Cretaceous, about 130 Ma. A maple has a tall upright stem made possible by an unusually strong composite material, wood. Wood appeared in the middle Devonian 386 Ma.

Maples are terrestrial plants. The earliest traces of terrestrial plants are spores recovered from the middle Ordovician 470 Ma and sterile stems from the early Silurian, 440–430 Ma, and there are fossil soils from the late Devonian, 360 Ma, with traces of vascular roots. Maples are multicellular organisms, which appeared about 1000 Ma. They are eukaryotes with endosymbiotic organelles, a spindle apparatus, and true meiosis. The eukaryote ancestors acquired the α proteobacterial ancestors of mitochondria between 3000 and 1700 Ma (Roger and Silberman 2002). Later the eukaryotes that became plants acquired chloroplasts from cyanobacteria between 1600 and 1000 Ma. Maple trees photosynthesize carbohydrates from water, carbon dioxide, and solar energy, and photosynthesis can be traced to fossil cyanobacteria (blue-green algae) much earlier, but exactly how early is controversial (Brasier et al. 2002, Schopf et al. 2002). Microbial life may have been flourishing on marine volcanoes by 3500 Ma (Furnes et al. 2004). Some molecules involved in photosynthesis are among the most ancient, conservative structures known to biochemists, relics, like membranes and RNA, of very early life. Life itself may have originated as early as 3900–3700 Ma, when the temperature of the cooling planet fell to the point where macromolecules could be stable in hot water under pressure.

Like all organisms, the maple tree is a mosaic of parts of different ages; some had their evolutionary origins at vastly different times.

Now consider the other organisms in the landscape (Figure 16.1 and Table 16.1). The oldest large plants are the liverworts, hornworts, and mosses, representatives of the group that colonized land about 470 Ma. Pines appeared in the fossil record 210 Ma, about the same time as dinosaurs, but gymnosperms, the group to which pines belong, appeared at the very end of the Devonian, about 350 Ma. They radiated in the Triassic and dominated from the mid-Triassic through the Cretaceous (225–65 Ma). The orchids, like the maple, are flowering plants, angiosperms. Angiosperms originated in the early Cretaceous and have dominated plant communities in the Cenozoic (65–0 Ma). The phylogeny of the seed plants is dominated by the conifers and their relatives; at that scale, the angiosperms are a small branch within which the monocotyledons

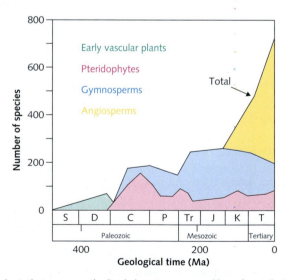

Figure 16.2 The plants that are currently alive belong to groups and have forms that originated in the different major eras of Earth history. The ferns and mosses dominated the late Paleozoic, the gymnosperms were characteristic of the Mesozoic, and the angiosperms are characteristic of the Cenozoic. S, Silurian; D, Devonian; C, Carboniferous; P, Permian; Tr, Triassic; J, Jurassic; K, Cretaceous; T, Cenozoic. (From Niklas et al. 1983. Reproduced by kind permission of the authors and *Nature*.)

(orchids, grasses, palms, lilies) are an insignificant twig nested within the dicotyledons.

The pteridophytes (ferns and their relatives), the gymnosperms, and the angiosperms each characterized and dominated, in that order, different major eras of Earth history (Niklas et al. 1983; Figure 16.2). The greatest diversity of major plant types is to be found among the liverworts, mosses, horsetails, and ferns, within which the seed plants form just one of many branches (Doyle 1998).

Among the terrestrial animals on this Alaska shoreline, the oldest forms are the dragonflies, which can be found as fossils 300 Ma. Frogs appear in the fossil record 190 Ma, loons 70 Ma, herons and rabbits 65 Ma.

Underwater we see forms that are much older, for life evolved for a long time in the sea before it came on to land. These also make a series that dominated successive ages (Figure 16.3). From just before the Cambrian we can still see sponges

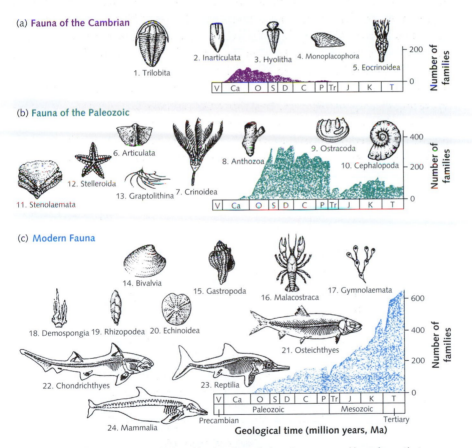

Figure 16.3 The animals currently alive, like the plants, belong to groups and have forms that originated in the different major eras of Earth history. The sponges and coelenterates remain from the Cambrian, the polychaetes and sea stars from the later Paleozoic, the clams, sea urchins, octopui, and bony fishes from the Mesozoic, and whales and teleost fishes from the Cenozoic. V, Vendian; Ca, Cambrian; O, Ordovician; S, Silurian; D, Devonian; C, Carboniferous; P, Permian; Tr, Triassic; J, Jurassic; K, Cretaceous; T, Cenozoic. (From Sepkoski 1984.)

and coelenterates (650–540 Ma). Many of the groups that characterize the later Paleozoic appear in the Cambrian, but more modern representatives appear in the Ordovician. Not much is left today that looks like the fauna of the Cambrian, which was characterized by trilobites, monoplacophoran mollusks, inarticulate brachiopods, and several echinoderm classes, all of which are wholly or mostly extinct. The Paleozoic fauna that originated primarily in the Ordovician dominated to the end of the Permian and has left many surviving forms: polychaete worms, sea stars, crinoids, articulate brachiopods, nautiloids, and ostracods (500–440 Ma). The Mesozoic fauna contains more familiar forms: clams, snails, sea urchins, crabs, lobsters, octopi, sharks, and bony fishes (230–65 Ma). Whales and teleost fish evolved 65–40 Ma.

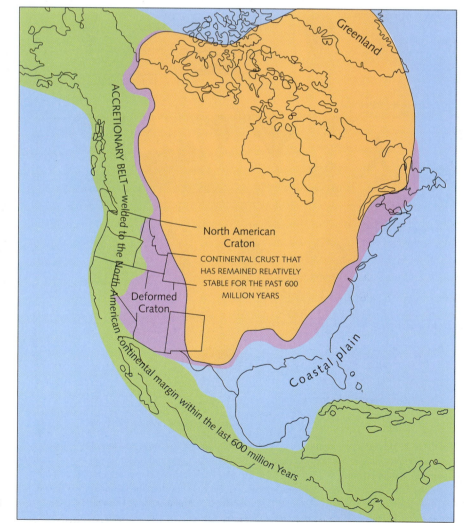

Figure 16.4 Western North America is built from terranes that accreted on to the core of the North American continent.

Many of the forms that dominated successive ages have disappeared, but there are survivors characteristic of a series of ancient communities. A modern community is a mosaic of forms of very different ages.

The land itself can have a mosaic history

Western North America is an accretionary belt of terranes (Figure 16.4), pieces of continental crust that do not belong to the original core of the North American continent, whose edge lies near the eastern edge of the Rocky Mountains. The terranes originated as island arcs or ribbon continents that drifted north and east towards North America on subducting plates while North America drifted west. They accreted into a broad belt thousands of miles long and hundreds of miles wide. Some pieces of Alaska seem to have come from as far away as present-day South America. Thus the rock on which the maple tree grows spent the Mesozoic and early Cenozoic—during which the flowering plants, birds, bees, wasps, and lizards originated, the dinosaurs and ammonites went extinct, and the mammals radiated—moving across the Pacific towards North America.

> **Terrane:** *A crustal block or fragment that preserves a distinctive geologic history that is different from the surrounding areas and that is usually bounded by faults.*

> **Accreted terrane:** *A terrane that has become attached to a continent (e.g. western North America) as a result of tectonic processes.*

How has the planet shaped life?

The universe, the Solar System, and planet Earth are billions of years old

The universe formed about 13 500 Ma; the leading hypothesis remains the Big Bang, which is plausible but not proven. According to that hypothesis, after a brief period of incredibly rapid expansion, a universe consisting of hydrogen and a little helium was formed. Gravitation then pulled concentrations of gas into primitive galaxies, within which stars formed, synthesizing heavier elements from hydrogen and helium nuclei as they burned with thermonuclear reactions until they consisted mostly of iron, at which point the larger stars exploded as novas and supernovas, forming the elements heavier than iron.

Our Solar System formed secondarily from the products of one or more cycles of star formation about 4600 Ma. There are two scenarios for the formation of the inner, Earth-like planets. In the first, they formed like the gas giants (Jupiter, Saturn, Uranus, and Neptune) of the outer system, as disks of gas that condensed into spheres consisting mostly of hydrogen and helium with a core of metal and rock debris. As the Sun began to burn, the solar wind blew off the primitive atmosphere of light molecules, leaving the small, heavy cores that we now know as Mercury, Venus, Earth, and Mars. In the second scenario the inner planets were never gas giants; they formed by accretion of heavy debris, planetoids, and meteorites. In both scenarios, the inner planets underwent an early period of intensive bombardment by planetoids and meteors, a bombardment

> ● **KEY CONCEPT**
> Continental drift, glaciation, local catastrophes, and other geological processes have repeatedly altered the surface, the oceans, and the atmosphere of the Earth, causing extinctions and changing the distributions of the species that survived.

that raised temperatures high enough to melt the crust and make life impossible. The traces of that bombardment, which ended about 4000–3800 Ma, can still be seen as giant impact craters on Mercury and the Moon.

The atmosphere began hot, toxic, reducing, sulfurous, and smelly

As the inner planets cooled, atmosphere formed by outgassing from the rocks. It was composed of compounds heavier than molecular hydrogen and helium, compounds that were more strongly held by gravity and less easily blown off in the solar wind. Water, carbon dioxide, methane, ammonia, and hydrogen sulfide were among them. Even this atmosphere was blown off Mercury, which is close to the Sun where the solar wind is strong, and it could not be completely retained by Mars, whose gravity is weaker than Earth's. Venus and Earth, further from the Sun than Mercury and larger and heavier than Mars, have retained thick atmospheres. The one on Venus, which has not been engineered by life, may resemble the early atmosphere of Earth. It consists mostly of carbon dioxide with some water and sulfuric acid, retains solar heat like a blanket, and raises temperatures on the surface of Venus well above the boiling point of water. The early atmosphere of the Earth was hot, toxic by our standards, reducing instead of oxidizing, and stank like rotten eggs from the sulfides.

Continents and oceans formed about 4000 Ma

In the molten mass of the cooling Earth, the heavier metals sank to the middle, forming a nickel–iron core, and the lighter rocks rose to the surface, forming a crust. Within this crust, which must have been driven by energy flows and wracked by movements more powerful and rapid than those in existence today, the cores of continents formed high ground. When the temperature of the surface fell below the boiling point of water, oceans formed. The continental cores and the first oceans formed about 4000 Ma. The oceans have been remodeled several times since then, but traces of the original continental cores can be found in South Africa, Brazil, the Canadian Shield, Greenland, Australia, and India. That is also where the oldest traces of carbon compounds specific to life occur.

Planet Earth provides the basic conditions for life

Our planet has characteristics important for the origin and maintenance of life as we know it. It circles a sun that is intermediate in size and energy output, as stars go, and at a distance where it can retain an atmosphere of moderate depth without losing it to the solar wind. The surface temperature of the Earth, the result of solar energy input and the blanketing effects of the atmosphere—the greenhouse effect—is a temperature at which water is a liquid. There is plenty of liquid water on Earth's surface, and water has some extraordinary properties. It dissolves many types of molecules and ions; it remains liquid over a broad temperature range; it has considerable buffering capacity, tending to maintain a pH near 7.0; and it has extraordinary heat capacity—it takes a lot of energy to heat it, and it loses heat slowly. The absorption and release of heat by water in the oceans and in the atmosphere plays a large role in determining climate and weather.

Over much of the Earth the surface temperatures have remained in a range where proteins and nucleic acids are stable—between 0 and 40°C—for the last 3.8 billion years. The Earth is supplied with carbon and nitrogen in the atmosphere and oceans, in many different molecular and ionic forms, as well as dissolved ions derived from the crust—phosphates, sulphates, and many metal ions. Such a chemical mixture allowed life to originate, and it was on the energy in inorganic molecules—such as sulfides—that early life fed, before photosynthesis evolved.

Continental drift has shaped the large-scale features of the planet

Since the formation of the continents, the crust of the Earth has consisted of a set of continental plates floating on the upper mantle, moved by the flow of heat rising in large convective cells from the interior. The plates either slide by one another, forming transverse faults, such as the San Andreas Fault in California, or move apart, as is happening in the Red Sea, the great Rift Valleys of Africa and along the Mid-Ocean Ridge, or collide with one another, forming mountain ranges (Figure 16.5). The great mountain arc containing the Himalayas that stretches from Iran to China formed in the Cenozoic collision of India with Asia. The Alps formed when Africa shoved Italy into Europe.

The energy source for the movement of the plates reaches the surface at the Mid-Ocean Ridge, the largest mountain range on the planet, stretching more than

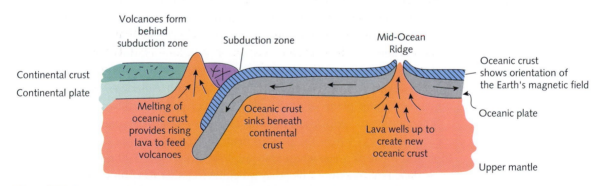

Figure 16.5 The crust of the Earth floats on the partially molten mantle. The crust is divided into plates which move by the flow of heat rising as convection currents from the molten interior. The continents are embedded into the plates and move along with them. At locations such as the Mid-Ocean Ridge, molten rock swells up in convection currents producing new crust and causing the plates to move apart. Where the expanding oceanic crust collides with the continental crust, the oceanic crust sinks beneath the continental crust at a subduction zone, where it remelts. The molten oceanic crust rises to the surface behind the subduction zone, forming volcanoes.

40 000 km around the globe. At the center of the ridge, which reaches the surface in Iceland, lava wells up as the plates move apart. When the lava solidifies, it preserves the orientation of the Earth's magnetic field. The field changes polarity irregularly every few million years. As the plates move apart, the space between them is filled by ocean bottom with magnetic stripes whose polarity and age are symmetrical around the ridge crest, younger nearer the ridge, older farther away. The plates move at about 2 cm/year, or 2000 km every 100 million years. Since the process began about 4 billion years ago, the continental plates have each traveled roughly 80 000 km, recycling the ocean bottoms completely several times over and scraping copious amounts of sea bottom on to the growing continental margins. The oldest part of the ocean floor, south-east of Japan, is less than 200 million years old—about 5% of the time since the continents started drifting.

The rate of sea-floor spreading has not been constant. At times the flow of heat from the Earth's interior speeds up, the mid-ocean ridges inflate with magma, the spreading from the ridges accelerates, the continents drift more rapidly, and the sea level rises, displaced by the increased volume of the mid-ocean ridges, flooding the continental plains. At such times shallow seas flooded large areas that are now dry land, providing increased habitat for marine organisms and opportunities for fossil deposition. At other times the flow of heat from the Earth's interior decreases, the mid-ocean ridges deflate, continental drift slows, and the seas retreat to the margins of the continental shelves. Habitat for marine organisms is limited, and there are not many places suitable for the deposit of fossils.

Because continental crust is lighter than oceanic crust, when ocean bottom collides with a continent, the oceanic crust is subducted beneath the continental plate (Figure 16.5). The lighter ocean-bottom sediment is scraped on to the continental margin to form an island arc or a coastal mountain range. The subducted oceanic crust forms a descending plate that melts 300–700 km deep and sends up plumes of lava, which form chains of volcanoes 100–300 km back from the leading edge of the subduction zone. The volcanoes and raised montane arches around the entire Pacific Ring of Fire—Indonesia, the Philippines, Japan, Kamchatka, the Aleutians, south-east Alaska, British Columbia, California, Mexico, Central America, and the Andes of South America—were so formed when the oceanic plates of the Pacific Basin were subducted beneath them.

The continents, which are drifting on the surface of a sphere, tend to drift away from areas where the flow of heat and the rate of sea floor spreading is high and towards areas where they are low. Every 300–500 million years, in a cycle called the Wilson Cycle, they all come together to form a supercontinent, with associated mountain building. Several hundred million years later they break up again, often along new rift zones, to form island continents and other configurations. The longer ago this happened, the more difficult it is to reconstruct the details, but we can recognize three repetitions of the process in the supercontinents Rodinia about 1 billion years ago, Pannotia about 570 Ma, and Pangaea about 250 Ma.

As they do their ponderous dance, the continents can move from the equator to the poles and back again. Thus any continent could, in principle, find itself in a position like that of Antarctica today, covered by ice, and then later in a position like that of Africa, covered by tropical forests, savannahs, and deserts. Several of them have done so. The positions of the continents also have profound effects on ocean currents and the turnover of surface and deep water. In some configurations, turnover is vigorous and the bottom waters are oxygenated; in others, turnover is weak, and the bottom waters can become anoxic. These processes have profound effects on the organisms living both on the continents and in the oceans.

Continental drift has left many traces in the distributions of living organisms

In the early Devonian, most of the continents were in the southern hemisphere: South America, Africa, and Australia were close enough to the South Pole to be covered by ice (Figure 16.6). By the end of the Permian (250 Ma), they had congealed into a supercontinent, Pangaea, and there was a single world ocean with some large islands in it. During the Mesozoic (250–65 Ma), that megacontinent began to break up. Huge rift valleys formed as plates pulled apart. They broadened into arms of the sea, as the Red Sea is doing today, then into oceans. First two major blocks were formed: Laurasia, consisting of North America and Eurasia, and Gondwana, consisting of South America, Africa, India, Antarctica, and Australia.

Fossils of the same terrestrial Triassic reptiles can be found on all four Gondwana continents, and the distribution of ratite birds (ostriches in Africa, rheas in South America, emus in Australia, kiwis in New Zealand), Southern Beeches (genus *Nothofagus*, from Chile and New Zealand), and *Araucaria* (conifers from South America and the islands off Australia and New Zealand) still reflect a time when those continents were joined in a single land mass.

By 200 Ma the Tethys Sea began to form (see Figure 14.8), separating Laurasia from Gondwana, and by the mid-Jurassic, 150 Ma, Gondwana began to fragment (Figure 16.6). India was by then already an island continent, but for a time South America remained joined to Africa and Antarctica to Australia. By about 70 Ma North and South America, Africa, India, Australia, and Eurasia were all unconnected island continents. The Atlantic finished opening, from south to north, about 60 Ma. Rocks found in Newfoundland match rocks in Scotland and Scandinavia, just as rocks in New England match rocks in Morocco and rocks in Brazil match rocks in Angola. North and South American continue to move away from Europe and Africa at 2–3 mm per year.

When North America began to separate from Europe, rift valleys opened, and in the bottoms of these valleys great rift lakes formed. As the climate changed, sometimes wetter, sometimes dryer, and as the landscape changed and with it the course of rivers, the lakes filled, dried up, and filled again. Each time

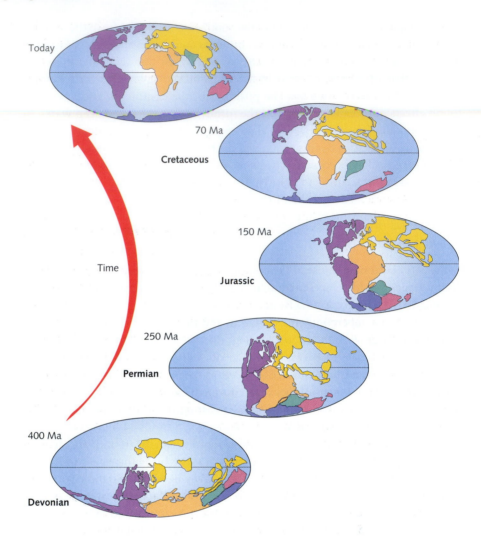

Today

70 Ma

Cretaceous

150 Ma

Jurassic

250 Ma

Permian

Time

400 Ma

Devonian

Figure 16.6 400 million years of continental drift. (From Marshak 2001.)

they filled with water, a now-extinct clade of teleost fish, the semionotids, underwent a radiation, filling the lakes with species that displayed a diversity of feeding types. Each time the lakes dried up, the species flocks went extinct (McCune 1996). Thus the evolution of the great species flocks of cichlid fishes in the African rift lakes was repeatedly foreshadowed in the Cretaceous lakes of eastern North America.

South America remained an island continent, like Australia and Antarctica today, until 2–10 Ma, when its northward drift into island arcs raised the Isthmus of Panama and parts of Central America, with major consequences. During its isolation a diverse endemic fauna and flora had evolved. Many endemic species went extinct when invaders from North America arrived; a few others, including armadillos and opossums, moved north. Like South America, India spent several tens of millions of years as an island continent and also evolved an endemic fauna and flora, most of which vanished after it collided with Asia during the middle Cenozoic. India continues to move north into Asia, and the Himalayas continue to rise. During the Cenozoic era Africa also moved north, closing the Tethys Sea at Suez 23 Ma and driving Italy into Europe, raising the Alps and the Jura.

Oceanic hot spots produce chains of isolated volcanic islands

The floor of the Pacific Basin, over which moved the terranes that docked along the west coasts of Central and North America, has had a more complex history than the Atlantic. The middle of the Pacific Plate contains another important geological mechanism: a hot spot. (Other hot spots can be found in French Polynesia, the Galápagos, Bermuda, the Azores, Iceland, St. Helena, Kerguelen, and Yellowstone National Park in Wyoming.) As the Pacific Plate moved north-west, the hot spot now located south-east of the youngest emergent island, Hawaii, repeatedly broke through the plate and built volcanic islands (see Figure 14.5). Such is the volume of lava produced that in less than a million years a volcano—Mauna Loa—about 10 km high and 300 km in diameter has formed. Only on Mars have larger volcanoes been built in this Solar System.

The plate carried the islands away, the conduits into the mantle were broken, the volcanoes stopped erupting, and their weight caused the crust to bend downwards, sinking the islands below the ocean surface. The Hawaiian shield volcanoes, larger in volume and higher from sea floor to summit than any other mountains on the planet, sink beneath the waves in 10–20 million years.

The Hawaiian Islands and their underwater continuation, the Emperor Seamounts, stretch 6500 km from the middle of the North Pacific into the subduction zone of the Kamchatka Trench. The oldest of the Emperor Seamounts, now about to be subducted, are about 170 million years old. Thus there have been islands above water near the present location of Hawaii at least since the Jurassic. About 85 Ma there was a major change in the direction of motion of the plate, from north-northwest to west-northwest, recorded in a change in the direction of the Emperor Seamounts north-west of Midway Island.

Glaciations have shaped Earth's landscape

There have been at least four major ice ages in Earth history—in the Vendian, in the late Ordovician, from the mid-Carboniferous to the mid-Permian, and in the Pleistocene and Holocene. There is some evidence for two more ice ages in the early Proterozoic (about 2.2 billion years ago) and in the Archaean (between 2.5 and 3.9 billion years ago).

In the Vendian, about 600–700 Ma, continental glaciers covered southern Africa, which was near the equator at that time. This suggested that the entire planet may have been covered in ice—the Snowball Earth hypothesis. The idea is controversial and not consistent with the mounting molecular evidence for complex multicellular animals and the lack of evidence of a mass extinction at that time. But it does appear that there were major continental glaciers that melted shortly before the Cambrian.

In the Ordovician, about 440 Ma, Africa formed the center of Gondwana and was located near the South Pole. The second major Ice Age was centered on northern and western Africa with glaciers growing to cover some of South America and Saudi Arabia. As water was withdrawn from the oceans and frozen in the continental glaciers, sea level dropped 50–70 m. The Ordovician event was relatively brief, perhaps no more than half a million years, and was associated with the Ordovician mass extinction.

The third major glaciation came in the late Carboniferous and early Permian, primarily in the southern hemisphere. At that time, South America, Africa, India, Australia, Madagascar, and Antarctica were joined together in Gondwana, much of which lay more than 60° south of the equator. The continental ice sheet spread outwards from a center over South Africa, western Australia, southern India, and Antarctica. Rocks that were scraped off Africa, for example, were carried by the ice and deposited as glacial erratics in South America. The direction of ice flow can still be seen scratched in the rocks on top of Table Mountain in Cape Town.

The fourth major Ice Age started in the Pleistocene and continues today. Over the last 2 million years, continental glaciers have advanced and retreated many times in the northern hemisphere. In so doing, they have wiped out the evidence on land of prior glaciations, but a record available in marine sediments suggests many more glaciations, 15–20 of them, than the four major events that can still be detected on land.

Glaciations caused major shifts in species distributions

Continental glaciation completely changes the distribution of life on the planet. Virtually nothing survives on or under the ice. Just south of the ice there is a barren strip, then tundra, then conifer forest, and only far to the south of the ice are there hardwood forests. In the tropics, ice ages are accompanied by shifts in rainfall. Savannah expands, forests contract. Sea levels drop as much as 100 m below current sea levels because so much water is locked up in continental ice sheets. While sea levels are low, reefs and beaches form that now are drowned beneath hundreds of metres of water. Glaciers and rivers cut canyons into the continental margins; and the continental shelf is partially exposed. When sea levels rise again, these canyons are flooded, producing, for example, the Norwegian fjords, the lower Hudson Valley, and Chesapeake Bay.

When sea levels were at their lowest at the peak of the last glaciation, 18 000–20 000 years ago, the Bering Strait was dry, Borneo, Java, and Sumatra were part of south-east Asia, New Guinea and Tasmania were part of Australia, Japan and Taiwan were connected to Korea and China, and England and Ireland were part of continental Europe (Figure 16.7). The Black Sea was cut off

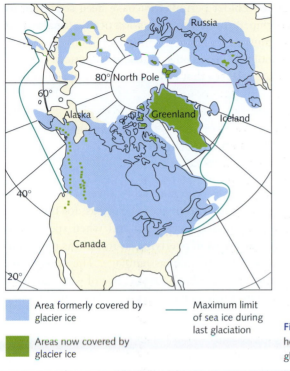

Area formerly covered by glacier ice

Areas now covered by glacier ice

_____ Maximum limit of sea ice during last glaciation

Figure 16.7 The northern hemisphere at the peak of the last glaciation.

from the Mediterranean Sea. Animals and people could walk from Siberia to Alaska, but only during recessions of the ice sheet could they walk from Alaska to Montana. Elephants and tigers could walk on to Borneo, but no snakes survived in Ireland, and none arrived before the rising sea cut it off from Europe again. South of the European ice cap saiga antelope, modern inhabitants of the Siberian taiga, roamed across the English Channel and northern France into Central Asia. Mammoths, woolly rhinoceroses, giant bison, giant beavers, dire wolves, lions, and other large mammals roamed North America. Large glacial lakes formed in the American West and in Siberia.

During the course of evolution, the surface of this planet has been extensively remodeled by continental drift, volcanism, and glaciation. The continents have had a variety of configurations. The level of the sea relative to the mean continental elevation has varied by several hundred meters. Island archipelagos have been repeatedly formed, and some of them have vanished, either by accretion on to continental margins or by sinking beneath the waves. During much of the Earth's history, the climate was warmer than it is today—as it was in all the Mesozoic and the Cenozoic up to the Pleistocene—but there have also been cold periods with continental glaciers in the late Vendian, the Permian, and the Quaternary. Temperatures were particularly high in the Jurassic, Eocene, and Pleiocene eras, and in the Jurassic they were accompanied by carbon dioxide levels in the atmosphere 10–16 times higher than those we experience today.

Because we live in one of the few colder periods that have been separated by much longer times of warmer temperatures, our impression of the climate of the globe is not representative of most of its history. These global changes in the geological theater have profoundly influenced the course of evolution.

We next discuss geological catastrophes that have had major impact on life.

Mass extinctions repeatedly changed the course of evolution

● **KEY CONCEPT**

The evolution of life has not been a constant increase in adaptation and biodiversity; it has been repeatedly altered by mass extinctions.

The Phanerozoic era (570 Ma–today) has been marked by several mass extinctions. It is no accident that they fall at the end of geological eras, for those eras are recognized as strata bearing characteristic fossils, and when species disappeared and were replaced by new forms in overlying strata, it was natural to invent a new name for the period with the new assemblage. There may have been as many as 20 major extinctions in the Phanerozoic, five of which are well documented—those at the end of the Ordovician, Devonian, Permian, Triassic, and Cretaceous periods. Three were truly massive.

Glaciation probably caused the mass extinction at the end of the Ordovician

At the end of the Ordovician, about 440 Ma, 22% of the families and nearly 60% of the genera of marine invertebrates vanished, including many trilobites, brachiopods, graptolites, echinoderms, and corals. The extinctions came in two waves, one about 10 million years before the end of the era, the other at the end. The mechanism is not clear. There is evidence of a glacial maximum associated with a drop in sea level that exposed the continental shelves, followed by deep flooding of the shelves with fresh water from melting ice caps and continental glaciers.

The end-Permian extinction was the most dramatic of all

The extinction at the end of the Permian 250 Ma was the largest ever. In it vanished about 49% of marine animal families, about 63% of terrestrial organisms, about 61% of all species (Benton and Twitchett 2003), and about 80% of the species of marine invertebrates, including all the trilobites, all the tabulate and rugose corals, about 70% of the brachiopod families, 65% of the bryozoan families, and 47% of the cephalopod families, among them many ammonites (Figure 16.8). Its intensity—numbers of genera disappearing per unit time—was the greatest ever. The event was massive and brief. The mechanism is not yet settled, but a consensus is starting to emerge (Benton and Twitchett 2003).

At that time the continents had just accreted into a single mass, Pangaea, with north China and Siberia finishing the process shortly before the end of the

(a)   (b)

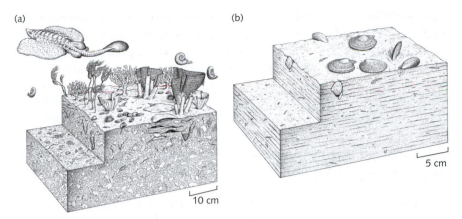

5 cm

10 cm

Figure 16.8 The marine community of Southern China immediately before (a) and after (b) the Permian extinction. There was rich reef life and a well-developed burrowing community before the extinction, but little after it. At that site a fauna of about 100 species, including rugose corals, ammonite molluscs, and crinoid echinoderms, was reduced to four or five bivalve molluscs and echinoid echinoderms. The fish on the left is a relative of skates and rays. (Artwork by John Sibbick.)

Permian. Ocean-circulation patterns probably could not supply enough oxygen to the deep basins, and except for a thin surface layer the oceans became anoxic. Surface photosynthesis continued to draw carbon dioxide from the atmosphere and export it to the anoxic sediments, as is happening in the Black Sea today. Deep water became enriched in methane and hydrogen sulfide. When vigorous circulation in the oceans started again, it brought deep water to the surface, poisoning the surface waters, and releasing vast quantities of carbon dioxide into the atmosphere. At the same time, right at the end of the Permian, massive lava flows erupted in Siberia, spewing out 2 million km³ of basalt and covering 1.6 million km² of eastern Russia—an area the size of the Gulf of Mexico—to a depth of 400–3000 m. These eruptions released additional huge quantities of carbon dioxide into the atmosphere. The resulting warming triggered the release of more methane from the oceans, and a runaway greenhouse effect raised the temperature of the surface of the planet by an average of 6°C according to the record of oxygen isotope ratios.

Consistent with this hypothesis are both the chemistry of the sediments and the selectivity of the extinctions. Marine animals with active circulation and gills compensate better for high carbon dioxide levels than those without them. Animals forming carbonate skeletons should have been particularly affected because of their sensitivity to disturbances of their internal acid–base balance. In fact, the groups most strongly affected were corals, articulate brachiopods, bryozoans, and echinoderms, all of which have carbonate skeletons, weak circulation, and low metabolic rates: they lost 65, 67, and 81% of extant genera in three waves of extinction. The mollusks, arthropods, and chordates, with gills, circulatory systems, and higher metabolic rates, lost 49, 38, and 38% of extant genera in each of those extinction waves.

It took 100 million years for global biodiversity at the family level to return to pre-extinction levels. Immediately after the extinction, microbial mats covered a sea floor in which deposit-feeding worms burrowed just below the sediment surface, a situation that lasted perhaps a million years. As oxygen levels rose and the food supply increased, more diverse communities with crinoids and bryozoans appeared. Complex communities—reefs and deep-burrowing sediment communities—were re-established within 10 million years. On land only one moderate-sized herbivore, *Lystrosaurus*, is known for several million years, and forest communities did not appear again until the mid-Triassic. Life was clearly tough in the 'post-apocalyptic greenhouse' (Benton and Twitchett 2003).

Massive volcanism and continental drift both contributed to the very strange chemistry of the late-Permian oceans and atmosphere. The immediate cause of the most massive extinction in Earth history was probably poisoning.

The end-Cretaceous extinction is associated with a meteorite impact and massive volcanism

In the mass extinction at the end of the Cretaceous, about 50% of extant genera disappeared. All marine invertebrates were affected. Prominent victims were foraminiferans, bivalves, bryozoans, all the remaining ammonites, gastropods, sponges, echinoderms, and ostracods. The dinosaurs appear to have been declining before the final extinction. However, many of the dinosaurs probably did die at the same time as the marine invertebrates.

The end-Cretaceous extinction also took place under unusual conditions. Widely scattered soot deposits—several centimeters thick in New Zealand—suggest fires on a hemispheric scale, as do spores from ferns that invaded habitats cleared by fire. Right at the Cretaceous–Paleocene boundary there is an enrichment of iridium, an element found at low concentrations on Earth and in higher concentrations in meteorites. The leading hypothesis is the impact of a meteorite 10 km in diameter that formed a crater 180 km in diameter. The ejected debris would have ignited hemispheric fires, and the impact would have injected so much dust into the atmosphere, both directly and through volcanic eruptions possibly induced by the impact, that the Earth would have been dark and cold for several years. That most seed plants survived suggests that the period of deepest crisis did not exceed the time that seeds can survive in the soil, which is about a decade. The timescales for destruction and recovery are given in Table 16.2. Both took a long time.

The leading candidate for the impact crater is Chicxulub on the Yucatan Peninsula of Mexico; it dates close to the end-Cretaceous boundary at 65 Ma, but may predate it by about 300 000 years (Keller et al. 2004). Shock quartz and tektite ejecta from that crater have been found both across the Caribbean and farther afield. Traces of huge tsunamis caused by the meteorite strike in the Yucatan have been found in Haiti. For several hours following a major meteorite impact, the entire Earth would experience intense heat from ejected

Table 16.2 The end-Cretaceous impact: destruction and recovery. (From Conway Morris 1998.)

Time	Effect
1 s	Annihilation around impact site ($c.30\ 000\ km^2$)
1 min	Earthquakes, Richter scale 10
10 min	Spontaneous ignition of North American forests
60 min	Impact ejecta cross North America
10 h	Tsunamis swamp the Tethyan coastal margins
1 week	First extinctions?
9 months	Dust clouds begin to clear
10 years	Severe climatic disturbance (cooling) ends
1000 years	Continental vegetation recovers; end of the fern spike
1500 years	Deeper-water benthic ecosystems start to recover
7000 years	Full recovery of benthic ecosystems
70 000 years	Ocean anoxia diminishes
100 000 years	Final extinction of dinosaurs (?)
300 000 years	Final extinction of ammonites (?)
500 000 years	Ocean ecosystems start to stabilize
1 000 000 years	Open ocean ecosystems partly recovered
2 000 000 years	Marine mollusk faunas mostly recovered
2 500 000 years	Global ecosystems normal

rock falling back through the atmosphere. That heat alone would have been enough to kill many terrestrial organisms that could not find shelter (Robertson et al. 2004).

The meteorite hypothesis is not universally accepted. We know that a large meteorite did smash into the Yucatan at about the right time, that its impact disturbed the entire globe, and that the disturbance lasted a long time. We also know that mass extinctions can occur without a meteorite impact, as at the end of the Permian, and that large impacts have occurred—the Montagnais impact structure, 45 km in diameter and 51 million years old, and a gigantic crater in the Kalahari, possibly 350 km in diameter and 145 million years old—without causing a mass extinction.

Mass extinctions repeatedly changed the course of evolution. It was probably not an accident that the amniotes radiated in the Triassic after the end-Permian mass extinction had eliminated many other groups and that the mammals radiated in the Paleocene after the end-Cretaceous mass extinction had eliminated most of the other groups. Molecular evidence indicates that the mammalian radiation started well before the end of the Cretaceous, but what form it might have taken if the dinosaurs had survived, we cannot say.

Other catastrophes have had dramatic local effects

The Mediterranean dried up, possibly several times

The Mediterranean Sea has been dry, probably several times, the last time about 18 Ma. This was discovered by ocean-bottom drilling that revealed salt deposits in the eastern Mediterranean that could only have formed if the entire sea had dried up. The mechanism requires that the Atlantic drop below the level of the rim of the basin at Gibraltar. Once it was cut off from the Atlantic, the Mediterranean would steadily evaporate, for the flow of rivers into it is not enough to compensate the evaporative loss, and it would dry up in about 15 000 years at a rate of 30–35 cm per year. When it was dry, the Rhone, the Nile, and the rivers flowing into the Black Sea (the Danube, Dnieper, and Don) cut deep canyons into its flanks, canyons that can still be traced underwater. Animals and plants could then disperse directly on to what are now islands—Cyprus, Crete, Sicily, Sardinia, Corsica, and the Balearics. Then the Atlantic would rise again, and the world's largest waterfall would be formed as the sea surged in through the Straits of Hercules. It would take less than 1000 years to fill the basin at a rate of about 1 cm per day.

Major volcanic outbreaks are regional catastrophes with impact on global climate

Some past volcanic outbreaks dwarf the major eruptions of historic times, like Krakatoa, in Indonesia, which sent tsunamis into Japan and the west coast of North America in August 1883. When the Phlegrean Fields, a suburb of Naples, erupted, they sent ash as far as the Ukraine. Oligocene and Miocene eruptions from volcanoes in the Cascades and Sierra Nevada of Oregon and California sent ash as far east as Nebraska, 2000 km away, burying a herd of woolly rhinoceroses. About 1600 BC the eruption of Santorini in the Cyclades sent falling ash, and the accompanying earthquakes sent giant waves, into the Greek islands and Crete, destroying many Mycenaean settlements and probably weakening the Cretan civilization, which fell not long after to Greek invaders.

Gigantic volcanic eruptions inject into the stratosphere a huge amount of ash that takes several years to settle out. During that period, it scatters radiation from the sun back into space, and the planet experiences a period of cooling. The eruption of Tambora Volcano in Indonesia in 1815, the largest eruption in historical times, lowered the global temperature by as much as 3°C; 1816 was known as the year without a summer.

Not all giant eruptions have been explosive; there have also been episodes of massive volcanic flooding. During the middle Cenozoic, 30–15 Ma, lava poured from cracks kilometers long in the Columbia Plateau of Washington and

Oregon and in the Ethiopean highlands, sterilizing hundreds of thousands of square kilometers with lava flows. Other flood basalts have been found in north-central USA (Proterozoic), central Siberia (as mentioned above, p. 391, right at the Permian–Triassic boundary), the Karoo of South Africa (Jurassic), and the Paraná plateau of Brazil and Uruguay and the Deccan plateau of India (both Cretaceous). Individual flows were typically 10–30 m thick, the area flooded was up to 1 million km^2, and the volume of lava in such an area could exceed half a million cubic kilometers, or fifty times the volume of the largest terrestrial shield volcanoes (Mohr 1983).

The breaking of glacial dams generates giant floods

During the Pleistocene glaciations, floods repeatedly swept across the State of Washington and south-central Siberia, down the valleys of the Columbia, Ob, and Yenisey Rivers.

When the continental glaciers grew, as they did at least seven times, an arm of the Columbia ice field moved south to dam the Clark Fork of the Columbia River near what is now Pend Oreille Lake in Idaho. The dam formed glacial Lake Missoula. As the water rose, it floated the ice that dammed it and swept out in a massive flood, carrying large icebergs with it, forming the Channeled Scablands of eastern Washington. Blocked by the Cascades, the flood could only drain through the Columbia River Gorge, where it was hundreds of meters deep. Blocked again between the Cascades and the coast, the water backed up in the Willamette Valley to a depth of 150 m over an area of several thousand square kilometers. Boulders lodged in the icebergs carried by the flood would have been scattered down the valley.

The same process was repeated in the Altay Mountains of Siberia. When ice-dammed lakes failed, they released floods with peak discharge rates of 18 million m^3 per second, slightly more than the rates achieved in the Missoula Floods. Other catastrophic floods occurred 15 000–20 000 years ago when water was released from beneath the melting continental ice sheets in Manitoba and Swedish Lapland (Baker et al. 1993). In 1996 a large flood devastated southern Iceland when a volcano erupted beneath that island's glacial cap.

Giant tsunamis are caused by submarine landslides and meteorite impacts

Extremely large waves can be generated by earthquakes, submarine landslides, and meteorite impacts. After the 1964 Anchorage in Alaska earthquake a local wave confined to a coastal Alaskan fjord destroyed a structure more than 200 m above sea level. A much larger wave went over the top of the Island of Lanai, in Hawaii, leaving sea water perched at an elevation of more than 1000 m. The cause was probably a massive landslide with a volume of tens of cubic kilometers. Remnants of such huge landslides are found on the ocean floor at the feet of the Hawaiian shield volcanoes. Massive undersea landslides also occur where unstable sediments are deposited at the mouths of rivers. Traces of such events have been found in the Mediterranean, off the Amazon, off the Carolinas, and off Norway; some moved 500 km^3 of material (Nisbet and Piper 1998).

When a meteorite impacted in the eastern South Pacific between Chile and Antarctica 2.15 Ma, it sent a tsunami up the western coast of South America that reached several hundred meters above sea level. The tsunami created by the end-Cretaceous meteorite impact was about 1 km high locally but was not nearly that large outside the Caribbean. Its force may have been dissipated when it inundated the islands in the Greater Antilles. Other large meteorites, producing craters 40–90 km wide, have impacted the continental shelf off Nova Scotia, between Norway and Spitzbergen, and in Chesapeake Bay. Theoretical calculations indicate that an impact in the ocean that creates a crater 150 km wide in water 5 km deep will generate a mega-tsunami 1300 m high near the impact and no lower than 100 m high anywhere on the planet (Dypvik and Jansa 2003). That is probably close to the largest meteorite impact that the Earth would experience every several hundred million years.

This selection of catastrophes shows that evolution has not occurred in a stable, predictable environment. It also shows that evolution has not been about constantly increasing adaptation and biodiversity. Major setbacks have repeatedly occurred when there was no opportunity for natural selection to produce adaptations that could help organisms to survive rare catastrophes. The catastrophes have varied in the geographical range of their impact. A very local catastrophe would only cause the extinction of endemic species occupying a limited area. As the range of a catastrophe increases, it would cause species with larger ranges to go extinct, and for major catastrophes, such as meteorite impacts and other causes of extreme climate change, most of the life on the plant would go extinct.

How has life shaped the planet?

Ancient bacteria changed the atmosphere and created mineral deposits

● KEY CONCEPT

Life has changed the composition of the atmosphere and the chemistry of the oceans, and has participated in the production of mineral deposits and ores.

As described above, a primitive atmosphere of hydrogen and helium was blown off the planet by the solar wind, after which a secondary reducing atmosphere of water, ammonia, and carbon dioxide formed, an atmosphere similar to that of Venus. The bacteria that evolved under those conditions utilized any available energy source accessible to anaerobic biochemistry. Among them are bacteria that use sulfates to reduce organic molecules, gaining chemical energy and excreting sulfur or hydrogen sulfide. The sulfur or sulfide then react with metal ions in solution, forming insoluble metallic sulfides in part responsible for the size of the copper, nickel, zinc, and mercury deposits of the world.

Other bacteria combine carbon dioxide with hydrogen, gaining chemical energy and excreting methane, which can accumulate in tremendous quantities in anoxic waters and soils. Still other bacteria ferment organic compounds, such as sugars and amino acids, gaining chemical energy and excreting carbon dioxide or ammonia, or oxidize metals to gain energy. Some anaerobic bacteria combine carbon dioxide with hydrogen or hydrogen sulfide in the presence of sunlight to gain chemical energy and synthesize organic molecules. These exotic metabolisms are all rapidly poisoned by oxygen, even at low concentration. The bacteria using them dominated the first 2 billion years of Earth's history, then survived in anaerobic pockets in the environment when cyanobacteria evolved the aerobic photosynthesis that now colors much of the planet green, combining carbon dioxide with water in the presence of sunlight to produce organic matter and excrete oxygen.

The oxygenation of the atmosphere was the most significant engineering done by life

After photosynthesis evolved, life began a major transformation of the atmosphere, removing carbon dioxide and adding oxygen (Figure 16.9). It took over 2 billion years for free oxygen to achieve significant concentrations. That point is marked all over the planet by deposits of iron ore, for when the atmosphere

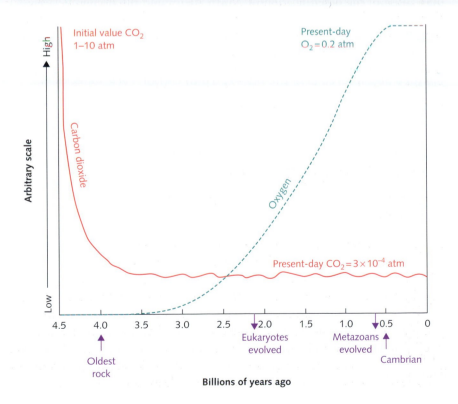

Figure 16.9 The history of oxygen and carbon dioxide in the atmosphere during Earth history. (Reproduced from Turekian (1996) *Global Environmental Change*, Prentice Hall, with permission.)

and the oceans switched from reducing to oxidizing, ferrous iron, which is soluble in water, was oxidized to ferric iron, which is insoluble, and all the iron dissolved in the world's oceans—about 600 trillion tons of it—fell out of solution. That happened between 2.5 and 1.8 billion years ago. Pink skies turned blue; brown seas turned blue-green.

The arrival of free oxygen in the atmosphere, caused by life, had two major consequences for life. The first was the conversion from anoxic to oxidative metabolism, which came with a considerable increase in efficiency and opened the opportunity to operate at much higher energy levels. The second was the creation of the ozone layer in the upper atmosphere, which shielded the surface of the Earth from ultraviolet radiation and substantially lowered the mutation rate of organisms living on or near the surface of the continents and oceans. Those two developments are thought to have played an enabling role in the evolution of the eukaryotes and the subsequent radiation of large, multicellular organisms. They were global in scale.

Life also has effects on the atmosphere that are local in scale. Forests and prairies take water from the soil and transpire it into the atmosphere, influencing local cloud formation and rainfall. A landscape covered by plants puts much more water into the atmosphere than a landscape of barren rock and sand. When goats and sheep grazed much of the Mediterranean vegetation to the ground, the climate changed from moderately humid to dry. This was not the only cause of the desertification of North Africa and the drying of Greece— post-glacial global climate trends also played a role—but it was an important contributing factor.

Bacteria also intervene in the nitrogen and carbon cycles and produce metal ores

In addition to oxidizing the atmosphere, life also affected the nitrogen cycle. Living things need nitrogen in many of the molecules of life, but they cannot acquire it directly from the molecular nitrogen, N_2, in the atmosphere. They require either oxidized nitrate or reduced ammonia as a nitrogen source. Molecular nitrogen moves from the atmosphere into organisms via two main paths, lightning and cyanobacteria in the ocean, which transform billions of tons of molecular nitrogen into nitrates each year. These processes would remove all the nitrogen from the atmosphere within 20 million years if there were not a return pathway, bacterial denitrification, in which some bacteria gain energy from the reaction of nitrates with organic matter, releasing molecular nitrogen in the process (Kump et al. 1999).

Life's adjustment of carbon levels in the atmosphere and oceans has not only lowered the level of carbon dioxide dramatically; it has also produced major mineral deposits. We owe to life considerable rock and liquid in the Earth's crust. Much of the carbon dioxide that was removed from the atmosphere by living organisms was deposited as the calcareous skeletons of countless algae,

Table 16.3 The fate of the original atmospheric carbon dioxide. (After Kump et al. 1999.)

Carbon reservoir	Amount (in gigatons)	Percentage of total
Limestone in sedimentary rocks	40 000 000	79.92709
Organic carbon in sedimentary rocks	10 000 000	19.98177
Oceanic bicarbonate ion	37 000	0.07393
Fossil fuels	4200	0.00839
Organic carbon in sediment and soil	1600	0.00320
Oceanic carbonate ion	1300	0.00260
Oceanic dissolved CO_2	740	0.00148
Living biomass	760	0.00152
Atmospheric CO_2	10	0.00002

protozoans, and corals—layers of calcium carbonate that can be thousands of meters thick and extend over hundreds of thousands of square kilometers. All of the chalk, limestone, and marble on the planet formed that way. Similarly, all of the oil, gas, and coal on the planet is the product of anoxic transformation of billions of tons of dead plants buried and heated under pressure. As a result, 99.9% of the carbon dioxide on the planet has been locked up in sedimentary rock (Table 16.3). Most of the rest exists as dissolved ions or dissolved gas in the world's oceans, and only a trace remains in the atmosphere. Life adds by photosynthesis and removes by respiration and decomposition equal amounts of carbon dioxide each year—about 60 billion tons, or 8% of the atmospheric total.

We owe to bacteria not only the iron mines that are the byproduct of free oxygen produced by photosynthesizing cyanobacteria. We can also thank them for copper, zinc, and other metals. Sulfur bacteria feed off sulfates and excrete sulfides. Sulfides react with many different metallic ions to form ores. It is thought that as much as 90% of some of the world's major metal deposits, including zinc and copper, were formed because of the presence of sulfur bacteria in the environment.

The planet is warmer because of greenhouse gases in the atmosphere

The long-term trend in atmospheric carbon dioxide, from a great deal in the early atmosphere to very little today, helps to explain a paradox. The Sun has been growing steadily warmer since the Earth formed, starting at about 70% of its present output. If there had been no greenhouse effect, the Earth would have been and would still be a very cold place indeed, with an average surface temperature today ($-15°C$) below the freezing point of water. But because the early atmosphere had a great deal of carbon dioxide, the greenhouse effect was strong, and it is thought that the average surface temperature of the planet rose above the freezing point of water about 2 billion years ago. Note that even before that, water was liquid in the oceans due to thermal input from sea-floor spreading and volcanoes.

Ever since plants invaded land and added their withdrawal of carbon dioxide from the atmosphere to that of marine invertebrates and algae, the fixation of carbon dioxide into biomass has been steady or increasing. It is therefore interesting to note that the rate of deposition of coal and oil appears to have been greater in the Carboniferous (360–286 Ma) than it has been subsequently (Figure 16.10). One possibility is that there was a huge area of tropical rain-forest because of the continental configuration, and as the continental glaciers waxed and waned, the sea level fell and rose, burying at the high stand of the sea the peat that was formed at the low stand of the sea. The amount of carbon dioxide present in the atmosphere and dissolved in the oceans has been so great, and the ability of plants to remove carbon dioxide from the atmosphere at low concentrations has been so efficient, that the burial of huge quantities of organic carbon probably has not affected the global rate of carbon fixation very much. The cumulative effect, however, has been to withdraw carbon from the atmospheres and oceans and bury it in solid form as fossil fuels and carbonate deposits, such as limestone.

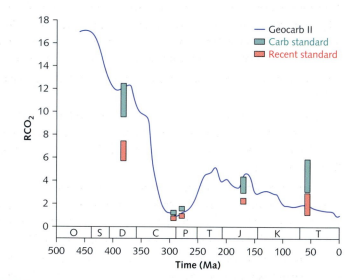

Figure 16.10 The Phanerozoic CO_2 curve (Geocarb II) estimated from a model compared to estimates from the stomatal ratios of fossil plants calibrated either to the Carboniferous (Carb. standard; where a stomatal ratio of 1 is equal to 600 ppm CO_2) or to the Recent (Recent standard; where a stomatal ratio of 1 is equal to 360 ppm CO_2). RCO_2 represents the *ratio* of atmospheric CO_2 to the pre-indtrial concentration of 300 ppm. All methods agree that there was a massive withdrawal of CO_2 from the atmosphere from the Ordovician to the Permian, then a reinjection in the Triassic, followed by a more gradual withdrawal to the current level of 360 ppm. (Reproduced from Briggs and Crowther (2003) *Paleobiology 2*, with permission from Blackwell Publishing.)

Conclusion

Life and the planet have interacted extensively, sequentially, and synergistically. Life has always been a thin layer wrapped around a massive ball of iron, nickel, and silicate, but it has had major influence on the composition of the atmosphere, the oceans, and the mineral deposits in the crust. The chemical composition of the atmosphere of the planet, and its average temperature, are far from the equilibrium that they would attain if there were no life on planet Earth—so far, in fact, that an extraterrestrial scientist would be able to conclude that something like life must exist on this planet just by examining the composition of the atmosphere.

● SUMMARY

This chapter describes how the planet has shaped life and how life has shaped the planet.

- Organisms, communities, and landscapes are mosaics of parts with very different ages: some ancient, some recent.

- The Earth has characteristics without which life as we know it could not have originated and continued: intermediate mass, intermediate distance from a medium-sized star, abundant surface water, a temperature at which that water is liquid, and a supply of elements and molecules out of which life can be built.

- Continental drift, volcanoes, and glaciers have profoundly altered the history of life.

- Mass extinctions have, at intervals of tens to hundreds of millions of years, killed as many as 60% of the species on the planet. The major mass extinctions were 250 Ma at the end of the Permian period and 65 Ma at the end of the Cretaceous period.

- The mechanism of the end-Permian extinction appears to have been continental drift causing glaciation coupled with massive volcanic eruptions in Siberia. The mechanism of the end-Cretaceous extinction appears to have been meteorite impact coupled with massive volcanic eruptions in India.

- Local catastrophes have also had an impact on life. They include the drying of the Mediterranean, gigantic volcanic eruptions, superfloods, and huge tsunamis.

- Life has shaped the planet by oxygenating the atmosphere, by removing carbon dioxide from the atmosphere and storing it in the Earth's crust as chalk, limestone, oil, and coal, and by creating soil and organic sediments.

With that as background, we proceed in the next chapter to consider the major patterns in the fossil record and the lessons that can be drawn from them.

● RECOMMENDED READING

Kump, L.R., Kasting, J.F., and Crane, R.G. (1999) *The Earth system*. Prentice Hall, Upper Saddle River, NJ.

Marshak, S. (2001) *Earth: portrait of a planet*. W.W. Norton, New York.

● QUESTIONS

16.1. Cyanobacteria are very small, yet they remove tens of billions of tons of carbon dioxide from the atmosphere each year. Describe what you would need to know to calculate the number of cyanobacteria that would have to live and die each year if a billion tons of carbon were to be stored in their corpses.

16.2. If all the ice stored in Antarctica and Greenland were to melt, the oceans would rise by about 100 m. Take a look at a relief map of the world. What would be some of the major consequences?

16.3. Review conditions on Mars and on the satellites of Jupiter. In what other parts of the solar system do you think life might exist? Why?

was marked by a mass extinction in which brachiopods and trilobites were hard hit but did not disappear completely.

The Silurian: 435–410 Ma; club mosses and fungi appear, jawless fish radiate

The Silurian deposits contain fossils of the first vascular plant, *Cooksonia*, a branching form about 10 cm long with neither roots nor leaves. The Silurian also saw the radiation of the club mosses and the jawless fish, and the first fossil fungi, which share ancestors with the cellular slime molds and other fungal-like amoebae (Lang et al. 2002). The arthropods on land in the Silurian included mites, springtails, scorpions, millipedes, and possibly insects (Engel and Grimaldi 2004).

The Devonian: 410–355 Ma; tetrapods invade land; seed plants appear

In the Devonian seas the first large, mobile invertebrate predators appear: eurypterids (related to scorpions) up to 2 m long and nautiloid cephalopods up to 3 m long. And they had many things to eat, for the seas filled with a diversity of jawed fish and cephalopods. During the Devonian terrestrial communities evolved from algae, small vascular plants, and small arthropods to forests that contained club mosses up to 40 m high and horsetails and tree ferns up to 15 m tall.

The late Devonian saw a key innovation in plant evolution. Up until then, plants had an alternation of independent generations, a sporophyte and a gametophyte. The sporophytes began to produce two kinds of spores, one developing into a female gametophyte that produced a food store, the other developing into a male gametophyte that produced many small sperm. The sporophyte then retained the gametophytes as part of its structure. The female gametophyte formed the seed and surrounding structures, and the male gametophyte produced sperm that swam to the female gametophyte through a film of water. We see the first evidence of this process in fossil progymnosperms, which were able to invade drier habitats than were their ancestors. Fertilization by relic motile sperm that hatch from pollen has been retained to the present by ginkgoes and cycads.

By the late Devonian most terrestrial invertebrates were probably already on land, but we do not have fossils of many of them. They would have included earthworms, leeches, flatworms, snails, centipedes, millipedes, scorpions, spiders, and mites.

When the first tetrapods came on to land in the late Devonian, they had to walk, and to walk they had to have limbs. The comparison of *Eusthenopteron*, a Devonian jawed fish, with *Ichthyostega*, one of the first amphibians, is instructive (Figure 17.2). The basic elements of the vertebrate limb evolved in the lobe-finned fishes, the ancestors of tetrapods, of which the coelecanth,

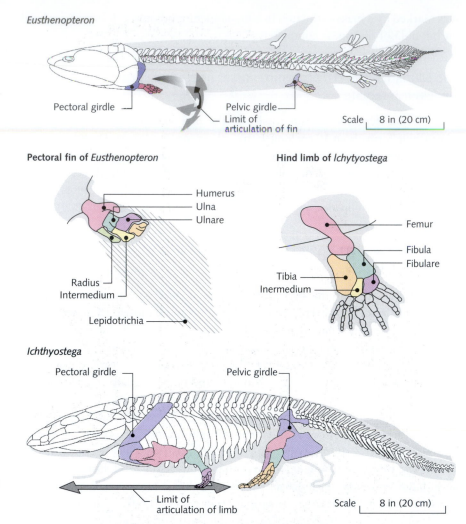

Eusthenopteron

Pectoral girdle

Pelvic girdle
Limit of
articulation of fin

Scale   8 in (20 cm)

Pectoral fin of *Eusthenopteron*

Humerus
Ulna
Ulnare

Radius
Intermedium

Lepidotrichia

Hind limb of *Ichytyostega*

Femur

Fibula
Fibulare

Tibia
Inermedium

Ichthyostega

Pectoral girdle

Pelvic girdle

Limit of
articulation of limb

Scale   8 in (20 cm)

Figure 17.2 The vertebrate limb originated in the late Devonian, about 370 Ma, but the basic elements of the limb were already in place in the lobe-finned fishes. The combinatorial control of limb development by the HOX D genes was arguably in place nearly 400 Ma. (From Gould 2001.)

Latimeria, is a living representative. The features of the vertebrate limb that appear to have been fixed from the origin of the tetrapods are a single proximal bone (humerus or femur) joined to two more distal bones (radius and ulna in the arm, tibia and fibula in the leg), joined to a complex of bones forming the wrist or ankle on which a variable number of digits are anchored.

Carboniferous: 355–295 Ma; coal deposits form; tetrapods freed from water

The Carboniferous is so named because much of the Earth's coal was deposited then. It was a warm period when the sea repeatedly rose over low-lying continental forests (Figure 17.3), burying layers of organic material beneath marine sediments to form coal seams in many of which fossils are found. In those deposits are the first fossils of earthworms, seed plants, and reptiles. Some

Figure 17.3 Late Carboniferous coal-swamp trees. From left to right, a calamite, a scrambling cordaite, a tree fern, a lycopsid, a seed fern, a lycopsid tree, and a mangrove cordaite. (By Mary Parrish, from Briggs and Crowther 2003.)

terrestrial arthropods were giants: there were millipedes up to 2 m long and dragonflies with 1 m wingspans.

Tetrapods liberated themselves from the amphibian aquatic larval stage early in the Carboniferous by repackaging offspring. The key innovation was the cleidoic egg, which has two important features. The first is a semipermeable outer shell that protects the internal fluids from evaporation and the embryo from physical harm. The second is a set of membranes within the egg that protect the embryo, enclose the yolk, and segregate waste. Tetrapods also evolved a scaly, impervious skin and the ability to produce near-solid urine, both of which greatly reduced water loss.

All the tetrapods that possess an amniotic membrane, either as part of an egg or as part of a viviparous embryo, are called *amniotes*, which includes all tetrapods except the amphibians. In the late Carboniferous the amniotes radiated into three clades with important futures.

1. The *anapsids* included amphibians and many extinct terrestrial amniotes. Anapsid means having no temporal openings in the skull, a condition also found in fishes.

2. The *synapsids* would later become the mammals. Synapsids were tetrapods with a single pair of temporal openings behind the eye socket.

3. The *diapsids* would later become the lizards, snakes, crocodiles, dinosaurs, and birds. Diapsids have two pairs of temporal openings, one above, the other behind the eye socket.

Much later, in the Mesozoic, a fourth group, the *euryapsids*, introduced a skull with a single pair of facial openings like the synapsids, but smaller and set higher behind the eye sockets. The euryapsids lived in the sea and included plesiosaurs and ichthyosaurs.

Permian: 295–250 Ma; gymnosperm forests, large land animals, largest mass extinction

The Permian saw major developments in plant biology. Forests of club mosses and horsetails were replaced by forests of seed plants: cycads, ginkgoes, and seed ferns. For the first time a terrestrial vertebrate took to the air—a gliding lizard with thoracic wings supported on expanded ribs—and another re-entered the sea. The synapsids included herbivores larger than rhinoceroses and meter-long saber-toothed carnivores. In the seas the ammonites underwent another major radiation, and the trilobites and the rugose corals played their final act before disappearing in the end-Permian mass extinction.

Triassic: 250–203 Ma; dinosaurs appear and radiate; Gondwana distributions develop

The Triassic began as an empty world with less life than had been seen since the Vendian 350 million years earlier. There were re-radiations of nautiloids, brachiopods, bivalves, gastropods, and sea urchins. A new type of coral appeared, the hexacorals, and a new type of bivalve, the oysters. The holostean fish radiated, with symmetrical tails and improved mouths, and the first teleost fish appeared, with improved skeletal support of the tail and more flexible jaws. The first ichthyosaurs swam in the seas; by the late Triassic some had become 15 m long.

Figure 17.4 *Coelophysis*, a small, late Triassic dinosaur from Ghost Ranch, New Mexico, was an active predator on small animals, including juveniles of its own species. Its ball-and-socket hip joint allowed it to run upright on its hind legs, freeing its strong, clawed forelimbs for prey capture. (Illustration by Marianne Collins from The Book of Life by Stephen Gould, published by Ebury Press. Reprinted by permission of The Random House Group Ltd.)

In the Triassic gymnosperm radiation, the genus *Araucaria* appeared. Its living representatives have a characteristically Gondwana distribution that stretches from Chile to islands near New Zealand. *Araucaria* are also the dominant fossils in the Petrified Forest of Arizona.

In the Triassic archosaur radiation, shoulder and hip girdles evolved that permitted an upright gait. The first true frogs, turtles, crocodiles, pterosaurs, and mammals appeared in the Triassic, and the dragonflies and damselflies radiated. But it would be the dinosaurs that would dominate the Earth, both in diversity and in size of the largest species, for the next 150 million years. The first dinosaur fossils date to 225 Ma; by the late Triassic they were radiating explosively. Triassic dinosaurs, such as *Coelophysis* (Figure 17.4), first had a ball-and-socket insertion of the femur into the hip. This, together with improvements in their ankles, allowed them to run upright on their hind legs at high speed.

The Triassic ended with two episodes of mass extinction, the first with impact primarily on land, where several branches of the early amniote radiation vanished, and the second, about 15–20 million years later, with impact primarily in the sea, where entire families of crinoids, sea urchins, bryozoans, scallops, and fishes disappeared. The dinosaurs survived.

Jurassic: 203–135 Ma; birds, pterosaurs, and insects take flight

The Jurassic saw the appearance of many familiar groups that survive today, including yews, crabs and lobsters, modern sharks and rays, newts and salamanders, birds, and three of the great insect clades—the diptera (flies), hymenoptera (bees, ants, and wasps), and lepidoptera (butterflies and moths). For the first time insects pollinated plants. There were major radiations of cephalopods and of teleost fish. The dominant trees in Jurassic forests were ginkgoes, cycads, and araucarias. In the air the pterosaurs and the birds joined the insects, which had been flying at least since the Carboniferous and possibly the Silurian (Engel and Grimaldi 2004).

The most striking terrestrial animals in the Jurassic were the dinosaurs, among them the famous stegosaurs, giant sauropods, and predatory allosaurs. The birds are directly descended from one group of dinosaurs, the theropods. The emergence of the birds marks the uncontested origin of vertebrate homeothermy—warm-bloodedness—that may also have emerged about then in the mammals and dinosaurs. Some pterosaurs had fur and were probably warm-blooded.

Two particularly exquisite Jurassic fossils must be mentioned. The first, from the Holzmaden quarry in Germany, is a pregnant ichthyosaur giving birth (Figure 17.5). She probably expelled one of her four offspring in her death throes.

The second is *Archaeopteryx*, from the Solnhofen quarry in Germany (Figure 17.6). *Archaeopteryx*, which lived about 155 Ma, is an intermediate step in the origin of modern birds. Its ancestral features include heavy jaws with short, spiky teeth; a hand that retains three long fingers; a long tail; and a wishbone similar to that of a small running dinosaur. Its reversed toe has been lengthened to aid in gripping a perch. There are non-flying dinosaurs from Chinese deposits that have feathers, perhaps for gliding or conserving heat, but the wing feathers of *Archaeopteryx* have been modified in a fashion characteristic of modern flying birds and lacking in the feathers of modern flightless birds: they have narrow leading edges and long trailing edges, adaptations that help to create lift.

Cretaceous: 135–65 Ma; angiosperms radiate; at end dinosaurs and ammonites vanish

The Cretaceous landscape took on a decidedly modern cast as the flowering plants appeared and diversified, reducing the conifers to about half their former

Figure 17.5 A pregnant ichthyosaur from Jurassic deposits in Holzmaden, Germany, where more than 50 pregnant ichthyosaurs have been found. The skeletons of three offspring lie within her rib cage, and one lies just outside her cloaca. (Photo copyright The Natural History Museum.)

Figure 17.6 *Archaeopteryx*, a key intermediate in the origin of birds, from the Solnhofen quarry in Germany, deposited 155 Ma. (Photo copyright Jim Amos/Science Photo Library.)

abundance. The oldest angiosperm fossils date to 130–120 Ma; they radiated about 100 Ma. Their special features include the reduction of the male gametophyte to the pollen tube, and the flower itself. The earliest flowers and fruit in the fossil record resemble those of sycamores and magnolias.

Other familiar groups making their first appearance in the Cretaceous include termites, lizards, snakes, loons, and marsupials. In the oceans the neogastropods, the modern snails, appeared.

Some Cretaceous animals were very large. There were marine lizards 10 m long, turtles 3 m long, clams 1.8 m wide, bivalves 1 m tall, and sharks 6 m long. Through Cretaceous skies flew giant pterosaurs, the largest of which, *Quetzalcoatlus*, had a wingspan of 11–12 m, four times the wingspan of a condor—the largest animal ever to fly. And dinosaur evolution continued. Near the end of the Cretaceous *Tyrannosaurus*, at 14 m in length, 6 m in height, and 5 tons, was the largest terrestrial predator in history, and *Triceratops*, at 9 m in length and 6 tons, was about twice the weight of a rhinoceros and worthy prey for tyrannosaurs, whose tooth marks have been found on its skeletons.

The mass extinction at the end of the Cretaceous is marked in the sediments by an iridium spike, soot deposits, shock quartz, and a spike of fern spores. There were fires of global scale followed by several years that were so cold and dark that many plants died. The dinosaurs and the ammonites disappeared, but few plants, which had seeds that could survive in the soil for several years.

Paleocene: 65–53 Ma; radiations of modern birds, mammals, and plants begin

The Paleocene was relatively empty of large land animals, but it rapidly filled with plants and birds. In Paleocene deposits we see the first fossil rhododendrons, grasses, maples and willows, herons, falcons, anhingas, avocets, parrots, and rollers. The rabbits and rodents appeared, as did mosquitoes.

Eocene: 53–34 Ma; mammal clades continue to form and radiate

Eocene deposits hold fossils of many modern trees, including beeches, elms, casuarinas, and tilias (lyme trees or basswoods). Both fossils and molecular systematics suggest that whales, primates, ungulates, canids, proboscids, and bats originated by the Eocene.

Oligocene: 34–23 Ma; grasslands form; the teeth of grazers get harder

When grasslands developed in the Oligocene, the teeth of grazers hardened to deal with the silicates in grass. All the major mammal clades expanded.

Miocene: 23–5.3 Ma; savannah communities converge

One of the most striking features of the mammal radiation is the repeated convergence of herbivores and carnivores in communities inhabiting savannah habitats. This convergence happened at least four times: in North America, South America, Australia, and Africa. The closest resemblance is that of the North American Miocene to present-day East Africa, where guilds of mega-herbivores, grazers, browsers, mixed feeders, omnivores, and carnivores evolved. While the species involved differed in morphological detail, they had roughly the same relative body size and played similar ecological roles. The savannah radiations were made possible by the appearance in the Miocene of all the subfamilies of grasses, a resource that was quickly exploited by the first grazers.

Pliocene: 5.3–1.75 Ma; hominids and orchids appear

During the early Pliocene hominids diverged from chimpanzees and gorillas. The Pliocene also saw the emergence and diversification of one of the largest families of flowering plants, the orchids.

Pleistocene: 1.75–0.012 Ma; repeated glacial periods; language and culture appear

During the Pleistocene, a relatively cool period of repeated continental glaciation, *Homo* moved out of Africa and colonized the planet. Language and culture emerged.

Holocene: 12 000 years ago to the present; mammal and bird extinctions; agriculture; writing

The Holocene refers to the period from the melting of the last continental glaciers to the present. It has seen the extinction of the Pleistocene mammal megafauna and about 25% of the birds, and the origin of agriculture with two major consequences—cities and many types of infectious disease; and the origin of writing.

The major radiations

A walk through the geological eras is one way to approach the past. Another is to take a major radiation and ask, when did this clade acquire its key characters? In this section the age of many events is estimated from molecular systematics as well as from fossils (Table 17.2).

The prokaryote radiation has three main branches: the Eubacteria, the Archaea, and the Eukaryota

The prokaryotes lie at the base of the Tree of Life and consist of three major clades: the Eubacteria, the Archaea, and the nuclei of the Eukaryota (Figure 17.7).

The Eubacteria are ancient and metabolically and digestively diverse. They include species that can use inorganic molecules, organic molecules, and sunlight for energy. There appears to have been a great deal of lateral gene transfer among the Eubacteria, making eubacterial relationships uncertain, including those reported in Figure 17.7. Notable Eubacteria include

- the Thermatogales, anaerobic heterotrophs living in hot springs, and close to the base of the tree;
- the green non-sulfur bacteria, anaerobic chemotrophs and photoautotrophs also living in hot springs;
- the Flavobacteria, a large group of anaerobic heterotrophs including the *Escherichia coli* in our large intestine;
- the cyanobacteria, the aerobic photosynthesizers that oxygenated the atmosphere and currently live in oceans, lakes, and soils;

Table 17.2 Fossil and molecular evidence of origins.

Group	First fossils (Ma)[1]	Molecules (Ma)	Comments
Prokaryotes			
Cyanobacteria	3500		Warrawarra Deposit, Australia, and Onverwacht Deposit, Africa
Split between prokaryotes and eukaryotes[2]		2000	Controversial, approximate
Sex	1100–900		
Multicellular algae	1000–700		Little Dal Deposit, Canada
Coevolutionary associations			
Arthropod herbivory[3]	c.400		Early Devonian, spore-feeding, piercing
Gall formation[3]	c.300		Mid- to Late Pennsylvanian
External foliage feeding[3]	c.280		Early Permian
Insect pollination[4]	c.180		Mid-Jurassic
Plants			
Mosses	400–280		Devonian, Carboniferous
Lycopodia, ferns	400–350		Devonian
Selagellinas (Equiseta)	280–230		Permian
Ancestor of seed plants[5]		300	Late Carboniferous
Gymnosperms	350		Early Carboniferous
Cycads	280		Early Permian
Podocarps	230		Early Triassic
Araucarias, pines	210		Mid-Triassic
Yews	190		Early Jurassic
Angiosperms	75–60		Late Cretaceous, Early Tertiary
Monocot–dicot split[6,7]		200–300	Late Carboniferous to Mid-Triassic
Walnut family	75–65		Late Cretaceous
Rhododendron family	65–60		Paleocene
Grass family	65–60		Paleocene
Maples, willows	60–55		Late Paleocene
Beeches, Casuarinas, elms, Tilias	50		Mid-Eocene
Orchids	10		Pliocene
Fungi	440–400		Silurian
Split between protists and other eukaryotes[2]		1230	Controversial, approximate
Split between plants, animals, and fungi[2]		1000	Controversial, approximate
Sponges	570		Early Cambrian
Coelenterates		650	A few Precambrian, mostly Ordovician
Split between sponges and coelenterates, and bilateral animals[7]			Much before the Cambrian—perhaps several hundred million years
Mollusks			
Amphineura	570		Cambrian
Gastropods	570		Cambrian
Scaphopods	500–440		Ordovician
Bivalves	500–440		Ordovician
Cephalopods			
Ammonites	570		Cambrian, mostly Ordovician, repeated radiations, vanish in end-Cretaceous mass extinction, 65 Ma

Table 17.2 (*Continued*)

Group	First fossils (Ma)[1]	Molecules (Ma)	Comments
Nautiloids	500–400		Ordovician–Silurian
Coleoids			
Squid	400–190		Devonian–Jurassic
Octopi	150–135		Late Jurassic
Cuttlefish	65–60		Paleocene, Early Tertiary
Annelids			
Polychaetes	570–440		Cambrian, but mostly Ordovician
Oligochaetes			
Tubifex	300		Late Carboniferous
Earthworms	15		Late Tertiary
Arthropods			
Trilobites	570		Cambrian, vanish in end-Permian extinction
Chelicerates	570–440		Cambrian, mostly Ordovician
Scorpions	440–400		Silurian, invasion of land
Harvestmen (Opilionids)	350		Early Carboniferous
Ticks	400		Early Devonian
Most spiders	40		Mid-Tertiary
Crustacea	570		Early Cambrian
'Shrimp'	400		Devonian, many groups Triassic–Cretaceous
'Crabs and lobsters'	190–130		Jurassic–Early Cretaceous
Daphnia	135		Early Cretaceous
Ostracods	570		Cambrian, but mostly Ordovician
Darwinulidae		350	
Insects	400–300		Some Devonian, mostly Carboniferous
Odonates (dragonflies)	300–200		Carboniferous, but mostly Permian–Triassic
Termites	100		Mid-Cretaceous
Mantises	40		Mid-Tertiary
Orthopterans (grasshoppers)	300–100		Carboniferous, but mostly Triassic–Cretaceous
Diptera	150		Jurassic, but mostly Cretaceous–Tertiary
Mosquitoes	65		Earliest Tertiary, important disease vectors
Hymenoptera	150		Jurassic, but mostly Cretaceous–Tertiary
Honeybees	70		Late Cretaceous
Ants	150		Mid-Jurassic
Wasps, parasitoid wasps	130		Early Cretaceous
Lepidoptera	150		Jurassic, but mostly Mid-Tertiary (see plants)
Brachiopods	570		Cambrian, peak in Ordovician, relics survive
Bryozoans	500–440		Ordovician
Echinoderms	570		Many now-extinct classes in Cambrian and Ordovician
Sea lilies (crinoids)	570		Cambrian, mostly Ordovician
Most surviving classes	500–440		Ordovician
Vertebrates	500–440		Ordovician
Agnatha (jawless fishes)	500–400	564[9]	Ordovician, mostly Silurian
Placoderms	400–350		Devonian
Chondrichthyes	400–350	538[9]	Devonian
Sharks	190–135		Jurassic
Rays	190–65		Jurassic, mostly Cretaceous

Table 17.2 *(Continued)*

Group	First fossils (Ma)[1]	Molecules (Ma)	Comments
Osteichthyes	440–280	450[9]	Silurian, mostly Devonian and Carboniferous
Teleosts	230–65		Triassic, mostly Jurassic and Cretaceous
Most modern fish	65–40		Early–Mid-Tertiary
Lungfishes	400–350		Devonian
Amphibians	360	360[9]	Late Devonian
Frogs, salamanders	190–135	197[9]	Jurassic
Amniotes	350–280		Carboniferous
Turtles	190–65		Jurassic, mostly Cretaceous
Lizards and snakes	135–65		Cretaceous, many first in Tertiary
Crocodiles	230–65		Triassic–Cretaceous
Crocodile–alligator split[8]		75–55[9]	Immunological distances from albumin
Dinosaurs	230–190		Triassic, vanish in end-Cretaceous extinction
Birds	155		Late Jurassic
Loons	135–65		Cretaceous
Herons, falcons, anhingas, Avocets, parrots, rollers	65–60		Early Tertiary
Many others	50–30		Mid-Tertiary
Mammals	c.145		Jurassic
Marsupials	90–60	173[9]	Jurassic–Early Tertiary
Placentals	65–60	129[9]	Cretaceous–Early Tertiary
Rabbits	65–60	91[9]	Cretaceous–Early Tertiary
Whales	65–60	58[9]	Early Tertiary
Rodents	65–60	66[9]	End-Cretaceous–Early Tertiary
Ungulates, carnivores	65–60	83–74[9]	End-Cretaceous
Bats, dogs, weasels, elephants	40–30		Mid-Tertiary

Sources: [1]Benton (1993); [2]Doolittle et al. (1996); [3]Labandeira (1998b); [4]Labandeira (1998a); [5]Savard et al. (1994); [6]Martin et al. (1993); [7]Laroche et al. (1995); [8]Hass et al. (1992); [9]Kumar and Hedges (1998).

- the purple bacteria, anaerobic photoautotrophs that use hydrogen sulfide and are free-living relatives of mitochondria;
- the Gram-positive bacteria, heterotrophs that eat organic molecules and produce, among other things, ethanol.

The Archaea appeared after the Eubacteria and are thought to be the sister group of the Eukaryota, with which they share several traits not found in the Eubacteria. They include:

- the pink Halobacteria, living on salt flats;
- the methanogens, an abundant group and the only one that excretes methane, a fact with consequences for global change;
- the eocytes, anaerobic chemotrophs that live in hot springs;
- the ancestor of the eukaryotic nucleus.

Striking in the Prokaryota are the many groups only capable of living in extreme environments that lack oxygen and have high temperatures, low pH, or significant

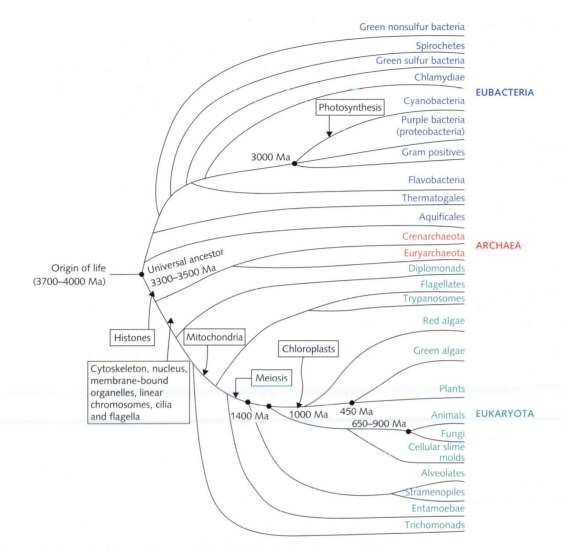

Figure 17.7 The three major clades of the Tree of Life. (Adapted from Offner 2001.)

osmotic stress. These living fossils persist in habitats that once formed the bulk of the Earth's surface but that declined in abundance with the oxygenation of the atmosphere, the lowering of the Earth's surface temperature, and the buffering of pH. Their biology hints at the conditions under which life originated, conditions that dominated for about 2 billion years.

The Eukaryotes: the basal radiation contains most biodiversity

The eukaryote radiation is immensely diverse and poorly resolved at present (Figure 17.8). The eukaryotes experienced two of the most important events in the history of life: the acquisition of symbiotic mitochondria and chloroplasts,

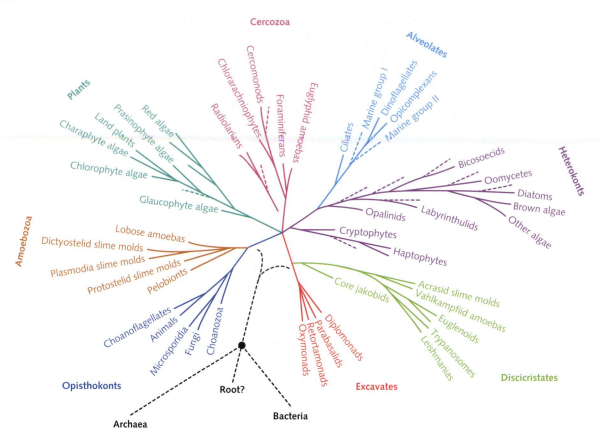

Figure 17.8 The eukaryotes are the clade with life's greatest macrodiversity. This consensus phylogeny is based mostly on the molecular phylogeny of small-subunit ribosomal sequences. The central branches depict events that occurred between 1 and 2 billion years ago. (From Baldauf 2003.)

and the origin of meiotic sex. Both major transitions increased the potential scope and rate of evolution.

Here is a brief summary of the key features of some of these groups, few as well known as is merited by their status as major branches in the Tree of Life, many with exotic genetics and cell structures.

The *opisthokonts* (*-kont* refers to flagellum) have a single basal flagellum and flat mitochondrial cristae. They include (Figure 17.9) the following.

- The *microsporidians*—unicellular intracellular parasites that once had but then lost mitochondria. Microsporidians cause diseases in both vertebrates and arthropods, including an important disease of silkworm larvae.

- The *fungi*—the biochemically and genetically talented sister group of the microsporidians.

- The *choanoflagellates*—either colorless heterotrophs or pigmented photosynthesizers. Many are fixed to the substrate by a cell stalk, and some are colonial. They look like sponge choanocytes, with whom they share ancestors.

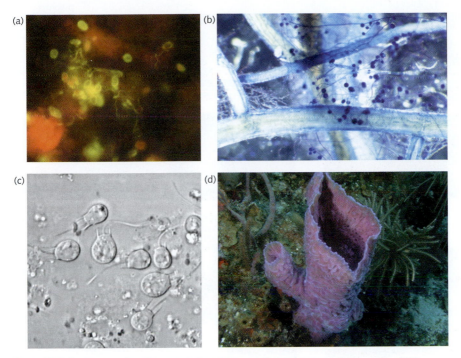

Figure 17.9 A collection of opisthokonts. (a) A microsporidian infection in the lung of an AIDS patient. (b) The arbuscular mycorrhizae of a fungus living symbiotically with plant roots. (c) Choanoflagellates. (d) A reef with a central glass sponge surrounded by cnidarian corals of various types. ((a) Photograph courtesy of DPDx: CDC's website for parasitology identification; (b) photograph courtesy of Philip Pfeffer, USDA/Agricultural Research Service; (c) photograph courtesy of Steve Paddock; (d) Photograph courtesy of Casey Dunn.)

- The *animals*—multicellular heterotrophs ranging from sponges to blue whales.

The *plants* (Figure 17.10) have the only plastids with just two membranes. All photosynthesize. They include the following.

- The *red algae*—large, multicellular organisms with a few tissue types. All reproduce sexually with separate egg- and sperm-producing organs. Life cycles are diverse; most are dominated by haploids; many alternate haploid and diploid stages. They are the source of agar, important in bacteriology and ice-cream production.

- The *green algae*—the sister group of the land plants, a morphologically diverse group of unicellular and multicellular photosynthesizers. They have symmetrical cells and complex chloroplasts. The prominent phase of the life cycle is haploid.

- The *land plants*—multicellular with diverse tissue types, they include liverworts, hornworts, mosses, ferns, and seed plants.

Figure 17.10 A collection of plants. (a) A filamentous red alga. (b) Single-celled green algae. (c) Green algae growing in the intertidal. (d) A liverwort growing on a stone. (Photo (a) courtesy of Mike Guiry at AlgaeBase, (b) and (c) courtesy of Rolf Hoekstra, and (d) courtesy of Nelson Wood, Currie Community High School, Edinburgh.)

The *heterokonts* (Figure 17.11) all have a unique flagellum decorated with hollow tripartite hairs; many also have a second, plain flagellum. They include the following.

- The *brown algae*—prominent seaweeds, nearly all marine, such as the giant kelp, which can grow to 100 m. The 1500 described species are all photosynthetic; most are sexual. Independent haploid and diploid phases alternate in the life cycle, and the haploid phases are often small and cryptic, looking like little more than splotches of tar.

- The *diatoms*—unicellular photosynthesizers that secrete silicate skeletons in the form of two nested valves. More than 10 000 species are known, some from Cretaceous fossils. They form much of the base of both the oceanic and the freshwater food chains and can remove silica from water down to a concentration of less than one part per million.

- The *oomycetes*—the water molds, white rusts, and downy mildews. They feed by extending fungus-like tubes into host tissue, excreting digestive

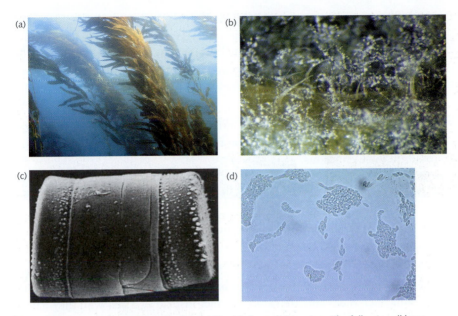

Figure 17.11 A collection of heterokonts. (a) Giant kelp, scale in meters. The following all have scales in micrometers. (b) *Phytophthera* spores on the surface of a soybean leaf. (c) Diatoms. (d) *Labyrinthula*, a slime net. ((a) Photograph courtesy of Brent Mardian, Marine Science Institute, University of California; (b) Photograph courtesy of William Fry, Cornell University; (d) Photograph courtesy of Mary Olsen, University of Arizona.)

enzymes, and absorbing the nutrients released. They have a sexual life cycle with motile zoospores. Some are important plant pathogens, including *Phytophthora*, the blight that caused the Irish potato famine. Others are ectoparasites of fish and fish eggs, often visible in home aquaria.

- The *labyrinthulids*—slime nets, colonies of cells that grow and move within self-manufactured slime tracks up to several centimeters long. They live in the ocean, in fresh water, and in moist soil. The nets can move toward a food source; digestion is extracellular; food uptake is by diffusion of small molecules through slime. Some have isogametic sex.

The *alveolates* have systems of alveoli directly beneath their plasma membranes. They include the following.

- The *dinoflagellates*—a group of several thousand species of mostly unicellular, mostly marine planktonic photosynthesizers. Many are bioluminescent, making for some magical night diving in the tropics. Some cause the red tides that produce powerful toxins in fish and shellfish. Others are farmed by reef-building corals, an association that limits reefs to the well-illuminated top 30 m of the ocean.

- The *apicomplexans*—heterotrophic unicellular microbes with a complex of fibrils, microtubules, and vacuoles at one end of the cell that gives them their name. They reproduce sexually, with alternating haploid and diploid generations. Both haploids and diploids can undergo schizogony, a type of mitosis without growth that rapidly produces many small, infective spores. They include *Plasmodium*, the pathogen causing malaria, and the coccidian gut parasites of domestic fowl and livestock.

- The *ciliates*—sexual heterotrophic unicells, covered with cilia, and having a micronucleus for inheritance and a macronucleus for RNA production. Famous ciliates include *Paramecium* and *Tetrahymena*. Marine ciliates left fossils in the Cretaceous.

Discicristates have mitochondrial cristae shaped like discs and in some cases a deep ventral feeding groove. They include the following.

- The *trypanosomes*—both free-living and the pathogens that cause sleeping sickness and Chagas' disease. They have a special, large mitochondrion, the kinetoplast, and remnants of chloroplast genes in their nucleur genome.

- The *euglenids*—motile unicells. Most have chloroplasts and photosynthesize but some lack them and are heterotrophic. Instead of a cell wall they have a flexible protein pellicle. Meiosis has never been observed.

- The *leishmanias*—pathogens transmitted by sand flies and causing skin sores, fever, weight loss, and enlarged spleen and liver.

The *Amoebozoa* are mostly naked amoebae without shells that move with pseudopodia. They include these groups.

- The *dictyostelid slime molds*—mobile unicells in fresh water, in damp soil, and on rotting vegetation, where they feed on bacteria. Their life cycle includes a stage in which independently moving and dividing amoebas feed separately, then aggregate into a single multicellular mass, called a slug, that transforms into a fruiting body bearing spores.

- The *plasmodial slime molds*—their feeding stage is a true plasmodium with many nuclei in a single, large cytoplasm, and unlike cellular slime molds they have the option of sexual outcrossing, although they do not always take it.

- The *lobose amoebas*—the familiar amoebas. They are highly mobile and surround and engulf food particles. Some form shells or tests that resemble Proterozoic fossils.

The *cercozoa* (Figure 17.12) are amoebae with thread-like podia that usually live in tests, or hard outer shells, that can be very complex. They include these groups.

- The *foraminifera*—single-celled amoebae living in shells ranging in size from 100 μm to 20 cm. Large forams farm algae within their cytoplasm; others eat bacteria, algae, and even copepods. They are useful indicator fossils; specific forms characterize geological eras since the Cambrian.

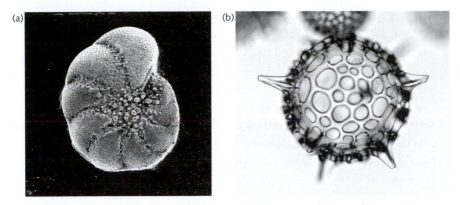

Figure 17.12 Two cercozoans. (a) A Holocene foraminiferan from California. (b) A Miocene radiolarian from Antarctica. Scale in micrometers. ((a) Photomicrograph by Mary McGann. Courtesy of the U.S. Geological Survey; (b) photograph courtesy of David Lazarus, Museum für Naturkunde Berlin.)

- The *radiolarians*—single-celled amoebae living in complex, beautiful silicate skeletons; mostly planktonic; filter feeders or predators. Like the forams, they have left a detailed fossil record.

Long as it is, this list of major basal groups of eukaryotes is incomplete. Basal eukaryote phylogeny is still in flux; for a recent view see Cavalier-Smith and Chao (2003).

Multicellularity, parasitism, amoebae, and fungal forms evolved several times in eukaryotes

The base of the eukaryotes is a huge, diverse universe of forms, life cycles, cell structures, and biochemistries. It does not fall neatly into the three traditional groups of plants, animals, and fungi. The terms algae, fungi, and amoebae are all misleading because they refer to groups that are polyphyletic. Long before the Cambrian, long before the first large deposits of hard-bodied fossils, the eukaryotic radiation was in full bloom, producing many types of sexuality, many life cycles, repeatedly and independently developing parasitic forms out of free-living forms, independently inventing multicellularity several times, and exploring the biochemical and morphological potential of the single cell.

The animals invent multicellular development with cell–cell interactions

The animal radiation (Figure 6.5) began with sponges evolving from choanoflagellates deep in the Precambrian. Fossil sponges from the Vendian about 630 Ma already coexisted with organisms that could have been cnidarians. Molecular evidence suggests that the sponges originated 700–900 Ma.

Key issues include the origin of cell layers, body cavities, mouth and anus, and segmentation

The following major steps were all taken before fossils of organisms with hard parts appeared in the Cambrian.

- Sponges have two important innovations in control over the extracellular environment: collagen for binding cells in an extracellular matrix, and two cell layers separated by a gelatinous extracellular mesenchyme.

- The cnidarians and ctenophores mark the earliest appearance of multicellular development with the formation of a well defined endoderm and ectoderm. Because they are characterized by these two basic cell layers, they are called the diploblasts. Some evidence suggests that the diploblasts are secondarily derived from a triploblastic state.

- The flatworms are the oldest group that clearly has a third basic cell layer, the mesoderm, which together with the endoderm and ectoderm forms the triploblastic state. The digestive, circulatory, and excretory systems are built from mesoderm.

- Body cavities and a circulatory system with blood originated after the flatworms branched off.

- Also in place before the Cambrian were two main groups of complex animals, the *protostomes*—whose mouth derives from the opening formed in gastrulation—and the *deuterostomes*—whose anus derives from that opening.

- Segmentation is ancestral to all the protostomes except the round worms and arrow worms.

Figure 6.5 represents only the groups with living representatives; some major clades may have disappeared without trace in the Vendian and Cambrian. If they were included, the phylogenetic tree of animals would be broader at the base and narrower at the top. The most striking feature of the tree is the Cambrian Explosion—the appearance of all major types in a brief period about 525 Ma, with no major types evolving since.

The relationships depicted represent major shifts of opinion driven by recent analysis of DNA sequences. The arrow worms are the sister group of the nematodes, not the cousins of the deuterostomes. The arthropods are the sister group of the velvet worms, not a branch of the annelid clade, segmentation being ancestral to all of them. Other relationships confirm earlier opinion. The phoronids (marine worms), brachiopods (lamp shells), and bryozoans (moss animalcules), all of which share a distinctive feeding appendage, the lophophore, still group together, and the echinoderms remain the sister group of the chordates and hemichordates.

Diploblastic: *Having two primary cell layers.*

Triploblastic: *Having three primary cell layers: endoderm, mesoderm, and ectoderm.*

Gastrulation: *Process by which cells of the blastula, a hollow ball, are translocated to new positions in the embryo, in part by invagination, producing the three primary cell layers.*

The land plants evolved alternation of generations, vasculature, wood, leaves, seeds, and flowers

Land plant phylogeny differs dramatically from animal phylogeny in several important ways (Figure 17.13). There was no Cambrian Explosion. The major groups emerged erratically in time, not all at once. The major innovations in land plant evolution include the following.

- Mid-Ordovician—origin of land plants, the nonvascular bryophytes, with *alternation of* sexual and asexual *generations* (Figure 17.14). In these groups the large, obvious individual is the haploid gametophyte; in mosses the diploid sporophyte grows out of the gametophyte. We see the sporophyte as the reproductive structure.

- Mid-Silurian—origin of vascular plants with the sporophyte dominant. Vascular tissues to move water and nutrients enabled the plants to move on to dry land and to grow large. The first group to possess such tissues were the Rhyniophyta, named for the Rhynie Chert in Scotland. They looked like the plant in Figure 17.14(b).

- Lower Devonian—origins of heterospory (large and small spores), allowing dryer environments to be invaded, and simple leaves.

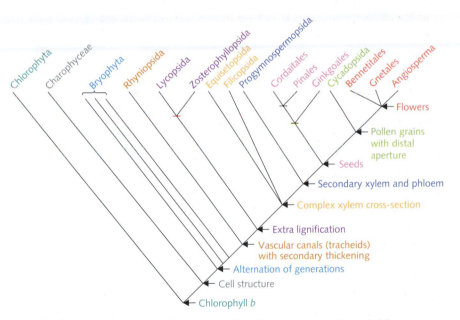

Figure 17.13 Plant phylogeny. The plants emerged from the green algae at the end of the Ordovician or the beginning of the Silurian about 435 Ma. Club mosses, horsetails, and various fern-like plants radiated in the Devonian. They were joined in the Carboniferous by cycads, ginkgoes, and early conifers, which formed the forests in the Permian, Triassic, and Jurassic. The flowering plants emerged in the Cretaceous and dominated plant communities in the Cenozoic. (Reproduced from Benton and Harper (1997) *Basic Paleontology*, Pearson Education Limited, with permission.)

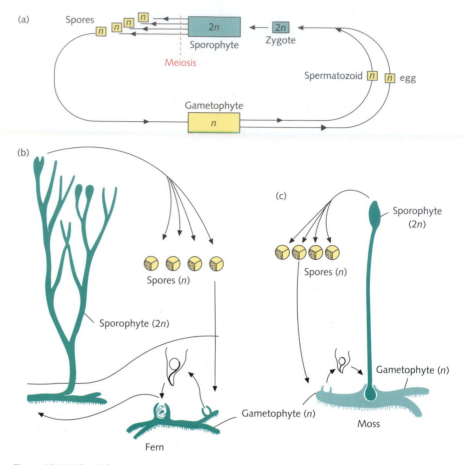

Figure 17.14 Plant life cycles (a) are all characterized by an alternation of haploid and diploid phases. Here green = diploid, yellow = haploid. What varies is the relative prominence of each phase. In mosses (c), the haploid gametophyte is the prominent individual, and the diploid sporophyte grows out of the gametophyte, forming a structure that looks like the reproductive organ of the moss. It is in fact a separate generation. In tracheophytes the diploid sporophyte is the prominent individual (b), and the haploid gametophyte is reduced in size and complexity. In angiosperms the male gametophyte has been reduced to the pollen tube. (Reproduced from Benton and Harper (1997) *Basic Paleontology*, Pearson Education Limited, with permission.)

- Middle Devonian—origin of things that looked like trees, made possible by the evolution of wood. The development of wood involves the targeted secretion of lignin, which acts as a resin, to produce a strong fiber-resin composite material.
- Upper Devonian—origin of compound leaves.
- Uppermost Devonian—origin of seeds.
- Early Cretaceous—origin of *flowers* (Figure 17.15) in the ancestors of angiosperms, gnetales, and bennettitales. The gnetales include *Welwitschia*, *Gnetum*, and *Ephedra*; the bennettitales are represented by *Williamsonia*, a

(a) *Williamsonia*

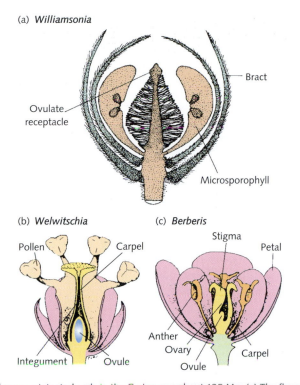

(b) *Welwitschia*   (c) *Berberis*

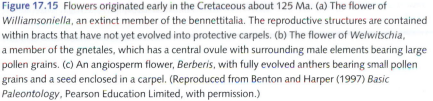

Figure 17.15 Flowers originated early in the Cretaceous about 125 Ma. (a) The flower of *Williamsoniella*, an extinct member of the bennettitalia. The reproductive structures are contained within bracts that have not yet evolved into protective carpels. (b) The flower of *Welwitschia*, a member of the gnetales, which has a central ovule with surrounding male elements bearing large pollen grains. (c) An angiosperm flower, *Berberis*, with fully evolved anthers bearing small pollen grains and a seed enclosed in a carpel. (Reproduced from Benton and Harper (1997) *Basic Paleontology*, Pearson Education Limited, with permission.)

Mesozoic fossil. The most primitive extant angiosperm is probably *Amborella*, a rare shrub found only in New Caledonia.

After their origin about 125 Ma, the angiosperms radiated massively into their current 250 000 species; only 550 species of conifers remain. Angiosperms enclose their seeds in an ovary that protects them from fungi, insects, and desiccation. The evolution of the flower involved precise developmental control over leaf homologues and reproductive structures to produce a better-nourished and better-protected seed. That control was achieved through the precisely targeted expression of MADS genes (Chapter 6, pp. 143 ff.).

The fungi have multinucleate cells, special reproductive structures, and associate with plants

Fungi are more closely related to animals than they are to plants. It appears that the terrestrial fungi diverged from the chytrids at the base of the Cambrian

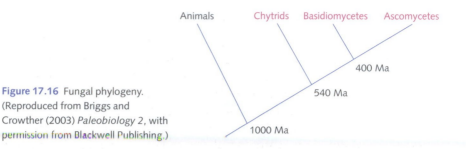

Figure 17.16 Fungal phylogeny. (Reproduced from Briggs and Crowther (2003) *Paleobiology 2*, with permission from Blackwell Publishing.)

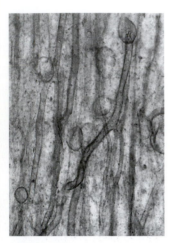

Figure 17.17 Hyphae and spores of a fungus within the tissue of a plant stem from the Lower Devonian, Scotland, 400 Ma. (From Briggs and Crowther 2003.)

Mycorrhizal association: *A symbiotic association between a fungus and a plant root; nutrients extracted from the soil by the fungus are transported into the plant.*

540 Ma, and that ascomycetes split from basidiomycetes about the time that plants invaded land in the Silurian 400 Ma (Figure 17.16). The first credible fossil fungi come from the Devonian and Carboniferous, where they appear to have been decomposers feeding on dead plant material (Figure 17.17). A Carboniferous fungus, *Palaeancistrus*, has hyphae forming a mycelium and structures characteristic of living basidiomycetes. Post-Carboniferous fungal fossils are often found in association with plant fossils, either as parasites or as mycorrhizae.

- The chytrids live in fresh water or soil, and like the higher fungi they form hyphae. They are sexual and have motile gametes. A chytrid causes brown-spot disease in maize.

- The basidiomycetes are the mushrooms, rusts, smuts, jelly fungi, puffballs, and stinkhorns. Their distinctive reproductive structure, the *basidium*, bears spores. This large radiation of about 25 000 species contains important plant pathogens as well as fungi eaten by humans.

- The ascomycetes—the yeasts, bread molds, morels and truffles—form a clade of more than 10 000 species. All form an *ascus*, a reproductive capsule, when the hyphae of compatible mating types fuse. Some form **mycorrhizal associations** with flowering plants; others are plant pathogens. Many aid in the production of beer, wine, and cheese. Others produce antibiotics, including penicillin.

Groups expanding, vanishing, or gone

Some groups of plants, insects, and mammals are still expanding

The process of evolution did not stop yesterday so that we could examine it: it continues all around us. Some groups are in the middle of a continuing expansion: the Coleoptera (beetles), Diptera (flies, mosquitos), Chiroptera (bats), Rodentia (mice, rats, squirrels), Asteraceae (sunflowers, asters), Orchidaceae (orchids), Poaceae (grasses), and the emergent viral diseases (influenza, SARS, Ebola).

Most species and many clades are extinct

Many other groups have vanished or are vanishing. Most of the large Pleistocene mammals outside Africa are gone, including most of the hominids. Most elephants and their relatives have vanished. In the last 12 000 years 25–35% of the world's birds have gone extinct, most of them on islands in the Pacific, probably because of human hunting. Other entire large clades vanished long ago—the trilobites at the end of the Permian, the tongueferns at the end of the Jurassic, the ammonites and dinosaurs at the end of the Cretaceous, the South American notoungulates when North American carnivores invaded about 10 Ma.

On the geological timescale major groups expand and contract, dominant types vanish and are replaced. Nothing is permanent. The world with which we are familiar is only one of many states in which this planet has found itself since life appeared.

Vanished communities and extraordinary extinct creatures

The Burgess Shale reveals the Cambrian Explosion 520 Ma

The most famous Cambrian Lagerstatt is the Burgess Shale in British Columbia, Canada. It formed 520 Ma when the western coast of North America was near what is now the border between British Columbia and Alberta. A rich community of diverse invertebrates inhabited a well-lit, shallow, muddy-bottomed shoreline near an underwater cliff. Occasionally the unstable mud slumped down the cliff, carrying with it the organisms that lived in and on it and depositing them into an anoxic environment, where they died quickly and were well preserved.

Lagerstatt: *German for resting place; plural Lagerstätte. A particularly complete fossil deposit representing much of an ancient community, often with exceptional preservation of detail.*

The Burgess Shale reveals an unfamiliar world. While some of its creatures have changed little since then, others looked like no living organism, including some strange creatures that have left no descendants and cannot be placed into any extant group. They represent vanished major body plans. Among the recognizable forms in the Burgess community are algae, sponges, lamp shells (brachiopods), animals with lophophores like those of brachiopods and phoronids, sea anemones, possibly ctenophores, polychaete worms, priapulids, arthropods, among them many trilobites, velvet worms, echinoderms, and a chordate.

Among the strangest Burgess denizens was *Opabinia regalis* (Figure 17.18), a predator up to 7 cm long. It had five eyes, a remarkable extensible feeding appendage, a flexible segmented body, and three pairs of upwardly pointed plates on its hind end. Nothing like it survives today. It would be more readily comprehensible if its feeding appendage were paired instead of singular. Multiple eyes are still found in spiders; many arthropods have flexible, segmented bodies; none have a single, central proboscis. In the Cambrian basic body plans that could never be inferred from organisms alive today originated and disappeared.

Figure 17.18 *Opabinia regalis*, a segmented predator with five eyes, a single central feeding appendage that brought food to a backward-facing mouth, and three pairs of upward-facing plates on its hind end. (Illustration by Marianne Collins from The Book of Life by Stephen Gould, published by Ebury Press. Reprinted by permission of The Random House Group Ltd.)

Messel records a European community 49 Ma with surprising species

A Lagerstatt in the oil shales at Messel, in Germany (Schaal and Ziegler 1992), reveals the community of a mid-Eocene (49 Ma) lake with an anoxic bottom layer. Organisms that fell into the anoxic layer were preserved virtually intact, including soft parts and stomach contents. Two finds at Messel are particularly surprising. Many of the plants belong to tropical families, for example the screw pines (Pandanaceae), which are currently confined to central and west Africa, south-east Asia, Australia, New Zealand, and Polynesia. And a South American ant-eater (Figure 17.19) from Messel suggests that central Germany was then sub-tropical or tropical. Plants now found only in the African and Asian tropics and animals now found only in South America were then living together in western Europe.

The message of Lagerstätte: things might have turned out very differently

Lagerstätte are perhaps the clearest evidence we have that the particular world we live in is only one of many possibilities, that many things that we take for granted are fairly arbitrary. Much might well have turned out differently.

Stasis may be selected

> ● **KEY CONCEPT**
> Evolutionary change is often not constant; there may be long periods when nothing happens, and short periods when a lot happens.

One of the most striking patterns in the fossil record is stasis—lack of morphological change over long periods of time. Every living fossil is an example of stasis. The following list of examples contains dates in parentheses indicating how long ago we can find evidence in the fossil record of organisms that closely resemble the living ones mentioned. Among the living

Figure 17.19 A tamandua, a South American ant-eater, from the 49 million-year-old Lagerstatt at Messel, Germany. (Photo courtesy of Senckenberg, Messel Research Department, Frankfurt a. M. (Germany).)

fossils are priapulids and oncychophorans (Cambrian, 530 Ma), liverworts (Ordovician, 500 Ma), club mosses and lungfish (Devonian, 400 Ma), and cycads (Permian, 270 Ma). Hundreds of other species have not changed significantly over tens of millions of years. Their long persistence without significant change is puzzling. One explanation is that stasis results from stabilizing selection on adults as a byproduct of precise habitat selection by larvae. By seeking out habitats that may shift in space, dispersing larvae track an environment that stays fairly constant for the adults that grow up in it. This explanation is particularly plausible for the marine invertebrates with larval stages that dominate the fossil record and provide some of the best evidence for stasis. There are other explanations for stasis, but they are not yet well tested. Stasis leads us to another striking feature of the fossil record, punctuation.

> **Stasis:** *A long period without evolutionary change. If the organisms involved survive to the present, we call them living fossils.*

Punctuational change is real but not universal

Eldredge and Gould (1972) observed that in many fossil lineages long periods of stasis were broken by brief periods of rapid change, and that these periods of rapid change—punctuations—seemed to be associated with apparent speciation

> **Punctuation:** *A short period of rapid change breaking a long period of stasis in the fossil record.*

events. They thought that most morphological change occurred during speciation and that during the rest of their existence most species did not change very much. The pattern to which they called attention is real, but it is not as general as they claimed. Many characters, and some species, are static for long periods, some of them lasting tens to hundreds of millions of years. The fossil record of many lineages is marked by brief periods of major change in which many new species appear and a great deal of morphological change occurs. But not all traits and lineages show this pattern of stasis and punctuation, and only in some lineages does most morphological change occur during or soon after speciation events.

Some groups—such as Pleistocene corals in New Guinea and bryozoans in tropical America—display stasis over several million years broken by near-simultaneous speciation events associated with major climatic change. Some groups—such as mollusks in Lake Victoria—display morphological change during speciation events and stasis between speciation events. Jackson and Cheetham (1999) found that 29 of 31 species with well-documented fossil histories displayed punctuated morphological change associated with cladogenesis. Other groups, including rodents, change as much between as during speciation events. And it is difficult to separate speciation in time from ecological replacement in space in a fossil record that is usually quite patchy in space even when it is fairly continuous in time.

Cope's Law—things get bigger—can be explained by either drift or selection

Another striking pattern in the fossil record is the tendency for the largest members of a clade to get larger over time. The invertebrates of the lower Cambrian were measured in millimeters and centimeters; those of the Permian were measured in meters. The first dinosaurs were the size of house cats; the last were the largest land animals that ever lived. The first mammals were the size of mice; today elephants and whales are the upper limits on land and sea. Two explanations have been offered for this trend, which is referred to as Cope's Law.

The *upwards drift hypothesis* notes that at the origin of a new group, all species in it are usually small, perhaps because new groups usually originate after mass extinctions have wiped out most of the large organisms. Subsequent changes in body size could be random. The lower limits would be rapidly reached, but the upper limits would be much farther away, and so as size changes at random, the largest members of a clade will tend to get larger in time. They drift upward. This idea is a useful null hypothesis, but it has not yet been possible to test it rigorously.

The *adaptive life-history hypothesis* notes that body-size evolution can go either up or down depending on the balance of costs and benefits of early versus late maturation and of growth versus reproduction. Coevolution will shape predators to improve their predation and prey to escape it. One way to become a better predator is to increase in body size. One way to escape predation is also to increase in body size. If the trend is towards higher mortality rates on smaller adults, and lower mortality rates on larger adults, life-history evolution may increase body sizes in both predators and prey. Evidence from changes in body sizes when animals colonize islands that lack predators supports this notion (Palkovacs 2003). When elephants colonized islands in the Mediterranean and were released from predation by lions and hominids, they rapidly evolved small body size, becoming no more than chest high. Such reduction in body size may have been driven by intraspecific competition for limited food: a smaller elephant can mature earlier and produce offspring that need less food to survive and mature than can a large elephant.

Thus some evidence supports the idea that long-term changes in body size seen in the fossil record are a byproduct of selection forces acting on life histories. One balance is between getting larger to resist predation and getting smaller to compete more effectively for limited food.

Evolution does not make progress; it simply continues to operate

An overview of evolution suggests that the most complex things in existence have gotten more complex through time. And it would certainly be satisfying to think that things that arrived late and are large and complicated, like ourselves, are somehow better than things that emerged earlier and were simpler and smaller. But there is little evidence of progress in evolution. A few remarks show just how misleading that notion can be:

- The Eubacteria and Archaea have existed for 3 billion years and remain among the most abundant organisms on Earth without much change in their basic biology.

- Most of the great Tree of Life consists of early branches, the great diversity of prokaryotes and basal eukaryotes, all of which have persisted to the present. At that scale, the more-complicated organisms, such as animals, are minor shoots.

- The dinosaurs were mighty impressive—large, complicated, diverse. They persisted for 100 million years, far longer than hominids have been in existence. If dinosaurs had been sentient, they might have spoken of progress and considered themselves the pinnacle of creation. They would have had better evolutionary reason to do so than we do. Yet they vanished completely 65 million years ago.

- The hominid radiation had more representatives living simultaneously in the past than it does at the present. In the early Pleistocene, five or six hominid species coexisted on the planet. That number declined; we are the only survivors of a once speciose lineage. If we were to give a prize for current radiation, we would not give it to hominids. We would give it to beetles, rodents, orchids, grasses, or viruses.

- Could we use probability of extinction to measure progress in evolution objectively? There is little evidence that the probability that a species will go extinct has changed significantly since the Cambrian; 1 to 10 million years is a good estimate of the duration of a species throughout the last 540 million years.

Other arguments can be laid on the table, but these should suffice to make the point. All we can say about the success of species is that some survived longer than others. Eventually all species join the more than 99.9% that have already gone extinct. To speak of progress in a system in which mass extinctions occur at long intervals, in which background extinctions are normal, in which increasing complexity has not brought with it decreased likelihood of extinction, seems inappropriate. The word progress is loaded with implicit cultural assumptions that do not help us to find and may distract us from the true nature of the evolutionary process. Evolution does not have a purpose. It does not have a goal. It simply is.

● SUMMARY

This chapter considered the history of life on Earth from four perspectives: the geological eras and what happened in them; the radiations of the major clades with emphasis on when they acquired their key features; insights from two ancient communities, Burgess and Messel; and major patterns in the entire history of life—stasis, punctuation, Cope's Law, and progress.

- The scope of geological time and the space encompassed by the planet's surface are the scale on which evolution has occurred. We must get familiar with deep time to understand macroevolution.

- The major divisions of geological time are the Archaean (4600–2500 Ma), the Proterozoic (2500–540 Ma), and the Phanerozoic (540 Ma–present).

- The Phanerozoic is divided into the Paleozoic (540–251 Ma), the Mesozoic (251–65 Ma), and the Cenozoic (65 Ma–present).

- Prokaryotes originated in the Archaean, eukaryotes in the Proterozoic, the major animal clades in the Cambrian, the major land plant clades from the Silurian through the Cretaceous, and the major fungal clades in the Cambrian and Silurian.

- There were major mass extinctions in the Permian and Cretaceous as well as smaller mass extinctions in several other eras.

- The radiation of the basal eukaryotes in the Proterozoic produced much of the fundamental diversity of life, including many quite different things inappropriately grouped as algae, fungi, and amoebae.

- The radiation of the animals, mostly in the Vendian and Cambrian, produced about 35 body plans that have persisted to the present. Two key innovations that made possible the animal colonization of land were the tetrapod limb and the cleidoic egg.

- In both plants and animals the evolution of multicellularity involved increased control over the extracellular environment—collagen and lignin being just two of many important innovations.

- Plants did not emerge suddenly, as did animals, in an explosive radiation. Their major groups appeared sequentially over several hundred million years. Their key innovations include alternation of generations, conductive tissue, apical meristem, roots, wood, the separation of xylem and phloem, seeds nourished and protected by parental tissues, and flowers.

- The fungi share with animals choanoflagellate ancestors. Key events in fungal evolution include the acquisition of hyphal organization with major changes in cell structure, the development of myorrhizal associations with plants, the origin of the basidium in basidiomycetes and of the ascus in ascomycetes, and the biochemical evolution that has permitted fungi to manipulate their environment with antibiotics and to digest cellulose.

- Some clades are currently expanding; others are currently contracting; many have vanished.

- The extraordinarily well-preserved communities of Lagerstätte suggest that ours is but one of many possible worlds, that things might well have turned out differently.

- Stasis is a prominent feature of the fossil record of many clades. It may result from stabilizing selection.

- Punctuation is also common in the fossil record, but it is not always associated with speciation, and it is difficult to separate speciation in time from ecological replacement in space in a fossil record that is quite patchy in space even when it is relatively continuous in time.

- The largest members of a clade do tend to get larger in time. One explanation is that they start small and simply drift randomly upward. Another is that body size is driven by coevolutionary interactions with predators balanced by intraspecific competition for food.

- There is no progress in evolution and no goal to attain. Evolution simply exists; it is not going anywhere in particular.

RECOMMENDED READING

Briggs, D.E.G., Erwin, D.H., Collier, F.J., and Clark, C. (1995) *The fossils of the Burgess Shale*. Smithsonian Institution Press, Washington.

Gould, S.J. (ed.) (2001) The book of life: an illustrated history of the evolution of life on earth. W.W. Norton, New York.

Margulis, L. and Schwartz, K.V. (1982) *Five Kingdoms: an illustrated guide to the phyla of life on earth*. W.H. Freeman & Co., San Francisco.

Schaal, S. and Ziegler, W. (1992) *Messel: an insight into the history of life and of the earth*. Oxford University Press, Oxford.

● QUESTIONS

17.1. What do the fossil record and molecular systematics tell us about when the HOX genes acquired their roles in organizing the body axis? What might they have been doing before that?

17.2. Similarly, what do the fossil record and molecular systematics tell us about when the MADS genes acquired their roles in organizing the flower? What might they have been doing before that?

17.3. What do the Scala Natura and the Great Chain of Being have to do with the notion of progress in evolution?

17.4. What important points about evolution would we never have known without fossils?

17.5. Both the malaria pathogen (an apicomplexan) and the sleeping sickness pathogen (a trypanosome) have chloroplast genes in their nuclei although they do not photosynthesize. What evidence do we have from the Tree of Life that some of their relatives are photosynthetic (see Figure 13.2).

PART 5

Integrating micro- and macroevolution

Several connections between micro- and macroevolution have already been made. Microevolutionary genetic drift within populations allows molecular systematists to reconstruct the macroevolutionary Tree of Life from DNA sequence data. Deeply conserved developmental control genes help to build the framework within which microevolution and phenotypic plasticity fine-tune the phenotypes of individual organisms. Microevolutionary processes produce speciation as a byproduct, generating macroevolutionary biodiversity.

We now take a deeper look at two of the many issues in which micro- and macroevolution are thoroughly intertwined: coevolution and evolutionary medicine. The same could be done for the evolution of life histories, sex, sexual selection, and other major themes.

Coevolution, in which two or more species interact with continued reciprocal change, introduces both an intricate dynamic and additional contingency to the history of life. This game of move and counter-move leads in some cases to intimate, peaceful, mutualisms, in other cases to escalating arms races between predators and prey, hosts and pathogens.

In evolutionary medicine the macroevolutionary context describes the origin and spread of *Homo sapiens* across the planet, the histories of contact of diverging populations with different diseases in different places, and the local evolution of differential susceptibility to drugs and disease. The microevolutionary processes relevant to medicine include the evolution of antibiotic resistance, of pathogen virulence, of selective abortion, and of mate choice for resistance genes.

Major issues are then highlighted in the essay of Chapter 20 with which the book concludes.

CHAPTER 18
Coevolution

Cleaner wrasses, the fishes they clean, and saber-toothed blennies

A large grouper approaches a cleaning station on a coral reef; a group of small brightly colored and strikingly patterned cleaner wrasses swim toward it with an unusual undulating motion—the so-called cleaning dance. The grouper adopts a rigid posture and opens its mouth, around and through which the wrasses swim, picking parasites off the skin, gums, and gills. The grouper, an impressive predator, does not eat the wrasses, instead remaining rigid with its mouth open while the wrasses glean ectoparasites from its body (Figure 18.1). Before the wrasses finish, another fish, about the same size as the wrasses, with a similar color pattern and unusual undulating motion, approaches the grouper. Instead of gleaning parasites, however, it bites a chunk out of the grouper and flees. This second fish is a saber-toothed blenny, an aggressive mimic that closely resembles cleaner wrasses in size, shape, coloration, and behavior. The blenny exploits the mutualism between the wrasse, the grouper, and the many other fish

● **KEY CONCEPT**
Coevolution involves reciprocal change in interacting species.

(a) (b)

Figure 18.1 (a) Tiger grouper being cleaned by several cleaner wrasses, Little Cayman Island, Caribbean. (Photo courtesy of David Matthews.) (b) Closeup of a saber-toothed blenny. (Photo courtesy of Robert Fenner.)

that the wrasse cleans. The blenny resembles a treacherous criminal: it takes selfish advantage of a community service, thereby threatening the stability of the service by directly harming a participant while pointing the finger of blame at the wrong party.

Co-evolution has molded these species into tightly specialized interactions involving striking coloration, unusual behavior, and puzzling consequences. How can the grouper benefit so much from being cleaned that it pays not to eat the wrasse? How did the coloration and behavior of the wrasse originate? Would not the first wrasse that tried to clean another fish simply be eaten? How can the aggressive mimicry of the blenny remain stable? Would its presence not erode the stability of the interaction between the wrasse and the grouper? And are such specialized mutualisms really typical products of the coevolutionary process? Such questions suggest some of the many intriguing puzzles posed by coevolution.

The types of coevolution

Coevolution is a process of reciprocal evolutionary change in interacting species (Thompson 1994). It moves the issue of specialization to center stage, for spectacular examples of specialized interactions—like the wrasse, grouper, and blenny—first attracted attention to the process. However, coevolution can also influence more diffuse interactions—for example, between guilds of many species of frugivores and clades of many species of plants, such as the guild of birds and mammals that eat figs and the many fig species that inhabit a single tropical forest, set fruit at different times of year, and provide a reliable source of food throughout the year. When will specialization evolve, when will interactions remain, or become, diffuse? Answering this question remains a challenge that is discussed further below.

While some coevolutionary interactions are specialized mutualisms, many are not. Interactions can be classified by their impact on the fitness of the partner species, as follows.

- ++: mutualisms, including cleaner wrasse–client fish, plant–pollinator, plant–seed disperser, insect–domesticated fungus, and Müllerian mimicry: both partners benefit.

- +−: parasite–host, pathogen–host, predator–prey: the parasites, pathogens, and predators benefit (+); the hosts and prey suffer (−).

- −−: competition: both species suffer from the presence of the other, as has been demonstrated within grasses, salamanders, and barnacles and between ants and rodents eating desert seeds.

Inquiline: *An animal living in a space provided by another; a guest.*

- 0+: commensalism, including **inquilines**: one partner benefits from living in close contact but has little or no impact on the other partner, such as scale worms (*Arctonoë vittata*) living in keyhole limpets (*Diodora aspera*)—the worm gets shelter, the impact on the limpet is obscure.

- 0−: one partner suffers but the interaction has no impact on the other partner, as is the case with many species now forced to live in contact with humans— they suffer reduced fitness, our fitness does not change.

Usually only the non-zero interactions, where evolutionary change is reciprocal, are considered coevolutionary. We consider the non-zero interactions to coevolve in the narrow sense, and the interactions containing a zero to coevolve in the broad sense. Broad-sense coevolution means that one species has evolved in response to changes in another species even though that second species has not responded to changes in the first. In this chapter we concentrate on narrow-sense coevolution, having noted here that coevolution can be defined more broadly.

The interaction signs can be hard to determine and controversial, for measuring the fitness consequences of an interaction in nature is not often easy. Interactions are often mixtures of positive and negative effects, and the methods used for measuring the signs and intensities of interactions may be biased towards the positive or towards the negative effects. It is therefore wise to remain skeptical about the nature of an interaction until it has been viewed by several methods from different angles.

Whether we recognize a coevolutionary interaction as tight or diffuse, specialized or generalized, is determined by the frequency and intensity of the interaction with the partner. Interaction frequency is determined by how often the two species encounter each other in space and time. Interaction intensity is determined by impact on the reproductive success of the two species, which can range weak to strong and from symmetrical to highly asymmetrical. The interaction may occur throughout the entire life cycle of both species, or only at particular stages of one or both. This in turn affects its intensity and the opportunity for specialization to evolve.

The plan of this chapter is next to illustrate the range of coevolution by discussing the spatial and temporal scales and the biological levels at which it can occur. Then we review the principles that govern coevolutionary interactions and some striking results of coevolution, including what happens when naïve hosts and prey encounter predators and pathogens with which they have not coevolved. We conclude the chapter by discussing the criteria we can use to judge a claim of coevolution, and how coevolution illustrates the interaction of micro- and macroevolution.

Scales of coevolution

Spatial: the geographic mosaic

In tight interactions, the geographical distributions of the partners overlap completely: both species are always found with each other. Such interactions are the product of an evolutionary process and do not represent either the conditions

● **KEY CONCEPT**

For an interaction to become specialized, the partners must occur at the same place, at the same time, at the appropriate stages in their life cycles.

under which interactions in general start to evolve or the majority of coevolving interactions in nature. Interspecific interactions differ among populations in frequency, symmetry, and intensity—and thus in impact and outcome—for many reasons. As a result, the interaction will affect the evolution of both partners in some populations, only one partner in others, and neither partner in the rest. In some places only one partner may be present and the interaction will not exist at all; in others both may be present and interacting intimately. In some populations the interaction may be specialized, in others, generalized. Among some populations gene flow is strong; among others, weak or nonexistent. Thus interspecific interactions—at least initially, and often for a long time—exist in a geographic mosaic. They are coevolutionary vortices in an evolutionary stream (Strong et al. 1984). This has some important consequences (Thompson 1994). (1) The overall coevolution of one species may be the product of interaction with several partners, even though the interaction is restricted to only one partner in any one population. (2) Character changes in some populations—coevolutionary hot spots—will dominate the coevolutionary response; other populations will contribute little or nothing. Therefore observations limited to only one or a few populations do not allow us to conclude that coevolution is probable or improbable, strong or weak—for that, we need a reliable sample of the overall pattern. (3) In many cases the coevolutionary response within the species as a whole will shift back and forth as the contributions of the various populations change over time. An escalating series of adaptations and counteradaptations may occur, but as a special case, not as the inevitable outcome.

If species start to coevolve diffusely in a geographical mosaic, then how can we account for the eventual production of specialized interactions? Here is one answer. Where local interactions are strong, symmetrical, and mutually beneficial, a series of reciprocal trait changes will occur to produce local coevolved adaptations. Once that has started to happen, there will be selection for assortative mating with individuals who interact well with the partner species and selection against mating with individuals, either locally born or recently arrived from other populations, who interact poorly with the partner species. Even with a fair amount of gene flow there is likely to be some steady improvement in selection of mates who perform well in the interaction. This process of ecological speciation, sketched in Chapter 12 (pp. 290 ff.), could result in the local speciation of both partners—cospeciation—that would cut off gene flow from the rest of the geographic mosaic and accelerate the evolution of the specialization of the interaction. Not all specialized interactions must have involved cospeciation, but in those that have, the mutual isolation of both interacting species from other gene pools will have made specialization easier and carried it further than would have otherwise been possible.

Temporal

To interact directly, as is the case with most coevolutionary interactions, organisms not only have to occur in the same place—they also have to be active in the same season of the year and the same time of day. Different plant species flower and set fruit at different times of year in the same habitat; different insect species emerge to lay eggs, eat leaves, and pollinate flowers at different times of year in the same habitat. For an interaction to occur and eventually lead to specialization, the seasonal patterns in the traits involved in the interaction must overlap at least to some extent at the start.

In most habitats there is a day shift of diurnally active organisms and a night shift of nocturnally active organisms. At the time the shifts change—dawn and dusk—other, crepuscular organisms, such as bat hawks, may be active. For example, flowers pollinated by birds, flies, and butterflies are usually open during the day and often closed at night; flowers pollinated by bats and moths are usually open at night and often closed during the day. In oceans and lakes planktonic organisms make regular diurnal migrations from deep to shallow water and back again; their predators follow suit; their parasites move with them. The biomass of marine organisms moving daily between deep and shallow water is in the billions of tons.

In many interactions co-occurrence in space and time involves coordination of specific life-history stages. Organisms with complex life cycles often change habitat with each stage in the life cycle, and each stage also usually occurs at a particular season, sometimes, for short-lived organisms, even at a particular time of day. Many coevolutionary interactions involve only a single stage of a life cycle; others involve several.

Thus to coevolve, two species must not only occur in the same place but be active at the same season of the year, the same time of day, and be in the stage of their life histories that is appropriate for the interaction.

Levels of coevolution

The study of coevolution was stimulated by interactions between species, but interactions that lead to evolutionary adjustments of both partners occur at many other levels. Life originated through the coevolution of molecules linked in metabolic nets (Chapter 15, pp. 356 ff.); cellular processes coevolve; the major cell organelles have coevolved with their cellular hosts; males and females coevolve, as do parents and offspring; species coevolve; genes and culture coevolve in ways that we are only beginning to understand; and even entire clades coevolve diffusely. We now briefly explore these levels of coevolution; books could be written about each.

● **KEY CONCEPT**

Coevolution occurs at all levels of the biological hierarchy, from interacting molecules to interacting clades.

Coevolution of elements within organisms

A particularly intricate coevolution of molecules was involved in moving from the RNA world to the DNA world early in the history of life. In translating the message written in DNA into the proteins that run the cell, the message transcribed in messenger RNA (mRNA) must be interpreted as a sequence of amino acids. This is done by matching triplet codons—sequences three nucleotides long—to amino acids. The molecules responsible for making these matches are the transfer RNAs (tRNAs). The origin of the genetic code involved extensive coevolution of the tRNAs with the triplet codons in the mRNA to which they bind specifically at one end, with the amino acids to which they bind specifically at the other end, with the structure of the ribosome where the protein is assembled, and with the structure of the enzyme, elongation factor, that catalyzes the addition of one amino acid to the growing protein. The sequence of nucleotides in the tRNAs had to be adjusted in both length and composition until each of the different tRNAs could fold into a three-dimensional structure just the right size to fit between the mRNA and the growing protein in the cleft at the entrance to the ribosome with just the right amount of space left, at just the right spot, for the elongation factor enzyme to do its work. During the process coordinated adjustments were doubtless made in the ribosome and in the elongation factor molecule as well (Maynard Smith and Szathmáry 1995). The result is a set of molecules that fit together with exquisite precision (Figure 18.2). Similar coevolutionary adjustments have occurred in all interacting biological macromolecules. They are used to predict the interacting domains of proteins from their amino acid sequences.

It is not only the amino acid sequences of interacting proteins that evolve; so do their expression levels, which are regulated so that the proteins that interact

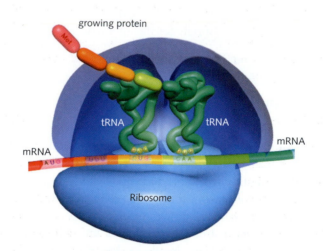

Figure 18.2 The elements of the machinery for translating the messages written in DNA into the sequences and hence the structures of proteins have coevolved to produce exquisitely precise fits, much more precise than any watchmaker is required to achieve. The ribosome moves along the mRNA molecule from left to right; it contains two chambers into each of which a tRNA fits and within which the amino acids carried by the tRNA are joined to each other while being released from the tRNAs. The genetic code is implemented in the match between the sequence on one end of a tRNA that lines up on a triplet codon in the mRNA and the sequence on its other end that binds to a particular amino acid. (Reproduced from *Virtual Cell*, North Dakota State University.)

are present in the appropriate proportions within the cell. A study of four species of yeast demonstrated that the expression levels of interacting proteins coevolve, changing in concert as the species diverge. This observation suggested that one could predict which proteins were interacting by studying the coordinated evolution of their expression levels, and in at least this case using expression levels turned out to be a better predictor of interactions than did studying the matches of the amino acid sequences (Fraser et al. 2004).

At a higher level of the biological hierarchy, organs and traits within organisms also coevolve with each other, for the environment in which a trait is evolving consists of the other traits in the organism as well as the external environment. For example, control over the ratio of root to leaf shoots in plants has evolved so that the construction and maintenance of the different major parts of plants is dynamically adjusted during growth and in response to damage. There must be enough root surface to provide the leaves with water and nutrients, and without a certain leaf surface there is no point in having a large root surface, for no use could be made of it. Similarly control over the ratio of the absorptive surface in lungs to body mass in tetrapods and the ratio of gill surface to body mass in aquatic animals has been adjusted so that there is an adequate flow of oxygen and carbon dioxide into and out of tissues without incurring the costs of producing more of an organ than is actually needed.

Life-history evolution (Chapter 10, p. 214) adjusts all of the major life-history traits to each other. If parental care increases so that juvenile mortality drops, then usually fecundity will decrease. If fecundity increases, the size of individual offspring often decreases. If age at maturity decreases and fecundity increases, then longevity often decreases. Such coordinated adjustments usually occur among all the life-history traits, not just two or three, keeping lifetime reproductive success as high as possible within the limits of constraints imposed by tradeoffs among the traits.

Ancient and intimate symbioses: mitochondria and chloroplasts

Some of the most extensive and intimate coevolutionary changes have occurred between the partners in intracellular symbioses, and the most ancient and most widespread of these symbioses are the evolution of α proteobacteria into mitochondria, which started about 1500–2000 Ma, and of cyanobacteria into chloroplasts, which happened several times about 800–1000 Ma. The free-living ancestors of both mitochondria and chloroplasts could both photosynthesize and respire. The closest free-living relatives of mitochondria cannot photosynthesize in the presence of oxygen, and the cyanobacteria, the closest free-living relatives of chloroplasts, use the same machinery both to photosynthesize and to respire, which compromises separate control of the two functions. Mitochondria lost the ability to photosynthesize and chloroplasts lost the ability to respire, and eukaryotic cells that do both need both types of organelle.

Mitochondria are about the size and shape of free-living bacteria, 1–2 μm long and 0.1–0.5 μm wide. Like bacteria they have an inner and an outer membrane, and they have their own genome, a single, circular DNA molecule. In many other respects, they have been extensively modified by their long history of living inside eukaryotic cells. Many mitochondrial genes have been transferred to the nucleus of their host cells. The ancestral free-living bacterium had at least 500 protein-coding genes; the mitochondrion with the most retains only 62 (it lives in a basal unicellular eukaryote). The mitochondria of animals have 13 such genes, and those of yeast have eight. Thus most of the proteins out of which mitochondria are built are now coded in the nucleus, manufactured in the cytosol, and imported into the mitochondrion.

Why were many mitochondrial genes transferred to the nucleus? The two main hypotheses are not mutually exclusive. The first is that it is more efficient during mitosis only to have to duplicate the two copies of the genes in the nucleus than the hundreds or thousands of copies of mitochondrial genomes in the cytoplasm. The second is that the asymmetrical inheritance of cytoplasmic and nuclear genes creates great potential for intragenomic conflict (see Chapter 9, p. 197, especially the discussion of the petite mutation in the mitochondria of yeast). The transfer of mitochondrial genes to the nucleus allowed the nucleus to control the rate of mitochondrial replication, thus preventing rogue mitochondria from taking over the cell.

While losing most of their protein-coding genes, mitochondria have retained the genes that code for tRNA and many of the genes that code for ribosomes. These genes have been retained for a striking reason. Mitochondria use a different genetic code than does the eukaryote nucleus, and this code differs among the mitochondria of mammals, flies, ascomyctes, basiodiomyctes, and plants. We do not know whether this diversity in genetic codes existed before the symbiotic event and thus reflects several independent acquisitions of mitochondria by eukaryote ancestors, or whether it evolved intracellularly after a single domestication. It may have been driven by specialization of mRNAs on the subset of tRNAs most frequently used in oxidative metabolism.

As the energy-producing organelle, mitochondria have become essential to the cells that contain them. Those cells have lost their ancestral ATP-producing machinery and have become dependent for their energy on their mitochondria. In turn the mitochondria have evolved from the generalized metabolism of free-living bacteria into specialized ATP factories, losing all non-essential functions and becoming dependent on their host cells for many materials and enzymes that their ancestors manufactured themselves.

Chloroplasts may have entered the eukaryote clade only once (Figure 18.3) or, according to another hypothesis not depicted, more than once. In the clade leading to the plants, their history was relatively simple, but in other clades there followed several events of secondary and tertiary ingestion, eventually leading to some dinoflagellates whose chloroplasts derive from red algae through haptophytes and have four membrane layers, other dinoflagellates whose chloroplasts also derive from red algae directly ingested by dinoflagellates

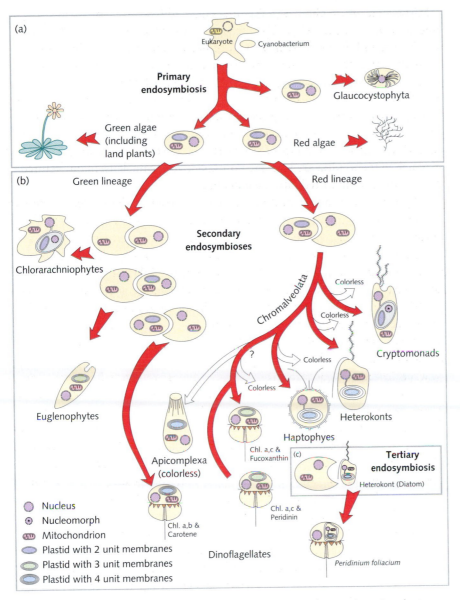

Figure 18.3 The chloroplasts found in eukaryotes have experienced a complex series of secondary and tertiary endosymbioses. In apicomplexans the photosynthetic machinery has been lost from the organelle, which remains as an unpigmented plastid, and the history can only be inferred from nuclear DNA sequences orthologous to chloroplast DNA sequences, thought to have resulted from transfer of genes from the chloroplast to the nucleus. (Reproduced from 'Assembling the Tree of Life'(2004) by Joel Cracraft & Michael J. Donoghue with kind permission from Oxford University Press.)

and have three membrane layers, and other dinoflagellates whose chloroplasts derive from green algae and have three membrane layers. In addition euglenids, haptophytes, and photosynthetic stramenopiles also have chloroplasts with three membrane layers, indicating secondary endosymbiosis. Because some dinoflagellates are symbiotic in reef-building corals, the maximum number of cell membranes through which a molecule of carbon dioxide must move to reach a chloroplast and be converted into sugar is five or six: one for the coral, one for the outer membrane of the dinoflagellate, and three or four surrounding the chloroplast. All chloroplasts have at least two membrane layers.

The chloroplast genome has not lost nearly as many genes as has the mitochondrial genome, and the chloroplast remains capable of synthesizing more of its essential molecules than does the mitochondrion. One explanation is that mitochondria have been intracellular symbionts longer than chloroplasts. This may play some role, but both have been in that state for more than a billion years. Another, more plausible, explanation is that because chloroplasts are surrounded by two to four membrane layers, it is more difficult to transport macromolecules in and out of them than mitochondria, and therefore chloroplasts continue to synthesize most of their own macromolecules.

Other ancient and intimate symbioses in which previously independent organisms live within the tissues of other, previously independent organisms include the algae and fungi that form lichens, the nitrogen-fixing bacteria that live in the root nodules of some flowering plants, the cyanobacteria that live in the stems of a large-leafed tropical plant called *Gunnera*, and the dinoflagellates mentioned above (pp. 450 ff.) that live in the tissues of corals (and giant clams). Exchange of nutrients is thought to be the basis of these mutualisms; in some cases it has been demonstrated.

Males and females

Whenever the reproductive success of one individual depends directly or indirectly upon interaction with another individual, those two individuals will coevolve. When the interaction is with mates or between parents and offspring, coevolution may be based on mutual self-interest that results in a cooperative mutualism, but it may also be based on a conflict that results in an arms race.

Males and females coevolve in many ways, the most striking of which centers around the sexual selection of male traits and female preferences in Fisher's runaway process (Chapter 11, pp. 260 ff.), but there are many others. For example, in some monogamous birds where both parents care for the offspring, the parents take turns sitting on the eggs and provisioning the young. In emperor penguins and wandering albatrosses, these tasks involve weeks of sitting on the egg without eating and long journeys to find food. Only by being reliable and returning to the nest to relieve the partner and feed the chick can birds like these enjoy any reproductive success. They have evolved elaborate, stereotypic displays thought to ensure partner recognition and to aid in bonding both partners to the task.

In the context of parenting in monogamous species, male and female interests coincide, but in many other contexts conflicts over mate choice, mating frequency, reproductive investment, and parental care are common. For example, in the fruit fly *Drosophila melanogaster* it is in the interest of males to mate frequently, and sperm from several males are often simultaneously in the reproductive tract of a single female. Males secrete proteins in their seminal fluid that decrease female receptivity and aid in sperm competition. These proteins, which increase the reproductive success of males, impose significant mating costs on females, reducing female survival and lifetime reproductive success—a substance that can poison a competitor's sperm can also poison a partner. Because the males mate with several females, they can realize a net benefit even though they damage each of their mates to a certain extent. In response, females exposed to frequent matings evolve greater resistance to the damaging effects of the seminal fluid proteins than do females exposed to single matings (Wigby and Chapman 2004).

Parents and offspring

Just as male–female interactions in mating and parenting are adjusted by coevolution, so are parent–offspring interactions. The parent is the dominant element in the environment of the offspring early in life in any species with internal fertilization. One key offspring trait that coevolves with parental traits is stage of development at birth. In altricial birds and mammals, offspring hatch or are born with eyes closed, unable to thermoregulate, and unable to fly or run. Parental behavior has coevolved to provide weeks, months, or even years of care before offspring reach independence. In **precocial** birds and mammals, offspring hatch or are born with eyes open and can thermoregulate, fly, or run within hours. Parental behavior has been adjusted accordingly.

> **Precocial:** *Born or hatched capable of seeing, moving, feeding, and thermoregulating.*

And just as males and females are often in conflict over mating and parenting, parents and offspring are often in conflict over reproductive investment. In iteroparous species, offspring born early in the life of the parent will want to extract more resources from the parent than the parent will want to give, for it is in the parent's interest to hold more resources in reserve for later offspring than the earlier-born offspring would want to concede. This produces conflicts at weaning and fledging, with offspring demanding care longer than parents are willing to give it (Trivers 1974).

Coevolving species

Coevolutionary interactions among species range from tightly, reciprocally coevolving pairs that live in intimate contact with each other, such as hosts and pathogens and specialized mutualists, to diffusely interacting guilds of predators, prey, and competitors. We discuss tight interactions between pathogens and hosts and diffuse interactions between predators and prey to give you some idea of the diversity that exists.

The tightest form of coevolution is the gene-for-gene interaction that occurs between some hosts and some pathogens: a gene that makes the pathogen more

Virulence: *The property of pathogens of being highly malignant or deadly, causing serious disease or death in the host.*

virulent is countered by a gene that makes the host more resistant, which in turn is countered by a new allele of the virulence gene, which then causes a new allele of the resistance gene to increase in frequency, and so forth. Some evidence for such interactions has been found in flax and flax rust and in other crop plants and their rust pathogens. The interactions appear to be occurring in a geographic mosaic, for the local match of virulence genes to resistance genes is imperfect (Thompson 1994).

The coevolutionary interactions of predators and prey are more diffuse, for normally a predator eats several species of prey and prey are eaten by several species of predators. For example, moths react immediately to the sounds emitted by bats; they have evolved sense organs that detect bat cries, nerve cells that fire specifically when a bat cry is detected, and evasive maneuvers to avoid being eaten. Most moths cannot afford to specialize on the characteristics of the cry of a single bat species, for several bat species usually hunt in any given area. Instead they detect very loud ultrasonic sounds over a broad range of frequencies. Some bats may use low-frequency calls and passive listening to reflections of cries uttered by other bats to overcome these moth defenses. One family of moths—the arctiids—appears to be distasteful. They actually emit clicks at frequencies that bats can detect, presumably to advertise their distastefulness and thus to avoid being attacked (Waters 2003). The mites that live in the ears of moths have also evolved behavior to avoid being eaten when the moth is eaten: they are typically found in only one of the moth's two tympanic organs, leaving at least one organ free to detect bat cries.

The coevolutionary arms race between aggressive arms in predators and defensive armament in prey is neatly exemplified in crabs and marine snails. Over the last several hundred million years snails have evolved thicker shells with morphologies that make them more difficult for crabs to hold and process, and snail-eating crabs have evolved larger, more powerful claws with more effective gripping surfaces (Figure 18.4). Crabs are not specialized on a single species of snail, but they still have been able to evolve very effective claws.

Figure 18.4 Crab claws have coevolved with snail defenses. *Carpilius maculatus*, one of several snail-eating crabs from Guam. (Photograph courtesy of Joseph Poupin, Institut de Recherche de l'Ecole Navale, France.)

Coevolving clades

The example of crabs and snails provides a transition to the issue of coevolving clades. Clades interact when the coevolutionary interactions between species persist through the speciation event and continue to characterize the interactions between the two radiating clades. This is the case for cospeciation of hosts and parasites, plants and pollinators, and plants and herbivores.

The phylogenetic pattern does not have to be as precise as is implied by cospeciation for the clades to continue to interact. Cospeciation simply represents the precise extreme of interactions that can be diffuse. For example, key innovations followed by radiations yield dominant clades that may retain their position of dominance for tens to hundreds of millions of years. Such radiations produced the geological eras informally called the Age of Fishes (the Devonian), the Age of Reptiles (the Mesozoic), and the Age of Mammals (the Cenozoic). When a radiation pre-emptively occupies ecological space, perhaps simply by getting there first, it may be able to suppress the radiation of other clades. This effect can only be seen when a mass extinction removes the dominant clade, following which another clade then radiates. There are other explanations for this pattern. This one is no more than plausible, for this is not the sort of system on which one can do experiments to sort rigorously among alternatives.

Many clades exhibit a long-term tendency for their largest members to become larger, a pattern known as Cope's Law (Chapter 17, pp. 434 ff.). Two explanations seem reasonable. One is that the organisms that found most clades are, for whatever reason, small. Random evolution of body size will then produce the pattern. Another is that prey can escape predation by becoming larger, and predators can get better at capturing larger prey if they also become larger. The first hypothesis is a useful tool to keep adaptive explanations honest; the second is just as plausible and appears to be confirmed by cases where prey colonize islands but their predators do not. Large prey released from predation pressure then may rapidly become smaller, as did elephants on islands in the Mediterranean before humans colonized those islands and hunted them to extinction.

When a clade participates in global change, such as the oxygenation of the planet by cyanobacteria, it changes the environment both for itself and for everything else on the planet. When the atmosphere switched from reducing to oxidizing, many clades went extinct; those that adapted to free oxygen seized what proved to be an enormous opportunity. The coevolutionary interactions between cyanobacteria and everything else on the planet were diffuse and complex, but they were exceedingly important.

Coevolution of genes and culture

The coevolution of genes and culture is another diffuse and complex interaction that may in the long run prove as significant for the history of life as did the oxygenation of the planet. It can be studied in its early stages in nonhuman organisms that have some elements of culture. The most cultured of these non-human species is the chimpanzee, where the seven longest-term studies have accumulated a total of 151 years of observation. In all, 39 different behavior patterns, including tool usage, grooming, and courtship, are customary or habitual in some communities. In other communities, they are absent, and ecological explanations for their absence can be discounted (Whiten et al. 1999). For example, in West Africa chimpanzees crack nuts using hammers and anvils west of a certain river; east of that river nuts and the raw materials for hammers and anvils all occur, but chimpanzees do not crack nuts.

The acquisition of a trait through cultural inheritance creates new selection pressures on genes whose products influence the ability of the organism to learn and to carry out the culturally inherited trait. In the case of nut-cracking by chimpanzees, the behavior is difficult, it takes years to learn, and those who do it well have access to an important high-energy, high-calcium food source that would otherwise be unavailable. Nut-cracking chimpanzees can be expected to evolve improved learning capacities and improved ability to store mental maps with the positions of nuts, hammers, and anvils. To the extent that such capacities increase ability to acquire and use other tools and to engage in other kinds of social interactions as well, they set in motion a coevolutionary escalation of culture and generalized intelligence.

In humans culture has been so well developed for so long that its complex history of interactions with genetic change has become difficult to disentangle. We have the simple example of the evolution of lactose tolerance in adults following the domestication of sheep, goats, and cattle (see Chapter 2, pp. 46 f.); that is relatively easy to understand. Many gene–culture interactions in humans are potentially much more important than the ability of some of us to digest milk, but they are also much more difficult to analyze and to demonstrate. One is the interaction of language, and its capacity for manipulation, with intelligence—the coevolution of signaling mind and receiving mind, where the capacities of the minds are mediated in part by genes. Another is the issue of whether the kinds of cultural interactions possible in large nation states can overcome the kinds of biological reactions that evolved in small hunter–gatherer groups. There are many others. Here we simply point to their existence and note that, as with male–female and parent–offspring interactions, they are likely to have elements of both cooperation, where genes and culture reflect similar interests, and of conflict, where culture is at odds with the genes.

Thus the levels of coevolution stretch from interactions between molecules within cells through interactions between species to interactions among entire clades and between genes and culture. This view broadens Thompson's (1994) definition from interacting species to any entities that interact in a process producing reciprocal evolutionary change. Our focus, however, remains on interacting species. We now discuss some general principles of coevolution before presenting some of the most striking consequences of the process.

Principles of coevolution

Much of coevolution occurs according to the broader principles of evolution in general. Selection on traits occurs when there is variation in reproductive success and when variation in the trait is correlated with reproductive success. A response to selection occurs when some of the variation in the trait is genetic. The types of variant that can be produced are constrained within lineages by developmental mechanisms specific to those lineages. The amount of mutational variation entering the population depends on its effective size, and the opportunity for mutations to affect a trait depends both on the number of genes coding for the trait and on the manner in which the trait is expressed. The capacity of a local population to adapt to local conditions depends upon the geographical structure of the meta-population in which the local population is embedded and the amount of gene flow it experiences. All that applies to the evolution of any trait, whether it is involved in a coevolutionary interaction or not.

What is new in coevolution is interaction with an evolving partner. Among the new effects that this introduces are those listed below.

- Reproductive success may now depend largely or in part on the reproductive success of the partner.
- The optimal values of interaction traits may now continue to shift from generation to generation as the partner changes, either in an endless, open-ended fashion, or towards some definite value in both partners.
- The geographical pattern of the interaction is determined by the overlap between the geographical occurrence of the focal species and the geographical occurrence of the partner. This overlap may be large or small, symmetric or asymmetric.
- Speciation may now result from speciation of the partner.

A central issue in coevolution is whether the interaction will evolve towards specialization or not. Some of the interactions that evolve towards specialization from a generalized starting point are mutualisms and parasite–host interactions. Predator–prey interactions and competition evolve towards specialization less often, but some of those interactions do become specialized. The direction in which an interaction will evolve depends both upon the co-occurrence of several conditions and upon the opportunities for selection that they create. We now discuss those conditions.

Frequency of interaction and impact on reproductive success

Whether an interaction is, or will become, tight and specialized or diffuse and generalized depends upon the frequency with which the partners interact and

● **KEY CONCEPT**

The intensity of a coevolutionary interaction depends on its frequency and its impact on the reproductive success of the partners.

on the impact the interaction has on their reproductive success. In the discussion of levels of coevolution above, we began with intimate interactions between molecules within cells, moved through intracellular symbioses and interspecific interactions, and ended with interactions between clades. Along that gradient the importance of other factors that could interfere with the coevolutionary process and prevent it from achieving a high degree of adaptation ranged from low for molecules to great for clades. If the focal species can devote its attention, so to speak, exclusively to the coevolving partner, without being 'interrupted by distractions,' then it is relatively easy for specialization to evolve. This will happen for the following interactions under the conditions mentioned that favor specialization.

- *Parasite–host interactions*: completion of the entire life cycle on a single host; feeding on the living host and having to deal with its induced responses; living within the host rather than on its surface.

- *Plant–herbivore and predator–prey interactions*: feeding on a species that is continuously available in edible form throughout the growing season of the consumer; the victim species is reliably available year after year; capturing, handling, or digesting the victim requires specialized adaptations that reduce the consumer's ability to eat other things; the victim is sessile, or moves slowly, and is easy to find (Thompson 1994).

- *Mutualists*: interactions either already are or—during the evolution of the interaction—become predominantly positive, with large, and fairly symmetrical, impacts on the reproductive success of both partners, this being facilitated by living in intimate contact for most or all of the life cycle.

Some of the remarkable cases of specialization in vertebrates include giant pandas and red pandas, which eat only bamboo; sage grouse, which eat almost exclusively one species of sagebrush; and aardwolves, the most specialized of the several mammal species that have converged on eating ants and termites. Aardwolves eat only termites of one genus, *Trinervitermes*, a genus that is protected from all other predators by the corrosive terpenes secreted by the soldier caste. Young aardwolves being weaned from a diet of bland milk to corrosive termites have been observed having difficulty accepting the food that will become their diet for life (T.H. Clutton-Brock, personal communication). In all three cases the consumer has to overcome significant defenses—silicate in bamboo leaves and defensive chemicals in sagebrush and termites. By being able to overcome these defenses, they gain access to a food resource largely free from exploitation by other consumers.

Relative evolutionary potential

Coevolutionary interactions can evolve towards stable coexistence; they can also evolve towards the extinction of one of the partners. There is no guarantee that both species will be able to stay in the game. The factors that determine which of the partners may be able to gain the upper hand in the interaction include generation time, genetic system, and genetic variation for the interaction traits.

(1) All else being equal, *the partner with the shorter generation time will evolve more rapidly.* Most pathogens have much shorter generation times than their hosts; part of the vertebrate defense against pathogens is achieved by having an immune system whose response time is much closer to the generation time of pathogens than is the generation time of the organisms containing it.

(2) All else being equal, *sexual partners will be able to evolve more rapidly than asexual partners.* Interactions with pathogens and parasites are one of the major reasons for the maintenance of sexual reproduction in hosts, and they have at times been countered with increased rates of recombination in parasites and pathogens (Chapter 8, pp. 192 f.).

(3) And, all else being equal, *the partner with more genetic variation for the interaction traits will evolve more rapidly.* For bacterial pathogens living in vertebrate hosts, the chief threat is the vertebrate immune system. The immune system rapidly sorts through a huge array of molecules until it finds one that binds specifically to the surface of the bacterium. It is therefore in the interest of bacteria to frequently change the characteristics of the surface they present to their hosts, and in fact the genes that code for surface proteins have the highest mutation rate and are the most rapidly evolving part of their genome (Moxon et al. 1994).

Unless it is in the interest of both partners that the interaction persist, and that interest is a top priority for both, one partner could win the arms race and drive the other to extinction. If a predator, herbivore, or parasite can feed on several different prey species, the disappearance of one of them may make little difference to its reproductive success. If a competitor can be eliminated without serious consequences, there will be no reason for the focal species to alter its characteristics. The array of coevolutionary interactions that we observe thus probably consists of those in which this has not been the case, where the interaction stabilizes in a manner permitting coexistence, and of those in which extinction has not yet happened but is in the process of doing so.

The Red Queen

There is some paleontological evidence to support the idea that coevolutionary interactions have kept the extinction rate fairly constant over a long period of time. We label such dynamics **Red Queen** interactions after the character in *Through The Looking Glass*, by Lewis Carroll. The Red Queen told Alice, 'Now here, you see, it takes all the running you can do, to keep in the same place.' To van Valen (1973), the Red Queen's message for evolution was that no matter how much one species adapted to what the other was doing, the other would change just as much in response, with the result that there would be no net long-term improvement in fitness. They would remain locked in an arms race with no exit.

Red Queen: *Coevolutionary dynamics in which the participants struggle forever against each other with no long-term reduction in extinction probability.*

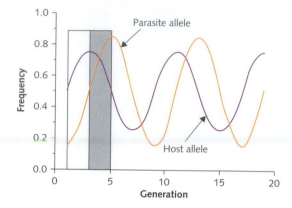

Figure 18.5 Cycling of parasite and host allele frequencies as a result of frequency-dependent selection (courtesy of Curt Lively). The dark rectangle indicates the portion of the cycle where the common host genotype is having serious problems with the parasite because the specialized parasite genotype is over 50%. The light rectangle indicates the portion of the cycle where the host is not having the problem because the specialized parasite genotype is at low frequency. (From Clay and Kover 1996.)

In Chapter 8 (pp. 192 f.) we discussed such arms races in the context of host–pathogen interactions for the maintenance of genetic variation in genes for virulence in the pathogen, for resistance in the host, and for sexual reproduction in both. We revisit that issue here because it is such a clear example of coevolution linking genetic changes in one species to genetic changes in another.

Frequency-dependent selection makes life difficult for common virulence and resistance types and should maintain offset cycles of frequencies of virulence and resistance alleles in host and pathogen (Figure 18.5).

Frequency-dependent selection does maintain very high levels of genetic diversity—many alleles at each of many loci—in the vertebrate immune system. Considerable genetic variation for genotype-specific virulence and resistance has also been found in crustacean and plant host–pathogen pairs, including the water flea *Daphnia* (bacterial pathogen *Pasteuria*; Carius et al. 2001), the ragwort *Senecio* (powdery mildew pathogen *Erysiphe*), the soybean *Glycine* (rust pathogen *Phakopsora*), the flax *Linum* (rust pathogen *Melampsora*), and the cress *Arabidopsis* (mildew pathogen *Perenospora*; Clay and Kover 1996).

There is thus suggestive genetic evidence for dynamic interactions well described as a coevolutionary arms race: each time the pathogen improves its attack the host changes in a way that improves its defense, which then causes a change in the pathogen that improves its attack. Whether those interactions are a sufficient explanation for the maintenance of sexual reproduction in host and pathogen remains a plausible hypothesis that has not yet been strongly confirmed (see Chapter. 8, pp. 192 ff.).

Evolution of niche width and location

The niche width of competitors can also be seen as the host range of pathogens, parasites, and herbivores—the number of different species they attack, and therefore the number of different species on which they could encounter competitors. How many different species should a consumer eat or infect? The answer is couched in terms of the balance of factors favoring specialization or generalization and the constraints on ability to evolve in one direction or the other.

The classical view of the evolution of niche width in competitors is that interspecific competition will cause competitors to evolve to reduce the impact of that competition. They are thought to do so by diversifying to use different resources. As they evolve to reduce interspecific competition, they specialize on using different parts of the available food supply. In so doing, they are at least not decreasing, and perhaps increasing, the impact of intraspecific competition, for as they avoid the other species they will encounter more of their own species. Thus the rate of this process, and the degree to which it will result in narrower, more-specialized niches, depends on the balance of inter- and intraspecific effects.

A pattern consistent with competition-driven niche shifts is called **character displacement**. Character displacement is seen by comparing closely related competitors in sympatry and allopatry. If they differ more in sympatry than in allopatry, then one explanation is that competition has caused the divergence. This explanation becomes more plausible if the traits that change are functionally related to competitive ability.

As competitors specialize, another important effect comes into play. It is often, but not always, the case that a jack of all trades is a master of none, and a master of one trade is poor at the others. A superb specialist is often a poor generalist: when the reproductive success of a species increases as it becomes adapted to one resource, its reproductive success on other resources decreases. This effect is strongly confirmed by serial transfer experiments on the evolution of virulence: as a pathogen adapts to a new host it loses its ability to function on the old host. And it is clearly present in cases where the consumer has had to evolve specialized adaptations to overcome specialized defenses in the resource. There is then a tradeoff in performance on different resources.

If the only effect determining niche width were a tradeoff in performance on different resources, one might expect all species to become specialized. There are at least four reasons why this does not happen:

- Intra-specific competition can limit the benefits of specialization.
- Organisms often have to eat different things at different stages of their development if only because they start smaller and end larger.
- Some resources are only intermittently present or exist in a geographical mosaic in which they do not overlap with the consumer everywhere; they must then be supplemented with alternatives if the consumer is to survive.

Character displacement: *Competing species differ in competition-related traits more in sympatry than they do in allopatry. Competition is then thought to have displaced the sympatric state from the allopatric state.*

• Another consequence of spatial heterogeneity is that it provides the contrasting selection pressures in different places that maintain both niche width and genetic variation within the focal species. Niche width usually, but not always, evolves to match the amount of variation present in the environment, with homogeneous environments containing specialists and heterogeneous environments containing generalists (Kassen 2002).

For all these reasons, we should expect specialization to evolve often, but not always. Both extreme specialists and extreme generalists are interesting cases, for they represent evolutionary histories with unusual combinations of effects.

Evolutionary transitions among parasitism, mutualism, and commensalisms

The nature of a coevolutionary interaction is not written in stone. The impact of the partners on each other can evolve, even to the extent of changing sign, at least from − to 0. Such evolution of the interaction is particularly well understood in the case of hosts and pathogens (see next chapter, pp. 488 ff.). If the pathogen is vertically transmitted from host parent to host offspring, then the reproductive success of the pathogen depends upon the survival and reproduction of its host. The expected result is that the virulence of the pathogen will evolve downwards, eventually to zero. In the process a host–pathogen interaction will have been converted into a commensalism, from +− into +0.

Such complete attenuation of virulence is only expected with strict **vertical transmission**. Where there is **horizontal transmission** of the pathogen from host to unrelated host, independent of host reproduction, then virulence is expected to stabilize at an intermediate level that balances reproductive success within one host with transmission probability between hosts. It does not pay a pathogen to kill a host so quickly that transmission is unlikely. These effects were seen clearly when a flea-transmitted viral disease, myxomatosis, was introduced from Europe to Australia to control the exploding population of introduced rabbits (Fenner and Ratcliffe 1965). Samples of the virus were frozen for later reference when it was introduced. At the beginning the disease was extremely virulent, but as it spread its virulence decreased because the strains that were most virulent killed their hosts before they could be transmitted. Comparison of the frozen with the evolved strains demonstrated that the reduction in virulence was due both to a decrease in the virulence of the virus when tested on a standard host and to an increase in the resistance of the rabbits. Virulence did not evolve to zero; it stabilized at an intermediate level.

There is another way for coevolution to convert a parasite–host relation into a commensal relation. A commensal +0 relation can evolve from a +− parasite–host interaction when genes in the host for resistance compete with genes for tolerance. A gene for resistance cannot take over the population completely because as it becomes more frequent, the incidence of infection declines, which

Vertical transmission:
Transmission of parasites or pathogens from host parent to host offspring, often during host reproduction.

Horizontal transmission:
Transmission of parasites or pathogens to other organisms at times and places not necessarily associated with host reproduction or relationship.

reduces the advantage of resistance. If there is any cost to resistance, resistance genes cannot be fixed, and resistance genes alone cannot eliminate diseases. In contrast, if a host gene for disease tolerance sweeps through a population, it increases the number of hosts that can serve as habitat for the disease, the incidence of the disease increases, and that in turn creates positive frequency-dependent feedback on the frequency of the tolerance gene, which rapidly goes to fixation. As predicted by this hypothesis, resistance traits tend to be polymorphic and tolerance traits tend to be fixed in field studies of interactions between diverse plant species and rust fungi (Roy and Kirchner 2000).

The evolution of mutualisms has attracted a great deal of attention because the principles involved also explain the evolution of cooperation between unrelated individuals. The principle theoretical tool used in these investigations has been game theory. In this kind of theory one imagines that both partners can exist in two versions, cooperators and defectors, and that there are certain pay-offs for cooperating and for defecting. The payoffs are delivered in the form of changes in reproductive success. The analysis then asks, under what conditions will a rare cooperator mutant be able to spread into a population composed mostly of defectors? If it can spread, then cooperation will increase, and if it can take over the population and go to fixation, which is a separate question, then a cooperation will have evolved uniformly throughout the population. There are two conditions that greatly facilitate the evolution of cooperation (Doebeli and Knowlton 1998):

- increased investments in the partner must yield increased returns, and
- the partners must be close to each other in space. Cooperation spreads more easily when contacts between cooperating partners are more frequent than random, and spatial contiguity is one effective way to achieve this.

Under these conditions, it is surprisingly easy for mutualism to evolve. Thus the barrier to the evolution of mutualisms may be not the evolution of the interaction but the initial penetration of host defenses. Once the two partners are reliably associated closely in space and exchanging some resource with even a very small initial mutual benefit, the intensity of the interaction will be selected to increase to the benefit of both. Such a process could take an interaction that was very close to 00 and convert it rapidly into one that was strikingly ++.

Striking outcomes of coevolution—and its absence

Volumes have been written on the outcomes of coevolution, and it is not easy to make an illustrative selection from the abundance of striking examples. We have chosen four classic interactions: mimicry in butterflies, the specialized active pollinators of yuccas and figs, leaf-cutter ants and their domesticated fungi, and the introduction of foreign pathogens and predators to naïve ecosystems. In

● **KEY CONCEPT**

Coevolution has produced striking examples of complicated adaptations embedded in ancient clades.

making this selection we have left out many equally fascinating and well-studied interactions between plants and herbivores (Karban and Baldwin 1997, Moraes et al. 2004), between seeds, fruit, and their dispersers and predators, including the striking phenomenon of mast fruiting (Curran and Webb 2000), between ant-plants and the ants that protect them from herbivores (Beattie and Hughes 2002), and between dinoflagellates and corals that make possible the construction of reefs (Baker 2003). We recommend them all to your attention.

Butterfly mimics

Mimicry is used to describe the remarkable morphological convergence between distantly related species originally discovered in butterflies. Bates (1862) and Müller (1879) proposed explanations that bear their names; both explanations are well supported. In addition, the term aggressive mimicry is used to describe cases like the saber-toothed blenny mimicking the cleaner wrasse.

Batesian mimicry

In Batesian mimicry an edible model evolves to resemble a warningly colored, noxious species. In butterflies the noxious model often acquires its defensive chemicals from the plants that it eats as a caterpillar. Harmless flies and beetles mimic the warning coloration of stinging wasps. Harmless king snakes mimic deadly coral snakes. Batesian mimicry is mediated by predators that do not eat the mimics because they are trying to avoid being harmed by the models. The convergence between model and mimic is therefore limited by the ability of the nervous systems of the predators to remember patterns and to perceive differences. That the convergences can be remarkably precise is testimony to the sophistication of predator sensory systems.

One of the most remarkable systems of Batesian mimicry can only be maintained if the harm caused to the predator by eating a model outweighs the benefit of eating a mimic. The nature of learning is weighted in favor of the mimics, for a predator that has had a bad first experience with a model tends to avoid anything that looks like it for a long time and does not re-sample soon to see whether the initial experience was a false negative. However, if mimics become more abundant than models, then the probability of a young predator having a first experience with a mimic increases. Such systems are therefore most likely to be stable in times and places where both the model and the mimic occur and where the model is more abundant than the mimic.

One of the most remarkable systems of Batesian mimicry is that of the African mocker swallowtail butterfly, *Papilio dardanus* (Figure 18.6), which feeds on *Citrus* plants. Throughout its extensive range, the male is a brightly colored butterfly with tails on its wings. The females, however, have diverged dramatically from the male morphology to mimic several different species of noxious model. This suggests interesting issues in the genetic control of the

Figure 18.6 The mocker swallowtails of Africa (*Papilio dardanus*) are one of the most remarkable cases of Batesian mimicry known. The females mimic different toxic models in different geographical regions, with the result that they look very different both from the males of their own species and from the females of their species in other geographical regions. The males are not mimics, and on Madagascar, where no toxic models are available, the females are not mimics and resemble the males. Top row: Left, male; right, female from Madagascar. In the remaining rows the mimicking female is on the left and the toxic model is on the right of each pair. Second row: left pair, left specimen, *P. dardanus* var. *planemoides* female (mimic); right specimen, *Bematistes poggei* (model) from Kenya; right pair, left specimen, *P. dardanus* var. *trophonius* female (mimic); right specimen, *Danaus chrysippus* (model) from South Africa. Third row: left pair, left specimen, *P. dardanus* var. *niobe* female (mimic); right specimen, *Bematistes tellus* (model) from Sudan; right pair, left specimen, *P. dardanus* var. *hippocoonides* female (mimic); right specimen, *Amauris albimaculata* (model) from Mozambique. Fourth row: left pair, left specimen, *P. dardanus* var. *hippocoon* female (mimic); right specimen, *Amauris naivius naivius* (model) from Great Lakes region; right pair, left specimen, *P. dardanus* var. *cenea* female (mimic); right specimen, *Amauris echeria* (model) from South Africa. The models are in the family Nymphalidae; the mimics are in the family Papilionidae. The papilionid mimics accurately reproduce the patterns that evolution elicits from the nymphalid ground plan (Figure 7.2). (Butterfly photos credit to Terry Dagradi; specimens courtesy of the Peabody Museum, Yale University, arranged by Raymond Pupedis, Curatorial Assistant in Entomology.)

development of complex morphology, the early investigation of which is summarized in Ford (1975) with a recent update given by Nijhout (2003).

Müllerian mimicry

Müllerian mimicry describes convergence between species all of which are distasteful; in this system all species are both models and mimics. In effect the entire set of species, which each initially had a different warning pattern, can be thought of as having converged on a method of cooperatively educating their predators in what is worth avoiding. They thereby reduce the number of deaths that each species must suffer in the process of informing naïve predators about their nature (Müller 1879).

A remarkable example of Müllerian mimicry is provided by the extensive formation of parallel geographic races in the butterfly genus *Heliconia* in the neotropics (Gilbert 1983; see Figure 18.7). Such a pattern need not be caused by reciprocal change; one species could converge on another, relatively unchanged species. Molecular systematics suggests that *H. erato* has historically had a much larger population size and more gene flow than *H. melpomene*, supporting the hypothesis that *H. erato* has served as the model on which *H. melpomene* converged (Flanagan et al. 2004).

Aggressive mimicry

Aggressive mimics are usually predators that have evolved to disguise themselves as something their prey will not recognize as dangerous. Many fish besides the saber-toothed blenny (pp. 443 f.) have evolved aggressive mimicry; most of them are in the families Serranidae (sea basses), Cichlidae (tilapia and their relatives), and Blenniidae (the blennies). This has usually happened in one of three contexts (Sazima 2002), as follows.

- Fish that feed on prey smaller than themselves mimic model species harmless to those prey and conceal themselves in schools of the models, from which they launch their attacks.

- Fish that feed on prey larger than themselves mimic cleaners or hide in schools of species harmless to their prey.

- Fish that feed on prey about their own size mimic their prey, becoming wolves in sheep's clothing.

The importance of the prey sensory system in determining the kinds of coevolutionary change that occur in the aggressive mimic is underlined by several examples that by no means exhaust the diversity of aggressive mimics.

- *Ant-eating spiders* in the family Zodariidae have evolved to resemble ants in size, color, and behavior (Pekar and Kral 2002). If a spider carrying a dead, captured ant is confronted by another ant, it taps the approaching ant with its front legs, transmitting a tactile cue that mimics an ant recognition signal, then exposes the corpse of the captured ant to the curious ant, transmitting a tactile cue.

Figure 18.7 In the neotropics geographical races of *Heliconius* species, and some other Nymphalidae, all of which are distasteful, form parallel sets of Müllerian mimics, and some of these Müllerian mimicry complexes are mimicked by edible species (Batesian mimicry). Column 1 (left, top to bottom): *Heliconius pachinus*, *H. sara*, *H. wallacei*, *Laparus doris astromache* (all nymphalid models), *Eurytides pausanios* (Batesian papilionid mimic). Column 2: *Heliconius ethilla*, *H. numata*, *Melinaea menophilus* (all nymphalid models), *Perrhybis pyrrha* (Batesian pierid mimic). Column 3: *Mechanitis lysimnnia* (nymphalid model), *Dismorpha amphione* (pierid mimic), *Lycorea cleobaea* (nymphalid model), *Papilio zagreus* (papilionid mimic). Column 4: *Tithorea harmonia martina* (nymphalid model), *Charonias eurytele* (pierid mimic), *Hypothyris lepeireui ninyas* (nymphalid model), *Ithomeis eulema* (rodinid mimic). (Butterfly photos credit to Terry Dagradi; specimens courtesy of the Peabody Museum, Yale University, arranged by Raymond Pupedis, Curatorial Assistant in Entomology.)

- Some *spider-eating spiders* enter the nets of their prey and mimic the signals of struggling insects by vibrating the web in an appropriate pattern, enticing the resident spiders to approach close enough to be attacked (Jackson 1992).

- *Blue butterflies* lay their eggs on thyme plants. The larvae fall to the ground and emit pheromones that mimic the pheromones given off by the larvae of specific species of ants. Foraging ants encounter the butterfly larvae, mistake them for ant larvae from their own colony, and carry them back to the colony.

There the butterfly larvae turn into predators, eat ant larvae, and emerge from months to years later for a brief life as adult butterflies (e.g. Thomas 2002).

- The females of certain species of *fireflies* (beetles in the family Lampyridae) emit the light signals characteristic of females of other species, attracting males of those other species, whom they then eat (Lloyd 1984).

Models and mimics thus range across several types of interactions. Batesian mimicry is usually a 0+ system with a positive impact on the mimic and little or no impact on the model. When the mimic becomes too abundant, however, the model may suffer from being mistaken for a mimic that has been identified as edible by naïve predators. Such effects are not likely to become large, because the effect itself implies that the mimic's population will soon decline. In Müllerian mimicry, the system is usually ++, with both partners benefiting. Aggressive mimicry is definitely a +− predator–prey or grazer–resource relation in which the reciprocal evolution of the prey is likely to take place in the discriminatory abilities of the sensory system.

Flowers with active pollinators

Plants have repeatedly coevolved with their pollinators. There are bird-flowers, bee-flowers, moth-flowers, and bat-flowers; in all those cases the morphology and physiology of the flower and the morphology and behavior of the pollinator have coevolved reciprocally. This is beautifully demonstrated in Darwin's orchid and its moth pollinator (Figure 2.8). In most of those cases, however, the pollination is a passive byproduct of the attempt of the pollinator to reach a food source in the flower, either pollen or nectar. In only a few cases has active pollination evolved, pollination in which the pollinator goes out of its way to make sure the flower is pollinated without gaining food from the act of pollination itself.

Two such cases are figs and their fig wasps and yuccas and their yucca moths. In both of these cases of active pollination, the pollinator not only pollinates the flowers; it lays its eggs in the seeds and fruit of the plant that it pollinates and thereby destroys some of the plant's reproductive potential. These pollinating seed predators exchange their pollination services for larval food. And in both of these cases the plant has some control over the degree to which such destruction of its seeds can evolve, for if too many of the seeds in any given fruit are destroyed, the plant will abort the fruit, thereby destroying the larvae of the pollinator and imposing negative selection on the further evolution of the tendency to destroy seeds.

A fig is a compound flower whose hundreds of floral elements are born on the inside of an enclosed sphere with a narrow opening. When the fig is receptive, a passage opens and volatile chemicals are exuded that attract the female wasps. The females arrive from neighboring figs bearing pollen with them. They enter the fig and lay their eggs into the endosperm of developing fig seeds. They also actively pollinate both the flowers into which they lay their eggs and neighboring flowers. The females then die, and the eggs develop into larvae that eat the fig seeds. Some figs produce two types of seed, those into which the wasps lay eggs and those that are available for pollination. When their offspring hatch, the flightless males fight each other for access to the newly hatched virgin females, which they inseminate. Then the males die inside the fig, and the inseminated females, coated with pollen acquired both passively and actively, fly off to the next receptive fig to initiate another cycle.

This microevolutionary co-adaptation has led to a high degree of specialization and to the macroevolutionary consequence of frequent cospeciation of the figs and their wasps. Each fig species has its own unique species of wasp. (Not all wasps found in figs are pollinators; some are gall-forming parasites on the figs, some are parasitoids on the fig wasps.) Not all members of the wasp subfamily *Agaoninae* pollinate figs; about 300 species do, about 400 species do not. Not all fig wasps are active pollinators; some are passive (Cook et al. 2003). And not all figs are monoecious, with male and female flowers in the same fruit. In dioecious figs, where a tree has either all-male or all-female flowers, the situation can be more complex, the conflicts are different, and the opportunities that arise for gall-forming and parasitoid wasps are different (Weiblen 2002).

The phylogenetic trees of figs and fig wasps are partially congruent, suggesting that cospeciation may have played a role during their evolution (Figure 18.8). The case for cospeciation is strengthened when one finds monophyletic groups of pollinators with conserved associations with their hosts (Weiblen 2002), and some are found in several segments of these two phylogenies. Where there are exceptions, one fig species hosts several wasp species, each in a different part of the fig's geographic range. Host switching appears to have been very rare, probably because of the extreme specialization of the interaction and the interdependence of the timing of reproductive events.

The other dramatic case of active pollination involves yuccas and yucca moths (Pellmyr 2003). Yuccas are large members of the agave family (Agavaceae) with huge inflorescences that live in south-western North America. They are pollinated by moths in the family Prodoxidae. The different species of moths are difficult to distinguish morphologically, and it was originally thought that several species of moth might be pollinating a single species of yucca, but molecular systematics has now shown that most moths pollinate and prey on a single species of agave, making this an example of specialized coevolution. Flower and fruit abortion by the yucca function to prevent over-exploitation by the moth. In two cases the + + mutualism has changed into a + − plant–herbivore relation with the evolution of non-pollinating moths that oviposit into fruits pollinated by other moth species. The phylogenetic data are as yet incomplete, but the data that are available do not suggest that cospeciation has been as frequent in yuccas and yucca moths as it has been in figs and fig wasps. The molecular estimate of the age of the interaction is about 40 million years; the mutualism appears to have evolved rapidly once it started.

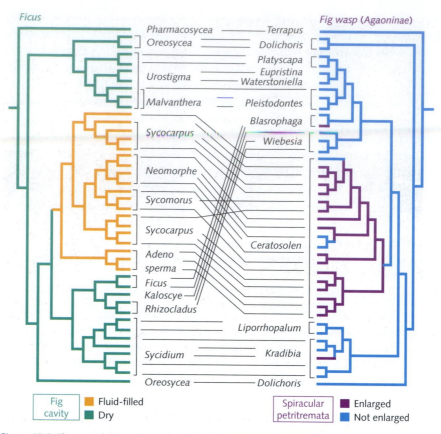

Figure 18.8 The coevolution of Indo-Australian fig wasps (*Agaoninae*) and figs (*Ficus*) based both on morphology and on DNA sequences. The colored branches indicate where the figs are filled with fluid and the wasps have enlarged organs allowing them to breathe in a fluid-filled environment. The morphology evolved once in the fig, and all but one instance of the co-adapted morphology evolved once in the wasp. (Reprinted, with permission, from the *Annual Review of Entomology*, vol. **47**. ©2002 by Annual Reviews.)

The co-evolution of yuccas and yucca moths was made possible by a key innovation in the moths. The female moths have complex tentacles that they use to collect pollen and actively pollinate the yucca flowers (Figure 18.9). These tentacles do not correspond to appendages on other insects; they are a novel limb. It appears that the tentacles are a morphological module related to the proboscis that develops at a new site, the first segment of the maxillary palp, through the expression of the genetic template for the proboscis. Both the tentacle and the proboscis are controlled by a shared hydraulic extension mechanism; thus the expression of many elements of the proboscis module at a new site meant that no new mechanism had to evolve to control tentacle function (Pellmyr and Krenn 2002). The surface of the female's pollinating tentacle has, moreover, been extensively modified by comparison with the proboscis (Figure 18.9b, lower right panel). The origin of the tentacles was a key innovation involving the

(a)

(b)

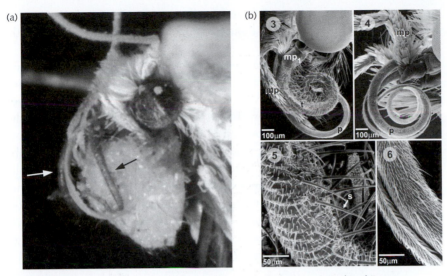

Figure 18.9 (a) Female yucca moth carrying a pollen load. Black arrow, tentacle; white arrow, proboscis. (b) The two upper panels contrast the female, who has both a proboscis and a pollinating tentacle, with the male, who has only a small protuberance on the first segment of the maxillary palp (mp) and no tentacle. The bottom two panels contrast the surface of the female's tentacle, on the left, with the surface of her proboscis, on the right, illustrating how microevolution modified the surface of the tentacle to make it more efficient at transferring pollen. (Photos courtesy of Olle Pellmyr, University of Idaho and Harald Krenn, University of Vienna.)

development of a new appendage at a new site; it has had the macroevolutionary consequence of enabling the radiation of many species of yucca moths. Following that innovation microevolution fine-tuned the surface of the tentacle to better adapt it to transferring pollen.

Leafcutter ants, their fungal cultivars, fungal enemies, and bacterial pharmacy

This example combines macroevolutionary cospeciation with a parasite–host arms race in one of the most remarkable biological objects on the planet. Leafcutter ants (*Atta* spp.) build huge nests in the New World tropics where they farm domesticated fungi for food. A leafcutter ant colony can strip the leaves from a large tree in a single day; it consumes as much as an adult cow. Specialized workers cut leaves from vegetation and carry them along cleared trails into the nest, where they are given to another worker caste that chew the leaf fragments into a mulch, which is then fed to the fungus garden. The fungus produces special structures, gongylidia, on which the ants and their larvae feed. The queen, many times larger than the nest workers, lays her eggs in the fungus garden. The larvae are cared for by a nursing caste. The colony is kept clean: dead ants, dead fungus, and exhausted mulch are placed in dump chambers. The entire colony is several meters across and several meters deep. Whether we consider it as a superorganism or an especially complex family, a leafcutter ant colony is one of the largest, most energy-demanding biological objects in the neotropical forest.

The ants provide the fungus with a protected habitat, a substrate for growth (the leaves cut and chewed by the ants), dispersal to new nests, and protection from enemies. One particular devastating enemy of the fungus is a specialized microfungal parasite, *Escovopsis*. To protect their fungal garden from this pathogen, the ants culture on specialized parts of their bodies a filamentous bacterium (an actinomycte) that produces antibiotics that specifically inhibit the microfungal parasite. There are thus four species involved in the coevolutionary interaction: ant, food fungus, pathogenic fungus, and bacterial antifungal agent. The fungus used by the ants for food is transmitted vertically from ant colony to ant colony by founding queens; the bacterium used by the ants to produce antibiotic is probably also vertically transmitted. The microfungal pathogen is transmitted horizontally (Currie et al. 2003).

The ant–fungus symbiosis is about 50 million years old

The ant–fungus symbiosis is an ancient coevolutionary domestication, at least 50 million years old. The phylogenies of the ant, the fungus, and the microfungal pathogen suggest that the tripartite association is also old. The more recent parts of the phylogeny contain some evidence for host switching, both between the food fungi and the ants and between the microfungal parasites and the food fungi (Figure 18.10).

Thus the phylogenies suggest a very long coevolutionary association. They also suggest the occasional acquisition of new fungal cultivars by the ants in the lower attine radiation where the closest relatives of the cultivars are free-living. That the ants have evolved body parts on which to grow the antibiotic-producing bacterium suggests that that interaction is also well-established. The ants' use of a bacterium to control a microfungus appears to be a microevolutionary solution to the coevolutionary virulence–resistance problems posed by the shorter generation time, larger population size, and more rapid rate of response of the sexual microfungal pathogen for the cultivated fungus, which is obligately asexual.

When things have not coevolved, catastrophe can result

Invasive species are often able to invade because the local community has not had an evolutionary experience with the invader. Disasters occur most impressively when the invader is a pathogen to which the local species have had no chance to adapt. Such was the case with rinderpest in Africa. Rinderpest changed the ecology of an entire continent for more than a century.

Bovidae: Cattle and their relatives, including antelope, buffalo, sheep, and goats.

Rinderpest is a directly transmitted viral disease of the Bovidae, cattle and their wild relatives, particularly buffalo, eland, kudu, giraffe, and bushbuck,

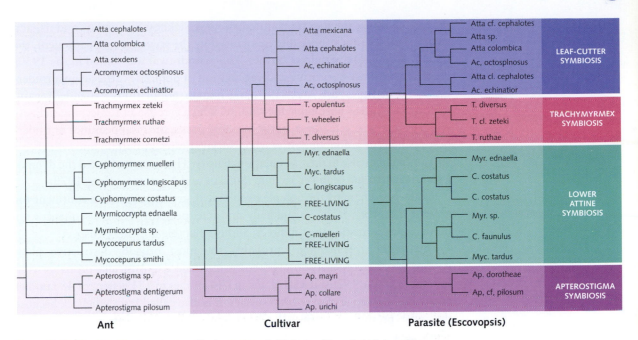

Figure 18.10 The tripartite coevolution of leafcutter ants (left), their cultivars (middle), and the pathogen of their cultivars (right) is ancient, as indicated by their partially congruent phylogenies and other evidence. (From Currie et al. 2003.)

but also infecting warthogs and bush pigs. It is endemic to the Asian steppes, from which it has been repeatedly introduced to Europe through human invasions. It was brought either to Italian Somaliland in cattle imported from Asia in 1889 or arrived in Russian cattle that accompanied the relief of General Gordon in Khartoum in 1884. Whatever the cause, by 1890 it had crossed the barrier of the Sahara Desert and invaded populations with neither evolved nor acquired resistance. Spreading over eastern, central, and southern Africa between 1890 and 1899, it eliminated most of the domestic cattle and buffalo and many of the related bovids. One species of antelope went extinct as a result of the epidemic, and the distributions of the other species remain altered to this day. The pastoral and nomadic peoples lost their food sources and, under the stress of starvation, the human populations succumbed to an outbreak of endemic smallpox (Spinage, 1962).

Subsequent **epizootics** occurred in 1917–18, 1923, and 1938–41. Over much of the infected area, tsetse flies disappeared after the epidemic. These flies require trees and bushes as their habitat. A second consequence of the disappearance of game was the appearance of man-eating lions. The man-eaters of Tsavo appeared in 1898, and in an outbreak in Uganda in the 1920s one individual killed 84 people. The presence of man-eating lions caused the farmers to abandon large areas, in which thickets of brush grew up. The wild ungulates developed immunity to rinderpest and moved back into the abandoned farming areas, where they became hosts for the tsetse flies that could now live in the new

Epizootic: *An epidemic of disease in an animal population.*

thickets. Because the flies transmitted sleeping sickness to humans and nagana to domestic livestock (both diseases caused by the trypanosome pathogen), the human population withdrew further. This withdrawal was halted in the 1930s by mechanical brush clearing and vaccination of domestic cattle against rinderpest, and in the 1960s and 1970s the human populations expanded again, making large areas unsuitable for wildlife, which is resistant to sleeping sickness and had been protected by it. In the national parks, the disappearance of rinderpest resulted in a doubling of yearling survival rates in buffalo and wildebeest and a rapid increase in their population density. This in turn affected the dynamics of the entire grassland ecosystem (Sinclair and Norton-Griffiths 1979).

Rinderpest changed the complex ecological structure of half a continent for at least a century. It did so suddenly and unexpectedly. The consequences for humans were drastic, often indirect, and predictable only in retrospect.

Phytophthora cinnamomi, probably the most broadly destructive plant pathogen ever recorded, is a root pathogen introduced to Australia, probably from Indonesia, in the 1920s. It kills plants in 444 species in 131 genera of 48 families and has a host range of over 1000 species. It is also called Jarrah dieback because the most prominently affected timber species is a eucalyptus called Jarrah. Three-quarters of the victims are in four families, the Proteaceae, Leguminosae, Epacridaceae, and Myrtaceae, some of which are valuable for their timber (eucalyptus) or their flowers (proteas). *Phytophthora* has devastated more than 300 000 hectares of forest, woodland, and heath in western Australia. In some places, more than half the plant species in the community have been destroyed.

Previous outbreaks of *Phytophthora* in the 19th century destroyed chestnuts in the southern USA, where the variety of the pathogen was much more specific. It has caused disease in *Nothofagus* forests in New Guinea, ohia forests in Hawaii, and montane forests in South Africa. Its foreign origin and remarkable host breadth testify to the potential disasters that lie in wait when there has been no opportunity for coevolution.

Lack of a history of coevolution is also one hypothesis for the virulence of emerging diseases, such as Ebola. When they first enter the human population, their virulence levels are far from equilibrium. It is not in the interests of the pathogens to kill humans as quickly as these do, for they then have little opportunity for transmission. In fact, Ebola is so virulent that it quickly dies out after having created an intense local outbreak. Similar selection forces were operating on the plague bacterium, *Yersinia pestis*, as it made its way northward through Europe in the 14th century: the initial virulence was high, with mortality over 70% in some towns in southern Europe, and then declined towards 30% in northern France and Scandinavia as the pathogen evolved to increase transmission opportunities. When measles and other diseases arrived with Europeans in the New World and Polynesia, they were extremely virulent in natives, who had no experience of them, having left those diseases behind them when they migrated into new areas in small groups, in which infectious diseases tend to go extinct.

In the fossil record, the history of South America suggests that when that continent, formerly an island, became connected to North America through the Isthmus of Panama from 13 to 2 Ma (Haug and Tiedemann 1998), there was a great exchange of faunas. Ant-eaters, armadillos, and possums moved north; predators, competitors, and possibly pathogens of the unique hoofed animals of South America—the litopterns and notoungulates—moved south. The arrival of North American species, with which the South American fauna had not coevolved, was followed by the rapid extinction of the litopterns and notoungulates and their replacement by horses, tapirs, peccaries, camels, and deer from the north (Simpson 1949). Thus patterns seen in microevolutionary time, such as the stories of rinderpest in Africa and *Phytophthora* in Australia, are reflected in the history of entire faunas in macroevolutionary time when naïve species encounter efficient predators, pathogens, and competitors that evolved elsewhere and whose arrival comes as a shock.

Discussion

How should we judge a claim that an interaction has coevolved?

The assertion that an interaction is coevolved is a hypothesis to be tested, not an obvious fact beyond dispute. Even in apparently obvious cases like the wrasse, grouper, and blenny, the process that led to the current state has not been observed. Could only one process have produced such a state, or could many have done it? Such issues are very similar to those associated with a claim of adaptation, and as with adaptation, we can judge support for a claim of coevolution with criteria ranked for rigor and reliability (Stearns and Ebert 2001). They are all variations on one theme: to be accepted as narrowly coevolved, an interaction must be shown to have resulted from a process of reciprocal evolutionary change.

● **KEY CONCEPT**

Microevolutionary processes produce macroevolutionary patterns.

- *The selection criterion.* This is the most stringent: we observe the evolution of the interaction itself and document changes in the reproductive success of both partners as the interaction evolves. Examples that achieve this for one but not both partners include the serial transfers of pathogens between hosts described in Chapter 19 (p. 489).

- *The perturbation criterion.* We perturb the interaction with a controlled manipulation and document changes in the reproductive success of the partners. Examples include the manipulation of orchid spurs that caused changes in pollination efficiency described in Chapter 2 (p. 38).

- *The functional criterion.* We could define an interaction as coevolved if it has a clear functional relationship to the reproductive success of both partners and only occurs when it increases the success of at least one partner. Under other conditions, where it might impose a cost rather than yield a benefit, it does not occur.

- *The design criterion.* We could claim that an interaction had coevolved if it were complicated, unusual, and precise, like something an engineer might design—for example, an interaction like that of the grouper, wrasse, and blenny. Such claims need to be checked experimentally against alternatives, including the alternative that the interaction is a byproduct of selection for something else.

The point of this list is simply that no scientific claim should be accepted until it has survived risky tests in which it is explicitly contrasted with plausible alternatives.

How coevolution illustrates the interaction of micro- and macroevolution

Coevolution connects micro- to macroevolution in several ways.

- The concept of the Red Queen connects coevolutionary arms races at the level of interacting populations to extinction patterns in deep time, as the previous section illustrated with examples of epizootics and extinctions caused by introductions of predators, pathogens, and competitors to naïve populations. So long as species must coevolve with enemies that are coevolving in response, we should not expect much increase in long-term fitness as measured by reductions in extinction probabilities.

- Ecological cospeciation in a geographical mosaic connects the concept of adaptive—ecological—speciation to congruent phylogenies. If positive reciprocal influence promotes assortative mating for efficient interactions in both partners, cospeciation becomes more likely.

- Significant portions of several major clades have coevolved fairly intimately. Angiosperms and their insect pollinators and herbivores have co-radiated; trematodes and lice have co-radiated with their vertebrate hosts; and viral, bacterial, and fungal pathogens have coevolved with their hosts, pathogens, predators, and competitors—in short, virtually all living things.

Thus microevolutionary process produces macroevolutionary pattern, here as elsewhere, but pattern does not imply one process. In the coevolutionary context, we need to remember that the pattern of cospeciation detected in clades does not necessarily imply reciprocal effects on interacting adaptations in partner species. We can use the criteria listed above to help us decide when reciprocal effects suggested by congruent phylogenies have in fact occurred within each of the interacting species.

● SUMMARY

- There are three types of *narrow* coevolutionary interaction—mutualism ($++$), parasite–host, pathogen–host, and predator–prey (all lumped as $+-$), and competition ($--$).

- There are two types of *broad* co-evolutionary interaction—commensalism ($0+$) and unreciprocated negative impacts ($0-$).

- Coevolution occurs in a geographic mosaic and in temporal patterns that may differ between the coevolving partners.

- Coevolution has occurred at many levels, including the adjustments of components within organisms to each other; the ancient symbioses of mitochondria and chloroplasts; the interactions of males and females and parents and offspring; interactions between species; interactions between clades; and the coevolution of genes and culture.

- The importance of coevolution in the overall balance of selective forces depends upon the frequency with which interactions with partners occur and the impact that those interactions have on the reproductive success of both species.

- The relative evolutionary potential of the evolving partners is determined by their generation time, whether they are sexual or asexual, and their standing genetic variation for interaction traits.

- Coevolving partners can become locked into an arms race in which each maintains genetic variation in interaction traits in the other; sexual reproduction may then be promoted in both.

- Microevolutionary arms races may be causing a macroevolutionary extinction pattern: times to extinction do not seem to get appreciably longer over hundreds of millions of years. Both the microevolutionary arms race and the macroevolutionary extinction pattern are called the Red Queen.

- Species evolve specialization to avoid interspecific competition and to reduce the tradeoff-imposed costs of generalization. They do so more easily in homogeneous environments.

- Species evolve generalization to reduce intraspecific competition, because they must eat different things as they inhabit a series of habitats as they develop, and because they cannot escape spatially and temporally heterogeneous habitats.

- Virulence increases in horizontally transmitted pathogens and in cases where two or more pathogen genotypes inhabit the same host. In those cases it is limited by the cost of transmission. Virulence decreases in vertically transmitted pathogens.

- Mutualism can evolve from commensalism when the two partners in the interaction each reward the other for increases in investment in the interaction and when the partners and their descendants maintain contiguity in space.

- Commensalism can evolve from parasitism when genes for tolerance are fixed in the host population.

- Mimics and their models are a striking class of products of coevolution. The interactions producing mimicry can be 0+ (e.g. Batesian mimicry), ++ (e.g. Müllerian mimicry), or +− (e.g. aggressive mimicry).

- Flowers and their pollinators are another striking class of coevolutionary product. Active pollination has evolved at least twice, in figs and fig wasps and in yuccas and yucca moths. In both cases the pollinator is also a seed predator, and in both cases the plant can control the degree of seed predation by abortion of fruit with unacceptable levels of seed predation.

- The oldest coevolutionary domestication that we know of involves leafcutter ants, the fungi that they farm, the microfungal pathogens that attack the farmed food fungi, and the bacteria that the ants raise to combat the pathogens that attack their crops. The congruence of the phylogenies of ants and their farmed fungi is the evidence suggesting an age for the association of at least 50 million years.

- When interacting species have no history of coevolution, invasion can be catastrophic, as it was with rinderpest in Africa, *Phythophthora* in Australia, the plague in Europe, measles in the New World and Polynesia, and the invasion of South America by North American predators, pathogens, and competitors as the Isthmus of Panama was formed 13–2 Ma.

● RECOMMENDED READING

Herrera, C.M. and Pellmyr, O. (eds) (2002) *Plant–animal interactions: an evolutionary approach*. Blackwell Publishing, Oxford.

Karban, R. and Baldwin, I.T. (1997) *Induced responses to herbivory*. University of Chicago Press, Chicago.

Page, R.D.M. (ed.) (2002) *Tangled trees: phylogeny, cospeciation, and co-evolution*. University of Chicago Press, Chicago.

Thompson, J.N. (1994) *The co-evolutionary process*. University of Chicago Press, Chicago.

Tollrian, R. and Harvell, C.D. (eds) (1999) *The ecology and evolution of inducible defenses*. Princeton University Press, Princeton, NJ.

● QUESTIONS

18.1. Is it possible to test the claim that the reason there are so many different kinds of insects is that there are so many different kinds of flowering plants? If not, why not? If yes, how?

18.2. van Valen noted that the average duration of a species, from origin to extinction, did not increase much over the last 600 million years. He explained this by noting that if partners in coevolutionary interactions evolved counter-measures equally rapidly, the result would be no net improvement in fitness and no reduction in time to extinction. Is this pattern evidence for coevolution, or is it merely consistent with coevolution, and might it be explained by something else? Does time to extinction have anything to do with the microevolutionary definition of fitness?

18.3. In what sense is coevolution usefully analogous to a chemical chain reaction? In what ways is it not?

18.4. What are the advantages to the leafcutter ant of using a bacterium as a partner in a coevolutionary arms race with a sexual pathogen that is attacking the ant's obligately asexual cultivar?

18.5. What are the reasons to consider you, your mitochondria, and your symbiotic gut bacteria an organism? What are the reasons not to do so?

CHAPTER 19

Human evolution and evolutionary medicine

Evolutionary medicine asks what difference evolution has made to how we think about human health and disease. Both of the main branches of evolutionary thought—the reconstruction of history and the analysis of the consequences of natural selection—shed light on these topics. The chief players are *Homo sapiens*, its pathogens, and how they interact with each other.

Aspects of evolutionary medicine have been discussed in other contexts: aging in Chapter 10 (pp. 233 ff.), host–pathogen coevolution in Chapter 18 (pp. 460 ff.), to mention the two most obvious. But in an important way everything we have discussed in the book is preparation for this chapter, for it has developed a way of thinking about issues that combines patterns in deep time with short-term dynamics to expose puzzles and problems to illumination from several angles as we rotate them in our minds.

This chapter has two main sections. In the first, we consider what the historical aspect of evolution has to say about some medical issues. In the second, we do the same for selection.

The evolution of humans

The Pleiocene radiation of hominids, chimpanzees, and gorillas

● **KEY CONCEPT**

Hominids evolved in Africa in the Pleiocene and immigrated into Eurasia in the Pleistocene. Many issues of timing and trait evolution remain uncertain and draw huge interest.

The reconstruction of the relationships of anthropoid apes and humans elicits intense interest and heated debate. Three decades ago the identity of our closest living relative was not yet settled; molecular systematics now tells us that we are more closely related to chimpanzees than to gorillas, to both of those than to orangutans, and we last shared ancestors with chimpanzees about 6 Ma. When the fossil data are combined with the molecular evidence, a picture emerges of a hominid radiation (Figure 19.1) in which our closest relatives are now all extinct.

Several issues are not yet resolved, including how many species of hominids radiated in the clade to which we belong after we parted company with

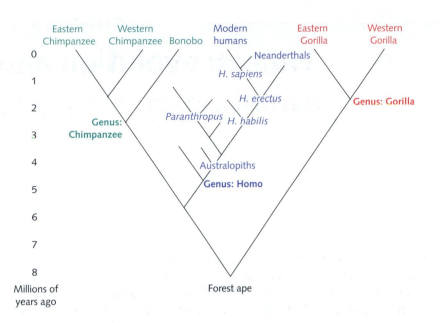

Figure 19.1 A simplified human phylogeny based both on fossils and on molecular systematics that places the last common ancestor with chimpanzees at about 6 Ma, the origin of genus *Homo* at about 4 Ma, and the origin of *H. sapiens* at about 0.5 Ma.

chimpanzees, and how many species of chimpanzees exist today (at least the common chimpanzee and the bonobo, and probably a third). The search for fossils in East Africa has yielded striking finds to which names have been assigned, among them *Australopithecus robustus*, *Homo erectus*, and *Homo habilis*: the robust southern ape, erect man, and tool-using man. There were others; not too long ago several hominid species coexisted in the same region.

Pleistocene emergence from Africa—about 150 000 years ago

The age of the genus *Homo* has been pushed back to about 4 Ma. Hominid fossils dating between 0.5 and 1.0 Ma have been recovered from Europe, Indonesia, and China. (Those in Spain are associated with evidence of systematic cannibalism.) Sometime in the last half million years *H. sapiens* appeared and diverged into two forms, *H. sapiens neanderthalensis* and *H. sapiens sapiens*. Thus Neanderthals were not our ancestors but our cousins. These coexisted for a while in Europe and south-west Asia; Neanderthals were probably extinct by 50 000 years ago. Anthropologists continue to debate two hypotheses for our origin: a single origin in Africa followed by spread to the other continents, or multiple origins in Africa and Asia (incipient speciation) followed by fusion upon secondary contact. The weight of the evidence at the moment favors an East African origin for *H. sapiens* about 150 000–250 000 years ago and emergence from Africa into the Middle East about 100 000 years ago (Figure 19.2).

This model is based on the evidence we have to date, but as new evidence comes in, it may change. Much of the picture of human origins given here will probably remain roughly correct, but experience suggests that future fossil

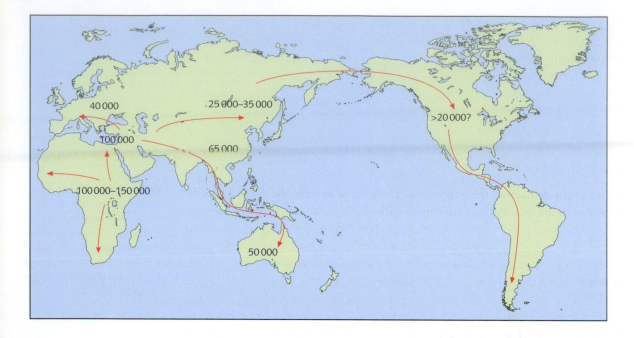

Figure 19.2 One recent view of the origin of humans and their occupation of the planet. (From Stearns 1999.)

discoveries and improved technology and methods will change our opinions:

- of when we diverged from the chimpanzees;
- of how many hominid species have existed and gone extinct since then;
- of when we emerged from Africa;
- of whether we also speciated in Asia; and
- of when we arrived in various parts of the globe.

There is great uncertainty about when language with grammatical structure evolved. The anatomy of the larynx suggests that Neanderthals may not have been capable of the refined modulation of sound production that we have, but that too is controversial. If it is true, then we may have had decent language capacity from about the time we diverged from the Neanderthals. Evidence of self-consciousness in the form of art and burial is fairly convincing for the last 30 000–60 000 years.

● **KEY CONCEPT**

As we colonized the planet we adapted to local variation in diseases, diet, and other aspects of lifestyle. Our genes still contain important traces of our history and not all of that information is currently adaptive, for cultural change is much faster than genetic change.

How our history has affected health and disease

We have spent a long time in hunter–gatherer groups and a short time in settlements and cities

During the period depicted in Figure 19.2 we existed in hunter–gatherer groups. Agriculture emerged in the Near East 10 000–12 000 years ago, with later, independent origins in China and the Americas. The origin of agriculture led to the growth of permanent settlements near cultivated fields, to cities, then to city states, regional powers, nations, and empires within the last 4000–5000 years: an evolutionary blink of the eye that contains within it all written history. That is the framework within which we have coevolved with many of our diseases, and that is the evolutionary legacy that our bodies and minds bring to life in the 21st century of the current era.

History has left genetic traces in disease resistance and drug response

As we expanded across the planet, we encountered specific diseases in specific locations. Some populations coevolved with one disease, others with other diseases. The evolutionary response often involved alleles with pleiotropic effects in the sense that they improved disease resistance but had detrimental effects when disease was absent. Thus not all humans are the same with respect to their ability to resist disease. We also differ in ability to metabolize drugs and other chemicals.

For example, there is variation among individuals within a given population, and among populations in different areas, in ability to metabolize carcinogens. One enzyme involved in such metabolism is N-acetyltransferase (NAT2), the gene for which exists in several alleles whose function is scored from slow to fast in terms of ability to catalyze the transfer of an acetyl group. Slow acetylation is inherited recessively; people homozygous for the slow allele are at increased risk for breast and bladder cancer if they smoke. Slow alleles are more common in Caucasians and Africans than they are in Japanese, Koreans, and American Indians. Why they remain so prevalent is a mystery; one supposes that it either currently has or at some time in the recent past had a beneficial function, but that function remains unknown (Meyer 1999).

Blood groups are associated with disease susceptibility and resistance

We all have an ABO blood type that helps to determine with whom we can have compatible blood transfers. ABO antibodies are secreted into saliva, where they form a first line of defense against infectious diseases. People carrying mutations that reduce such secretion are at higher risk of contracting diseases such as meningitis. One's blood group also influences susceptibility to cholera, plague, and some kinds of diarrhea. This suggests that the frequency of a given blood type in a given human population may depend on the history of exposure to those sorts of infectious disease. Blood group O is more common in isolated populations, such as natives of Central and South America, that have not shared exposure to such diseases as much as have Europeans and Asians. The reason for this geographical pattern may be that blood group O is associated with susceptibility to peptic ulcers, whose causal agent, *Helicobacter pylori*, adheres to gastric epithelial cells using antigens associated with the O blood group. The isolated populations mentioned may never have been colonized by *H. pylori*, or colonized by it only recently, after contact with Europeans. If there were no infectious diseases, we might all have blood group O or no blood groups at all.

Malaria incidence and sickle-cell frequency overlap in space

The best-documented example of how geographic variation in exposure to disease has shaped human genetic variation is that of sickle-cell anemia, which is found at higher frequency in areas where malaria is a serious health problem. The sickle-cell mutation changes a single amino acid in one of the two protein subunits of hemoglobin. That change alters the binding properties of the molecule in such a way that red blood cells are distorted, making them unsuitable for habitation by *Plasmodium*, the malaria parasite. In consequence, *Plasmodium* has to live in the liquid component of the blood rather than inside the red blood cells in which it normally hides from the immune system of its host. There it is exposed to attack by the human immune system and suffers accordingly. Humans homozygous for the sickle-cell allele are anemic and suffer high mortality from anemia early in life. Those homozygous for the normal allele are susceptible to malaria, which infects hundreds of millions and kills millions of people each year, most of them children. It is the heterozygotes that benefit in the presence of malaria: they are not anemic, and they are resistant. This genetic mechanism of resistance imposes strong costs, for the only way to continue to produce resistant heterozygotes is also to produce homozygotes that suffer either anemia or malaria.

Phylogenetic methods yield evidence in medical cases involving criminal activity

In Chapter 13 (p. 303) we saw how systematics allows us to reconstruct phylogenetic trees from nucleic acid sequences; there they were illustrated with detective work on a rape victim (Figure 13.7). These methods have been applied to a medical problem: who was the person responsible for infecting others with the HIV virus (Hillis and Huelsenbeck 1994)? The methods work because HIV is an RNA virus, and RNA viruses evolve so rapidly that each host has a new, unique strain soon after infection. HIV samples were recovered from each person and sequenced. The phylogenetic tree of HIV, with a different branch for each infected human (Figure 19.3), was constructed and used to identify both the Florida dentist who transmitted HIV to his patients and the sailor who was among the first to introduce HIV into Sweden. (Both died before the analyses were published.)

Thus one contribution of systematics has been the development of reliable methods that allow us to assign probabilities to particular historical scenarios. The same methods are used to reconstruct the human evolutionary tree.

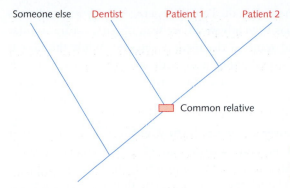

Figure 19.3 The phylogenetic tree of the HIV virus identifies who transmitted it.

Human genetic diseases may have provided resistance to infectious diseases that have disappeared

Infectious diseases have probably altered the genetic constitution of most human populations. Exactly how they have done so, to which genes, to what extent, and what might be the costs of resistance—the detrimental effects of resistant alleles in the absence of the disease—are all active research topics in medical genetics. It is often thought that human genetic diseases that currently exist at higher frequencies than their detrimental effects would lead us to predict are as common as they are because in the recent past, before infectious diseases were brought under control by hygiene and antibiotics, they conferred disease resistance. For example, there are genes on the short arm of chromosome 5 associated with allergies and asthma that appear to protect us from infections of parasitic worms (Hill and Motulsky 1999). If that is the case, they continue to exist only because we have not yet come into genetic equilibrium with the modern environment. This idea forms a natural transition to the next topic.

Lifestyle diseases: are we stuck in the Paleolithic?

There can be no doubt that many of us no longer have the lifestyle of a hunter–gatherer. We are more sedentary and have to set aside periods for exercise, if we exercise at all, rather than having exercise forced upon us by the need to find food, shelter, and clothing and to escape from and defend ourselves against predators. We eat more carbohydrates, refined sugar, high-fat meat, and processed foods, fewer fresh vegetables, fruits, and nuts, and less low-fat meat. We no longer live in small, cohesive groups; many of us no longer live in extended families; we encounter one another at high population densities; we can no longer simply move to another place to find open land, food to harvest, and clean water.

These contrasts suggest the hypothesis that the chronic, non-infectious diseases of modern life are the product of poor adaptation to our lifestyle, that our bodies and minds remain stuck in the Paleolithic era. Among the conditions for which this explanation has been advanced are war, obesity, cardiovascular disease, nearsightedness, and some of the mental illnesses. While the idea is plausible, two things must be said about it immediately. The first is that it is much easier to tell a story, to assemble a set of correlated observations, than it is to establish that lack of adaptation to the modern environment is the only possible cause of the problem. The second is that our bodies and minds have been evolving continuously over our entire history, not just in one particular environment, and this makes the choice of a certain set of environmental conditions as the putative ancestral environment a bit problematic.

Of the many conditions for which this kind of explanation has been advanced (see Eaton and Eaton 1999, Trevathan et al. 1999), we have chosen to discuss two, female reproductive cancers and autoimmune diseases.

The more menstrual cycles per lifetime the higher the probability of reproductive cancers

In naturally cycling populations that do not practice contraception, a normal human female life history would be roughly as follows. A woman would start menstruating at 14–16 years of age, have her first child before 20, experience several years when she did not ovulate while nursing her child, become pregnant again soon after weaning a child, have a child about every 3–5 years until she was 35–40, and have a completed family size of 4–6 children, of which on average half would survive to adulthood. During her entire life she would experience 110–130 menstrual cycles (Strassmann 1997).

In contrast, modern women start menstruating earlier in life, practice contraception, have fewer children, and experience many more menstrual cycles per lifetime: 350–450 before menopause. Each time a woman goes through a menstrual cycle, cells divide in many of her organs—ovaries, uterus, breasts, and cervix. Every cell division is an opportunity for mutation, and cancer is a disease that results after between seven and nine mutations have been accumulated in a somatic cell line. There are billions of cells in the organs mentioned, and even though a mutation in any one of them is unlikely, the chance of a mutation somewhere in the entire organ once every few menstrual cycles is fairly probable (Greaves 2000). In fact, most cancers, not just reproductive cancers, occur in cell lines that are undergoing regular mitosis throughout adult life: in the skin, in the intestinal epithelium, in the lung epithelium—as well as in breasts and uteri.

Quadrupling the number of menstrual cycles per lifetime greatly increases the probability that some cell, in some organ, will sequentially acquire all the mutations needed to let it first slip out of cell-cycle control, multiplying to form a local tumor, and then, after acquiring further mutations, start to move about the body, forming metastases in other tissues. Female reproductive cancers do occur at higher frequency in populations where women cycle more frequently; this does seem to be a case where our bodies have not adapted to modern conditions by evolving better defenses against mutational damage.

The hygiene hypothesis: do clean environments increase autoimmune diseases?

The post-industrial human environment is relatively clean and disease-free. Hunter–gatherers had a higher incidence of worm infections (trematodes and nematodes), and pre-modern city dwellers had a much higher incidence of infectious diseases. It is thus likely that our immune systems are set up to deal with a much higher level of exposure to parasites and pathogens than we currently encounter (Stearns and Ebert 2001).

The immune system responds to infection by producing a huge random array of antibodies, then recruiting those cell lines that produce the antibodies that match the antigens presented by a particular invader. That huge initial array of antibodies is a double-edged sword. If, by mistake, an antibody against your own tissues is recruited, the immune system will mount an immune response against part of your own body. There are fail-safe devices built into the system, most prominently in the thymus gland, one of whose functions is to screen out antibodies against self. But sometimes a mistake is made, antibodies against self are recruited and amplified, and an autoimmune disease is the result. The autoimmune diseases include multiple sclerosis, lupus, Type 1 diabetes, rheumatoid arthritis, Crohn's disease, Grave's disease, psoriasis, asthma, and allergies.

Childhood allergies have been increasing in frequency over the last few decades in industrialized countries, with allergies to milk, eggs, peanuts, tree nuts, fish, shellfish, soy, and wheat among the most prominent. The symptoms include sneezing, vomiting, and inflammation. When the lungs are affected, patients suffer asthma attacks. Children of families that use antibiotics and vaccinations appear to have more allergies than children of families that do not (Bjorksten 1999), suggesting that children exposed to a dirty environment and a variety of pathogens early in life develop fewer allergies.

Autoimmune disease may develop more frequently when an immune system that has evolved to deal with multiple invaders fails to adapt to a more sterile environment. Pre-industrial humans may never have experienced life without infections by worms that tend to reduce our immune response, probably to their own benefit. When worms were a normal part of our environment, our immune system evolved to overexpress immune function to counter the suppressive effect of the worms. Without worms, the released immune system overproduces antibodies, resulting in various autoimmune diseases. One therapy suggested for these diseases, among which are found some syndromes for which there is not yet any other effective therapy, is inoculation with the surface proteins of parasitic worms to re-create the way the immune system experienced the ancestral environment. Here evolutionary thinking suggests a concrete benefit from an unexpected source. The idea is plausible and is starting to be tested.

Thus the historical perspective on human health and disease suggests two main points. First, as we colonized the continents, we encountered and adapted to different pathogens in different places. The various human populations of the planet diverged genetically and now differ in their genetic capacity to resist specific diseases. They also differ in the ability to metabolize given drugs. Second, we have not yet adapted to the environment in which we now live, an environment that has been changing rapidly. Cultural evolution has outpaced the ability of biology to keep up with it. Modern medicine and hygiene have improved the quality of life in many ways, but they have also had unintended consequences with serious effects. Methods of birth control that allow menstruation appear to increase the frequency of women's reproductive cancers, and the unusually clean modern environments in which children are raised may increase the frequency of autoimmune diseases.

How selection shapes virulence and atresia

> **● KEY CONCEPT**
> Some important medical issues appear in a dramatically different light when we view them as shaped by natural selection.

Having looked at a few of the ways in which history sheds light on medical issues, we now turn to ways in which thinking about selection does the same. Of the many topics that could be discussed, the evolution of virulence and antibiotic resistance in pathogens stand out. We also discuss some of the subtle ways in which selection has shaped human physiology, focusing on selective atresia of oocytes and selective abortion of zygotes.

Virulence evolves to intermediate levels set by tradeoffs

> **Virulence:** *The property of pathogens of being highly malignant or deadly, causing serious disease or death in the host.*

The basic tradeoff that a pathogen faces when it infects a host is the tradeoff between immediate reproductive success—rapid growth—and impacts on the host that potentially reduce long-term reproductive success. In many cases this tradeoff will result in the evolution of an intermediate level of **virulence**, for it often does not pay pathogens to kill their host before they can be transmitted. Exactly where that optimal level of intermediate virulence will be established depends on the mode of transmission.

Diseases that are primarily vertically transmitted, from parent to offspring, such as the nematode worms that parasitize fig wasps (Herre 1993), should evolve lower levels of virulence than diseases that can also be horizontally transmitted, from host to unrelated host. If symptoms associated with virulence alter the probability that the pathogen will be transmitted, then selection will shape the pathogen to *increase* the intensity of those symptoms if it can thereby increase its lifetime transmission probability. Similarly, selection will shape the pathogen to *decrease* that intensity if it can thereby increase its lifetime transmission probability. Thus the level of virulence actually encountered depends on how symptoms affect transmission probability (Ewald 1984).

Virulence also responds to competition among strains within hosts

The second mechanism for virulence evolution focuses on competition of pathogens within the host. Virulence will be selected to increase if several pathogen genotypes infect the host at once. In multiple infections, competition within the host will be won by the pathogen genotype that exploits the host completely and rapidly before the other competing pathogens can rob it of resources. The level of virulence that will evolve depends on the relative importance of competition within and between hosts and thus on the type and frequency of multiple infections and the transmission costs associated with early host death caused by the pathogens. In vertically transmitted parasites, where between-host competition is important, virulence should be low. If we experimentally remove the costs of horizontal transmission, virulence should increase, and it does (Ebert 1998). These cases are the extremes of a continuum: vertical transmission largely excludes within-host competition, and serial-passage experiments exclude between-host selection. Most diseases evolve under a mixture of conditions, and it is not clear that interpolating between the extremes will predict the virulence of most diseases. The ecological conditions that allow some pathogens to increase and others to decrease in virulence are only partially understood.

Virulence appears to be acquired in a series of steps that are adaptive for the pathogen

The evolution of virulence also has an historical dimension that can be explored with molecular phylogenetics. Such analyses reveal that virulent organisms are 'constructed' through the stepwise accumulation of virulence determinants. The pattern suggests that each step represents an adaptive advance for the pathogen; the accumulation of steps is compelling evidence that virulence is adaptive. For example, the evolution of pathogenic *Escherichia coli* strains began about 9 Ma, and the virulent pathogen responsible for epidemics of food poisoning, *E. coli* O157: H7, branched off from a common ancestor of *E. coli* K-12 about 4.5 Ma. Lineages of *E. coli* acquired the same virulence factors in parallel, including genes involved in sticking to the gut wall, breaking open blood cells, and poisoning tissues. The parallel evolution suggests that selection has favored an ordered acquisition of genes that progressively built up the molecular mechanisms that increase virulence (Reid et al. 2000). *E. coli* virulence genes have levels of nonsynonymous change five to ten times greater than housekeeping genes. Such changes may help the individual organism to escape host immune response or a variant to spread in a host population that previously acquired immunity to other variants. *E. coli* is thus a rapidly evolving species capable of generating new pathogenic variants that can escape host protective mechanisms and produce new disease syndromes (Donnenberg and Whittam 2001).

A high rate of nonsynonymous substitutions in genes associated with virulence has also been found in the molecular phylogeny of influenza A viruses (Bush et al. 1999). The branch with the highest proportion of nonsynonymous substitutions is the one that takes the virus from epidemic to epidemic, presumably because it has the greatest selective advantage. Such information could predict which flu strain in this year's flu season will cause next year's epidemic, and vaccination policy could be adjusted accordingly.

Thus both selection and history shed light on the evolution of virulence.

Increased mutation rates help bacteria deal with antibiotics but hinder transmission

In Chapter 2 (pp. 38 ff.) we discussed how pathogenic bacteria rapidly evolve resistance to antibiotics. The mechanisms include horizontal acquisition of resistance genes carried by plasmids or transposons, recombination of foreign DNA into the chromosome, and mutations in chromosomal loci. The rapid evolution of antibiotic resistance is some of the most convincing evidence that we have for adaptive evolution. Moxon et al. (1994) gave the story a new twist by suggesting that bacteria have two kinds of genes: housekeeping genes that mutate at low frequency, and highly mutable genes whose mutability helps bacteria adapt to changing environments. Many populations of bacteria contain some cells with a mutation rate from 10 to 10 000 times that of normal cells, usually because they have a defective repair system. In combination mutable regions of the genome and cell lineages with increased overall mutation rates would allow a bacterial lineage rapidly to accumulate many alleles, some of which could evade stressful environments such as host defenses or antibiotics and give bacterial pathogens living in humans under treatment for infectious disease a selective advantage.

There is some evidence that the mutation rate of bacteria challenged with antibiotics does increase in human hosts (Martinez and Baquero 2000), but experiments on mice suggest that the advantage may be confined to a single host and may not be able to accumulate because the higher mutation rate, while helpful within the host, causes problems with transmission to new hosts (Giraud et al. 2001). This example illustrates the formulation of a potentially important hypothesis, its initial support, and its subsequent qualification through experimental tests—a standard story in much of science that is becoming more common in evolutionary biology as the scope of experimental evolution expands.

Selection arenas are screening mechanisms that eliminate defective gametes and zygotes

Selection arena: *An adaptation produced by natural selection that uses natural selection to produce its effect.*

A selection arena is a selection process occurring inside an entity that is a unit of selection in its own right at a higher level (see Ch. 9, p. 200, multilevel selection). Here natural selection has produced an adaptation that uses natural selection to achieve its effect. Selection arenas have been invoked to explain the vertebrate immune system, the overproduction of zygotes (Stearns 1987), the elimination of defective mitochondria from the germ line by oocytic atresia (Krakauer and Mira 1999), and selection among pollen tubes growing down flower styles.

Selection on variation in reproductive performance of the organisms—the higher level—has shaped the internal selection process within the organisms—the lower level.

Discarding oocytes may cleanse the germ line of mutationally damaged mitochondria

The process that leads to total oocyte loss—menopause—usually finishes by the time a woman is in her 50s, but it starts much earlier. By the third month of gestation, ovaries containing about 7 million oocytes have developed in the female human embryo. By birth that number has been reduced to about 1 million and by the onset of menstruation at puberty to a few thousand. This oocyte destruction, called oocytic atresia, is found in most mammals. It poses a puzzle. Why make so many oocytes and then destroy most of them? And why destroy 40–60% of fertilized zygotes, as appears to be the case, with early natural miscarriage without the knowledge of the woman being the outcome of many human pregnancies?

One idea is that selective atresia eliminates mitochondria damaged by mutations (Jansen and de Boer 1998, Krakauer and Mira 1999). If mitochondria reproduce asexually and pass regularly through population bottlenecks, they cannot avoid accumulating deleterious mutations by Muller's Ratchet; eventually all will be damaged. The problem can be avoided, however, if the mitochondria with the deleterious mutations can be isolated and discarded, which would happen if only a few mitochondria were introduced into each of many oocytes and if each oocyte signaled the performance of its mitochondria, allowing the maternal tissue to identify the oocytes to be discarded.

This hypothesis makes at least three predictions. (1) If ovaries are sampled from fetuses from 3 months onwards, the percentage of defective mitochondria should decline. This has not been confirmed for humans. (2) The number of mitochondria allocated to a new oocyte should be small, ideally just one. The number of mitochondria with which an oocyte starts life is not yet known precisely, but a review of all published microphotographs of primordial oocytes suggests that the number is less than 10 (Jansen and de Boer 1998). If the number of mitochondria entering a primordial oocyte were large, the mechanism would not work, because then most oocytes would get some defective mitochondria, the biochemical signals given off by the oocytes with defective mitochondria would not be distinctive, and oocytes could not be eliminated selectively. Estimating the number of mitochondria with which an oocyte starts life is thus an important issue. (3) The signal that initiates the process that selectively destroys an oocyte should have a functional relationship to mitochondrial performance; it should reliably indicate the presence of damaged mitochondria. Whether such signals exist is not yet known.

Atresia could be selective but not necessarily have anything to do with damaged mitochondria; it could also eliminate nuclear mutations. Later, selective

> **Oocytic atresia:** *The elimination of oocytes from the developing ovary.*

miscarriages could eliminate unfortunate diploid gene combinations with serious enough effects to be detected in the performance of embryos early in development. One such unfortunate gene combination occurs when the two partners have similar alleles for immune response—similar HLA genes. The evidence comes from a long-term medical study of the Hutterite communities in South Dakota.

Humans spontaneously abort early-stage embryos with genetically deficient immune systems

The Hutterites, who moved to North America from Switzerland in the 19th century, are a small community that has become inbred. Hutterites are endogamous—they only marry other Hutterites. They prohibit contraception, and while they do not have to marry, if they do marry, they are not allowed to divorce. Thus a mistake in mate choice has life-long consequences. Very importantly, the Hutterites cooperate with the research team that has been investigating their reproductive biology. Because of that cooperation, a complete genealogy is available, and the limited number of five-locus HLA haplotypes is known for 411 Hutterite couples.

Some Hutterite women suffer from recurrent spontaneous abortions; those whose husbands have similar HLA loci are more likely to suffer spontaneous abortions than women whose husbands have different HLA alleles (Ober et al. 1992). Women who do not share such genes with their husbands achieve a family size of five children within 8 years; those who share one or more than one such gene take 14–15 years to have five children. The distributions of completed family sizes reveal the very strong fitness cost associated with choosing the wrong partner (Figure 19.4). Women without such problems have an average of nine children per lifetime; women suffering spontaneous abortions have an average of 6.5.

Ober et al. (1997) then examined mate choice to see whether potential mates avoided partners with similar immune genes. Significantly fewer matches of HLA haplotypes between spouses were observed than would be expected at random. Thus Hutterites avoid mating with partners with the same HLA haplotype. Whether the mechanism of mate choice is physiological—through scent or taste, as it is in mice—or cultural, through avoidance of partners from families where older siblings have experienced problems, or both, is not yet clear. Avoidance reduces the problem of spontaneous abortions, but some errors are made unwittingly, and those who choose a partner with similar immunity genes suffer greatly.

Spontaneous abortion based on immune deficiency may reflect evolution under inbreeding

In our post-industrial society, an adaptation for discarding early-stage embryos likely to die of infectious disease would not evolve for two reasons. First, infectious

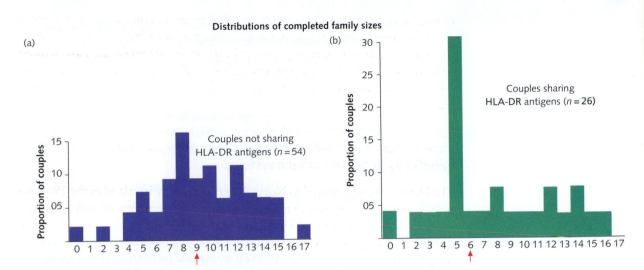

Figure 19.4 Completed family sizes in Hutterite couples who do and do not share HLA alleles. Those not sharing HLA alleles (a) have larger families, often ending up with 8–15 children; those sharing such alleles (b) experience frequent spontaneous abortions and tend to stop attempting to have children when they have five. (Courtesy of Carole Ober.)

diseases are, at least for the moment, not a significant problem in the developed world. Second, we now live in a very outbred society. The invention of the bicycle, the train, automobiles, and airplanes has greatly increased both general mobility and the average distance between the birthplaces of married couples. The probability that both partners would share immune genes is low. But in hunter–gatherer groups of 50–100 people, especially in those that are endogamous, inbreeding levels would have been high and the number of HLA haplotypes circulating within the group would have been small. It is that combination of circumstances—frequent exposure to infectious disease and high probability that a sexual partner will have a similar HLA haplotype—that would select for the mechanisms that make spontaneous abortions of the type observed in the Hutterites not only possible but advantageous. An important difference in a hunter–gatherer society would be the opportunity to select a new mate if there were repeated spontaneous abortions with the current mate: where divorce is possible and children can be born more quickly with a new partner, the strength of selection to improve the mechanism would be greater.

● SUMMARY

Both history and selection shed light on the evolution of human health and disease.

- Systematics can clarify the recent history of pathogens; in the case of HIV, it can identify who it was that transmitted the disease to a particular person.

- Humans colonized the continents over roughly the last 150 000 years and arrived in the last large habitable islands within the last 1000–2000 years.

- As they colonized the continents, human populations diverged genetically, in part due to adaptation to local encounters with particular infectious diseases. They also diverged in the genetic capacity to metabolize drugs.

- Many human genetic diseases are currently found at higher incidence than one would predict given their detrimental effects. They are thought to have had some beneficial effect in the past, such as conferring resistance to an infectious disease that has since become unimportant.

- The rapid development of civilization has outrun the ability of biology to keep pace. Lifestyle diseases are the result. They include obesity, cardiovascular disease, and autoimmune diseases.

- An unintended consequence of birth-control methods that do not suppress menstruation may be an increased incidence of female reproductive cancers.

- Oocytic atresia may reduce mutational buildup; spontaneous abortions may eliminate embryos with little chance of surviving to reproduce.

- Humans may choose their mates in part for complementary immune genes that would result in more resistant offspring.

The application of evolutionary thinking to medical issues has yielded important insights. Some have withstood critical tests; others await such tests. In the future, both clinicians and medical researchers will find increasing reason to be aware of evolutionary interpretations of the problems they confront.

● RECOMMENDED READING

Nesse, R.M. and Williams, G.C. (1995) *Why we get sick: the new science of Darwinian medicine*. Times Books, New York.

Stearns, S.C. (ed.) (1999) *Evolution in health and disease*. Oxford University Press, Oxford.

Trevathan, W.R., Smith, E.O., and McKenna, J.J. (eds) (1999) *Evolutionary medicine*. Oxford University Press, Oxford.

● QUESTIONS

19.1. Deviations from the environment to which we are adapted have been used to explain a huge variety of conditions, including sudden infant death syndrome (SIDS), the persistent crying described as colic, lower back pain, drug addiction, and depression. Consult the recommended readings for details and then evaluate the quality of evidence for an evolutionary interpretation of one such condition.

19.2. If one is feeling a bit under the weather in Vietnam and purchases some pills to help with headache, one gets as a matter of course an aspirin, an antimalarial, and an antibiotic. Discuss the selective impact of widespread sublethal doses of antimalarials and antibiotics on the evolution of drug resistance in infectious diseases. Discuss why doctors insist that patients on antibiotics complete the regimen and take all their pills.

19.3. Design a research project that could detect selective oocytic atresia and establish whether the object of selection was defective mitochondria, defective nuclear genes, or both.

CHAPTER 20

Conclusion and prospect

In this concluding chapter we comment on the status, nature, and preoccupations of evolutionary biology and describe some unsolved problems.

Key conclusions about evolution

● KEY CONCEPT

Evolution is real, well tested, and reliable.

Evolution is as well tested as any major idea in science. It did happen, and it is continuing. Natural selection is always at work; gene frequencies do drift; both have often been observed in the laboratory and in the field. The fossil record, molecular systematics, and the mechanisms shared by all living cells clearly demonstrate the ancient origin of life and the continuity of descent with modification from shared ancestors. Evolution is a reality that cannot be ignored.

● KEY CONCEPT

Everything in biology has both proximate and ultimate causes.

Evolutionary theory resembles theoretical physics in its attempt to derive empirically sufficient predictions from simple first principles. Evolution has an historical element, however, not easily incorporated into the theory and shared with astronomy and geology, where unique events have also had important consequences. Thus evolutionary causation is complex. Everything in biology has both proximate, or mechanistic, and ultimate, or evolutionary, causes. The ultimate causes are themselves a mixture of history, accidents, and adaptation.

● KEY CONCEPT

Evolution is happening now and can be fast.

Evolution is not something that takes millions of years. It can produce results quickly. Bacteria evolve resistance to antibiotics within a few months, and insects evolve resistance to insecticides within a few years. Evolution has not stopped in humans: it occurs in all populations in which individuals vary in their reproductive success and in which some heritable traits are correlated with reproductive success. Both conditions are true for humans, at least for some heritable traits. The key question is, which ones? In most cases, we do not know.

Applying evolution to humans remains controversial

Science assumes that everything has a material explanation. This assumption has been so successful that it is now often regarded as a result. Evolution and molecular biology extend the scope of materialistic explanation to all things biological. As long as the things explained do not include humans, the implications of materialism have not been too controversial, but at least since the publication of Darwin's *Origin of species* in 1859, people inside and outside science have struggled with materialistic explanations of humans.

How far does evolutionary explanation apply to humans? The lack of a generally accepted answer continues to be an important reason why evolution is not more widely understood and accepted. Most scientists accept evolution as a fact, and virtually all biologists use it in their thinking, but in society as a whole opinions are much more diverse. They range from complete rejection by fundamentalist Christians and Muslims, through the position of the Catholic Church, which is that evolution did occur, and the variable positions of other religions, to the situation of many informed lay people, who think that evolution probably did occur but do not think through its implications. Thus evolution continues to encounter resistance because it contradicts some aspects of some religions. And some who think deeply about it find that its implications challenge their self-conceptions. This can be uncomfortable.

Within the academic community, evolutionary explanations are controversial in disciplines that focus on humans—anthropology, psychology, medicine, economics, political science, and sociology—for quite different reasons. Here it is not materialism that causes the resistance, but the competition of evolution with explanatory paradigms pitched at other levels. Those raised in other disciplines defend their intellectual territory and are not happy to see it threatened. Nevertheless, evolutionary explanations have gained some ground, and there are now sub-disciplines of evolutionary anthropology, evolutionary psychology, and evolutionary epistemology. In the analysis of optimality, tradeoffs, game theory, and conflicts there is considerable overlap between evolution, economics, and political science. Even so, the spread of evolutionary explanations into related disciplines continues to meet with resistance, some of it justified.

How does evolution interact with cultural change? Does anything divide humans from animals?

At least two important and interrelated questions about the scope of evolutionary explanation remain unanswered. First, how does biological evolution interact with cultural change? How should we analyze phenomena with both biological and cultural causes? Second, are humans different from other animals in some essential respect, or are the boundaries between humans and other animals diffuse? Do other animals have culture, consciousness, and language? By some criteria they do, by others they do not. Because several borders between humans and chimpanzees that had previously seemed impervious have recently proven to be unclear, we think it unwise to draw hard borders between humans and animals.

Thus the scope of evolutionary explanation is not yet determined. It has taken our culture more than 140 years to partially digest the implications of evolution, and it will take at least as long before they are fully digested. The scope of evolutionary explanation will continue to expand, will continue to meet resistance, and will continue to inject new ideas and insights into cognate disciplines, but we would not be surprised if a book written at the end of the 21st century also concluded that the implications of evolution had not yet been fully assimilated, for they go to the heart of what we are. People fight very hard over self-definitions.

Evolutionary biology focused on genetics and will focus on development

That organisms might be related by descent was recognized by Aristotle, and evolutionary ideas were widespread before Darwin, but Darwin changed everything by proposing a *mechanism* to produce adaptation: natural selection of heritable traits. His lack of a plausible mechanism for inheritance that could maintain the genetic variation needed to sustain a response to selection was so striking that when Mendel's laws were rediscovered in 1900, and it became accepted that genes were material particles located on chromosomes whose behavior followed Mendel's laws, genetics became the main preoccupation of evolutionary biology. The whole first half of the 20th century was devoted to the assimilation of genetics into evolutionary theory. This assimilation consisted primarily of the demonstration that phenomena from the population through the species to higher taxonomic groups and the fossil record were *consistent* with the newly discovered genetic mechanisms.

There are three criteria used in logic to judge how well we understand the causation of any phenomenon: consistency, necessity, and sufficiency. Here is what they mean:

- *consistency*: A does not rule out B, and there is some reason to think that A might be associated with B;
- *necessity*: B only happens when A is present;
- *sufficiency*: B only happens when A is present, and A is all that is needed to elicit B—no other condition is needed.

The strongest criterion is *sufficiency*. If one can show that a cause is by itself sufficient to elicit an effect, then the cause must be present for the phenomenon to exist and nothing else is necessary. It was the consistency of all evolutionary phenomena with genetics that was established, not the *sufficiency* of genetics to explain all evolutionary phenomena. At the time that consistency was established, some exaggerated the logical state of affairs and claimed sufficiency when only consistency had been demonstrated.

The second half of the 20th century has seen the emergence of molecular biology, the assimilation into evolutionary biology of molecular genetics, the reinvigoration of the interaction between evolutionary and developmental biology, and the emergence of evolutionary and behavioral ecology. The influx of these alternative approaches to explanation—molecular, developmental, and ecological—and the recognition of the complexity of the determination of phenotypes—through the interaction of developmental mechanisms with the environment—have brought evolutionary biology closer to the goal of logical sufficiency.

But it is not there yet.

At the turn of the 21st century, it is clear that we will see rapid advances made in the use of developmental molecular genetics in a comparative context to understand the mechanistic basis of body plans. One consequence will be the clarification of the genotype–phenotype relationship, whose obscurity has blocked progress in 20th century evolutionary biology just as surely as the lack of a genetic mechanism blocked progress in 19th century evolutionary biology. In the process, two major puzzles may be solved: how traits are fixed and how constraints evolve.

How are traits fixed? How do constraints evolve?

Some traits are fixed within large lineages whereas others remain variable within populations. For example, all tetrapod vertebrates have four limbs, two eyes, and one backbone, but quite variable adult body weights, numbers of offspring, and lifespans, and all arthropods have chitinous exoskeletons and grow discontinuously by molting, but the morphology of their appendages and the color of their bodies varies dramatically among lineages and species. Because all genes are exposed to mutation, and because environments change considerably over long periods of time, it is unlikely that extrinsic stabilizing selection has maintained so many traits in so many lineages in a fixed state for hundreds of millions of years. It seems reasonable to look for additional reasons for long-term maintenance of traits in fixed states in the genotype–phenotype relation, in development, and in the constraints inherent in multi-trait evolution (Stearns 1994).

Microevolution does not yet explain how varying traits evolve into fixed traits, nor does it explain how fixed traits become constraints on the further evolution of the still-varying traits. Microevolutionary theories of phenotypic evolution appear to be ahistorical, as is physics: they deal with processes that occur at all times and in all places and that do not require information about a particular, concrete history to be understood. To make successful predictions about phenotypes, however, they must assume, usually implicitly, that certain traits are fixed and that other traits vary. The traits that happen to be fixed in the lineage are historical particulars that appear as boundary conditions or empirically fitted parameters in the models. This is not satisfactory. It should be possible to understand the transformation of variable traits into fixed ones. Doing so would help to connect macro- to microevolution.

The two major puzzles concern the relation of the traits that are fixed within clades to the traits that still vary among individuals.

- The first is, how did the fixed traits—which are not just fixed genes—become invariant? How did trait fixation evolve? It must have had something to do with the developmental control of gene expression, for all DNA sequences mutate, all genes become variable.

- The second puzzle is, what are the effects of the fixed traits on the further evolution of the variable ones? Is the expression of the genetic variation of the variable traits affected by which other traits happen to be fixed in that clade? Do the fixed traits affect the further evolution of the traits that remain genetically variable, thus producing clade-specific patterns of response to selection?

Vavilov (1922) documented clade-specific patterns of variation in crop plants, and theoretical and experimental analyses of the causes of clade-specificity in patterns of variation have been made by Alberch (1989) for amphibian limbs, by Nijhout (1991) for butterfly wings, and by Ebert (1994) for crustacean life histories. Their work (see Chapter 7, pp. 152 ff.) provides starting points for the solution of these two puzzles.

Other unsolved problems

If evolution can be fast, why is it often so slow?

Two remarkable and apparently contradictory facts have emerged in the last 30 years. On the one hand, there is stasis documented in the fossil record, where one frequent evolutionary pattern is for morphology not to change over periods of several million years. On the other hand, field studies and laboratory experiments have shown that evolutionary change can happen with surprising speed when selection is strong and genetic variation is available. If evolution can be fast, why is it often so slow? Getting an answer will not be easy, for we do not live long enough to see many changes in our lifetimes, and fossils are beyond the reach of experiment.

What is the nature of species?

Variation in life forms is not continuous. Organisms form discrete clusters that we call species. The variation within species is generally small compared to differences between species; although there is a continuum of variation, there are clusters within the continuum. We do not yet fully understand why this is so. It seems plausible that sexual reproduction plays an important role in maintaining relative uniformity within species, but it cannot be the whole explanation, because discrete species also exist among organisms where sexual recombination is absent or very rare, as in many imperfect fungi. Do ecological conditions only allow a finite number of discrete forms, with intermediate forms less well adapted? Or do clusters simply emerge when evolving lineages lose species at random to extinction? Such ideas are currently being developed.

Do developmental constraints exist and, if they do, what is their nature?

The importance of broadly shared developmental control systems was highlighted in Chapter 6 (pp. 124 ff.). What is not clear is the extent to which deeply conserved developmental control systems constrain evolutionary change, or the extent to which they enable such change, for they appear to do both. Developmental switches and position indicators were evolved early on. In some clades they retained their specific function. In others they were co-opted to control new morphologies, such as the radial symmetry of echinoderms. When duplicated in the vertebrate line, one set of HOX genes continued to control the body axis; a duplicate set was co-opted to control limb development. It is thus not at all clear that one can point to deeply conserved developmental control genes and assert that they are a source of phylogenetic constraint on the set of phenotypes that evolution can attain. They appear to enable as much as they appear to constrain.

One consequence of developmental constraints should be some of what we call parallelism—the tendency for the same traits to evolve repeatedly in the same clade. For example, the ancestor of the arthropods had many pairs of legs. The number of pairs has been repeatedly reduced. Some arthopods (insects) have three pairs; some (spiders) have four pairs; others (crabs and lobsters) have five pairs. It would be a triumph of evolutionary thought to demonstrate that these numbers of legs have repeatedly evolved from ancestors with pairs of legs on most segments and that other numbers of pairs—for example two, six, or seven—are not possible. Work on this issue has begun (Salazar-Ciudad and Jernvall 2004).

What are the limits to evolutionary prediction?

Evolution may share a theoretical element with physics and historical elements with astronomy and geology, but it also shares an element of unpredictability with meteorology. Evolutionary theory makes some successful predictions—so prediction is not impossible—but evolution is a complex, nonlinear, dynamic process. Such processes are often characterized by unpredictability: tiny differences in initial conditions can lead to huge differences in outcomes. Just how much can be predicted in principle is not yet clear. To illustrate the scope and limits of evolutionary predictions, we describe two case studies.

In the evolution of RNA molecules we can see the genotype–phenotype fitness relation

The nucleotide sequence of an RNA molecule is its genotype, and the spatial structure of the molecule is its phenotype. Schuster and colleagues have made a detailed study of the relationships between genotype and phenotype of RNA molecules (Schuster 1993, Schuster et al. 1994). The result is the first complete characterization of the genotype–phenotype relationships of any system. Their findings can be summarized as follows:

1. There are far fewer possible shapes (phenotypes) than there are sequences (genotypes).

2. A few of the shapes produced by a large collection of RNA sequences are common; many are rare.

3. The same phenotype (shape) may result from quite different genotypes.

4. All the main phenotypes can be realized from a relatively small set of similar genotypes.

5. The adaptive phenotypes are sometimes connected by networks of neutral mutations, rather than intervening forms with reduced fitness. This makes the evolutionary exploration of phenotype space much easier.

RNA molecules, linear chains of four nucleotides, are among the simplest entities that can evolve by natural selection, as is apparent from the test-tube evolution experiments described in Chapter 2 (pp. 32 ff.). In those experiments the precursors of RNA were placed together with an enzyme that replicates RNA. This system is simple enough that one can in principle compute which phenotypes (molecular shapes) allow the highest replication efficiency with that particular enzyme. This means that we can rank all genotypes with respect to replication efficiency in an environment where this enzyme is available, and thus predict the evolution of a population of RNA molecules, given a particular starting composition.

It would be interesting to know to what extent conclusions from this study apply to the evolution of traits in organisms, but for no trait in a real organism do we have such complete information on mutations and development as in the case of the RNA molecules. Perhaps a few genetic diseases in humans, where we know that specific mutations in a single gene can produce a specific disease phenotype, come closest. But for most traits studied by evolutionary biologists there is very little knowledge of the genes affecting the trait, of the phenotypic consequences of the possible mutations in these genes, or of the fitness differences between the possible phenotypes.

The dynamics of a lethal gene agree with quantitative prediction

A more limited evolutionary prediction would concern just one aspect of the evolution of a population. For example, one can establish a population of

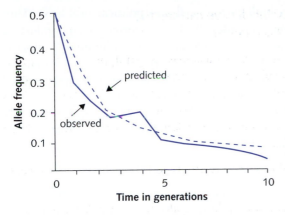

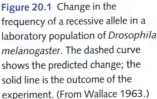

Figure 20.1 Change in the frequency of a recessive allele in a laboratory population of *Drosophila melanogaster*. The dashed curve shows the predicted change; the solid line is the outcome of the experiment. (From Wallace 1963.)

known genetic composition for some trait under controlled conditions, and ask the theory to predict how fast and how far the trait will change. Wallace (1963) did just that. He measured the rate at which a recessive lethal trait disappears from a population (Figure 20.1). A *Drosophila melanogaster* population was started from flies that were all heterozygous for a recessive lethal allele. Therefore the initial allele frequency was $q = 0.5$. Each generation a random sample of the surviving flies were used as parents. The frequency of heterozygotes was determined each generation by test matings, from which the allele frequency was estimated. The expected change in the frequency of the lethal allele can be predicted with the model in Box 4.2. Substituting $s = -1$ and $h = 0$ yields the case of a recessive lethal allele and gives us the formula for the expected change in allele frequency $q = 1/(1+q')$, which predicts the change in allele frequency given by the dashed curve in Figure 20.1. The data fit the prediction reasonably well.

Both these cases are examples of experimental evolution, and in both evolutionary change can be predicted quantitatively. The quantitative precision could only be achieved with detailed knowledge of genetics. Genetic details, however, often do not matter for qualitative predictions. For example, genetic details are not required to predict that an initially non-resistant population of insects fed food containing insecticide will evolve resistance to the pesticide.

It is taking a long time to assimilate Darwin's insights

Neither science nor culture has found it easy to assimilate evolution. The implications of evolution are often thought to be in conflict with traditional religions and philosophies. Evolution has also been falsely interpreted to serve unpopular, even evil, political ends, in Social Darwinism and in Nazi Germany. That experience has left many people extremely sensitive to and deeply cautious about any attempt to apply biology to social and political problems. Thus although evolutionary biology provides the best scientific understanding of where we came from and what we are, the messages that evolutionary biologists try to communicate to society at large must be very carefully formulated, for even the best formulations are liable to be misinterpreted by many and misused by some. It will be a long time before Darwin's insights are fully assimilated.

● SUMMARY

This chapter discusses the status, scope, and prospects of evolutionary explanation.

- The mechanisms of evolution are well understood. It is strongly supported by many kinds of data, from fossils to molecules. It is as well confirmed as any major idea in science.

- Evolution has a complex causal structure that combines historical influences with selection and drift.

- Evolutionary change can be rapid. It is happening all around us.

- The scope of evolutionary explanation remains controversial. How far it can be applied to humans remains open, both because of resistance from established religions and because the scientific problem of understanding the interaction of biological evolution with cultural change has not been resolved.

- The major preoccupation of evolutionary biology in the first half of the 20th century was demonstrating its consistency with Mendelian genetics. In the second half of the 20th century, the major preoccupation was the integration of evolution with molecular biology. In the 21st century, it is likely to be the developmental connection of genotype with phenotype.

- Two outstanding puzzles concerning the connection of genotype and phenotype are these. How did traits fixed within lineages become invariant? What are the effects of the fixed traits on the further evolution of the variable ones?

- Other unsolved problems are the huge variation in evolutionary rates, from very fast to almost zero, and the nature of species—the reasons for the formation of clumps in phenotype space.

- Evolutionary biology has made successful predictions, but the limits to what can be predicted, in principle, are not clear. They will probably be pushed back. They usually are.

- There are good reasons why it will take a long time for evolutionary thinking to be fully assimilated, both in academic circles and by the lay public.

● RECOMMENDED READING

Betzig, L. (1997) *Human nature: A critical reader*. Oxford University Press, Oxford.

Dunbar, R.I.M. (ed.) (1995) *Human reproductive decisions*. MacMillan, London.

Hrdy, S.B. (ed.) (1998) *Mother nature*. Pantheon, New York.

● QUESTIONS

20.1. Is the evolution of canalization necessary to account for the fixation of traits within lineages? Or is such fixation sufficiently explained by shared developmental control genes?

20.2. Was Darwin's prediction of a long-tongued moth needed to pollinate a deep-nectaried orchid just as impressive as Wallace's prediction of the change in frequency of a recessive lethal gene in a *Drosophila* population cage? Do qualitative and quantitative predictions have different value?

20.3. How much can evolution predict without genetics?

20.4. Discuss the nature and limits of materialistic explanations of human behavior. How does biological evolution interact with cultural evolution? Is there any danger that the sciences might replace the humanities as the central sources of insight on what it means to be human?

GENETIC APPENDIX

This appendix provides a simple summary of basic genetics. Because evolution can only occur if variation is heritable, some knowledge of genetics is necessary to understand it. Many genetic terms are used in the book; most of them are explained in this appendix. Concepts introduced in bold are also listed in the Glossary.

The blueprint of an organism is encoded in DNA molecules

All organisms, from unicellular bacteria to multicellular plants, fungi, and animals, contain a blueprint of their form and function in the form of DNA. In some viruses— for example those that cause influenza and Aids—a related molecule, RNA, replaces DNA. DNA stands for deoxyribonucleic acid, RNA for ribonucleic acid. These are long macromolecules, consisting of a sugar-phosphate backbone with nucleotides attached to it. There are four different nucleotides in DNA, adenine (A), guanine (G), cytosine (C), and thymine (T). Each DNA molecule consists of a double chain of linked nucleotides (see Figure A.1). The two strands are twisted around each other to form a double helix and connected by chemical pairing of complementary nucleotides. Adenine pairs with thymine (A–T) and guanine pairs with cytosine (G–C). Each strand is constructed from millions of nucleotides. The sequential order of the nucleotides determines the messages, just as the letters of the alphabet can be combined in different ways to form distinct words and sentences. The entire sequence of DNA in an organism is called its *genome*.

> **Nucleotide:** *A molecule consisting of a nitrogen base, a sugar and a phosphate group; the basic building block of DNA and RNA.*

Genotypes consist of information; phenotypes consist of matter

In general every cell of an organism contains the genetic blueprint for the complete organism. The information specified by an organism's DNA is called its genotype. The material organism itself (or the part of it under study), built according to genotypic instructions, is called the phenotype. Thus the genotype contains information; the phenotype contains matter.

DNA is embedded in chromosomes

In eukaryotes—organisms with cells containing a nucleus—the DNA resides in linear *chromosomes* in the cell nucleus. Chromosomes are long structures consisting of a central scaffold around which the DNA molecule is wrapped together with associated proteins. In prokaryotes (unicellular organisms lacking a nucleus, such as bacteria) the DNA forms a continuous loop, called a circular chromosome. The number of chromosomes in a cell nucleus is usually constant within a species and often differs between species. For example, in humans the complete genetic information is carried by 23 chromosomes, in *Drosophila melanogaster* by four chromosomes. A somatic cell contains two sets of chromosomes (in humans, this number is 46 chromosomes), one set from the father and one from the mother. Such a cell is called diploid, in contrast to gametes (eggs and sperm) which are haploid, i.e. they contain a single set of chromosomes.

> **Somatic cell:** *A 'body cell' in nonreproductive tissue.*

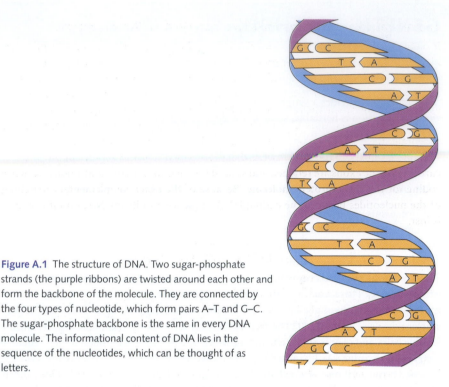

Figure A.1 The structure of DNA. Two sugar-phosphate strands (the purple ribbons) are twisted around each other and form the backbone of the molecule. They are connected by the four types of nucleotide, which form pairs A–T and G–C. The sugar-phosphate backbone is the same in every DNA molecule. The informational content of DNA lies in the sequence of the nucleotides, which can be thought of as letters.

Genes, the units of heredity, code for the proteins that build the organism

Organisms are built from thousands of proteins and other materials whose construction is controlled by protein enzymes. The unique nature of a species is thus determined by the specification in its DNA of the precise structure of each protein in a cell. A segment of DNA that instructs a cell to produce a particular protein is called a gene. In addition to the genes that encode proteins, some genes code for RNA molecules involved in protein synthesis. A single chromosome contains about a thousand genes.

Each gene occupies a specific position on its chromosome. This site is called the locus of the gene. For example, the locus of the gene coding for the enzyme alcohol dehydrogenase (ADH) in humans is located on the long arm of chromosome 4. Within a species, a gene may occur in different variants, called alleles. Typically, different alleles show small variations in their nucleotide sequence, some of which may change the encoded protein. Thus, a person may carry a different ADH allele on each of its two chromosomes 4. Such a person is called heterozygous for the ADH gene. Someone possessing two identical alleles is homozygous.

A typical eukaryotic gene consists at its endpoints of sequences that indicate the beginning of the gene, marked by a start codon, and the end of the gene, marked by a stop codon. Between the start and stop codons, the sequence coding for the protein is interrupted by non-coding segments called introns. The functions of introns are still unclear. The coding sequences interdigitated between the introns are called exons.

The central dogma: information flows from DNA to RNA to proteins

Protein-coding genes are first transcribed into RNA then translated into protein (see Figure A.2). Transcription is the process by which the DNA sequence of a gene from one of the two DNA strands is copied into a single-stranded RNA molecule, the so-called messenger RNA (mRNA). Transcription is guided by complementary nucleotide pairing: at every position where the DNA sequence contains a G, the RNA sequence contains the complementary nucleotide C, and similarly with the other nucleotides. There is one exception: RNA does not contain thymine (T); it is replaced with uracil (U).

While the mRNA is being transcribed, it is processed. The introns are cut out and discarded, and the exons are retained and spliced together to form the continuous sequence coding for the entire protein molecule. Because of the exact complementary matching of the nucleotides between the coding DNA sequence and its mRNA, no information is lost.

The genetic code is implemented by tRNA molecules

The meaningful units in the nucleotide sequence of an mRNA molecule are the *codons*, groups of three consecutive nucleotides. Each codon represents either a particular amino acid or a start or stop signal. Because there are $4 \times 4 \times 4 = 64$ different codons available to encode only about 20 different amino acids, the genetic code is redundant, meaning that several codons can code for the same amino acid.

The mRNA carries the encoded information for a specific protein out of the nucleus into the cytoplasm where it is translated into protein at cytoplasmic organelles called ribosomes. Here an essential role is played by transfer RNA (tRNA) molecules. Each amino acid is brought to the ribosome by a tRNA molecule that binds on one end to a specific amino acid and on its other end to a specific mRNA codon. In this way a string

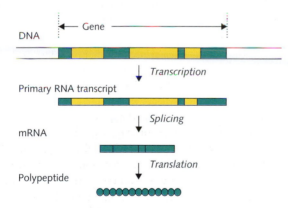

Figure A.2 A protein-coding gene in a eukaryotic organism consists of an alternation of exons, the coding segments (green), and introns (yellow). The DNA code of the gene is transcribed into RNA, the introns are spliced out and a messenger RNA (mRNA) molecule is created that carries the information of the gene outside the nucleus into the cytoplasm where it is transcribed into polypeptide. This figure also demonstrates the so-called central dogma: genetic information flows from DNA to RNA to protein, not in the reverse direction.

of amino acids (a polypeptide) is created, the primary structure of the protein, that precisely reflects the sequence of codons in the mRNA.

Transmission of genetic material during cell division

Replication and reproduction precisely regulate the transmission of chromosomes

In a multicellular organism cells must divide so that the organism can grow and replace damaged cells. Cell division is also essential for reproduction, both in the direct production of daughter cells by single-celled organisms and in the production of gametes by multicellular organisms. Whenever cells divide, the daughter cells must receive a complete, correct copy of the genetic blueprint encoded in the DNA packaged in chromosomes. Therefore a DNA copying process must exist. It is called *replication*.

The double-helix structure of DNA is eminently suitable for replication. During replication, the two complementary strands of the DNA molecule unwind; then each of the two exposed nucleotide chains acts as a template for the construction of a new strand (Figure A.3), which is built by attaching complementary nucleotides (A to T, G to C) to the template. An enzyme called DNA polymerase moves along each growing DNA branch to catalyze the formation of the new double helices. In this way a single DNA molecule produces two identical daughter copies.

DNA replication is the first step in cell division. It is followed by replication of the chromosomes and other cellular structures. There are two types of cell division in eukaryotes, defined by the behavior of the chromosomes. They share the mechanism for aligning the two copies of the chromosomes at the central plane where the cell will divide, then pulling them apart, one set into each daughter cell (Figure A.4). The chromosomes are pulled apart by the **spindle apparatus**, which itself divides before cell division, the two copies moving to opposite ends of the cell. One end of the spindle apparatus anchors itself to the cell wall; the other sends out microtubules that attach to a special structure on the chromosome called the *centromere*. The microtubules then contract, pulling the chromosomes apart.

Figure A.5 illustrates mitosis, the nuclear division that produces two identical daughter nuclei. Mitosis occurs at cell division during growth and also when single-celled propagules (spores) are produced asexually, as in many fungi.

Figure A.6 illustrates meiosis, the process that reduces the number of copies of the chromosome set from two to one and produces reproductive cells (gametes and sexual spores). Meiosis starts in a diploid parent cell in the gamete- or spore-producing tissues; its end products are haploid cells.

The behavior of chromosomes at meiosis explains inheritance patterns

The foundations of modern genetics were laid in 1865 by Gregor Mendel, who formulated the basic principles of inheritance from the results from crossing experiments with garden peas (*Pisum sativum*). Mendel did not know about the role of chromosomes in heredity and derived his ideas purely from the regularities he observed in the distribution of visible hereditary characteristics among parental plants and their offspring. Among the traits he studied were flower color, seed color and seed shape. His theory can now readily be derived from our present insights into the location of genes on chromosomes and the chromosomal behavior at meiosis. The main principles are referred to as Mendel's laws.

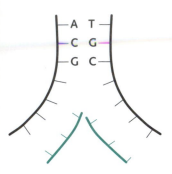

Figure A.3 DNA replication. The newly formed nucleotide strands are in green. Note that each daughter DNA molecule has an old strand and a new strand.

Spindle apparatus: Pulls the chromosomes apart during meiosis and mitosis. It replicates, divides, and each copy anchors itself. Microtubles grow from the spindle to the centromeres of the chromosomes, attach, contract, and pull a set of chromosomes to one end of the dividing cell.

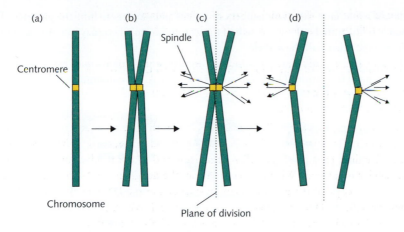

Figure A.4 (a) A single eukaryote chromosome prior to replication; the *centromere* is a region in the DNA important at chromosome replication. (b) The chromosome has replicated to form two chromatids, each containing one of the two DNA daughter molecules. (c) The replicated chromosome is oriented (together with the other chromosomes) in the plane of division of the cell (the equatorial plane); the spindle (consisting of contractile fibers) is attached to the centromeres. (d) The centromeres split and the daughter chromosomes are pulled apart into the daughter nuclei by the action of the spindle fibers. The figure shows what happens in mitosis. In the first division of meiosis the two homologous chromosomes line up in the equatorial plane, but the centromeres of the chromatids do not split. Consequently, the two chromatids in each chromosome remain attached to each other, and the spindle pulls the complete (i.e already replicated but not yet divided) chromosomes apart (details in Figure A.6).

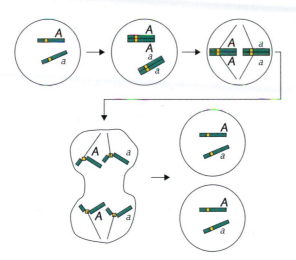

Figure A.5 Mitosis in a diploid cell with one pair of chromosomes. *A* and *a* are a pair of alleles at a locus on this chromosome pair, for which the cell is heterozygous. After replication, each chromosome consists of two sister chromatids. Both pairs of chromatids come to lie in the equatorial plane of the cell. Then the sister chromatids are pulled to opposite ends of the cell by fibers of the nuclear spindle. Finally the cell divides into two daughter cells, with a genetic composition identical to the parent cell.

Mendel's first law: the two members of a gene pair segregate into the gametes. This principle is illustrated in Figure A.6: half the gametes from a heterozygote *Aa* carry the *A* allele and the other half the *a* allele.

Mendel's second law: gene pairs located on different chromosomes assort independently in gamete formation. This is illustrated in Figure A.7: a consequence of independent

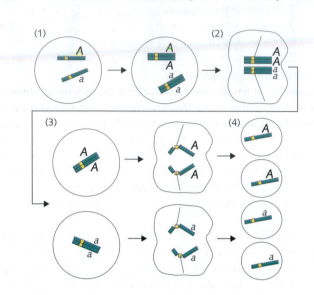

Figure A.6 Meiosis occurs in a special diploid cell (a meiocyte), here shown with only one pair of chromosomes (1). The cell is heterozygous at a gene locus with alleles *A* and *a*. Meiosis consists of two successive divisions. In the first meiotic division—called reduction division—the homologous chromosomes, after replication consisting of two sister-chromatids, pair in the equatorial plane of the cell, forming a group of four chromatids (2). Each of the two pairs of sister chromatids is then pulled into a different daughter nucleus, yielding two haploid daughter cells (3), which go through a subsequent mitotic division. The haploid end products of meiosis (4) become the gametes. The two members of the gene pair *A/a* present in the parent cell segregate from each other in the gametes. Half of the gametes carry *A* and the other half *a*.

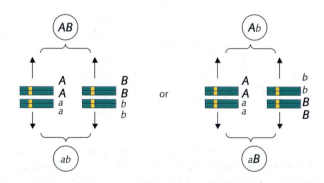

Figure A.7 The principle of independent assortment of two gene pairs (*A/a* and *B/b*) located on different pairs of chromosomes. At meiosis each member of a chromosome pair has an equal chance of being pulled into the upper or into the lower daughter cell. Consequently, four different gamete types are produced, *AB*, *Ab*, *aB*, and *ab*, each with probability 0.25.

assortment is that a double heterozygote *Aa Bb* will produce gametes carrying *AB*, *Ab*, *aB*, and *ab* in equal proportions (each with a frequency of 0.25).

From our insights into meiosis as reflected in Mendel's laws we can deduce which gametic genotypes a parent of known genotype is expected to produce. For example, considering only the genotype at a single locus: an *AA* parent will only produce *A* gametes, and an *Aa* parent will produce 50% *A* and 50% *a* gametes. If we consider a genotype consisting of two loci located on different chromosomes, an *AABB* parent will only produce *AB* gametes, an *AABb* parent will produce 50% *AB* and 50% *Ab* gametes, while the gametes from an *AaBb* parent will consist of 25% *AB*, 25% *Ab*, 25% *aB*, and 25% *ab* (the case illustrated in Figure A.7).

We may assume that at fertilization a gamete from the mother is combined randomly with a gamete from the father. Given that assumption, we can predict the genotypes of the progeny emerging from a cross. For example, if both the mother and the father are heterozygous for a gene pair *A/a*, the two female gametic types will be fertilized randomly by the two male gametic types to produce the offspring. This can conveniently be shown in a 2 × 2 grid (see Figure A.8). When the genotypes are specified at two loci, the same method requires a 4 × 4 grid. It is clear that working out the expected genotypes of the offspring from a cross quickly becomes very laborious with an increasing number of loci.

What we see when observing the offspring from a cross are their phenotypes, not their genotypes. It follows that for comparing the actual outcome of a cross with the calculated prediction, we should know the relationship between genotype and phenotype. It may happen that the number of different phenotypes observed is smaller than the number of genotypes predicted. For example, the cross worked out in Figure A.8 may produce only two different *phenotypes*, while three different *genotypes* are predicted. This phenomenon was known to Mendel, who used the terms dominant and recessive to describe it. For example, he deduced from his experiments with the garden pea that purple flower color is dominant over white flower color. If in Figure A.8 the allele *A* determines purple color and allele *a* white color, and allele *A* is dominant over allele *a*,

Female gametes	Male gametes: *A* 0.50	Male gametes: *a* 0.50
A 0.50	*AA* 0.25	*Aa* 0.25
a 0.50	*Aa* 0.25	*aa* 0.25

Figure A.8 Predicted genotypes of the offspring from a cross between two parents heterozygous for a gene pair *A/a*. Both parents produce gamete types *A* and *a* in equal proportions. An offspring resulting from random fertilization is expected to be *AA* with probability 0.25, *Aa* with probability 0.50, and *aa* with probability 0.25. Thus among a large number of offspring from a cross between two heterozygotes we expect offspring of genotype *AA*, *Aa*, and *aa* in proportions 1:2:1. If allele *A* is dominant over allele *a*, the offspring are predicted to occur in phenotypic proportions of 3:1.

heterozygotes *Aa* have the same purple phenotype as the homozygotes *AA*. Allele *a* is then called recessive to *A*. In this example the cross worked out in Figure A.8 involves two parents with purple flowers producing offspring with phenotypic proportions of 3 purple:1 white.

Recombination produces offspring genotypes that differ from the parental genotypes

One of the most fundamental characteristics of meiosis is that it results in offspring that differ genetically from their parents. To see how it does this, consider a zygote formed from a fusion between a sperm with genotype *AB* and an egg with genotype *ab*. This *AaBb* zygote develops into an adult individual, which will itself produce four different types of gametes: *AB*, *Ab*, *aB*, and *ab*. Two of those, *AB* and *ab*, are called *parental types*, because they contain the same combinations of alleles as the individual had received from his parents. The other two, *Ab* and *aB*, are called *recombinant types*, because they contain a new combination of alleles, derived from each of the original parents.

This key aspect of meiosis, called recombination, can result from two different processes.

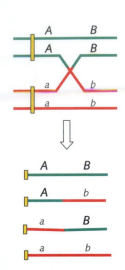

Figure A.9 Crossing-over involving a pair of homologous chromosomes during meiosis. During pairing of the chromosomes two non-sister chromatids come into close contact, break at homologous positions and rejoin.

- From the independent assortment of different chromosome pairs (Mendel's second law). In that case the alleles forming the new combinations are located on different chromosomes, as in Figure A.7.
- From a process called crossing-over, the exchange of chromosome parts between the two homologous members of a chromosome pair by breakage and reunion. In that case the alleles forming the new combinations are located on the same chromosome. Figure A.9 shows a schematic representation of crossing-over.

Figure A.10 illustrates both types of meiotic recombination. Recombination is important in evolution because it generates genotypes consisting of new combinations of alleles and thus increases the amount of genetic variability on which natural selection can act.

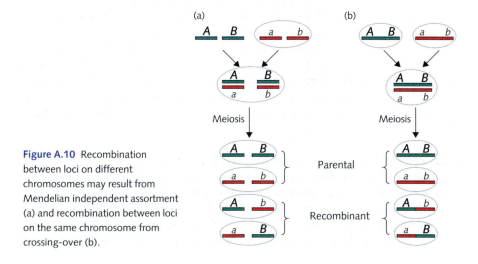

Figure A.10 Recombination between loci on different chromosomes may result from Mendelian independent assortment (a) and recombination between loci on the same chromosome from crossing-over (b).

```
A T C T A T G T A G C C A A G     Original sequence

A T C C A T G T A G C C A A G     Substitution of
                                  single nucleotide
                                  pair

A T C T A T G T A C C A A G       Deletion of single
                   ↑              nucleotide pair;
                                  causes frameshift

A T C T A T G C C A A G           Deletion of three
               ↑                  nucleotide pairs; no
                                  frameshift

A T C T A T G T C A G C C A A G   Addition of single
                 ↑                nucleotide pair;
                                  causes frameshift
```

Figure A.11 Common types of gene mutation. So-called point mutations involve the substitution, deletion, or addition of a nucleotide pair or of a few adjacent nucleotide pairs. If the changed codon resulting from a substitution of a single nucleotide pair still codes for the same amino acid, the substitution is called *silent* or synonymous. If it codes for a different amino acid the functionality of the protein involved will be changed (often impaired). If the deletion or addition of nucleotide pairs is not a multiple of 3, the reading frame in coding segments is changed because the amino acids are coded by codons consisting of three consecutive nucleotides. Such a frameshift mutation results in different amino acids from that point on, leading frequently to a chain termination and a nonfunctional protein.

DNA has a tendency to change by mutation

A mutation is a hereditary change in the DNA. Mutations can be divided into two categories, gene mutations and chromosome mutations.

Gene mutations are mostly small alterations in the nucleotide sequence of a gene and are responsible for the formation of new alleles. Thus mutation is responsible for the origin of genetic variation, which in turn is a prerequisite for evolution to occur. The most important causes of gene mutation are errors made during the replication of the DNA and spontaneous damage to the DNA. The mistakes and damage that occur are far greater than the mutations that result, for DNA maintenance and repair systems in the cell prevent most of the errors and repair most of the damage. As a result, DNA replication is a very reliable process. A rough estimate of the average mutation rate in DNA per nucleotide pair per replication is about 10^{-10}—just one mistake per 10 billion copying events.

Chromosome mutations involve changes in chromosome structure or in chromosome number. They may occur as a consequence of chromosome breaks or when mistakes occur in chromosome separation during cell division. Some serious human genetic diseases result when one chromosome fails to segregate at meiosis, producing an offspring that has a normal complement of chromosomes for all but one chromosome, which is then present in three, not two, copies.

Mutations can occur in any cell for a variety of reasons. Only mutations occurring in tissue from which gametes are derived, the germ line, can be transmitted to progeny, and are therefore of direct relevance to evolutionary processes. Mutations occurring in somatic cells can also have important consequences, however. For example, cancer is the result of somatic mutations, often because a DNA maintenance or repair system is not working properly. A germ-line mutation in a DNA-repair gene can lead to an elevated rate of somatic mutation, giving some people a genetic predisposition to develop cancer.

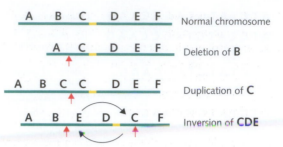

Figure A.12 Some common types of chromosome mutation. **A, B**, etc. refer not to single gene loci but to relatively large segments of the chromosome. When a chromosome breaks in one or more places a portion may get lost (*deletion*), or the fragment between two breaks may rotate 180° before rejoining (*inversion*). A deletion may have serious consequences, because genes are missing from that chromosome, while an inversion need not cause abnormalities because no DNA has been lost. Irregularities during replication may cause a doubling of part of a chromosome (duplication). Duplications may cause phenotypic changes due to an imbalance in the cell between the amounts of proteins produced.

Mutations can be classified in different ways. Figure A.11 shows the most common types of gene mutation involving one or a few nucleotides.

Figure A.12 shows the most common types of chromosomal mutation. They involve entire genes or sets of genes—thousands of nucleotides. In addition to the chromosomal mutations shown which involve structural changes of a chromosome, other chromosomal mutations cause changes in chromosome number. Particularly important in evolution is polyploidy, the multiplication of the entire chromosome set. If the haploid set has *n* chromosomes and the diploid set has *2n* chromosomes, then *triploids* have *3n*, *quadruploids* have *4n*, and so forth.

The genetic composition of a population

The genetic composition of a population is described by allele frequencies

Evolution involves change over time within and between species. To be inherited, and thus to endure, evolutionary change must be reflected at the genetic level as changes in the DNA. Thus it is important to understand how a species can change genetically. The simplest situation is to focus on a single population of individuals from the same species, to describe its genetic composition and ask how this genetic composition may change over time. Every individual of the population can in principle be described by its genotype. One possible genetic description of the population would be a list of the genotypes of all its individuals. However, because each individual consists of thousands to tens of thousands of genes, following entire genotypes would be extremely cumbersome and impractical. Instead suitable statistics are used.

The most important concept is gene frequency, or more precisely *allele frequency*, which is the frequency at which a particular allele occurs at a given locus. Suppose there are 100 individuals in a population of a diploid species. Concentrating on a locus *A*, the genotype of each individual is specified by the two alleles it carries at locus *A*, one on each of the two homologous chromosomes. To describe this population we now forget about individual genotypes and view the population as a set of 200 homologous chromosomes, each carrying one allele at locus A. If all chromosomes carry the same allele, there is no genetic variation in this population at this locus A. More interesting is the case where different alleles occur at locus A, let us say three: A_1, A_2, and A_3. If 50 chromosomes carry A_1, 70 chromosomes have A_2, and 80 chromosomes A_3, the respective allele frequencies are $50/200 = 0.25$, $70/200 = 0.35$ and $80/200 = 0.40$. With these three numbers the genetic composition of the population at locus A is specified.

In practice we cannot readily inspect all the chromosomes and count how often they carry A_1, A_2, and A_3. Instead, the genotype of a representative random sample of the population are characterized. In our example of 20 randomly chosen individuals, we might have observed the following distribution:

Genotype	A_1A_1	A_1A_2	A_1A_3	A_2A_2	A_2A_3	A_3A_3
Individuals	0	4	5	2	6	3

The chromosome types are directly reflected in the notation of the genotypes. Therefore we can simply count how often they occur in this sample, and compute the allele frequencies. Thus the allele frequency of A_1 is $(0 + 4 + 5)/40 = 0.225$; of A_2 $(4 + 2 \times 2 + 6)/40 = 0.35$, and of $A_3(5 + 6 + 2 \times 3)/40 = 0.425$. As expected these values inferred from a smaller sample are close to, but not exactly the same as, the values inferred from a larger sample.

Random mating results in a stable genotype distribution, the Hardy–Weinberg equilibrium

We now consider how the principles of genetic transmission, expressed in Mendel's laws, affect the genetic composition of a population. Will genetic transmission during reproduction in this population change the allele frequencies? You may think that the answer to this question requires a full specification of the genotypes of all mating partners, so that we can deduce the expected genotypes of the offspring and hence the new allele frequencies among them. This again would be extremely laborious and perhaps impossible. A much simpler approach uses the concept of *random mating*, which assumes that males and females mate with each other irrespective of their genotype. This is equivalent to mixing all sperm and all the eggs in the population together and combining them to form zygotes in proportion to their relative frequencies.

For example, if at locus A two alleles are present in the population, A and a, with respective frequencies in the gametes of 0.7 and 0.3, the probability of an AA zygote produced by random mating is $0.7 \times 0.7 = 0.49$. Similarly, an aa zygote will be formed with probability $0.3 \times 0.3 = 0.09$. In contrast to the homozygotes, heterozygotes can be formed in two ways: by a fusion of an A sperm with an a egg, or of an a sperm with an A egg. Thus heterozygotes will occur with probability $2 \times 0.7 \times 0.3 = 0.42$. Note that $0.49 + 0.42 + 0.09 = 1.0$, as it must if we have calculated correctly, for the probabilities of all independent events must sum to 1.0.

The following more general analysis leads us to important conclusions. We assume that at a locus A only two different alleles occur in a population, A and a, with respective allele frequencies p and q, where $0 < p < 1$, $p + q = 1$, and therefore $q = 1 - p$.

	Male gametes	
	A	a
	p	q

<table>
<tr><td rowspan="2">Female gametes</td><td>A p</td><td>AA
p^2</td><td>Aa
pq</td></tr>
<tr><td>a q</td><td>Aa
pq</td><td>aa
q^2</td></tr>
</table>

Figure A.13 Random mating in a population, specified by a single locus at which two alleles occur, results in the *Hardy–Weinberg distribution* of genotypes:

AA	Aa	aa
p^2	$2pq$	q^2

This result implies that Mendelian genetic transmission in a random-mating population preserves the genetic variation which is present and does not change allele frequencies when other disturbing forces are absent.

The consequences of random mating combined with the Mendelian inheritance principles are summarized in a 2 × 2 table (Figure A.13). Random mating produces an off-spring generation in which the three genotypes *AA*, *Aa*, and *aa* occur in the proportions p^2:$2pq$:q^2. By counting *A* and *a* chromosomes we compute the allele frequency of *A* among the offspring as $(2p^2 + 2pq)/2 = p^2 + pq = p(p + q) = p$ (because $p + q = 1$). By similar reasoning, the new allele frequency of *a* becomes *q*. Therefore, Mendelian genetic transmission in a random mating population *does not change the allele frequencies*. It follows that under continued random mating, each generation the frequencies of the three genotypes will again be p^2:$2pq$:q^2. Therefore continued random mating does not change the genetic composition of the population. This composition is called the *Hardy–Weinberg distribution*, after the two scientists who independently discovered it early in the 20th century.

Just as DNA replication provides faithful copying of genetic information at the molecular level, random mating combined with Mendel's laws provide faithful copying of the genetic information at the population level. This means that the genetic system itself has no tendency to cause evolutionary change, leaving maximal room for the selection process.

GLOSSARY

ACCRETED TERRANE. A terrane that has become attached to a continent (e.g. western North America) as a result of tectonic processes.

ACQUIRED CHARACTERISTICS. The states of traits acquired by individuals as they develop through their interaction with the environment; they are not inherited and do not respond to natural selection.

ADAPTATION. A state that evolved because it improved relative reproductive performance; also the process that produces that state.

ADAPTIVE EVOLUTION. The process of change in a population driven by variation in reproductive success that is correlated with heritable variation in a trait.

ADDITIVE GENETIC VARIANCE. The part of total genetic variance that can be modeled by allelic effects whose influence on the phenotype in heterozygotes is additive— halfway between the effects caused in the phenotype by the two homozygotes. This part of genetic variance determines the response to selection by quantitative traits.

ADDITIVE TREE. A phylogenetic tree in which the branch lengths between two species add up to the distance between those species.

ALLELE. One of the different homologous forms of a single gene; at the molecular level, a different DNA sequence at the same place in the chromosome.

ALLOPATRY. Occurring in two or more geographically separate places.

ALLOPOLYPLOIDY. An increase in the number of chromosomes involving hybridization followed by chromosome doubling. The offspring then contain chromosomes from both parental species.

ALTRICIAL. Born or hatched helpless, blind, and needing parental care for warmth and food.

ALTRUISM. Behavior that increases the reproductive success of others while reducing one's own reproductive success.

ANAGENESIS. Evolutionary change occurring within a species without speciation. Usually applied to fossils.

ANALOGY. Convergent traits whose similarity is caused by shared selection.

ANCESTRAL STATE. A state characteristic of an ancestor shared with related groups.

ANISOGAMY. Having gametes of different sizes; large eggs and small sperm.

ANTAGONISTIC PLEIOTROPY. One gene has positive effects on fitness through its impact on one trait but negative effects on fitness through its impact on another trait.

APOMIXIS. Asexual formation of seeds in plants without a genetic contribution from a male gamete. Apomictic seeds are genetically identical to the mother plant.

APOMORPHIC. Derived, relative to an ancestral, or plesiomorphic, state.

ARITHMETIC AND GEOMETRIC MEAN. The arithmetic mean of $(2 + 3 + 2 + 3 + 2 + 3)/6 = 2.5$ and the geometric mean is $(2 \times 3 \times 2 \times 3 \times 2 \times 3)^{1/6} = 2.45$.

ARTIFICIAL SELECTION. Selection carried out by humans with the aim of changing particular trait values, widely used in plant and animal breeding to improve desired traits and in evolutionary research to test hypotheses.

ASEXUAL. A type of reproduction in which a parent produces an offspring either by dividing in two or by producing an egg that develops without fertilization. In most cases offspring produced asexually are genetically identical to the parent.

AUTOPOLYPLOIDY. A doubling in the number of chromosomes when a gametocyte fails to divide in meiosis; all the resulting chromosomes come from a single parent.

AUTOSOME. A chromosome that does not function as a sex chromosome.

BAYESIAN INFERENCE. A method that focuses on the likelihood of observing one thing given that another thing has already occurred.

BIOLOGICAL SPECIES CONCEPT (BSC). Species are actually or potentially interbreeding natural populations that are reproductively isolated from other such groups.

BOOTSTRAPPING. A statistical method to estimate our confidence in a pattern. One takes random samples of the original data (with replacement to get a data set of the same size) and repeats the calculations with the new, artificial data sets many times. The patterns that do not change, or only rarely change, receive our confidence.

BOOTSTRAP VALUE. The proportion of times a pattern is repeated in a bootstrapping procedure.

BOVIDAE. Cattle and their relatives, including antelope, buffalo, sheep, and goats.

BOXES. DNA sequences that specify protein structures that bind to the DNA double helix. They characterize families of transcription factor genes, each family having its own box. Shared boxes suggest shared origins.

BREEDING TRUE. Displaying the same character state in the offspring as in the parents, an indication that the trait is genetically determined and that the parents are genetically similar.

BROAD-SENSE HERITABILITY. The proportion of total phenotypic variation that can be ascribed to all the genetic differences among individuals in the population. For example, the amount of variation caused by differences among full-sib families.

CANALIZATION. The limitation of phenotypic variation by developmental mechanisms. It can be demonstrated by disturbing developmental control to reveal the underlying genetic variability that had been canalized.

CENTRAL DOGMA. A key concept in molecular biology stating that DNA makes RNA, and RNA makes proteins. Thus information flows from DNA out to the cell, not from the cell back into the DNA.

CENTROSOME. The cell organelle that becomes the spindle apparatus during mitosis and meiosis in eukaryotes.

CHARACTER. A trait that varies among taxa and that in any given taxon takes one out of a set of two or more different states.

CHARACTER DISPLACEMENT. Competing species differ in competition-related traits more in sympatry than they do in allopatry. Competition is then thought to have displaced the sympatric state from the allopatric state.

CHARACTER STATE. A specific value taken by a character in a specific taxon.

CHEMOHETEROTROPH. A bacterium that feeds on energy stored in molecules, inorganic or organic.

CLADE. A natural group of related species containing all descendants of the most recent common ancestor of the group.

CLADOGENESIS. The diversification of a lineage through splitting (speciation); usually applied to fossils.

CLINE. A spatial gradient in trait values or gene frequencies.

CLOCK-LIKE TREE. A phylogenetic tree in which the length of the branches is proportional to the time elapsed between branching events.

CLONAL INTERFERENCE. The inhibiting effect on the spread of advantageous mutations through an asexual population caused by the presence of clones that lack the mutations but still have a temporary fitness advantage from genes at other loci.

COALESCENCE. A process of inference from existing DNA sequences in related species back to the ancestral sequence in the shared ancestor. The process yields an estimate of the age at which the different daughter sequences coalesce into the ancestral sequence.

CODON BIAS. Synonymous codons do not occur with equal frequency.

COEVOLUTION. Evolutionary changes in one thing—genes, sexes, species—induce evolutionary changes in another which in turn induce further evolutionary changes in the first, and so forth.

COLINEARITY. A property of HOX genes: they have the same relative position on the chromosome as the part of the body the gene affects.

COMBINATORIAL CONTROL. The effects elicited by developmental control genes depend upon the *combination* of such genes expressed in a given tissue.

COMMENSALISM. One species lives in close proximity to and benefits from another without having much impact on its partner, either positive or negative.

COMPARATIVE TREND ANALYSIS. The relations of two or more traits among higher taxa are analyzed with proper control for phylogeny and covariates like body weight.

CONFLICT. Evolutionary conflict arises when two genes that interact with each other have different transmission patterns and therefore different evolutionary interests.

CONJUGATION. A form of bacterial sex involving DNA transfer from a donor to a recipient cell, which emerges from the process with a recombinant genome. The cells do not reproduce: cell number does not increase.

CONSERVED FUNCTION. A property of paralogous genes with high DNA sequence homology that code for proteins with similar function in distantly related organisms.

CONTRASTS. Differences in trait states between two taxa.

CONVERGENCE. Two species resemble each other not because they shared common ancestors but because evolution has adapted them to similar ecological conditions.

CROSSING-OVER. An exchange of material between non-sister chromatids, thus between maternal and paternal chromosomes, occurring in early prophase of the first division of meiosis. It results in new combinations of alleles for different genes and is the mechanism of recombination within chromosomes.

CRYPTIC. Disguised to look like the background.

CRYPTIC SPECIES. Species that resemble each other so closely that they cannot be distinguished by external morphology. Synonym of sibling species.

CYCLICAL PARTHENOGENESIS. A life cycle typical of aphids, rotifers, cladocerans and some beetles in which a series of asexual generations is interrupted by a sexual generation. The offspring of the sexual generation are often adapted to resist extreme conditions and to disperse.

DARWINIAN DEMON. An organism that matures instantaneously, gives birth to infinitely many offspring, and lives forever.

DARWINIAN DOLT. An organism that delays maturity, reproduces asexually, has one offspring that does not disperse, and dies immediately after reproducing.

DENSITY-DEPENDENT SELECTION. Selection that favors different things at different population densities.

DENSITY-INDEPENDENT SELECTION. Selection that favors the same things at all population densities, or that has always occurred at the same population density so that density effects did not occur.

DERIVED STATE. A state that evolved after the ancestral state in this lineage.

DEVELOPMENT. The processes of growth and differentiation that build the organism and allow it to survive and reproduce.

DEVELOPMENTAL CONTROL GENE. A gene that controls the expression of other genes, often in combination, through the production of proteins called transcription factors that bind to the control regions of the genes whose expression is being regulated.

DIOECY. Having separate sexes; individuals are either males or females; used for plants. See GONOCHORISM.

DIPLOBLASTIC. Having two primary cell layers.

DIPLOID. Diploid cells have two copies of each chromosome, usually one from the father and one from the mother.

DIPLONTIC LIFE CYCLE. A life cycle in which diploid somatic adults produce haploid gametes by meiosis that fuse to form diploid zygotes that develop into somatic adults.

DIRECTIONAL SELECTION. Selection that always acts in a given direction, for example, always to increase the value of a trait.

DISJUNCT DISTRIBUTION. Geographical distribution in which the closest relatives of a group are found far away with more distantly related groups occupying the intervening territory.

DISRUPTIVE SELECTION. Selection that favors the extremes and eliminates the middle of a frequency distribution of trait values, for example, increasing the frequency of small and large individuals and reducing the frequency of medium-sized individuals.

DISTAL. Located away from the body.

DIVERGENCE. Related species no longer resemble each other because evolution has adapted them to different ecological conditions.

DNA. Deoxyribonucleic acid is the hereditary material, the substance of genes, in all organisms except the RNA viruses.

DOMINANT. An allele is dominant if it is expressed in the phenotype in the heterozygous diploid state.

DOWNSTREAM GENE. A gene under the control of a regulatory gene; the genes downstream from a regulatory gene constitute a regulatory pathway.

DRIFT. The random walk of gene frequencies that occurs in both large and small populations when variation in genes is not correlated with variation in reproductive success.

DUPLICATION. Copying of a DNA sequence without loss of the original, increasing the size of the genome by the size of the sequence copied.

ECLOSION. Emergence from a pupa. Birds hatch from eggs; insects eclose from pupae.

ECOMORPH. A morphology associated with a particular ecological habitat.

ECTOPIC. In an abnormal position or place.

EFFECTIVE POPULATION SIZE. The size N_e of an abstract population that would experience the same amount of genetic drift as a real population of size N.

EPIGENETIC INHERITANCE. Somatic inheritance of the differentiated state of the cell through cycles of cell division.

EPIZOOTIC. An epidemic of disease in an animal population.

EUKARYOTE. An organism with a cell nucleus surrounded by a nuclear membrane, usually with organelles, such as mitochondria and chloroplasts, that have their own circular DNA genome. Eukaryotes include the protists, fungi, plants and animals.

EUSOCIALITY. A social system with nonreproductive workers.

EXON. The part of a eukaryotic gene whose DNA sequence is preserved in post-transcriptional splicing and is represented in the spliced mRNA and in the resulting amino acid sequence of the protein product. Exons occur in eukaryotes but not in prokaryotes.

EXTRINSIC MORTALITY. Mortality caused by environmental factors, such as predators, disease, and weather.

FECUNDITY. A synonym for offspring production used for all organisms, not all of which give birth.

FITNESS. Relative lifetime reproductive success, which includes the probability of surviving to reproduce. In certain situations, other measures are more appropriate. The most important modifications to this definition include the inclusion in the definition of the effects of age-specific reproduction and of density-dependence.

FIXATION. The process by which an allele increases in frequency until it becomes the only version of the gene in the population, at which point it is referred to as fixed.

FIXATION PROBABILITY. The probability that a new mutation will reach fixation.

FIXATION TIME. The time that it takes new mutations that get fixed to reach fixation, in generations.

FIXED EFFECTS. Biological features that do not vary and that are shared by all organisms within a lineage.

FOUNDER EFFECT. Major changes in gene frequencies that occur in a population founded with a small sample of a larger population.

FREQUENCY-DEPENDENT SELECTION. A mode of natural selection in which either rare types (negative frequency-dependent selection) or common types (positive frequency-dependent selection) are favored.

GAMETE. A reproductive cell, usually haploid. In anisogamous organisms, an egg or sperm cell. Gametes produced by different sexes or mating types fuse to form zygotes.

GASTRULATION. Process by which cells of the blastula, a hollow ball, are translocated to new positions in the embryo, in part by invagination, producing the three primary germ layers.

GENE. A gene is a unit of heredity, a segment of DNA transcribed to produce a messenger RNA that is translated to produce a protein.

GENEALOGY. A tree describing the history of a single gene, as opposed to a phylogeny, which uses information from many genes or traits to reconstruct the history of a set of species.

GENE FLOW. Genes flow from one place or situation to another when organisms born in one place or situation move to another where they have offspring that survive to reproduce there.

GENE FREQUENCY. The frequency of an allele in a population. If there are 100 individuals in a population of diploid individuals, and we consider one locus (one gene) that is present in two forms (two alleles), A and a, then if 20 of the individuals carry two copies of A (they are AA homozygotes), 60 of the individuals are Aa heterozygotes, and the remaining 20 individuals are aa homozygotes, then the gene frequencies are calculated as the number of each allele divided by the total number, in this case $(40+60)/200 = 0.50$ for both alleles.

GENE SUBSTITUTION. The process by which a new mutation becomes fixed in a population.

GENETIC BOTTLENECK. A reduction in population size to a low-enough level for long enough that many alleles are lost and others are fixed.

GENETIC CORRELATION. The portion of a phenotypic correlation between two traits that can be attributed to additive genetic effects.

GENETIC DIVERSITY. The probability that two homologous alleles chosen at random from a population differ.

GENETIC DRIFT. Random change in allele frequencies due to chance factors.

GENETIC IMPRINTING. The silencing of certain genes in the germ lines of parents prior to gamete production, accomplished by methylating the DNA sequence. Imprinting is usually removed after the early developmental stages of the offspring.

GENETIC SEPARATION. The separation of gene pools during speciation.

GENOMIC CONFLICT. Occurs when genes affecting the same trait experience different selection pressures because they obey different transmission rules or experience opposing selection at different levels of a nested hierarchy.

GENOTYPE. In evolutionary biology, the information stored in the genes of one individual; in population genetics, the diploid combination of alleles at one locus present in an adult prior to meiosis.

GERM LINE. The cell lineage of gametes, zygotes, stem cells, and gonadal cells that stretches continuously back from each organism to the origin of life.

GONOCHORISM. Having separate sexes; individuals are either males or females, not both; used for animals. See DIOECY.

GROUP SELECTION. Selection generated by variation in the reproductive success of groups.

GYNODIOECIOUS. Plant populations containing some individuals whose flowers only produce seeds and other individuals whose flowers produce both seeds and pollen.

HAPLOID. An organism or cell having a single set of chromosomes, in contrast to diploids, which have two sets, one from each parent. In humans the haploid set contains 23 chromosomes and the diploid set contains 46.

HAPLONTIC LIFE CYCLE. A life cycle in which haploid adults produce haploid gametes by mitosis; the diploid zygotes immediately undergo meiosis to produce haploid individuals.

HAPLOTYPES. Groups of closely linked genes that tend to be inherited together.

HEMIMETABOLOUS. Insects that undergo incomplete metamorphosis.

HEMIPARASITE. An organism that can live either independently or as a parasite.

HERITABILITY. The fraction of total phenotypic variance in a trait that is accounted for by additive genetic variance; measures the potential reponse to selection.

HERITABLE VARIATION. Differences in traits between individuals that are determined by genes and can be passed from parents to offspring.

HETEROGAMETIC. The sex having two different sex chromosomes; for organisms with chromosomal sex determination; males are XY in humans.

HETEROPLASMY. A cell that contains genetic variants of a cytoplasmic genome.

HETEROSIS. The heterozygote is more fit than either homozygote.

HETEROZYGOSITY. The proportion of a population which is heterozygous at a locus; also the average proportion of loci heterozygous per individual.

HETEROZYGOTE. A diploid cell that has two different alleles of the same gene.

HITCHHIKING. Changes in the frequencies of neutral traits that are pleiotropically linked to other traits that are under selection, or changes in the frequencies of neutral genes that are linked on chromosomes to changes in other genes that are under selection.

HOLOMETABOLOUS. Insects that undergo a complete metamorphosis, changing from larvae to pupae to adults (imagos).

HOMEOBOX. A 180-base-pair sequence in important regulatory genes that codes for a highly conserved protein segment that is a key part of a transcription factor.

HOMEOBOX GENE. These include the HOX genes as well as other genes that share the DNA sequence motif that translates into a protein sequence, part of which binds the transcription factor to the control region of the genes whose expression is being regulated.

HOMOGAMETIC. The sex having two similar sex chromosomes; for organisms with chromosomal sex determination; females are XX in humans.

HOMOLOGY. An *hypothesis* that similarity of a trait in two or more species indicates descent from a common ancestor.

HOMOPLASMY. A cell that contains only one type of cytoplasmic genome, which may occur in multiple copies.

HOMOPLASY. Similarity for any reason other than common ancestry. The commonest cause of homoplasy in morphological traits is probably convergence, in DNA sequences simple mutation.

HOMOZYGOTE. A diploid cell with the same alleles for a given gene.

HORIZONTAL TRANSMISSION. Transmission of parasites or pathogens to other organisms at times and places not necessarily associated with host reproduction or relationship.

HOX GENE. A gene that produces a transcription factor that controls the expression of genes that determine the major parts of the body axis; similar genes determine similar parts of the bodies of flies and mice.

HYDROSTATIC SKELETON. A cavity containing an incompressible fluid surrounded by a flexible covering to which muscles passing through the fluid attach.

HYPHAE. Long, cylindrical, multinucleate fungal cells (singular hypha).

INBREEDING. The mating of related organisms.

INBREEDING DEPRESSION. The reduction in the survival or reproduction of offspring of related parents caused by the expression as homozygotes of deleterious recessive genes that were present in the parents as heterozygotes.

INCLUSIVE FITNESS. Genetic contribution to the next generation through one's own reproduction plus the *increment* to the reproductive success of relatives caused by aid given to them.

INDEPENDENT CONTRASTS. Contrasts that do not contain any confounding influence of shared history because they evolved after that history was no longer shared.

INDIRECT SELECTION. When a gene increases in frequency not because it is being selected but because it is on a chromosome near a gene that is being selected.

INDIVIDUAL SELECTION. Selection generated by variation in the reproductive success of individual organisms, affecting all their genes and traits.

INDUCED RESPONSE. Developmental response to a specific environmental signal that has a functional relationship to that signal, resulting in improved growth, survival, or reproduction.

INFINITE ALLELE MODEL. A model in the neutral theory of molecular evolution that assumes that every mutation is unique in the sense that it does not already exist in the population, plausible for long DNA sequences.

INGROUP. An assumed monophyletic group, normally made up of the taxa of primary interest.

INNOVATION. A trait in descendants not present in ancestors.

INSTAR. The period between molts, usually used for arthropods.

INTERACTOR. The organism in its ecological role, in which it develops, grows, acquires food and mates, survives and reproduces.

INTRINSIC MORTALITY. Mortality caused by neglect of maintenance leading to degradation of essential functions.

INTRON. A sequence within a gene that is removed after transcription and before translation by gene splicing; its DNA sequence is not represented in the RNA sequence of the spliced mRNA or the amino acid sequence of the resulting protein; introns occur in eukaryotes but not prokaryotes.

INQUILINE. An animal living in a space provided by another; a guest.

ISOGAMY. Mating partners have gametes of the same size.

ITEROPAROUS. Having several discrete reproductive events per lifetime.

KEY EVENT. An event involving the origin of something fundamentally new with a large impact on subsequent evolution.

KIN SELECTION. Adaptive evolution of genes caused by relatedness; an allele causing an individual to act to benefit relatives will increase in frequency if that allele is also found in the relatives and if the benefit to the relatives more than compensates the cost to the individual.

LACK CLUTCH. The intermediate clutch size in altricial birds that produces that largest number of fledglings.

LAGERSTATT. German for resting place; plural *Lagerstätte*. A particularly complete fossil deposit representing much of an ancient community, often with exceptional preservation of detail.

LEK. A traditional display site where males gather to defend mating territories and females come to choose mates, often used for generations.

LIFE. A process involving metabolism and hereditary replication.

LIFE-HISTORY TRAIT. A trait directly associated with reproduction and survival, including size at birth, growth rate, age and size at maturity, number of offspring, frequency of reproduction, and lifespan.

LINEAGE-SPECIFIC DEVELOPMENTAL MECHANISM. A developmental mechanism found within all organisms of one lineage but not in other lineages, responsible for the morphology that characterizes the lineage. Limits the genetic variation that can be expressed in the lineage.

LOCAL MATE COMPETITION. The offspring of one parent compete with each other for mates.

LOCUS. The site of a gene on a chromosome, often used as a synonym for gene.

MACROEVOLUTION. The pattern of evolution at and above the species level, including most of fossil history and much of systematics.

MADS GENE. A family of genes coding for transcription factors present in plants, animals, and fungi. They are not colinear, but their action is combinatorial.

MALADAPTATION. The state of a trait that leads to demonstrably lower reproductive success than an alternative existing state. Maladaptations can arise when local populations are swamped by gene flow.

MATING TYPES. Sets of potential mating partners. Matings can occur between partners of different type but not with partners of the same type.

MAXIMUM LIKELIHOOD. A statistical method of inferring the process that would make the data actually observed the most likely of all possible data sets.

MEIOSIS. Reduction division of diploid germ cells to yield haploid gametes.

MEIOTIC DRIVE. Genes distort meiosis to produce gametes containing themselves more than half the time.

MENDELIAN LOTTERY. A particular allele will or will not be represented in the offspring because of the segregation of alleles at meiosis and the random chance that any particular gamete will form a zygote. Most easily seen with small family sizes. Think about single children.

MENDELIAN RATIO. The ratio of offspring of given genetic types expected when homozygous lines are crossed. If A and a denote two alleles at one locus, and if an AA homozygote is crossed with an aa homozygote, the expected Mendelian ratios in the offspring are 25% AA : 50% Aa : 25% aa.

METHOD OF INDEPENDENT CONTRASTS. A comparative method that controls for the fact that character states in related organisms are not statistically independent because of shared ancestors. The basic idea is that *differences* between one pair of species are independent of *differences* between another pair of species even if both pairs are related.

MICROEVOLUTION. The process of evolution within populations, including adaptive and neutral evolution.

MITOCHONDRIA. Intracellular organelles derived from bacterial ancestors with their own genomes. The energy factories of the cell where ATP and the intermediate products of the Krebs cycle, used in the cytoplasm for energy and biosynthesis, are made.

MITOSIS. Cell division by exact copying in which two daughter cells are created from one parental cell with no change in chromosome number or DNA content.

MOLECULAR CLOCK. The approximately constant rate of nucleotide substitution for particular genes and classes of genes within particular lineages. The constancy of the rate depends on the randomness with which particular nucleotides mutate and then drift to fixation.

MONOECY. Individuals reproduce both as males and as females; hermaphrodites; used for plants.

MONOGAMY. Each sex has a single mate for life.

MONOPHYLETIC. All species in a monophyletic group are descended from a common ancestor that is not the ancestor of any other group and no species descended from that ancestor are not in it.

MULLER'S RATCHET. A mechanism operating in finite asexual populations whose effect is that the number of deleterious mutations can only increase over time.

MULTIGENE FAMILY. Sets of multiple copies of genes derived by duplication from a common ancestor gene and retaining the same function.

MULTILEVEL EVOLUTION. Adaptive evolution occurring simultaneously at several levels of a biological hierarchy, e.g. nuclear and cytoplasmic genes.

MUTAGENESIS. The production of mutations in an experimental population using either chemicals or radiation.

MUTATION. A hereditary change in the DNA sequence or in chromosome number, form or structure.

MUTUALISM. Two species interact to their mutual benefit.

MYCELIUM. A network of fused hyphae.

MYCORRHIZAL ASSOCIATION. A symbiotic association between a fungus and a plant root; nutrients gathered in the soil by the fungus are transported into the plant.

NARROW-SENSE HERITABILITY. The fraction of total phenotypic variance in a trait that is accounted for by additive genetic variance; measures the potential response to selection.

NATURAL SELECTION. The correlation of a **trait** with variation in reproductive success.

NEUTRAL. Variation in state is not correlated with variation in reproductive success: states are equally fit.

NEUTRAL EVOLUTION. Changes in the genetic composition of populations that occur because genes that are not correlated with reproductive success are still influenced by random processes.

NUCLEOTIDE. A molecule consisting of a nitrogen base, a sugar and a phosphate group: the basic building block of DNA and RNA.

NUCLEOTIDE DIVERSITY. The average number of nucleotide differences per site between randomly chosen pairs of sequences.

OOCYTIC ATRESIA. The elimination of oocytes from the developing ovary.

OPERATIONAL SEX RATIO. The local ratio of sexually active males to receptive females.

ORTHOLOGOUS GENES. Genes similar enough in DNA sequence to justify the inference that they are descended from a common ancestor.

ORTHOLOGY. An *hypothesis* that the similarity of DNA sequences is explained by ancestry.

OUTGROUP. One or more taxa assumed to be phylogenetically outside the ingroup.

OUTGROUP COMPARISON. A method used to root phylogenetic trees and thus to establish the direction of character change.

PARALLELISMS. Parallel patterns of evolution in related species caused by shared developmental constraints, not by shared experience of selection.

PARALOGOUS. Similarity in the DNA sequence of two genes because one is a duplicate of the other.

PARAPATRIC. Occurring in adjacent areas sharing a border.

PARAPHYLETIC. A group that does not contain all species descended from the most recent common ancestor of its members; some of those species are outside it.

PARSIMONY. A principle of logic called Ockham's razor stated by William of Ockham (d. 1347): 'Entities should not be multiplied unnecessarily': keep it as simple as possible. In cladistic tree-building: the best tree has the fewest changes in character states and the least homoplasy.

PARTHENOGENESIS. Asexual reproduction from an egg cell that usually does not involve recombination. In most cases the daughters are exact genetic copies of the mothers.

PHAGE. The viruses that infect bacteria.

PHENOTYPE. The material organism, or some part of it, as contrasted with the information in the GENOTYPE that provides the blueprint for the organism.

PHENOTYPIC DIFFERENTIATION. The differentiation of phenotypes in separated gene pools during and after speciation.

PHENOTYPIC PLASTICITY. Sensitivity of the phenotype to differences in the environment. Less precise than reaction norm.

PHYLOGENETICS. The branch of biology that reconstructs the evolutionary history of species.

PHYLOGENETIC SPECIES CONCEPT (PSC). A monophyletic group composed of 'the smallest diagnosable cluster of individual organisms within which there is a parental pattern of ancestry and descent.'

PHYLOGENETIC TRAIT ANALYSIS. A comparative method in which one constructs a phylogenetic tree, plots character states (traits) on the tree, and infers transitions in character states from their position on the tree. Geographical locations of taxa can be plotted onto the tree to infer the location of ancestors.

PHYLOGENY. The history of a group of taxa described as an evolutionary tree with a common ancestor as the base and descendent taxa as branch tips.

PHYTOCHROMES. Light-detecting molecules that function as the 'eyes' of plants. The information they receive influences germination and growth.

PLASMID. A small piece of circular DNA that may exist in multiple copies in bacterial cells. Plasmids are involved in inducing conjugation, during which they carry genetic information from one bacterial cell to another.

PLASTICITY. Sensitivity of the phenotype to changes in the environment; the slope of a reaction norm of a trait measures the plasticity of that trait to the environmental factor to which it is reacting.

PLASTID. A cell organelle with its own bacteria-like genome: mitochondria and chloroplasts.

PLEIOTROPY. One gene has effects on two or more traits.

PLESIOMORPHIC. Ancestral, relative to a derived, or apomorphic, state.

POINT MUTATION. A change in a single DNA nucleotide, e.g. adenine mutates to thymine, or an insertion or deletion of a single nucleotide.

POLYANDROGYNY. Both sexes are variable in their mate numbers, but females are more variable than males.

POLYANDRY. Males mate with a single female for life, but females may mate with more than one male.

POLYGAMY. Both sexes have several partners and approximately equal variation in mate numbers.

POLYGYNANDRY. Both sexes are variable in their mate numbers, but males are more variable than females.

POLYGYNY. Females mate with a single male for life, but males may mate with more than one female.

POLYMORPHISM. The existence in a population of two or more alternative forms of a trait or gene.

POLYPHENISM. In contrast to continuous reaction norms, distinctly different, *discrete* phenotypes expressed by the same genotype in reaction to an environmental signal.

POLYPHYLETIC. A group that contains species descended from several ancestors from which members of other groups also descended.

POLYPLOID. A cell with more than two copies of the entire chromosome set; those with three copies are triploid, those with four, quadriploid, and so forth.

POLYPLOIDIZATION. A doubling of the complete chromosome set.

POPULATION GENETICS. The discipline that studies changes in frequencies of alleles in populations; issues include mutation, selection, inbreeding, assortative mating, gene flow, and drift; suitable when genetic differences at one or more loci can be detected as phenotypic differences.

POSITIONAL INFORMATION. Information used by cells during development that establishes where they are relative to other cell types.

POSTZYGOTIC. After fertilization, e.g. early development.

PRECOCIAL. Born or hatched capable of seeing, moving, feeding, and thermoregulating.

PREZYGOTIC. Before fertilization, e.g. mate choice.

PROKARYOTES. Single-celled organisms lacking a nucleus and organelles; they include Eubacteria and Archaea.

PROTANDRY. Individuals are born as males, reproduce as males, then change sex and reproduce as females. In plants, individuals express male function prior to female function, producing pollen before being pollinated.

PROTEIN. Proteins are macromolecules with roles both as catalysts of reactions (enzymes) and as building blocks of cell structures. They consist of chains of amino acids, usually several hundred long. Twenty different amino acids are common in proteins.

PROTOGYNY. Individuals are born as female, reproduce as females, then change sex and reproduce as males. In plants, individuals express female function prior to male function, being pollinated before producing pollen.

PROXIMAL. Located towards the body.

PROXIMATE CAUSATION. The mechanical determination of traits during the lifetime of an organism, couched in terms of chemistry and physics.

PSEUDOGENE. A nonfunctional copy of a gene; it is not expressed.

PUNCTUATED EQUILIBRIUM. A pattern seen in many but not all lineages in the fossil record in which a long period of stasis is broken by a short period of rapid change. In some cases the rapid change is associated with speciation.

PUNCTUATION. A short period of rapid change breaking a long period of stasis in the fossil record.

QUANTITATIVE GENETICS. Studies changes in traits in populations when genetic differences at one locus are too small to detect in phenotypes and when many genes affect one trait; common themes are heritability, genetic covariance, and response to selection.

RADIATION. The diversification and divergence of species within a group with a single common ancestor (a clade).

REACTION NORM. A property of a genotype that describes how development maps the genotype into the phenotype as a function of the environment.

RECESSIVE. An allele is recessive if it is not expressed in the phenotype in the heterozygous diploid state.

RECOMBINATION. The production of gametes during meiosis that differ from the haploid genomes of either parent. It consists of two processes: independent segregation of chromosomes to produce new combinations of chromosomes, and crossing over of nonsister chromatids of the same chromosome to produce new combinations of genes along a single chromosome.

RED QUEEN. Coevolutionary dynamics in which the participants struggle forever against each other with no long-term reduction in extinction probability. Named after a character in *Through the Looking Glass*, by Lewis Carroll.

REGULATORY GENE. A gene that turns another gene, or group of genes, on or off. Small changes in regulatory genes can cause large changes in phenotypes.

REPLICATOR. The organism in its role as information copier, the mechanism that copies the DNA sequence of the parent and passes it to the offspring.

REPRODUCTIVE SUCCESS. A measure of fitness defined as the number of offspring produced per lifetime. It can be extended through several generations; for example, one could define it as the number of grandchildren that survive to reproduce.

REPRODUCTIVE VALUE. The expected contribution of organisms in that stage of life to lifetime reproductive success.

RESIDUAL REPRODUCTIVE VALUE. The remaining contribution to lifetime reproductive success after the current activity has made its contribution.

RNA. Ribonucleic acid is a macromolecule that, like DNA, has a sugar-phosphate backbone to which are attached nucleotides, in this case adenine, uracil, guanine, and thymine. Thus uracil in RNA replaces thymine in DNA. RNA preceded DNA as the genetic molecule; it still is the genetic molecule of RNA viruses. In other organisms it occurs as messenger RNA (mRNA), transfer RNA (tRNA), and ribosomal RNA (rRNA).

SECONDARY REINFORCEMENT. The reinforcement of prezygotic barriers to hybridization after secondary contact between isolated populations.

SEGREGATION DISTORTION. Deviation from the Mendelian ratios that give equal chances to homologous alleles in meiosis; unfair ratios can be caused by nuclear genes that interfere with meiosis or with the products of meiosis to improve their own chances at the expense of their homologs.

SELECTION ARENA. An adaptation produced by natural selection that uses natural selection to produce its effect.

SELECTION DIFFERENTIAL. The difference between the population mean and the mean phenotype of the parents that produce the next generation.

SELECTION RESPONSE. The difference between the mean of the parental population and the offspring mean.

SELFISH CYTOPLASMIC GENE. A gene located in an organelle, plasmid, or intracellular parasite that modifies reproduction to cause its own increase at the expense of the fitness of the cell or organisms that carries it.

SEMELPAROUS. Reproducing once, then dying.

SESSILE. Permanently attached, therefore immobile.

SEX ALLOCATION. The allocation of reproductive effort to male versus female function in hermaphrodites and to male versus female offspring in species with separate sexes.

SEXUAL. A mode of reproduction in which the genetic material from two parents is mixed to produce offspring that differ genetically from both parents.

SEXUAL DIMORPHISM. Males and females have different phenotypes.

SEXUAL SELECTION. The component of natural selection that is associated with success in mating.

SIBLING SPECIES. Species that resemble each other so closely that they cannot be distinguished by external morphology.

SIMULTANEOUS HERMAPHRODITE. An organism with fully functional male and female reproductive organs producing both eggs or ovules and sperm or pollen.

SKELETON. A structure that permits muscles to be stretched back to their original length following contraction.

SOMA. The cells of the body that carry out all the functions of life except gamete production.

SOMATIC CELL. A 'body cell' in nonreproductive tissue.

SPECIATION. The process by which new species originate and thereafter remain separate.

SPECIES. Either a set of organisms that could share grandchildren (the biological species concept), or the smallest diagnosable cluster of individual organisms within which there is a parental pattern of ancestry and descent (the phylogenetic species concept), or a cluster of similar DNA sequences.

SPECIES CONCEPT. How we define what a species is.

SPECIES CRITERIA. Criteria used to recognize species when we see them.

SPINDLE APPARATUS. Pulls the chromosomes apart during meiosis and mitosis. It replicates, divides, and each copy anchors itself. Microtubules grow from the spindle to the centromeres of the chromosomes, attach, contract, and pull a set of chromosomes to one end of the dividing cell.

STABILIZING SELECTION. Selection that eliminates the extremes of a distribution and favors the center.

STASIS. A long period without evolutionary change; if the organisms involved survive to the present, we call them living fossils.

STRICT CONSENSUS TREE. A phylogenetic tree derived from a set of equally parsimonious trees and constructed by only including the groups that are supported in all the equally parsimonious trees.

SYMPATRY. Occurring in the same geographic area.

SYMPLESIOMORPHY. A character state shared by all members of this group as well as with members of related groups.

SYNAPOMORPHY. A shared, derived character state indicating that two species belong to the same group.

SYNERGISM. A nonadditive interaction between two or more factors.

SYNONYMOUS MUTATION. A point mutation (change in a single nucleotide) that does not change the amino acid for which the DNA triplet codes.

SYSTEMATICS. The branch of biology that investigates relationships among species to understand the history of life.

TERRANE. A crustal block or fragment that preserves a distinctive geologic history that is different from the surrounding areas and that is usually bounded by faults.

TRADEOFF. A tradeoff occurs when a change in one trait that increases fitness causes a change in the other trait that decreases fitness.

TRANSCRIPTION. The process by which the DNA sequence of a gene is copied into the complementary sequence of a messenger RNA.

TRANSCRIPTION FACTOR. The protein product of a developmental control gene; binds to the regulatory sequence of genes, where it interacts with other transcription factors to determine whether the genes will be turned on or switched off.

TRANSDUCTION. A virus that infects bacteria picks up some bacterial DNA from one host and transfers it to the next host, which may incorporate the DNA if it survives the infection.

TRANSFORMATION. Bacteria take up DNA from the medium and incorporate it into their circular chromosome.

TRANSLATION. The process by which mRNA produces proteins. It occurs when mRNA is fed through a ribosome and matched up with tRNAs carrying the amino acids that will form the protein. The amino acids are joined together, the tRNAs are released, and the mRNA continues to be fed through the ribosome until the process is completed.

TRANSPOSON. A segment of DNA that can move from one site on a chromosome to another. If it leaves a copy behind, it can multiply, increasing in copy number. Transposons are also called jumping genes. They do not code for proteins that are expressed in the cells of their hosts, behaving instead as genetic parasites. Their movements are responsible for some mutations, for when they insert into the DNA at a new site they can disturb local gene expression.

TRIPLOBLASTIC. Having three primary cell layers: endoderm, mesoderm, and ectoderm.

TRUNCATION SELECTION. Artificial selection in which only individuals with a value of a trait above (or below) some threshold are allowed to breed.

ULTIMATE CAUSATION. The evolutionary causes of a biological effect, couched in terms of selection and history.

VERTICAL TRANSMISSION. Transmission of parasites or pathogens from host parent to host offspring, often during host reproduction.

VIRULENCE. The property of pathogens of being highly malignant or deadly, causing serious disease or death in the host.

VIVIPARITY. Giving birth to fully formed offspring, as opposed to eggs.

WILD TYPE. A term used in classical genetics to designate the standard genotype in the population from which mutations formed rare deviations. Modern molecular data have destroyed the concept by revealing so much variation that it has become meaningless.

ZYGOTE. The cell resulting from the union of an egg and a sperm.

ANSWERS TO QUESTIONS

Chapter 2

2.1. Necessary: variation in a trait; variation in reproductive success; non-zero correlation between the two. Sufficient: combination of all three.

2.2. The correlation between the trait—or gene—and reproductive success must be close to zero.

2.3. Gene flow, insufficient time, tradeoffs, constraints. To check for gene flow, two methods are available: (1) isolate the population and see if it declines in numbers—if it does, it is a sink; (2) genotype enough individuals to be able to test the hypothesis that many loci are not in Hardy–Weinberg equilibrium (see Genetic appendix)—if they are not, one of several explanations is gene flow.

2.4. Tinkering produces short-term solutions that take no account of overall design and introduce many details and contingencies that complicate an overall solution. For example, the previous owners may have altered the plumbing in the bathroom so that the only way to re-route the pipes is to remove the outer wall of the house, making the house uninhabitable until the reconstruction is complete. The analogy with evolution would be, for example, that the light-sensitive cells in the retina of the eye cannot simply be moved to the eye's inner surface because they have to have a particular relationship with other cell layers during development to induce the formation of critical connections. Simply moving them without also changing the entire mechanism of eye development would destroy visual function.

2.5. This is a deep question with a complicated answer best pursued in discussion or essay. See Dawkins (1999) and Richerson and Boyd (2004) for an entry into the literature.

Chapter 3

3.1. In the developed world humans have been largely released from selection for resistance to infectious diseases and parasitic worms. We are likely to lose our genetic resistance to those diseases and parasites, especially if the genes involved have costly effects on other traits that are still important.

3.2. Yes, because the effective population size is strongly influenced by the small number of males that breed. The allele frequencies in the offspring generation are equally affected by the maternal and the paternal contribution. The latter will reflect genetic drift because of the small number of fathers.

3.3. It affects both: all genes are subject to a balance between selection and drift.

3.4. Cultural drift and founder effects are very important in the evolution of language. They lead to the rapid divergence of dialects.

3.5. See Chapter 4, but think about it before looking there.

Chapter 4

4.1. A_1 is the common allele that is selected against, therefore A_2 is a rare advantageous allele. A_2 will increase in frequency faster when it is dominant because it then experiences the effects of selection in both homozygotes and heterozygotes. When A_2 is dominant, A_1 is recessive, therefore A_1 decreases faster when it is recessive, as in case a.

4.2. The heritabilities within populations establish that variation in seed weight within each population has an important genetic component, but they do not tell us whether the differences between populations have an important genetic component. The best way to test that idea is to grow the plants in common gardens—in the same environments—to see if they still differ in seed weight when environmental differences among environments have been controlled.

4.3. We can use the formula from Box 4.2:

$$p' = (p^2 + pq(1 - hs))/(p^2 + 2pq(1 - hs) + q^2(1 - s))$$

If we assume the mutation is dominant, then $h = 1.0$; s is given as 0.1. We also need to make an assumption about effective population size. If we assume a population size of 500, then the initial frequency of the mutant is 1 in a 1000 in a diploid population, or $p = 0.001$. With those numbers, we need to repeat the calculation in the equation above 9089 times for the frequency to rise to 0.95. If we assume a human generation is roughly 30 years, then the answer is that it takes 272 670 years for the frequency to increase from 0.001 to 0.95. The factors affecting this calculation are the degree of dominance and population size. If we assumed a population of 50 and a starting frequency of 0.01, the answer would be 966 generations, for example.

4.4. He was assuming that the combined future reproductive success of two brothers, or of eight cousins, would be greater than his own.

4.5. Rate $= p(1 - p) = p - p^2$. The rate will be maximum when the first derivative $= 0$ and the second derivative <0. The first derivative $= 1 - 2p$. When that $= 0$, $1 - 2p = 0$, implying that $p = 0.5$. The second derivative is -2, which satisfies the inequality.

Chapter 5

5.1. If there is no genetic diversity, recombination cannot form any new haplotypes. If there is high genetic diversity, recombination will form many new haplotypes.

5.2. The equilibrium frequency of a recessive mutation under mutation–selection balance is $q = \sqrt{u/s}$. The frequency of a recessive gene is the square root of the frequency of recessive homozygotes. In this case, the frequency of recessive homozygotes is about the same—10^{-5}—as the equilibrium which would hold when s is about 1, the value of a selection coefficient for a lethal gene. So the diseases can be explained as a mutation–selection equilibrium if recessive homozygotes usually die before reproducing.

5.3. The square of the frequency of the allele for cystic fibrosis is 0.00025 (the frequency of the allele is about 0.0158). If the mutation rate is $10^{-6} = 0.000001$, then the equation used in question 5.2 implies, after squaring both sides, that $0.00025 = 0.000001/s$, i.e. $s = 0.004$. That is a very low selection coefficient for a disease as serious as cystic fibrosis; mutation–selection balance is not likely to be maintaining it at such a high frequency. One explanation is that the allele might have a positive effect in the heterozygote, perhaps through disease resistance. There is indeed evidence suggesting that it may provide protection against typhoid fever.

5.4. At equilibrium under overdominance $p = (1 - h)/(1 - 2h)$ (see Box 5.2). In that model p is the frequency of the normal allele, and q is that of the disease allele. The frequency of the cystic fibrosis allele is 0.0158 (see question 5.3), therefore $p = 1 - 0.0158 = 0.9842$. The equilibrium allele frequency equation $p = (1 - h)/(1 - 2h)$ can be reworked to yield $h = (1 - p)/(1 - 2p)$. Substitution of $p = 0.9842$ into this equation yields $h = -0.0163$. We may put $s = 1$, because cystic fibrosis patients are assumed not to reproduce. Therefore the heterozygote fitness $1 - hs$ becomes 1.0163. This implies that a small fitness advantage between 1 and 2% in heterozygotes is sufficient to explain the observed disease frequency.

5.5. (a) See Box 5.2 for the formulae used in this problem.

$$\Delta p = p' - p = p\left[\frac{p + q(1 - hs)}{W} - \frac{W}{W}\right];$$

we can write W in a different form as follows: $W = p^2 + 2pq(1 - hs) + q^2(1 - s) = p[p + q(1 - hs)] + q[q(1 - s) + p(1 - hs)]$. Using this expression, and remembering that $q = 1 - p$, the first equation becomes

$$\Delta p = p\frac{\left[p + q(1 - hs) - p(p + q(1 - hs)) - q(q(1 - s) + p(1 - hs))\right]}{W}$$
$$= pq\frac{p + q(1 - hs) - q(1 - s) - p(1 - hs)}{W} = pqs\frac{1 - h - p(1 - 2h)}{W}$$

(b) It is clear from this last expression that $\Delta p = 0$ for three values of p: $p = 0$, $p = 1$ (corresponding to $q = 0$), and $p = (1 - h)/(1 - 2h)$. The first two are so-called trivial allele frequency equilibria, because in both cases it is obvious that the allele frequency does not change if there is only one allele present in the population. The other value for which the allele frequency does not change is—as expected—the equilibrium value we derived earlier in another way (eqn 5.2). It can easily be verified that this allele frequency equilibrium only takes values in the interval $(0,1)$—that is, represents a real relative allele frequency—if $h < 0$ or if $h > 1$. Limiting ourselves to these two cases, we can check by trying a few values for h and p in the formula for Δp that the graph of Δp behaves qualitatively as follows:

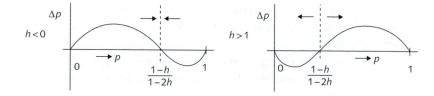

In the left graph (the case of $h < 0$) the internal equilibrium is stable, because for $p < (1 - h)/(1 - 2h)$ the value of Δp is positive, so p moves towards the equilibrium point, and for $p > (1 - h)/(1 - 2h)$ the value of Δp is negative, so in that case also p moves towards the equilibrium value. In the right graph ($h > 1$), the situation is reversed, and the allele frequency p moves away from the equilibrium value, causing the equilibrium to be unstable.

Note that the stable equilibrium corresponds to heterozygote superiority, and the unstable equilibrium to heterozygote inferiority.

(c) An unstable allele frequency equilibrium is not expected to be commonly observed, because the allele frequency will have the tendency to approach the values 0 or 1. A polymorphic situation (two alleles co-occurring at the locus in a population) is therefore unlikely when the heterozygotes have a lower fitness than either homozygote. One of the rare instances of such a polymorphism is the Rhesus blood-group polymorphism in European populations. Until medical intervention became possible to solve the problem, Rhesus-positive children from a Rhesus-negative mother ran the risk of the so-called hemolytic disease of the newborn. This is caused by an immunological incompatibility between mother and child. Such children are necessarily heterozygotes, because the Rh+ allele is dominant. Thus this is an example of selection against heterozygotes, and we do not expect to observe stable polymorphism. The likely explanation in this case is a historic one: the original population that lived in Europe before the large immigrations from Asia occurred following the invention of agriculture was probably Rhesus negative, and the people who came from the east were Rhesus positive. The resulting mixture created the polymorphism, which is unstable as long as selection operates against heterozygotes. The 10 000 or so years of selection against heterozygotes that have elapsed since then have apparently been insufficient to completely remove the minority allele.

Chapter 6

6.1. Both the elephant and the bacterium have to undergo a complex series of developmental changes to complete their life cycle. The bacterium must grow, differentiate within the cell, and divide. The elephant must grow, multiply greatly in cell number, and differentiate many different cell types. They differ in that all the bacterial differentiation occurs within a single cell, and all bacterial cells are capable of reproduction. The elephant differentiates primarily among cells, and only the cells in its germ line remain capable of reproduction.

6.2. The duplication of the entire set of HOX genes made available a set of developmental switches already specialized at defining different regions of a cylinder. Evolution was opportunistic in using a duplicate of something previously invented and applying it in a new context rather than re-inventing it from scratch.

6.3. Lipid bilayers spontaneously assemble from polar fatty acid precursors in aqueous solution, forming membranes. The spontaneous assembly does not need to be coded in the genes.

Chapter 7

7.1. No, intelligence is then clearly determined *both* by genes *and* by environment. The question is not posed properly; it inappropriately assumes that genes and environment can act on the phenotype independently.

7.2. The differences between basic features of arthropods and vertebrates probably result from changes in the sets of downstream genes controlled by the developmental control genes.

7.3. One could go on at great length. Our inclination is to argue that both are necessary and neither is sufficient, and to illustrate that using the evolution of RNA in test tubes (see Chapter 2, pp. 32 ff.).

7.4. One approach to answering this question is to construct the answer around a figure on which two reaction norms are sketched. One can then use that to argue that the differences between the two reaction norms have evolved, and that the particular point on either of the reaction norms that described an individual organism in a particular environment is due to the interaction of the genotype of that organism with that particular environment. Thus one can equate the position and shape of the reaction norm with nature and the position of any particular point on a reaction norm with nurture, making clear the sense in which organisms and their traits are always the products of an interaction between the two.

Chapter 8

8.1. This pattern is consistent with the idea that sex evolved as a method of spreading the risk of not having any offspring that could survive in a changed environment, that the function of sex is to produce offspring that differ genetically from their parents in ways that matter ecologically. In this kind of life cycle, the costs of sex are reduced in the many generations of asexual reproduction that occur when the environment is favorable. Sex only occurs when it has a benefit.

8.2. What one thinks of this argument depends upon whether sex is seen as a cause or a consequence. It is very unlikely to be a consequence of selection to keep species from going extinct, for such events happen at very long intervals, once in tens of thousands to millions of generations, and in the intervening period asex has many opportunities to invade. If that were the only reason for sex, then asex would dominate. However, if sex is seen as caused for other reasons, but having reduced probability of extinction as a consequence, then the argument makes sense. That is not, however, how old texts and popular articles present the case—they usually confuse cause with consequence.

8.3. Williams' argument is correct. The asexual alternative is always present but does not take over completely. That implies that it is being held at an equilibrium, probably by frequency-dependent selection.

8.4. It is clear that imprinting must be reset in the developing organism. Otherwise the imprinted condition would be inherited from grandparents, not from parents, or even from distant ancestors if it were never erased. The evidence needed is thus whether or not the genes in the adult somatic cells used to clone sheep and mice have had their germ-line imprinting erased.

Chapter 9

9.1. Both cases are examples of two-level genomic conflict. But there is a difference: yeast cells that have become petite can reproduce to form petite daughters, while cancer—fortunately—is not hereditary at the individual level.

9.2. We may use the model developed in Box 4.2. The present case corresponds with $s = 1$ and $h = 0$. However, due to the assumed segregation distortion the formulae for the allele frequencies among the gametes have to be modified. For A_1 this becomes $p' = (p^2 + 2pq(1 - k))/W = [p(p + 2q(1 - k))]/W$ and for A_2 $q' = (2pq.k)/W = q(2pk/W)$

In case of an allele-frequency equilibrium we must have $p' = p$ and $q' = q$, which implies (combining both equations) that $p + 2q(1 - k) = 2pk$, or (remember $p = 1 - q$), $\hat{q} = 2k - 1$. This is the expected allele-frequency equilibrium, which indeed requires $k > 0.5$ in order to take a value in the interval $(0,1)$.

9.3. When A_2 is fixed in the population, all individuals will have genotype A_2A_2. Then segregation distortion will not be observed, because it only occurs in heterozygotes.

9.4. If a trait is selectively favored at the individual level but selected against at the group level, the timescale at which the evolution proceeds at both levels becomes an important factor. Usually evolution proceeds much faster at the individual level, so that the individually selected trait will dominate most of the time.

Chapter 10

10.1. List: external fertilization, random mating, no parental care, no social structure, no size-dependence, or other factors affecting sex-specific reproductive success. Mammalian sex ratios are close to 50:50 at birth because they have sex chromosomes: XX females and XY males. Males produce 50% X-bearing sperm and 50% Y-bearing sperm. Why, then, did genetic sex determination using sex chromosomes (rather than some other mechanism, such as a single locus) evolve? Conflict resolution is one possible answer: a 50:50 sex ratio is in the interests of most of the genome, and chromosomal sex determination is a mechanism for achieving it that exploits the pre-existing fairness of meiosis while keeping together on one chromosome the several genes needed to initiate sex differentiation in development.

10.2. The argument is based on the idea of constraint, and constraint can enter into development in this case in at least two places. First, given chromosomal sex determination, the developing embryos get signals fairly early in development that they will be either male or female but not both. Those signals set in motion a sequence of events that lead to strikingly divergent differentiation in the male and female reproductive tracts. Thus, second, later in development the reproductive tract of mammals is not plastic with respect to sex. Unlike fish gonads, a mammalian ovary cannot function as a testis, nor a testis as an ovary. And after a certain early point in development, there is no tissue in a male embryo whose fate can be altered to become a uterus.

10.3. It would take about 260 generations for a dominant mutant to increase to 90% and about 1780 generations to increase to 99% starting at $p = 0.01$ with a selection differential of 0.06. If clutch size was polygenic, selection was directional, and heritability was 0.3, then $R = h^2 s = 0.3 \times .06 = 1.8\%$ per generation, or $(100/1.8 =) 55.55$ generations to change clutch size by one egg. This answer increases when selection is stabilizing because the selection differential gets smaller and smaller as the clutch size approaches the optimum, slowing the approach.

10.4. Under these assumptions, on average all four offspring of the first type find mates and breed, yielding 16 grandchildren. On average one of the two daughters of the second type fails to find a mate; her two brothers and sister mate successfully; this type can expect 12 grandchildren.

Chapter 11

11.1. Disease resistance should decrease in captivity because females are no longer able to select for it in their mates. Disease resistance might also decrease in captivity because it is negatively correlated with traits under artificial selection, such as weight gain, without any influence from female choice.

11.2. Mate choice can explain striking patterns and colors in sexually monomorphic species if both partners are choosing the same traits. If the alternative explanation is correct, it is not at odds with the first, for it just specifies that both partners are choosing a mate of their own species rather than some other species. And in that case, there is no difference between mate choice and species recognition.

Chapter 12

12.1. Witnessing the complete process of speciation requires observations extending over all stages involving increasing degrees of reproductive isolation and phenotypic differentiation. Limiting attention to the genetic side of speciation (Figure 12.7), this would take on the order of a million years, far exceeding the time span of a few generations of biologists. Only very rarely may speciation happen so rapidly that it can be observed, and then it involves atypical mechanisms, like polyploidization.

12.2. When populations are separated, differential selection among the populations and independently occurring random drift may cause differentiation to such an extent that the populations appear to be genetically separated upon secondary contact (allopatric speciation). Within a population, disruptive selection combined with assortative mating, sexual selection, and host shifts may cause genetic separation and phenotypic differentiation (sympatric speciation).

12.3. In this experiment gene flow between the two strains is effectively prevented by allowing only the offspring from homotypic matings to enter the next generation.

Reinforcement, in contrast, presupposes ongoing gene flow while prezygotic isolation evolves. Therefore, this experiment does not test the possibility of speciation by reinforcement. It would, if some fraction of the offspring from heterotypic matings were allowed to participate in forming the next generation.

Chapter 13

13.1. Because most species are extinct and most clades have vanished, only a very small proportion of fossils belong to species that have left living descendants.

13.2. No, it cannot. However, we can judge the reliability of inferring any particular tree structure, and some of them are much more likely than others.

13.3. The Linnean system is not useful if we want names to indicate relationship as depicted by a phylogenetic tree. A name that reflected all the information in a tree would be technically useful but cumbersome to use in everyday discourse, as well as being difficult to remember. To take a simplified example, humans belong (roughly) to the Eukaryote, Opisthokont, Deuterostome, Vertebrate, Tetrapod, Mammal, Primate, and Hominid clades (branches have been left out). We might want to name them something like *Eu.Op.De.Ve.Te.Ma.Pr.Ho.Ho.sapiens* for technical purposes, put that tag in a place from which we could easily retrieve it, and continue to use *Homo sapiens* for everyday discourse. If we use some such scheme, then it is possible for the name of a species to reflect the structure of a phylogenetic tree if we restrict ourselves to naming major clades. If we want the name to reflect every branch in the tree, then names will grow longer and longer as trees become more and more detailed, until they grow so long that the detailed tree can be reconstructed from the complete set of names. That can be done in principle and presents no great problem if the long technical names are stored in computers, but they would not be convenient objects to refer to for any reason other than knowing details of relationship.

Chapter 14

14.1. The choice of approach would depend on the question to be answered. If one wanted to know whether a pattern was consistent with an hypothesis about the impact of mating systems on sexual dimorphism, one would do the comparative study, perhaps discovering that males were more brightly colored in a certain clade (ancestral effect) or in species that displayed on tree trunks but not on the ground (response to selection based on habitat). If one wanted to see whether skin color was an honest signal of resistance to parasites, and whether females paid any attention to that signal, one would do an experiment.

14.2. Because these estimates change every few years, you should use ISI Web of Knowledge or some other literature search engine to find the articles containing the most recent estimates for the breakup of Gondwana and the common ancestor of marsupials and eutherians. If Gondwana broke up *before* marsupials and placentals diverged, then placentals arose on other continents and marsupials went extinct there. If Gondwana broke up *after* marsupials and placentals diverged, then most placentals went extinct on Australia but not on South America. These are only two of many scenarios involving several land masses, including Antarctica.

Chapter 15

15.1. Genomic conflict creates suboptimal states at one or more levels at which selection occurs, thus creating a reason for the evolution of mechanisms that suppress, resolve, or prevent the genomic conflict. In some cases this may have been the driving force for evolutionary novelty, in particular in the genetic system itself. For example, Hurst et al. (1996) suggest that genomic conflicts may have played a role in the evolution of chromosomes, sexual reproduction, meiosis, recombination, and other key aspects of genetic systems.

15.2. It depends on how strictly you want to limit the term organism to a physiologically independent unit. Separate individual bacterial cells are capable of independent functioning and reproduction, and can be considered as separate organisms, but a colony consisting of many millions of genetically identical cells, showing division of labor and functional specialization of its cells, comes very close to a multicellular organism. Because probably the genetical relatedness of the cells in such colonies is a crucial factor, the concept of kin selection is highly relevant.

15.3. Language, broadly interpreted as an information-transmission system, is extremely important in the biology of many—if not all—species. To mention but a few examples, it plays a key role in sexual reproduction where males and females recognize and select each other using specific signals, host–parasite recognitions depend on specific signals, and within social species communication is essential for much of individual behavior.

Chapter 16

16.1. For average cyanobacteria one would need to know the carbon content of an individual, the time between cell divisions, and the proportion that die that are sequestered in geological deposits rather than recycled through the food chain. One would make the assumption that half the products of a cell division die (this corresponds to assuming zero population growth). The greatest unknown is in the proportion recycled in the food chain—very large—and in the proportion sequestered in geological deposits—very small.

16.2. Bangladesh, the southern Ukraine, much of Micronesia, the eastern seaboard and Gulf Coast and Mississippi Valley of the United States, parts of Eastern China, southwest Alaska, the Central Valley of California, the Willamette Valley of Oregon, and large sections of Argentina, Uruguay, Paraguay, and Brazil would be submerged, as would the Nile Valley. Some of the world's most productive agricultural land would be lost; huge numbers of refugees would seek new homes. War and famine would be likely.

16.3. Life might exist on Mars because it has had free water on its surface and portions of its surface experience temperatures at which water is sometimes liquid.

Of Jupiter's moons Europa, Ganymede, and Callisto all have water ice, and Europa may have seas of water kept liquid by a molten core beneath its ice cap. Europa is probably the moon of Jupiter most likely to have life in its oceans. Ganymede and Callisto may be too cold. Io is rocky, volcanically active, and heated by Jupiter's magnetic field. Its extreme environment does not at first sight seem suited for life. The moons of Saturn have chemical precursors of life—for example on Titan—but are probably too cold for metabolism.

Chapter 17

17.1. HOX genes play a role in multicellular development in animals, including Cnidaria, and Cnidaria evolved in the pre-Cambrian, roughly 650 Ma. Recently HOX genes have been found in sponges, which may have evolved as long ago as 1000 Ma. In sponges HOX genes may play a role in simple tissue differentiation and reaggregation.

17.2. MADS genes are present in fern-like plants, which evolved in the Devonian about 400 Ma. One of their functions in ferns is to contribute to the control of leaf development.

17.3. The Scala Natura and the Great Chain of Being both refer to the notion that there is a hierarchy in nature, mammals, for example, being more advanced than fish, with humans being above the animals and just a little lower than the angels. This is a non-evolutionary world view, but elements of it were retained when evolution was discovered. They were then taken from a religious framework and applied to evolution in an attempt to maintain the notion of progress and to keep humans at the pinnacle of creation. The key point is this: progress is not something that one can derive from the pattern of evolution. It is an idea that comes from our cultural and intellectual history; it does not belong in evolutionary explanation.

17.4. There are many. We would have lacked key evidence for change over time; we would have lacked evidence for intermediate forms connecting extant groups; we would not have realized that the spatial distributions of species and clades was quite different in the past than it is today; we would not have estimates of times of divergence independent of those estimated from DNA sequences; we would not realize that most things that have existed are now extinct; we would be unaware of mass extinctions.

17.5. Trypanosomes are related to euglenoids, and apicomplexans are related to dinoflagellates. Both euglenoids and dinoflagellates are free-living, photosynthetic organisms.

Chapter 18

18.1. It is certainly possible to check whether the timing of the insect and angiosperm radiations is consistent with coevolution and perhaps cospeciation using fossils and molecular phylogenies. It is possible to use ecological information to estimate what percentage of insects feed on or otherwise utilize flowering plants, and what is the average range of species utilized by an average insect. Thus consistency checks are possible. But we cannot use such patterns to establish the necessity or sufficiency of the angiosperm radiation to explain insect diversity, for we do not know how many insect species there would be on the planet if the angiosperms had never evolved. There would still be quite a few.

18.2. The pattern is consistent with coevolution, but there is an alternative. Perhaps most microevolutionary adaptations have little influence on the extinction probability of species. van Valen's argument assumes that the concept of fitness has meaning in macroevolution. There may be a role for species selection in determining which species are currently present and which clades are currently diverse, but it does not seem likely that species selection could shape adaptations as precise as those shaped by microevolution as rapidly as microevolution can produce them. Extinction can sort among variants among species that have arisen as a result of microevolution, but it is unlikely to shape adaptations that reduce extinction probability.

18.3. Coevolution is analogous to a chemical chain reaction in the sense that once it starts, it continues to produce change for a long time. Chemical chain reactions, however, usually come to an end when they run out of energy or reach an absorbing state. Coevolution continues without limit because the interacting partners independently acquire free energy and convert it into offspring.

18.4. One advantage is that the defense used—a bacterium—can respond with change faster than the enemy—a fungus. Although the pathogenic fungus is sexual, the population of bacteria is many orders of magnitude larger and has a much shorter generation time. Thus even though it is asexual, it can generate enough variation, rapidly enough, to hold the sexual fungus in check.

18.5. We share with our mitochondria and our gut bacteria a pattern of vertical transmission that causes our genetic interests to be shared. Our mitochondria have virtually no opportunity for horizontal transmission, but our gut bacteria do. To the extent that they can behave independently in horizontal transmission, they can be considered separate organisms. Another way to analyze this question would be to ask, over what range of phenotypes does a genome have control? The nuclear and mitochondrial genomes interact in development to produce the integrated organisms that we think of as ourselves; our gut bacteria can have an existence independent of that development and do not have strong influence on it.

Chapter 19

19.1. There is no single answer. It would be wise to remain skeptical of explanations that cannot be tested or that have not been compared to alternatives.

19.2. Long-term, widespread use of low doses of antibiotics is probably the most effective way to cause resistance to evolve, for it eliminates the susceptible strains and is not stringent enough to kill the resistant strains, which increase rapidly in frequency. A proper antibiotic prescription is calculated to be intense enough, and to last long enough, to eliminate virtually all strains, including those that are partially but not fully resistant, and to do so quickly enough so that strains do not have time to exchange resistance genes on plasmids through conjugation.

19.3. There are many possible approaches. Mice are the logical model system. It would be helpful to have mutants to mitochondrion- and nucleus-specific DNA-repair mechanisms. It would be helpful to have methods for detecting DNA damage early and late in atresia, either directly in the DNA sequences or by isolating oocytes from early and late in the process and raising offspring from them with *in vitro* methods.

Chapter 20

20.1. Clearly the sharing of developmental control genes is not sufficient to account for the fixation of traits within lineages, for there have been many times in evolution when the downstream processes under such control have evolved while the upstream genes have remained similar. That does not mean that canalization is the answer, at least not in the sense of a trait that was selected directly. Canalization of traits may be a byproduct of patterns of gene regulation selected for stable function at the level of the cell, for example, or a byproduct of other processes that have not yet been discovered.

20.2. In its historical context, Darwin's prediction was more impressive, for it used ideas that had only recently been discovered. Wallace's prediction of a change in frequency of a recessive lethal in a *Drosophila* population cage comes very close to using fruit flies as analog computers to validate logic mathematically proven to be correct by other people.

20.3. Among other things, it can predict how phenotypes are likely to change when the environment changes, but not how fast or how far, and it can predict the probable equilibrium states of organisms if tradeoffs are assumed.

20.4. No brief answer could do justice to these questions. They may prove useful in stimulating discussions.

LITERATURE CITED

Abele, L.G., Kim, W., and Felgenhauer, B.E. (1989) Molecular evidence for inclusion of the Phylum Pentastomida in the Crustacea. *Molecular Biology and Evolution* 6, 685–91.

Aguinaldo, A.M.A., Turbeville, J.M., Linford, L.S., Rivera, M.C., Garey, J.R., Raff, R.A., and Lake, J.A. (1997) Evidence for a clade of nematodes, arthropods, and other moulting animals. *Nature* 387, 489–93.

Akam, M. (1998) Hox genes: From master genes to micromanagers. *Current Biology* 8, R676–8.

Alberch, P. (1989) The logic of monsters: evidence for internal constraint in development and evolution. *Geobios, mémoire spécial no. 12*, 21–57.

Alberch, P. and Gale, E.A. (1985) A developmental analysis of an evolutionary trend: digital reduction in amphibians. *Evolution* 39, 8–23.

Albert, V.A., Williams, S.E., and Chase, M.W. (1992) Carnivorous plants: phylogeny and structural evolution. *Science* 257, 1491–5.

Alexander, R.D., Hoogland, J.L., Howard, R.D., Noonan, K.M., and Sherman, P.W. (1979) Sexual dimorphisms and breeding systems in pinnipeds, ungulates, primates, and humans. In *Evolutionary biology and human social behavior, an anthropological perspective*. (Chagnon, N.A. and Irons, W., eds), pp. 402–35, Duxbury Press, North Scituate, MA.

Allen, J.F. (2003) The function of genomes in bioenergetic organelles. *Philosophical Transactions of the Royal Society of London B Biological Sciences* 358, 19–37.

Andersson, M. (1982) Female choice selects for extreme tail length in a widow bird. *Nature* 299, 818–20.

Andersson, M. (1994) *Sexual selection*. Princeton University Press, Princeton, NJ.

Andersson, D.I. and Levin, B.R. (1999) The biological cost of antibiotic resistance. *Current Opinion in Microbiology* 2, 489–93.

Antolin, M.F. and Herbers, J.M. (2001) Perspective: Evolution's struggle for existence in America's public schools. *Evolution* 55, 2379–88.

Antonovics, J. and Bradshaw, A.D. (1970) Evolution in closely adjacent populations, VIII. Clinal patterns at a mine boundary. *Heredity* 25, 349–62.

Antonovics, J. and Ellstrand, N.C. (1984) Experimental studies of the evolutionary significance of sexual reproduction. I. A test of the frequency-dependent selection hypothesis. *Evolution* 38, 103–15.

Avise, J.C. (1994) *Molecular markers, natural history and evolution*. Chapman and Hall, London.

Baker, A.C. (2003) Flexibility and specificity in coral–algal symbiosis: diversity, ecology, and biogeography of *Symbiodinium*. *Annual Review of Ecology, Evolution, and Systematics* 34, 661–89.

Baker, V.R., Benito, G., and Rudoy, A.N. (1993) Paleohydrology of late Pleistocene superflooding, Altay Mountains, Siberia. *Science* 259, 348–50.

Bakker, T.C.M. (1993) Positive genetic correlation between female preference and preferred male ornament in sticklebacks. *Nature* 363, 255–7.

Baldauf, S.J. (2003) The deep roots of eukaryotes. *Science* 300, 1703–6.

Barbujani, G. and Excoffier, L. (1998) The history and geography of human genetic diversity. In *Evolution in health and disease* (Stearns, S.C., ed.), pp. 27–40, Oxford University Press, Oxford.

Barraclough, T.G., Birky, C.W. Jr., and Burt, A. (2003) Diversification in sexual and asexual organisms. *Evolution* 57, 2166–72.

Bateman, A.J. (1948) Intra-sexual selection in *Drosophila*. *Heredity* 2, 349–68.

Bates, H.W. (1862) Contributions of an insect fauna of the Amazon Valley. *Transactions of the Linnean Society of London* 23, 495–566.

Beattie, A.J. and Hughes, L. (2002) Ant–plant interactions. In *Plant–animal interactions* (Herrera, C.M. and Pellmyr, O., eds), pp. 185–210. Blackwell, Oxford.

Beldade, P. and Brakefield, P.M. (2002) The genetics and evo-devo of butterfly wing patterns. *Nature Reviews Genetics* 3, 442–52.

Beldade, P., Brakefield, P.M., and Long, A.D. (2002a) Contribution of distal-less to quantitative variation in butterfly eyespots. *Nature* **415**, 315–18.

Beldade, P., Koops, K., and Brakefield, P.M. (2002b) Developmental constraints versus flexibility in morphological evolution. *Nature* **416**, 844–7.

Bell, G. (1982) *The masterpiece of nature. The evolution and genetics of sexuality*. University of California Press, Berkeley.

Bell, G. (1985) The origin and early evolution of germ cells as illustrated by the Volvocales. In *The origin and evolution of sex* (Halvarson, H. and Mornoy, A., eds), pp. 221–56. Alan Liss, New York.

Bentolila, S., Alfonso, A.A., and Hanson, M.R. (2002) A pentatricopeptide repeat-containing gene restores fertility to cytoplasmic male-sterile plants. *Proceedings of the National Academy of Sciences USA* **99**, 10887–92.

Benton, M.J. (ed.) (1993) *The fossil record 2*. Chapman and Hall, London.

Benton, M. and Harper, D. (1997) *Basic paleontology*. Prentice-Hall, Upper Saddle River, NJ.

Benton, M.J. and Twitchett, R.J. (2003) How to kill (almost) all life: the end-Permian extinction event. *Trends in Ecology and Evolution* **18**, 358–65.

Berbee, M.L. and Taylor, J.W. (1993) Dating the evolutionary radiations of the true fungi. *Canadian Journal of Botany* **71**, 1114–27.

Berglund, A., Bisazza, A., and Pilastro, A. (1996) Armaments and ornaments: an evolutionary explanation of traits of dual purpose. *Biological Journal of the Linnean Society* **58**, 385–99.

Bernasconi G., Ashman, T.L. Birkhead, T.R. et al. (2004) Evolutionary ecology of the prezygotic stage. *Science* **303**, 971–5.

Betzig, L. (1997) *Human nature: A critical reader*. Oxford University Press, Oxford.

Bierzychudek, P. (1987) Resolving the paradox of sexual reproduction: a review of experimental tests. In *The evolution of sex and its consequences* (Stearns, S.C., ed.), pp. 163–74. Birkhäuser, Basel.

Bjorksten, B. (1999) Allergy priming early in life. *Lancet* **353**, 167–8.

Blondel, J., Perret, P., Maistre, M., and Dias, P. (1992) Do harlequin Mediterranean environments function as source sink for Blue Tits (*Parus caeruleus* L.)? *Landscape Ecology* **7**, 213–19.

Boraas, M.E., Seale, D.B., and Boxhorn, J.E. (1998) Phagotrophy by a flagellate selects for colonial prey: a possible origin of multicellularity. *Evolutionary Ecology* **12**, 153–64.

Brakefield, P.M. (2000) Structure of a character and the evolution of butterfly eyespot patterns. In *The character concept in evolutionary biology* (Wagner, G.P., ed.), Academic Press, San Diego. Reprinted in *Journal of Experimental Zoology (Mol. Dev. Evol.)* **291**, 93–104.

Brakefield, P.M. and Larsen, T.B. (1984) The evolutionary significance of dry and wet season forms in some tropical butterflies. *Biological Journal of the Linnean Society* **22**, 1–12.

Brasier, M.D., Green, O.R., Jephcoat, A.P. et al. (2002) Questioning the evidence for Earth's oldest fossils. *Nature* **416**, 76–81.

Bried, J., Pontier, D., and Jouventin, P. (2003) Mate fidelity in monogamous birds: a re-examination of the Procellariiformes. *Animal Behaviour* **65**, 235–46.

Briggs, D.E.G. and Crowther, P.R. (2001) *Paleobiology II*. Blackwell Publishing, Oxford.

Briggs, D.E.G., Erwin, D.H., Collier, F.J., and Clark, C. (1994) *The Fossils of the Burgess Shale*. Smithsonian Press, Washington.

Brunetti, C.R., Selegue, J.E., Monteiro, A., French, V., Brakefield, P.M., and Carroll, S.B. (2001) The generation and diversification of butterfly eyespot color patterns. *Current Biology* **11**, 1578–85.

Bull, J.J. (1983) *Evolution of sex determining mechanisms*. Benjamin/Cummings, Menlo Park.

Bulmer, M.G. (1989) Structural instability of models of sexual selection. *Theoretical Population Biology* **35**, 195–206.

Bush, G.L. (1994) Sympatric speciation in animals, new wine in old bottles. *Trends in Ecology and Evolution* **9**, 285–8.

Bush, G.L. and Smith, J.J. (1998) The genetics and ecology of sympatric speciation: A case study. *Researches on Population Ecology* **40**, 175–87.

Bush, R.M., Bender, C.A., Subbarao, K., Cox, N.J., and Fitch, W.M. (1999) Predicting the evolution of human influenza A. *Science* 286, 1921–5.

Cann, R.L., Stoneking, M., and Wilson, A.C. (1987) Mitochondrial DNA and human evolution. *Nature* 325, 31–6.

Carius, H.J., Little, T.J., and Ebert, D. (2001) Genetic variation in a host–parasite association: potential for coevolution and frequency-dependent selection. *Evolution* 55, 1136–45.

Carroll, S.B. (1994) Developmental regulatory mechanisms in the evolution of insect diversity. *Development* 1994 (suppl.), 217–23.

Carroll, S.B., Weatherbee, S.D., and Langeland, J.A. (1995) Homeotic genes and the regulation and evolution of insect wing number. *Nature* 375, 58–61.

Carroll, S.B., Grenier, J.K., and Weatherbee, S.D. (2001) *From DNA to diversity: Molecular genetics and the evolution of animal design*. Blackwell Science, Oxford.

Cavalier-Smith, T. (2000) Membrane heredity and early chloroplast evolution. *Trends in Plant Science* 5, 174–82.

Cavalier-Smith, T. and Chao, E.E.Y. (2003) Phylogeny of choanozoa, apausozoa, and other protozoa and early eukaryote megaevolution. *Journal of Molecular Evolution* 56, 540–63.

Cavalli-Sforza, L.L. (1969) Genetic drift in an Italian population. *Scientific American* 221, 30–7.

Cavalli-Sforza, L.L. and Bodmer, W.F. (1971) *The genetics of human populations*. Freeman, San Francisco.

Cavalli-Sforza, L.L., Menozzi, P., and Piazza, A. (1994) *The history and geography of human genes*. Princeton University Press, Princeton, NJ.

Charlesworth, B. (1980) *Evolution in age-structured populations*. Cambridge University Press, Cambridge.

Charlesworth, B. and Williamson, J.A. (1975) The probability of the survival of a mutant gene in an age-structured population and implications for the evolution of life-histories. *Genetical Research* 26, 1–10.

Chen, J-Y., Oliveri, P., Li, C-W., Zhou, G-Q., Gao, F., Hagadorn, J.W., Peterson, K.J., and Davidson, E.H. (2000) Precambrian animal diversity: Putative phosphatized embryos from the Doushantuo Formation of China. *Proceedings of the National Academy of Sciences USA* 97, 4457–62.

Clay, K. and Kover, P. (1996) The Red Queen hypothesis and plant/pathogen interactions. *Annual Review of Phytopathology* 34, 29–50.

Clutton-Brock, T.H. (1991) *The evolution of parental care*. Princeton University Press, Princeton, NJ.

Clutton-Brock, T.H. (ed.) (1988) *Reproductive success*. University of Chicago Press, Chicago.

Clutton-Brock, T.H. and Iason, G.R. (1986) Sex ratio variation in mammals. *Quarterly Review of Biology* 61, 339–74.

Clutton-Brock, T.H. and Vincent, A.C.J. (1991) Sexual selection and the potential reproductive rates of males and females. *Nature* 351, 58–60.

Coen, E. (2001) Goethe and the ABC model of flower development. *Comptes Rendus de l'Academie des Sciences Serie III Sciences de la Vie Life Sciences* 324, 523–30.

Coen, E., Rolland-Lagan, A.G., Matthews, M., Bangham, J.A., and Prusinkiewicz, P. (2004) The genetics of geometry. *Proceedings of the National Academy of Sciences USA* 101, 4728–35.

Cohen, M.L. (1992) Epidemiology of drug resistance, implications for a post-antimicrobial era. *Science* 257, 1050–5.

Colegrave, N. (2002) Sex releases the speed limit on evolution. *Nature* 420, 664–6.

Conway Morris, S. (1998) The evolution of diversity in ancient ecosystems, a review. *Philosophical Transactions of the Royal Society B*, 353, 327–45.

Cook, J.M., Bean, D., Power, S.A., and Dixon, D.J. (2003) Evolution of a complex coevolved trait: active pollination in a genus of fig wasps. *Journal of Evolutionary Biology* 17, 238–46.

Cooper, A., Lalueza-Fox, C., Anderson, S., Rambaut, A., and Ward, R. (2001) Complete mitochondrial genome sequences of two extinct moas clarify ratite evolution. *Nature* 409, 704–7.

Coyne, J.A. and Orr, H.A. (1989) Patterns of speciation in *Drosophila*. *Evolution* 43, 362–81.

Cracraft, J. (1983) Species concepts and speciation analysis. In *Current ornithology* (Johnston, R.F., ed.), pp. 159–87. Plenum Press, New York.

Crow, J.F. (1986) *Basic concepts in population, quantitative, and evolutionary genetics*. Freeman, New York.

Literature cited 545

Crow, J.F. (1997) The high spontaneous mutation rate: is it a health risk? *Proceedings of the National Academy of Sciences USA* **94**, 8380–6.

Crow, J.F. and Kimura, M. (1965) Evolution in sexual and asexual populations. *American Naturalist* **99**, 439–50.

Crow, J.F. and Kimura, M. (1979) Efficiency of truncation selection. *Proceedings of the National Academy of Sciences USA* **76**, 396–9.

Curio, E. (1973) Towards a methodology of teleonomy. *Experientia* **29**, 1045–59.

Curran, L.M. and Webb, C.O. (2000) Experimental tests of the spatiotemporal scale of seed predation in mast-fruiting Dipterocarpaceae. *Ecological Monographs* **70**, 129–48.

Currey, J. (1984) *The mechanical adaptations of bones*. Princeton University Press, Princeton, NJ.

Currie, C.C., Wong, B., Stuart, A.E., Schultz, T.R., Rehner, S.A., Mueller, U.G., Sung, G.H., Spatafora, J.W., and Straus, N.A. (2003) Ancient tripartite coevolution in the attine ant–microbe symbiosis. *Science* **299**, 386–8.

Daan, S., Dijkstra, C., and Tinbergen, J.M. (1990) Family planning in the kestrel (*Falco tinnunculus*): the ultimate control of covariation of laying date and clutch size. *Behavior* **114**, 83–116.

Darwin, C. (1859) *On the origin of species by means of natural selection or the preservation of favoured races in the struggle for life*. John Murray, London.

Davies, N.B. (1992) *Dunnock behaviour and social evolution*. Oxford University Press, Oxford.

Dawkins, R. (1986) *The blind watchmaker*. Longman, London.

Dawkins, R. (1999) *The extended phenotype. The long reach of the gene*, revised edition. Oxford University Press, Oxford.

de Queiroz, K. (1998) The general lineage concept of species, species criteria, and the process of speciation. In *Endless forms: species and speciation* (Howard, D.J. and Berlocher, S.H., eds), pp. 57–75. Oxford University Press, Oxford.

Dean, G. (1972) *The porphyrias. A story of inheritance and environment*, 2nd edn. J.B. Lippincott, Philadelphia.

Delwiche, C.F., Andersen, R.A., Bhattacharya, D., Mishler, B.D., and McCourt, R.M. (2004) Algal evolution and the early radiation of green plants. In *Assembling the tree of life* (Cracraft, J. and Donoghue M.J., eds), pp. 121–37. Oxford University Press, Oxford.

Denholm, I., Devine, G.J., and Williamson, M.S. (2002) Evolutionary genetics. Insecticide resistance on the move. *Science* **297**, 2222–3.

Dieckmann, U. and Doebeli, M. (1999) On the origin of species by sympatric speciation. *Nature* **400**, 354–7.

Dobzhansky, T. (1937) *Genetics and the origin of species*. Columbia University Press, New York.

Dodson, S.I. (1989) Predator-induced reaction norms. *BioScience* **39**, 447–52.

Doebeli, M. and Knowlton, N. (1998) The evolution of interspecific mutualisms. *Proceedings of the National Academy of Sciences USA* **95**, 8676–80.

Donnenberg, M.S. and Whittam, T.S. (2001) Pathogenesis and evolution of virulence in enteropathogenic and enterohemorrhagic *Escherichia coli. Journal of Clinical Investigation* **107**, 539–48.

Donoghue, M.J., Bell, C.D., and Li, J. (2001) Phylogenetic patterns in Northern Hemisphere plant geography. *International Journal of Plant Sciences* **162** (suppl.), S41–52.

Doolittle, R.F., Feng, D., Tsang, S., Cho, G., and Little, E. (1996) Determining divergence times of the major kingdoms of living organisms with a protein clock. *Science* **271**, 470–7.

Dorit, R.L., Akashi, H., and Gilbert, W. (1995) Absence of polymorphism at the ZFY locus on the human Y chromosome. *Science* **268**, 1183–5.

Dowton, M., Austin, A.D., and Antolin, M.F. (1998) Evolutionary relationships among the Braconidae (Hymenoptera, Icheumonoidea) inferred from partial 16S rDNA gene sequences. *Insect Molecular Biology* **7**, 129–50.

Dowton, M., Austin, A.D., Dillon, N., and Bartowsky, E. (1997) Molecular phylogeny of the apocritan wasps, the Proctotrupomorpha and Evaniomorpha. *Systematic Entomology* **22**, 245–55.

Doyle, J.A. (1998) Phylogeny of vascular plants. *Annual Review of Ecology and Systematics* **29**, 567–99.

Drake, J.W., Charlesworth, B., Charlesworth, D., and Crow, J.F. (1998) Rates of spontaneous mutation. *Genetics* **148**, 1667–86.

Dudley, J.W. (1977) 76 generations of selection for oil and protein percentage in maize. In *Proceedings of the international conference on quantitative genetics* (Pollack, E., Kempthorne, O., and Bailey, T.B., eds), pp. 459–73. Iowa State University Press, Ames, IO.

Dunbar, R.I.M. (ed.) (1995) *Human reproductive decisions*. MacMillan, London.

Dypvik, H. and Jansa, L.F. (2003) Sedimentary signatures and processes during marine bolide impacts: a review. *Sedimentary Geology* **161**, 309–37.

Eaton, S.B. and Eaton, S.B., III (1999) The evolutionary context of chronic degenerative diseases. In *Evolution in health and disease* (Stearns, S.C., ed.), pp. 251–9. Oxford University Press, Oxford.

Ebert, D. (1994) A maturation size threshold and phenotypic plasticity of age and size at maturity in Daphnia magna. *Oikos* **69**, 309–17.

Ebert, D. (1998) Experimental evolution of parasites. *Science* **282**, 1432–5.

Eigen, M. (1971) Self-organization of matter and the evolution of biological macromolecules. *Naturwissenschaften* **58**, 465–523.

Eldredge, N. and Gould, S.J. (1972) Punctuated equilibria: an alternative to phyletic gradualism. In *Models in paleobiology* (Schopf, T.J.M., ed.), pp. 82–115. Freeman, Cooper and Co, San Francisco.

Elgar, M.A. (1992) Sexual cannibalism in spiders and other invertebrates. In *Cannibalism, ecology and evolution among diverse taxa* (Elgar, M.A. and Crespi, B.J., eds), pp. 128–55. Oxford University Press, Oxford.

Elinson, R.P. (1989) Egg evolution. In *Complex organismal functions, integration and evolution in vertebrates* (Wake, D.B. and Roth, G., eds), pp. 251–62. Dahlem Conference Report, John Wiley and Sons, New York.

Emelyanov, V.V. (2003) Mitochondrial connection to the origin of the eukaryotic cell. *European Journal of Biochemistry* **270**, 1599–618.

Enattah, N.S., Sahi, T., Savilahti, E., Terwilliger, J.D., Peltonen, Ll, and Jarvela I. (2002) Identification of a variant associated with adult-type hypolactasia. *Nature Genetics* **30**, 233–7.

Endler, J.A. (1986) *Natural selection in the wild*. Princeton University Press, Princeton, NJ.

Engel, M.S. and Grimaldi, D.A. (2004) New light shed on the oldest insect. *Nature* **427**, 627–30.

Ewald, P.W. (1994) *Evolution of infectious diseases*. Oxford University Press, Oxford.

Ewens, W.J. (1993) Beanbag genetics and after. In *Human population genetics. A centennial tribute to J.B.S. Haldane* (Majumder, P.P., ed.), pp. 7–29. Plenum Press, New York.

Eyre-Walker, A. and Keightly, P.D. (1999) High genomic deleterious mutation rates in hominids. *Nature* **397**, 344–7.

Fay, J.C., Wyckoff, G.J., and Wu, C.I. (2001) Positive and negative selection on the human genome. *Genetics* **158**, 1227–34.

Fay, J.C., Wyckoff, G.J., and Wu, C.I. (2002) Testing the neutral theory of molecular evolution with genomic data from Drosophila. *Nature* **415**, 1024–6.

Felsenstein, J. (1985) Phylogenies and the comparative method. *American Naturalist* **125**, 1–15.

Felsenstein, J. (1988) Phylogenies from molecular sequences, inference and reliability. *Annual Review of Genetics* **22**, 521–65.

Fenner, F. and Ratcliffe, F.N. (1965) *Myxomatosis*. Cambridge University Press, London.

Finch, C.E. and Austad, S.N. (2001) History and prospects: symposium on organisms with slow aging. *Experimental Gerontology* **36**, 593–7.

Fisher, R.A. (1918) The correlation between relatives on the supposition of Mendelian inheritance. *Transactions of the Royal Society of Edinburgh* **52**, 399–433.

Fisher, R.A. (1930) *The genetical theory of natural selection*. Oxford University Press, Oxford.

Flanagan, N.S., Tobler, A., Davison, A., Pybus, O.G., Kapan, D.D., Planas, S., Linares, M., Heckel, D., and McMillan, W.O. (2004) Historical demography of Müllerian mimicry in the neotropical *Heliconius* butterflies. *Proceedings of the National Academy of Sciences USA* **101**, 9704–9.

Foerster, K., Delhye, K., Johnsen, A., and Kempenaers, B. (2003) Females increase offspring heterozygosity and fitness through extra-pair matings. *Nature* **425**, 714–17.

Ford, E.B. (1975) *Ecological genetics*, 4th edn. Chapman and Hall, London.

Forster, L.M. (1992) The stereotyped behaviour of sexual cannibalism in *Latrodectus hasselti* Thorell (Araneae, Theridiidae), the Australian redback spider. *Australian Journal of Zoology* 40, 1–11.

Franco, M.G., Rubini, P.G., and Vecchi, M. (1982) Sex determinants and their distribution in various populations of *Musca domestica* L. of Western Europe. *Genetical Research* 40, 279–93.

Frank, S.A. (1998) *Foundations of social evolution*. Princeton University Press, Princeton, NJ.

Frank, S.A. (2000) Polymorphism of attack and defense. *Trends in Ecology and Evolution* 15, 167–71.

Fraser, H.B., Hirsh, A.E., Wall, D.P., and Eisen, M.B. (2004) Coevolution of gene expression among interacting proteins. *Proceedings of the National Academy of Sciences USA* 101, 9033–8.

Friedman, W.E., Moore, R.C., and Purugganan, M.D. (2004) The evolution of land plant development. *American Journal of Botany* 91, 1726–41.

Furnes, H., Banerjee, N.R., Muehlenbachs, K., Staudigel, H., and de Wit, M. (2004) Early life recorded in Archean pillow lavas. *Science* 304, 578–81.

Gerrish, P.J. and Lenski, R.E. (1998) The fate of competing beneficial mutations in an asexual population. *Genetica* 102–3, 127–44.

Ghiselin, M.T. (1969) The evolution of hermaphroditism among animals. *Quarterly Review of Biology* 44, 189–208.

Gilbert, L.E. (1983) Coevolution and mimicry. In *Coevolution* (Futuyma, D.J. and Slatkin, M., eds), pp. 263–81. Sinauer Associates, Sunderland, MA.

Gillespie, J.H. (1991) *The causes of molecular evolution*. Oxford University Press, Oxford.

Gillham, N.W. (1994) *Organelle Genes and Genomes*. Oxford University Press, New York.

Gingerich, P.D. (1983) Rates of evolution: effects of time and temporal scaling. *Science* 222, 159–61.

Giraud, A., Matic, I., Tenaillon, O., Clara, A., Radman, M., Fons, M., and Taddei, F. (2001) Costs and benefits of high mutation rates: adaptive evolution of bacteria in the mouse gut. *Science* 291, 2606–8.

Givnish, T.J., Sytsma, K.J., Smith, J.F., and Hahn, W.J. (1995) Molecular evolution, adaptive radiation, and geographic speciation in *Cyanea* (Campanulaceae, Lobelioideae). In *Hawaiian Biogeography* (Wagner, W.T. and Funk, V.A., eds), pp. 288–337. Smithsonian Institution Press, Washington.

Glesener, R.R. and Tilman, D. (1978) Sexuality and the components of environmental uncertainty: Clues from geographic parthenogenesis in terrestrial animals. *American Naturalist* 112, 659–73.

Godfray, H.C.J. (1994) *Parasitoids. Behavioral and evolutionary ecology*. Princeton University Press, Princeton, NJ.

Goldsmith, T.H. (1990) Optimization, constraint, and history in the evolution of eyes. *Quarterly Review of Biology* 65, 281–322.

Gould, J.L. and Keeton, W.T. (1996) *Biological Science*, 6th edn. Norton and Co. New York.

Gould, S.J. (1999) *Rocks of ages. Science and religion in the fullness of life*. Ballantine Books, New York.

Gould, S.J. (ed.) (2001) *The book of life: an illustrated history of the evolution of life on earth*. W.W. Norton, New York.

Grant, B.R. and Grant, P.R. (1989) *Evolutionary dynamics of a natural population. The large cactus finch of the Galápagos*. University of Chicago Press, Chicago.

Grant, P.R. (1986) *Ecology and evolution of Darwin's finches*. Princeton University Press, Princeton, NJ.

Greaves, M. (2000) *Cancer: the evolutionary legacy*. Oxford University Press, Oxford.

Greenwood, P.H. (1974) *Cichlid fishes of Lake Victoria, East Africa*. Natural History Museum, London.

Griffin, D.R. (1958) *Listening in the dark*. Yale University Press, New Haven.

Griffiths, A.J.F., Miller, J.H., Suzuki, D.T., Lewontin, R.C., and Gelbart, W.M. (1996) *An introduction to genetic analysis*, 6th edn. W.H. Freeman, New York.

Grubb, P.J. and Metcalfe, D.J. (1996) Adaptation and inertia in the Australian tropical lowland rain-forest flora: contradictory trends in intergeneric and intrageneric comparisons of seed size in relation to light demand. *Functional Ecology* 10, 512–20.

Gwynne, D.T. and Simmons, L.W. (1990) Experimental reversal of courtship roles in an insect. *Nature* **346**, 171–4.

Haig, D. (2002) *Genomic imprinting and kinship*. Rutgers University Press, New Brunswick, NJ.

Haig, D. and Graham, C. (1991) Genomic imprinting and the strange case of the insulin-like growth factor II receptor. *Cell* **64**, 1045–6.

Halder, G., Callaerts, P., and Gehring, W.J. (1995) Induction of ectopic eyes by targeted expression of the *eyeless* gene in *Drosophila*. *Science* **267**, 1788–92.

Hamilton, W.D. (1964a) The genetical evolution of social behaviour, I. *Journal of Theoretical Biology* **7**, 1–16.

Hamilton, W.D. (1964b) The genetical evolution of social behaviour, II. *Journal of Theoretical Biology* **7**, 17–52.

Hamilton, W.D. (1980) Sex versus non-sex versus parasite. *Oikos* **35**, 282–90.

Hamilton, W.D. and Zuk, M. (1982) Heritable true fitness and bright birds. A role for parasites? *Science* **218**, 384–7.

Hartl, D.L. (1988) *A primer of population genetics*, 2nd edn. Sinauer Associates, Sunderland, MA.

Hartl, D.L. (1994) *Genetics*, 3rd edn. Jones and Bartlett, Boston.

Hartman, H. and Fedorov, A. (2002) The origin of the eukaryotic cell: a genomic investigation. *Proceedings of the National Academy of Sciences USA* **99**, 1420–5.

Hass, C.A., Hoffman, M.A., Densmore, III, L.D., and Maxson, L.R. (1992) Crocodilian evolution, insights from immunological data. *Molecular Phylogenetics and Evolution* **1**, 193–201.

Haug, G.H. and Tiedemann, R. (1998) Effect of the formation of the Isthmus of Panama on Atlantic Ocean thermohaline circulation. *Nature* **393**, 673–6.

Haukioja, E. and Neuvonen, S. (1985) Induced long-term resistance of birch foliage against defoliators, defensive or incidental? *Ecology* **66**, 1303–8.

Herre, E.A. (1993) Population structure and the evolution of virulence in the nematode parasites of fig wasps. *Science* **259**, 1442–5.

Hewitt, G.M. (2001) Speciation, hybrid zones and phylogeography—or seeing genes in space and time. *Molecular Ecology* **10**, 537–49.

Hill, A.V.S. and Motulsky, A.G. (1999) Genetic variation and human diseases: the role of natural selection. In *Evolution in health and disease* (Stearns, S.C., ed.), pp. 50–61. Oxford University Press, Oxford.

Hillis, D.M. and Huelsenbeck, J.P. (1994) Support for dental HIV transmission. *Nature* **369**, 24–5.

Hillis, D.M., Moritz, C., and Mable, B.K. (eds.), (1996) *Molecular Systematics*, 2nd edn. Sinauer Associates, Sunderland, MA.

Hoekstra, R.F. (1987) The evolution of sexes. In *The evolution of sex and its consequences* (Stearns, S.C., ed.), pp. 59–91. Birkhäuser, Basel.

Hoekstra, R.F. (1990) Evolution of uniparental inheritance of cytoplasmic DNA. In *Organizational constraints on the dynamics of evolution* (Maynard Smith, J., ed.), pp. 269–78. Manchester University Press, Manchester.

Hoffmann, W.A. and Franco, A.C. (2003) Comparative growth analysis of tropical forest and savannah woody plants using phylogenetically independent contrasts. *J. Ecology* **91**, 475–84.

Holder, M. and Lewis, P.O. (2003) Phylogeny estimation: traditional and Bayesian approaches. *Nature Reviews Genetics* **4**, 275–84.

Holloway, G.J., Brakefield, P.M., and Kofman, S. (1993) The genetics of wing pattern elements in the polyphenic butterfly, *Bicyclus anynana*. *Heredity* **70**, 179–86.

Horai, S., Hayasaka, K., Kondo, R., Tsugane, K., and Takahata, N. (1995) Recent African origin of modern humans revealed by complete sequences of hominoid mitochondrial DNAs. *Proceedings of the National Academy of Sciences USA* **92**, 532–6.

Hori, M. (1993) Frequency-dependent natural selection in the handedness of scale-eating cichlid fish. *Science* **260**, 216–19.

Houle, D., Hoffmaster, D.K., Assimacopoulos, S., and Charlesworth, B. (1992) The genomic mutation rate for fitness in *Drosophila*. *Nature* **359**, 58–60.

Hrdy, S.B. (ed.) (1998) *Mother nature*. Pantheon, New York.

Hudson, R.R. (1990) Gene genealogies and the coalescent process. *Oxford Surveys in Evolutionary Biology* **7**, 1–44.

Hughes, A.L. and Nei, M. (1989) Nucleotide substitution at major histocompatibility complex class II loci: evidence for overdominant selection. *Proceedings of the National Academy of Sciences USA* **86**, 958–62.

Hurst, L.D., Atlan, A., and Bengtsson, B.O. (1996) Genetic conflicts. *Quarterly Review of Biology* **71**, 317–64.

Hurst, L.D. and Hamilton, W.D. (1992) Cytoplasmic fusion and the nature of sexes. *Proceedings of the Royal Society of London B* **247**, 189–94.

Ingman, M., Kaessmann, H., Pääbo, S., and Gyllensten, U. (2000) Mitochondrial genome variation and the origin of modern humans. *Nature* **408**, 708–13.

IUGS (International Union of Geological Sciences) (1989) *Global stratigraphic chart*. Episodes, June 1989.

Jablonka, E. and Szathmáry, E. (1995) The evolution of information storage and heredity. *Trends in Ecology and Evolution* **10**, 206–11.

Jackson, J.B.C. and Cheetham, A.H. (1999) Tempo and mode of speciation in the sea. *Trends in Ecology and Evolution* **14**, 72–7.

Jackson, R.R. (1992) Predator–prey interactions between web-invading jumping spiders and a web-building spider, *Holocnemus pluchei* (Araneae, Pholcidae). *Journal of Zoology* **228**, 589–94.

Jacob, F. (1977) Evolution and tinkering. *Science* **196**, 1161–6.

Jaenike, J. (1978) A hypothesis to account for the maintenance of sex within populations. *Evolutionary Theory* **3**, 191–4.

Jansen, R.P.S. and de Boer, K. (1998) The bottleneck: mitochondrial imperatives in oogenesis and ovarian follicular fate. *Molecular and Cellular Endocrinology* **145**, 81–8.

Jenkins, T. (1996) The South African malady. *Nature Genetics* **13**, 7–9.

Jockusch, E.L. and Wake, D.B. (2002) Falling apart and merging: diversification of slender salamanders (Plethodontidae: *Batrachoseps*) in the American West *Biological Journal of the Linnean Society* **76**, 361–91.

Jones, J.S., Ebert, D., and Stearns, S.C. (1992) Life history and mechanical constraints on reproduction in genes, cells and waterfleas. In *Genes in ecology* (Berry, R.J., Crawford, T.J., and Hewitt, G.M., eds), pp. 393–404. Blackwell Scientific, Oxford.

Juchault, P., Rigaud, T., and Mocquard, J.P. (1993) Evolution of sex determination and sex ratio variability in wild populations of *Armadillidium vulgare* (Latr.) (Crustacea, Isopoda), a case study in conflict resolution. *Acta Œcologia* **14**, 547–62.

Judson, O.P. and Normark, B.B. (1996) Ancient asexual scandals. *Trends in Ecology and Evolution* **11**, 41–6.

Kaneda, H., Hayashi, J., Takahama, S., Taya, C., Lindahl, K.F., and Yonekawa, H. (1995) Elimination of paternal mitochondrial DNA in intraspecific crosses during early mouse embryogenesis. *Proceedings of the National Academy of Sciences USA* **92**, 4542–6.

Kaneshiro, K.Y. and Boake, C.R.B. (1987) Sexual selection and speciation, issues raised by Hawaiian drosophilids. *Trends in Ecology and Evolution* **2**, 207–12.

Karban, R. and Baldwin, I.T. (1997) *Induced responses to herbivory*. University of Chicago Press, Chicago.

Karn, M.N. and Penrose, L.S. (1951) Birth weight and gestation time in relation to maternal age, parity and infant survival. *Annals of Eugenics* **16**, 147–64.

Kassen, R. (2002) The experimental evolution of specialists, generalists, and the maintenance of diversity. *Journal of Evolutionary Biology* **15**, 173–90.

Keller, G., Adatte, T., Stinnesbeck, W., Rebolledo-Vieyra, M., Fucugauchi, J.U., Kramar, U., and Stuben, D. (2004) Chicxulub impact predates the K-T boundary mass extinction. *Proceedings of the National Academy of Sciences USA* **101**, 3753–8.

Kempenaers, B., Verheyen, G.R., Van den Broeck, M., Burke, T., Van Broeckhoven, C.V., and Dhondt, A. (1992) Extra-pair paternity results from female preference for high-quality males in the blue tit. *Nature* **357**, 494–6.

Kenrick, P. and Crane, P.R. (1997) The origin and early evolution of plants on land. *Nature* **389**, 33–9.

Kimura, M. (1968) Evolutionary rate at the molecular level. *Nature* **217**, 624–6.

Kimura, M. (1989) The neutral theory of molecular evolution and the world view of the neutralists. *Genome* **31**, 24–31.

Kirkwood, T.B.L. (1987) Immortality of the germ line versus disposability of the soma. In *Evolution of longevity in animals* (Woodhead, A.D.H. and Thompson, K.H., eds), pp. 209–18. Plenum, New York.

Klein, J., Gutknecht, J., and Fischer, N. (1990) The major histocompatibility complex and human evolution. *Trends in Genetics* **6**, 7–11.

Klein, J., Sato, A., Nagl, S., and O'hUigin, C. (1998) Molecular trans-species polymorphism. *Annual Review of Ecology and Systematics* **29**, 1–21.

Klein, J., Takahata, N., and Ayala, F. (1993) MHC polymorphism and human origins. *Scientific American*, December, 46–51.

Knowlton, N. (1993) Sibling species in the sea. *Annual Review of Ecology and Systematics* **24**, 189–216.

Koelewijn, H.P. (1993) *On the genetics and ecology of sexual reproduction in* Plantago coronopus. PhD Thesis, University of Utrecht.

Kondrashov, A.S. (1982) Selection against harmful mutations in large sexual and asexual populations. *Genetical Research* **40**, 325–32.

Kondrashov, A.S. (1988) Deleterious mutations and the evolution of sexual reproduction. *Nature* **336**, 435–40.

Kondrashov, A.S. (1993) Classification of hypotheses on the advantage of amphimixis. *Journal of Heredity* **84**, 372–80.

Kondrashov, A.S. and Kondrashov, F.A. (1999) Interactions among quantitative traits in the course of sympatric speciation. *Nature* **400**, 351–4.

Kozlowski, J. (1992) Optimal allocation of resources to growth and reproduction, implications for age and size at maturity. *Trends in Ecology and Evolution* **7**, 15–19.

Krakauer, D.C. and Mira A. (1999) Mitochondria and germ-cell death. *Nature* **400**, 125–6.

Kumar, S. and Hedges, S.B. (1998) A molecular time scale for vertebrate evolution. *Nature* **392**, 917–20.

Kump, L.R., Kasting, J.F., and Crane, R.G. (1999) *The Earth system*. Prentice Hall, Upper Saddle River, NJ.

Kurtz, J., Kalbe, M., Aeschlimann, P.B., Häberli, M.A., Wegner, K.M., Reusch, T.B.H., and Milinski, M. (2004) Major histocompatibility complex diversity influences parasite resistance and innate immunity in sticklebacks. *Proceedings of the Royal Society of London B* **271**, 197–204.

Labandeira, C.C. (1998a) How old is the flower and the fly? *Science* **280**, 57–9.

Labandeira, C.C. (1998b) Early history of arthropod and vascular plant associations. *Annual Review of Earth and Plantary Science* **26**, 329–77.

Lack, D. (1947) *Darwin's finches*. Cambridge University Press, Cambridge.

Lan, R. and Reeves, P.R. (2001) When does a clone deserve a name? A perspective on bacterial species based on population genetics. *Trends in Microbiology* **9**, 419–24.

Lanciotti, R.S., Roehrig, J.T. et al. (1999) Origin of the West Nile virus responsible for an outbreak of encephalitis in the Northeastern United States. *Science* **286**, 2333–7.

Lander, E.S., Linton, L.M. et al. (2001) Initial sequencing and analysis of the human genome. *Nature* **409**, 860–921.

Lang, B.F., O'Kelly, C., Nerad, T., Gray, M.W., and Burger, G. (2002) The closest unicellular relatives of animals. *Current Biology* **12**, 1773–8.

Laroche, J., Li, P., and Bousquet, J. (1995) Mitochondrial DNA and monocot-dicot divergence time. *Molecular Biology and Evolution* **12**, 1151–6.

Law, R. and Hutson, V. (1992) Intracellular symbionts and the evolution of uniparental cytoplasmic inheritance. *Proceedings of the Royal Society of London B* **248**, 69–77.

Li, W.-H. and Graur, D. (1991) *Fundamentals of molecular evolution*. Sinauer Associates, Sunderland, MA.

Li, W.-H. and Graur, D. (2000) *Fundamentals of molecular evolution*, revised edn. Sinauer Associates, Sunderland, MA.

Li, W.-H., Gu, Z. et al. (2001) Evolutionary analyses of the human genome. *Nature* **409**, 847–9.

Lively, C.M. (1986) Predator-induced shell dimorphism in the acorn barnacle *Chthamalus anisopoma*. *Evolution* **40**, 232–42.

Lively, C.M. (1992) Parthenogenesis in a freshwater snail: Reproductive assurance versus parasitic release. *Evolution* **46**, 907–13.

Lloyd, J.E. (1984) Occurrence of aggressive mimicry in fireflies. *Florida Entomologist* **67**, 368–76.

Losos, J.B., Jackman, T.R., Larson, A., de Queiroz, K., and Rodriguez-Schettino, L. (1998) Contingency and determinism in replicated adaptive radiations of island lizards. *Science* **279**, 2115–18.

Lowe, C.J. and Wray, G.A. (1997) Radical alterations in the roles of homeobox genes during echinoderm evolution. *Nature* **389**, 718–20.

Luo, C.C., Li, W.H., and Chan, L. (1989) Structure and expression of dog apolipoprotein A-I, E, and C-I mRNAs: implications for the evolution and functional constraints of apolipoprotein structure. *Journal of Lipid Research* **30**, 1735–46.

Lyttle, T.W. (1991) Segregation distorters. *Annual Review of Genetics* **25**, 511–57.

Mallet, J. (1995) A species definition for the Modern Synthesis. *Trends in Ecology and Evolution* **10**, 294–9.

Margulis, L. (1970) *Origin of eukaryotic cells*. Yale University Press, New Haven, CT.

Margulis, L. (1981) *Symbiosis in cell evolution*. W.H. Freeman, San Francisco.

Margulis, L. and Schwartz, K.V. (1982) *Five kingdoms: an illustrated guide to the phyla of life on earth*. W.H. Freeman, San Francisco.

Marshak, S. (2001) *Earth: portrait of a planet*. W.W. Norton, New York.

Martin, W., Lydiate, D., Brinkmann, H., Forkmann, G., Saedler, H., and Cerff, R. (1993) Molecular phylogenies in angiosperm evolution. *Molecular Biology and Evolution* **10**, 140–62.

Martinez, J.L. and Baquero, F. (2000) Mutation frequencies and antibiotic resistance. *Antimicrobial Agents and Chemotherapy* **44**, 1771–7.

Mason, S.F. (1992) *Chemical evolution*. Oxford University Press, Oxford.

May, R.M. and Anderson, R.M. (1983) Epidemiology and genetics in the coevolution of parasites and hosts. *Proceedings of the Royal Society of London B* **219**, 281–313.

Maynard Smith, J. (1998) *Evolutionary genetics*, 2nd edn. Oxford Univesity Press, Oxford.

Maynard Smith, J., Dowson, G.C., and Spratt, B.G. (1991) Localized sex in bacteria. *Nature* **349**, 29–31.

Maynard Smith, J. and Szathmáry, E. (1995) *The major transitions in evolution*. W.H. Freeman, New York.

Maynard Smith, J. and Szathmáry, E. (1999) *The Origins of Life: From the birth of life to the origin of language*. Oxford University Press, Oxford.

Mayr, E. (1963) *Animal species and evolution*. Harvard University Press, Cambridge, MA.

McCune, A.R. (1996) Biogeographic and stratigraphic evidence for rapid speciation in semionotid fishes. *Paleobiology* **22**, 34.

McCune, A.R., Fuller, R.C., Aquilina, A.A., Dawley, R.M., Fadool, J.M., Houle, D., Travis, J., and Kondrashov, A.S. (2002) A low genomic number of recessive lethals in natural populations of bluefin killifish and zebrafish. *Science* **296**, 2398–401.

McDonald, J.H. and Kreitman, M. (1991) Adaptive protein evolution at the Adh locus in Drosophila. *Nature* **351**, 652–4.

McMillan, W.O., Monteiro, A., and Kapan, D.D. (2002) Development and evolution on the wing. *Trends in Ecology and Evolution* **17**, 125–33.

Metzker, M.L., Mindell, D.P., Liu, X-M., Ptak, R.G., Gibbs, R.A., and Hillis, D.M. (2002) Molecular evidence of HIV-1 transmission in a criminal case. *Proceedings of the National Academy of Sciences USA* **99**, 14292–7.

Meyer, U. (1999) Medically relevant genetic variation of drug effects. In *Evolution in health and disease* (Stearns, S.C., ed.), pp. 41–9. Oxford University Press, Oxford.

Michod, R.E. (1979) Evolution of life histories in response to age-specific mortality factors. *American Naturalist* **113**, 531–50.

Miles, D.B. and Dunham, A.E. (1993) Historical perspectives in ecology and evolutionary biology: The use of phylogenetic comparative analyses. *Annual Review of Ecology and Systematics* **24**, 587–619.

Milinski, M. and Bakker, T.C.M. (1990) Female sticklebacks use male coloration in mate choice and hence avoid parasitized males. *Nature* **344**, 330–3.

Miller, S.L. (1953) Production of amino acids under possible primitive earth conditions. *Science* **117**, 528.

Miralles, R., Gerrish, P.J., Moya, A., and Elena, S.F. (1999) Clonal interference and the evolution of RNA viruses. *Science* **285**, 1745–7.

Mohr, P. (1983) Ethiopian flood basalt province. *Nature* **303**, 577–84.

Monteiro, A., Prijs, J., Bax, M., Hakkaart, T., and Brakefield, P.M. (2003) Mutants highlight the modular control of butterfly eyespot patterns. *Evolution and Development* **5**, 180–7.

Monteiro, A.F., Brakefield, P.M., and French, V. (1994) The evolutionary genetics and developmental basis of wing pattern variation in the butterfly *Bicyclus anynana*. *Evolution* 48, 1147–57.

Moore, T. and Haig, D. (1991) Genomic imprinting in mammalian development: a parental tug-of-war. *Trends in Genetics* 7, 45–9.

Moraes, J.C., Goussain, M.M., Basagli, M.A.B., Carvalho, G.A., Ecole, C.C., and Sampaio, M.V. (2004) Silicon influence on the tritrophic interaction: Wheat plants, the greenbug *Schizaphis graminum* (Rondani) (Hemiptera: Aphididae), and its natural enemies, *Chrysoperla externa* (Hagen) (Neuroptera: Chrysopidae) and *Aphidius colemani viereck* (Hymenoptera: Aphidiidae). *Neotropical Entomology* 33, 619–24.

Mousseau, T.A. and Roff, D.A. (1987) Natural selection and the heritability of fitness components. *Heredity* 59, 181–97.

Moxon, E.R., Rainey, P.B., Nowak, M.A., and Lenski, R.E. (1994) Adaptive evolution of highly mutable loci in pathogenic bacteria. *Current Biology* 4, 24–33.

Mukai, T. (1964) The genetic structure of natural populations of *Drosophila melanogaster*. I. Spontaneous mutation rate of polygenes controlling viability. *Genetics* 50, 1–19.

Mukai, T., Chigusa, S.I., Mettler, L.E., and Crow, J.F. (1972) Mutation rate and dominance of genes affecting viability in *Drosophila melanogaster*. *Genetics* 72, 335–55.

Müller, F. (1879) *Ituna* and *Thyridis*: a remarkable case of mimicry in butterflies. *Proceedings of the Entomological Society of London* 1879, 20–9.

Muller, H.J. (1964) The relation of recombination to mutational advance. *Mutation Research* 1, 2–9.

Nanney, D.L. (1982) Genes and phenes in *Tetrahymena*. *BioScience* 32, 783–8.

Nei, M. (1987) *Molecular evolutionary genetics*. Columbia University Press, New York.

Neu, H.C. (1992) The crisis in antibiotic resistance. *Science* 257, 1064–73.

Newman, R.A. (1988) Genetic variation for larval anuran (*Scaphiophus couchii*) development time in an uncertain environment. *Evolution* 42, 763–73.

Newton, I. (1988) Age and reproduction in the Sparrowhawk. In *Reproductive success* (Clutton-Brock, T.H., ed.), pp. 201–19. University of Chicago Press, Chicago.

Nijhout, H.F. (1991) *The development and evolution of butterfly wing patterns*. Smithsonian Institution Press, Washington.

Nijhout, H.F. (2003) Polymorphic mimicry in *Papilio dardanus*: mosaic dominance, big effects, and origins. *Evolution and Development* 5, 579–92.

Nikaido, M., Rooney, A.P., and Okada, N. (1999) Phylogenetic relationships among cetartiodactyls based on insertions of short and long interspersed elements: hippopotamuses are the closest extant relatives of whales. *Proceedings of the National Academy of Sciences USA* 96, 10261–6.

Niklas, K.J., Tiffney, B.H., and Knoll, A.H. (1983) Patterns in vascular land plant diversification. *Nature* 303, 614–16.

Nilsson, L.A. (1988) The evolution of flowers with deep corolla tubes. *Nature* 334, 147–9.

Nisbet, E.G. and Piper, D.J.W. (1998) Giant submarine landslides. *Nature* 392, 329–30.

Nordskog, A.W. (1977) Success and failure of quantitative genetic theory in poultry. In *Proceedings of the international conference on quantitative genetics* (Pollack, E., Kempthorne, O., and Bailey, T.B., eds), pp. 569–85. Iowa State University Press, Ames, IO.

Nublerjung, K. and Arendt, D. (1994) Is ventral in insects dorsal in vertebrates? A history of embryological arguments favoring axis inversion in chordate ancestors. *Roux's Archives of Developmental Biology* 203, 357–66.

Ober, C., Elias, S., Kostyu, D.D., and Hauck, W.W. (1992) Decreased fecundability in Hutterite couples sharing HLA-DR. *American Journal of Human Genetics* 50, 6–14.

Ober, C., Weitkamp, L.R., Cox, N., Dytch, H., Kostyu, D., and Elias, S. (1997) HLA and mate choice in humans. *American Journal of Human Genetics* 61, 497–505.

Offner, S. (2001) A universal phylogenetic tree. *The American Biology Teacher* 63, 164–70.

O'Neil, P. (1997) Selection on genetically correlated phenological characters in *Lythrum salicaria* L. (Lythraceae). *Evolution* 51, 267–74.

Oring, L.W., Colwell, M.A., and Reed, J.M. (1991) Lifetime reproductive success in the spotted sandpiper (*Actitis macularia*)—sex differences and variance components. *Behavioral Ecology and Sociobiology* 28, 425–32.

Palkovacs, E.P. (2003) Explaining adaptive shifts in body size on islands: a life history approach. *Oikos* **103**, 37–44.

Parker, G.A., Baker, R.R., and Smith, V.G. (1972) The origin and evolution of gamete dimorphism and the male–female phenomenon. *Journal of Theoretical Biology* **36**, 529–53.

Pekar, S. and Kral, J. (2002) Mimicry complex in two central European zodariid spiders (Araneae: Zodariidae): How *Zodarion* deceives ants. *Biological Journal of the Linnean Society* **75**, 517–32.

Pellmyr, O. (2003) Yuccas, yucca moths, and coevolution: a review. *Annals of the Missouri Botanical Garden* **90**, 35–55.

Pellmyr, O. and Krenn, H.W. (2002) Origin of a complex key innovation in an obligate insect–plant mutualism. *Proceedings of the National Academy of Sciences USA* **99**, 5498–502.

Peters, A.D., Halligan, D.L., Whitlock, M.C., and Keightley, P.D. (2003) Dominance and overdominance of mildly deleterious induced mutations for fitness traits in *Caenorhabditis elegans*. *Genetics* **165**, 589–99.

Pfeiffer, T. and Bonhoeffer, S. (2003) An evolutionary scenario for the transition to undifferentiated multicellularity. *Proceeding of the National Academy of Sciences, USA* **100**, 1095–8.

Poole, A., Jeffares, D., and Penny, D. (1999) Early evolution: prokaryotes, the new kids on the block. *Bioessays* **21**, 880–9.

Price, G.R. (1972) Extension of covariance selection mathematics. *Annals of Human Genetics* **35**, 485–90.

Prince, V.E. and Pickett, F.B. (2002) Splitting pairs: the diverging fates of duplicated genes. *Nature Reviews Genetics* **3**, 827–37.

Reid, S.D., Herbelin, C.J., Bumbaugh, A.C., Selander, R.K., and Whittam, T.S. (2000) Parallel evolution of virulence in pathogenic *Escherichia coli*. *Nature* **406**, 64–7.

Reusch, T.B.H., Häberli, M.A., Aeschlimann, P.B., and Milinski, M. (2001) Female sticklebacks count alleles in a strategy of sexual selection explaining MHC polymorphism. *Nature* **414**, 300–2.

Reznick, D.A., Bryga, H., and Endler, J.A. (1990) Experimentally induced life-history evolution in a natural population. *Nature* **346**, 357–9.

Rice, W.R. and Hostert, E.E. (1993) Laboratory experiments on speciation, what have we learned in 40 years? *Evolution* **47**, 1637–53.

Richerson, P.J. and Boyd, R. (2004) *Not by genes alone: how culture transformed human evolution.* University of Chicago Press, Chicago.

Ridley, M. (1996) *Evolution*, 2nd edn. Blackwell, Oxford.

Robertson, D.S., McKenna, M.C., Toon, O.B., Hope, S., and Lillegraven, J.A. (2004) Survival in the first hours of the Cenozoic. *Geological Society of America Bulletin* **116**, 760–8.

Roca, A.L., Georgiadis, N., Pecon-Slattery, J., and O'Brien, S.J. (2001) Genetic evidence for two species of elephant in Africa. *Science* **293**, 1473–7.

Roderick, G.K. and Gillespie, R.G. (1998) Speciation and phylogeography of Hawaiian terrestrial arthropods. *Molecular Ecology* **7**, 519–31.

Roff, D.A. (1981) On being the right size. *American Naturalist*, **118**, 405–22.

Roger, A.J. and Silberman, J.D. (2002) Cell evolution: mitochondria in hiding. *Nature* **418**, 827–9.

Romer, A.S. (1962) *The vertebrate body*. W.B. Saunders, Philadelphia.

Rose, M.R. (1991) *Evolutionary biology of aging.* Oxford University Press, Oxford.

Roy, B.A. and Kirchner, J.W. (2000) Evolutionary dynamics of pathogen resistance and tolerance. *Evolution* **54**, 51–63.

Rozen, D.E. and Lenski, R.E. (2000) Long-term experimental evolution in *Escherichia coli*. VIII. Dynamics of a balanced polymorphism. *American Naturalist* **155**, 24–35.

Rozen, D.E., de Visser, J.A., and Gerrish, P.J. (2002) Fitness effects of fixed beneficial mutations in microbial populations. *Current Biology* **12**, 1040–5.

Ryan, M.J. (1985) *The tungara frog, a study in sexual selection and communication.* University of Chicago Press, Chicago.

Sacks, O.W. (1997) *The island of the colorblind.* Knopf, New York.

Salazar-Ciudad, I. and Jernvall, J. (2004) How different types of pattern formation mechanisms affect the evolution of form and development. *Evolution & Development* **6**, 6–16.

Sale, P.F. (1977) Maintenance of high diversity in coral fish communities. *American Naturalist* 111, 337–59.

Savard, L., Li, P., Strauss, S.H., Chase, M.W., Michaud, M., and Bousquet, J. (1994) Chloroplast and nuclear gene sequences indicate late Pennsylvanian time for the last common ancestor of extant seed plants. *Proceedings of the National Academy of Sciences USA* 91, 5163–7.

Sazima, I. (2002) Juvenile snooks (Centropomidae) as mimics of mojarras (Gerreidae) with a review of aggressive mimicry in fishes. *Environmental Biology of Fishes* 65, 37–45.

Schaal, S. and Ziegler, W. (eds) (1992) *Messel. An insight into the history of life and of the earth*. Oxford University Press, Oxford.

Schliewen, U., Rassmann, K., Markmann, M., Markert, J., Kocher, T., and Tautz, D. (2001) Genetic and ecological divergence of a monophyletic cichlid species pair under fully sympatric conditions in Lake Ejagham, Cameroon. *Molecular Ecology* 10, 1471–88.

Schluter, D. (2000) *The ecology of adaptive radiation*. Oxford University Press, Oxford.

Schmitt, J. and Wulff, R.D. (1993) Light spectral quality, phytochrome and plant competition. *Trends in Ecology and Evolution* 8, 47–51.

Schopf, J.W., Kudryavtsev, A.B., Agresti, D.G., Wdowiak, T.J., and Czaja, A.D. (2002) Laser-Raman imagery of Earth's earliest fossils. *Nature* 416, 73–6.

Schuster, P. (1993) RNA based evolutionary optimization. *Origin of Life and Evolution of the Biosphere* 23, 373–91.

Schuster, P., Fontana, W., Stadler, P.F., and Hofacker, I.L. (1994) From sequences to shapes and back, a case study in RNA secondary structures. *Proceedings of the Royal Society of London B* 255, 279–84.

Seehausen, O., Van Alphen, J.J.M., and Witte, F. (1997) Cichlid fish diversity threatened by eutrophication that curbs sexual selection. *Science* 277, 1808–11.

Selker, E. (1990) Premeiotic instability of repeated sequences in *Neurospora crassa*. *Annual Review of Genetics* 24, 579–613.

Sepkoski, J.J., Jr. (1984) A kinetic model of Phanerozoic taxonomic diversity. III. Post-Paleozoic families and mass extinctions. *Paleobiology* 10, 246–67.

Shaw, R.F. and Mohler, J.D. (1953) The selective advantage of the sex ratio. *American Naturalist* 87, 337–42.

Shine, R. (1985) The evolution of viviparity in reptiles, an ecological analysis. In *Biology of the Reptilia* (Gans, C. and Billet, F., eds), pp. 605–94. Academic Press, New York.

Shuster, S.M. and Wade, M.J. (2003) *Mating systems and strategies*. Princeton University Press, Princeton, NJ.

Simmons, J.A. (1973) The resolution of target range by echolocating bats. *Journal of the Acoustical Society of America* 54, 157–73.

Simons, A.M. and Johnston, O. (2003) Suboptimal timing of reproduction in Lobelia inflata may be a conservative bet-hedging strategy. *Journal of Evolutionary Biology* 16, 233–43.

Simoons, F.J. (1978) The geographic hypothesis and lactose malabsorption. A weighing of the evidence. *Digestive Diseases* 23, 963–80.

Simpson, G.G. (1949) *The meaning of evolution*. Yale University Press, New Haven.

Sinclair, A.R.E. and Norton-Griffiths, M. (1979) *Serengeti: dynamics of an ecosystem*. University of Chicago Press, Chicago.

Singh, N., Agrawal, S., and Rastogi, A.K. (1997) Infectious diseases and immunity, special reference to Major Histocompatibility Complex. *Emerging Infectious Disease* 3, http//www.cdc.gov/ncidod/EID/vol3no1/singh.htm.

Skelton, P.W. (ed.) (1993) *Evolution, a biological and palaeontological approach*. Addison-Wesley, New York.

Slack, J.M.W., Holland, P.W.H., and Graham, C.F. (1993) The zootype and the phylotypic stage. *Nature* 361, 490–2.

Smith, H. (1995) Physiological and ecological function within the phytochrome family. *Annual Review of Plant Physiology* 46, 289–315.

Smith, N.G. and Eyre-Walker, A. (2002) Adaptive protein evolution in *Drosophila*. *Nature* 415, 1022–4.

Spinage, C.A. (1962) Rinderpest and faunal distribution patterns. *African Wildlife* 16, 55–60.

Stearns, S.C. (1987) The selection arena hypothesis. In *The evolution of sex and its consequences* (Stearns, S.C., ed.), pp. 299–311. Birkhäuser Verlag, Basel.

Stearns, S.C. (1992) *The evolution of life histories.* Oxford University Press, Oxford.

Stearns, S.C. (1994) The evolutionary links between fixed and variable traits. *Acta Paleontologica Polonica* **38**, 215–32.

Stearns, S.C. (ed.) (1999) *Evolution in health and disease.* Oxford University Press, Oxford.

Stearns, S.C. and Ebert, D. (2001) Evolution in health and disease: work in progress. *Quarterly Review of Biology* **76**, 417–32.

Stearns, S.C. and Koella, J. (1986) The evolution of phenotypic plasticity in life-history traits: predictions for norms of reaction for age- and size-at-maturity. *Evolution* **40**, 893–913.

Stearns, S.C., Kaiser, M., Blarer, A., Ackermann, M., and Doebeli, M. (2000) The evolution of intrinsic mortality, growth, and reproduction in fruitflies. *Proceedings of the National Academy of Sciences USA* **97**, 3309–13.

Stegemann, S., Hartmann, S. Ruf, S., and Bock, R. (2003) High-frequency gene transfer from the chloroplast genome to the nucleus. *Proceedings of the National Academy of Sciences USA* **100**, 8828–33.

Stouthamer, R., Luck, R.F., and Hamilton, W.D. (1990) Antibiotics cause parthenogenetic *Trichogramma* (Hymenoptera/Trichogrammatidae) to revert to sex. *Proceedings of the National Academy of Sciences USA* **87**, 2424–7.

Strassmann, B.I. (1997) The biology of menstruation in H. sapiens: total lifetime menses, fecundity, and nonsynchrony in a natural fertility population. *Current Anthropology* **38**, 123–9.

Strong, D.R., Lawton, J.H., and Southwood, R. (1984) *Insects on plants: community patterns and mechanisms.* Blackwell, Oxford.

Stryer, L. (1988) *Biochemistry*, 3rd edn. W.H. Freeman, New York.

Sultan, S. and Stearns, S.C. (2005) Environmentally contingent variation. In *Variation: a hierarchical examination of a central concept in biology* (Hallgrimsson, B. and Hall, B.K., eds). Academic Press, New York (in press).

Swallow, J.G. and Wilkinson, G.S. (2002) The long and short of sperm polymorphisms in insects. *Biological Reviews of the Cambridge Philosophical Society* **77**, 153–82.

Swanson, W.J. and Vacquier, V.D. (2002) The rapid evolution of reproductive proteins. *Nature Reviews Genetics* **3**, 137–44.

Tarr, C.L. and Fleischer, R.C. (1995) Evolutionary relationships of the Hawaiian Honeycreepers (Aves, Drepaniidae). In W.L. Wagner and V.A. Funk (eds.), *Hawaiian Biogeography*, pp. 147–59. Smithsonian Institution Press.

Taylor, D.R., Zeyl, C., and Cooke, E. (2002) Conflicting levels of selection in the accumulation of mitochondrial defects in Saccharomyces cerevisiae. *Proceedings of the National Academy of Sciences USA* **99**, 3690–4.

Thomas, J.A. (2002) Larval niche selection and evening exposure enhance adoption of a predacious social parasite, *Maculinea arion* (large Blue Butterfly) by *Myrmica* ants. *Oecologia* **132**, 531–7.

Thompson, J.N. (1994) *The co-evolutionary process.* University of Chicago Press, Chicago.

Travisano, M., Vasi, F., and Lenski, R.E. (1995) Long-term experimental evolution in *Escherichia coli.* III. Variation among replicate populations in correlated responses to novel environments. *Evolution* **49**, 189–200.

Trevathan, W.R., Smith, E.O., and McKenna, J.J. (eds) (1999) *Evolutionary medicine.* Oxford University Press, Oxford.

Trivers, R. (1974) Parent–offspring conflict. *American Zoologist* **14**, 249–64.

Trivers, R.L. and Willard, D.E. (1973) Natural selection of parental ability to vary the sex ratio of offspring. *Science* **179**, 90–2.

Turekian, K. (1996) Global environmental change: past, present, and future. Prentice-Hall, Upper Saddle River, NJ.

Turgeon, J. and McPeek, M.A. (2002) Phylogeographic analysis of a recent radiation of Enallagma damselflies (Odonata: Coenagrionidae). *Molecular Ecology* **11**, 1989–2001.

Van Doorn, G.S., Luttikhuizen, P.C., and Weissing, F.J. (2001) Sexual selection at the protein level drives the extraordinary divergence of sex-related genes during sympatric speciation. *Proceedings of the Royal Society of London B* **268**, 2155–61.

van Hinsberg, A. (1997) Morphological variation in *Plantago lanceolata* L., effects of light quality and growth regulators on sun and shade populations. *Journal of Evolutionary Biology* **10**, 687–701.

van Valen, L. (1973) A new evolutionary law. *Evolutionary Theory*. **1**, 1–30.

Vavilov, N.I. (1922) The law of homologous series in variation. *Journal of Genetics* **12**, 47–89.

Venable, D.L. and Levin, D.A. (1985) Ecology of achene dimorphism in Heterotheca latifolia. I. Achene structure, germination, and dispersal. *Journal of Ecology* **73**, 133–45.

Vermeij, G.J. (1978) *Biogeography and adaptation: patterns of marine life*. Harvard University Press, Cambridge, MA.

Via, S. (2001) Sympatric speciation in animals: the ugly duckling grows up. *Trends in Ecology and Evolution* **16**, 381–90.

Vigilant, L., Stoneking, M., Harpending, H., Hawkes, K., and Wilson, A.C. (1991) African populations and the evolution of human mitochondrial DNA. *Science* **253**, 1503–7.

Vogel, F. and Motulsky, A.G. (1979) *Human Genetics*. Springer Verlag, Berlin.

Vulic, M., Dionisio, F., Taddei, F., and Radman, M. (1997) Molecular keys to speciation, DNA polymorphism and the control of genetic exchange in enterobacteria. *Proceedings of the National Academy of Sciences USA* **94**, 9763–7.

Wagner, G.P. (1989) The biological homology concept. *Annual Review of Ecology and Systematics* **20**, 51–69.

Wagner, W.E. Jr., Kelley, R.J., Tucker, K.R., and Harper, C.J. (2001) Females receive a life-span benefit from male ejaculates in a field cricket. *Evolution* **55**, 994–1001.

Wake, D.B. and Larson, A. (1987) Multidimensional analysis of an evolving lineage. *Science* **238**, 42–8.

Wallace, B. (1963) The elimination of an autosomal recessive lethal from an experimental population of *Drosophila melanogaster*. *American Naturalist* **97**, 65–6.

Wasserthal, L.T. (1997) The pollinators of the Malagasy Star Orchids *Angraecum sequepedale*, *A. sororium*, and *A. compactum* and the evolution of extremely long spurs by pollinator shift. *Botanica Acta* **110**, 343–59.

Waters, D.A. (2003) Bats and moths: what is there left to learn? *Physiological Entomology* **28**, 237–50.

Weiblen, G.D. (2002) How to be a fig wasp. *Annual Review of Entomology* **47**, 299–330.

Weigel, D. and Meyerowitz, E.M. (1994) The ABCs of floral homeotic genes. *Cell* **78**, 203–9.

Weigensberg, I. and Roff, D. (1996) Natural heritabilities: Can they be reliably estimated in the laboratory? *Evolution* **50**, 2149–57.

Weiss, K.M. (1993) *Genetic variation and human disease*. Cambridge University Press, Cambridge.

Welch, D.M. and Meselson, M. (2000) Evidence for the evolution of bdelloid rotifers without sexual reproduction or genetic exchange. *Science* **288**, 1211–15.

Wertz, J.E., Goldstone, C., Gordon, D.M., and Riley, M.A. (2003) A molecular phylogeny of enteric bacteria and implications for a bacterial species concept. *Journal of Evolutionary Biology* **16**, 1236–48.

West, S.A., Lively, C.M., and Read, A.F. (1999) A pluralist approach to sex and recombination. *Journal of Evolutionary Biology* **12**, 1003–12.

Whitelam, G.C., Patel, S., and Devlin, P.F. (1998) Phytochromes and photomorphogenesis in *Arabidopsis*. *Philosophical Transactions of the Royal Society of London B* **353**, 1445–53.

Whiten, A., Goodall, J., McGrew, W.C., Nishida, T., Reynolds, V., Sugiyama, Y., Tutin, C.E.G., Wrangham, R.W., and Boesch, C. (1999) Cultures in chimpanzees. *Nature* **399**, 682–5.

Wigby, S. and Chapman, T. (2004) Female resistance to male harm evolves in response to manipulation of sexual conflict. *Evolution* **58**, 1028–37.

Wilkins, A.S. (1993) *Genetic analysis of animal development*, 2nd edn. Wiley-Liss, New York.

Wilkins, A.S. (2002) *The evolution of developmental pathways*. Sinauer Associates, Sunderland, MA.

Williams, G.C. (1966) *Adaptation and natural selection*. Princeton University Press, Princeton, NJ.

Williams, G.C. (1992) *Natural selection. Domains, levels, and challenges*. Oxford University Press, Oxford.

Windig, J.J. (1993) The genetic background of plasticity in wing pattern of *Bicyclus* butterflies. PhD Thesis, University of Leiden.

Woolhouse, M.E.J., Taylor, L.H., and Haydon, D.T. (2001) Population biology of multihost pathogens. *Science* **292**, 1109–12.

Wu, C.-I. (2001) The genic view of the process of speciation. *Journal of Evolutionary Biology* **14**, 851–65.

Wynne-Edwards, V.C. (1962) *Animal dispersion in relation to social behaviour*. Oliver and Boyd, Edinburgh.

Yoder, A.D., Bums, M.M., Zelw, S., Delafossa, T., Veron, G., Goodman, S.M., and Flynn, J.J. (2003) Single origin of Malagasy Carnivora from an African ancestor. *Nature* **421**, 734–7.

Yoo, B.H. (1980) Long-term selection for a quantitative character in large replicate populations of *Drosophila melanogaster*. I. Response to selection. *Genetical Research* **35**, 1–17.

Zeyl, C., Mizesko, M. et al. (2001) Mutational meltdown in laboratory yeast populations. *Evolution* **55**, 909–17.

Zimmermann, W. (1931) Arbeitsweise der Botanischen Phylogenetik und anderer Gruppierungswissennschaften. In *Handbuch der biologischen Arbeitsmethoden*, (Abderhalden, E. ed.), pp. 941–1053. Urban und Schwarzenberg, Berlin.

Zöllner, S., Wen, X., Hanchard, N.A., Herbert, M.A. Ober, C., and Pritchard, J.K. (2004) Evidence for extensive transmission distortion in the human genome. *American Journal of Human Genetics* **74**, 62–72.

INDEX

Note: Page numbers in *italics* refer to figures and tables, and those in **bold** refer to marginal definitions of terms.